Master Electrician's Exam Preparation

Electrical Theory
National Electrical Code®
NEC® Calculations
Contains 2,300 Practice Questions

Michael Holt

Delmar Publishers
I(T)P An International Thomson Publishing Company

Albany • Bonn • Boston • Cincinnati • Detroit • London • Madrid
Melbourne • Mexico City • New York • Pacific Grove • Paris • San Francisco
Singapore • Tokyo • Toronto • Washington

NOTICE TO THE READER

Publisher does not warrant or guarantee any of the products described herein or perform any independent analysis in connection with any of the product information contained herein. Publisher does not assume, and expressly disclaims, any obligation to obtain and include information other than that provided to it by the manufacturer.

The reader is expressly warned to consider and adopt all safety precautions that might be indicated by the activities herein and to avoid all potential hazards. By following the instructions contained herein, the reader willingly assumes all risks in connections with such instructions.

The publisher makes no representation or warranties of any kind, including but not limited to, the warranties of fitness for particular purpose or merchantability, nor are any such representations implied with respect to the material set forth herein, and the publisher takes no responsibility with respect to such material. The publisher shall not be liable for any special, consequential, or exemplary damages resulting, in whole or part, from the readers' use of, or reliance upon, this material.

Cover Image by Mick Brady

Delmar Staff
Publisher: Susan Simpfenderfer
Acquisitions Editor: Paul Shepardson
Developmental Editor: Jeanne Mesick
Project Editor: Patricia Konczeski
Production Coordinator: Toni Bolognino
Art/Design Coordinator: Michael Prinzo
Marketing Manager: Lisa Reale

PNB Graphics
Design, Layout, and Typesetting: Paul Bunchuk
Internet email: pnbgraph@pnbgraphics.com

Graphic Illustrations: Mike Culbreath

COPYRIGHT © 1996 Charles Michael Holt Sr.

The ITP logo is a trademark under license

Printed in the United States of America
For more information, contact:

Delmar Publishers
3 Columbia Circle, Box 15015
Albany, New York 12212-5015

International Thomson Editores
Campos Eliseos 385, Piso 7
Col Polanco
11560 Mexico D F Mexico

International Thomson Publishing Europe
Berkshire House 168 - 173
High Holborn
London WC1V 7AA
England

International Thomson Publishing GmbH
Königswinterer Strasse 418
53227 Bonn
Germany

Thomas Nelson Australia
102 Dodds Street
South Melbourne, 3205
Victoria, Australia

International Thomson Publishing Asia
221 Henderson Road
#05 - 10 Henderson Building
Singapore 0315

Nelson Canada
1120 Birchmount Road
Scarborough, Ontario
Canada M1K 5G4

International Thomson Publishing Japan
Hirakawacho Kyowa Building, 3F
2-2-1 Hirakawacho
Chiyoda-ku, Tokyo 102
Japan

All rights reserved. No part of this work covered by the copyright hereon may be reproduced or used in any form or by any means—graphic, electronic, or mechanical, including photocopying, recording, taping, or information storage and retrieval systems—without the written permission of the publisher..

NEC, NFPA, and National Electrical Code are registered trademarks of National Fire Protection Association. This logo is a registered trademark of Mike Holt Enterprises, Inc.

    4 5 6 7 8 9 10 XXX 01 00 99 98 97

Library of Congress Cataloging-in-Publication Data
Holt, Charles Michael.
    Master electrician's exam preparation : electrical theory,
    National Electrical code, NEC Calculations : contains 2,000 practice questions / Michael Holt.
       p.   cm.
    Includes index.
    ISBN 0–8273-7623–5 (paperback)
    1. Electric engineering—United States—Examinations, questions, etc.
    2. Electricians—Licenses—United States.
    3. Electric engineering—Problems, exercises, etc. I. Title.
    TK 169.H653 1996
                               95–49219
    621.3 ′ 076-dc20   CIP

I dedicate this book to my family, and the Lord.

Contents

Preface	xi
About The Author	xi
Acknowledgments	xii
Getting Started	xii
How To Get The Best Grade On Your Exam	xiii
How To Take An Exam	xiv
Checking Your Work	xv
Changing Answers	xv
Rounding Off	xv
Summary	xv
Things To Be Careful Of	xv
Delmar Publishers' Internet Address	xvi
Chapter 1 Electrical Theory And Code Questions	1
Unit 1 Electrician's Math And Basic Electrical Formulas	3
Part A – Electrician's Math	3
1–1 Fractions	3
1–2 Kilo	4
1–3 Knowing Your Answer	4
1–4 Percentages	4
1–5 Multiplier	5
1–6 Percent Increase	5
1–7 Percentage Reciprocals	5
1–8 Squaring	6
1–9 Square Root	6
1–10 Rounding Off	7
1–11 Parentheses	7
1–12 Transposing Formulas	7
Part B – Basic Electrical Formulas	8
1–13 Electrical Circuit	8
1–14 Electron Flow	8
1–15 Power Source	8
1–16 Conductance And Resistance	9
1–17 Electrical Circuit Values	9
1–18 Ohm's Law I = E/R	10
1–19 PIE Circle Formula	12
1–20 Formula Wheel	13
1–21 Power Changes With The Square Of The Voltage	14
1–22 Electric Meters	15
Unit 1 – Electrician's Math And Basic Electrical Formulas Summary Questions	18
Challenge Questions	22
Getting The Most	25
Understanding The Terms And Theories	25
Understanding A Code Section	27
How To Use The NEC	27
Finding Information In The Nec	27
Customizing Your Code Book	28

NEC Questions	30

Unit 2 Electrical Circuits — 41
Part A – Series Circuits — 41
Introduction To Series Circuits — 41
2–1 Understanding Series Calculations — 42
2–2 Series Circuit Summary — 45
Part B – Parallel Circuit — 45
Introduction To Parallel Circuits — 45
2–3 Practical Uses Of Parallel Circuits — 45
2–4 Understanding Parallel Calculations — 46
2–5 Parallel Circuit Resistance Calculations — 47
2–6 Parallel Circuit Summary — 48
Part C – Series–Parallel And Multiwire Circuits — 49
Introduction To Series-Parallel Circuits — 49
2–7 Review Of Series And Parallel Circuits — 49
2–8 Series-parallel Circuit Resistance Calculations — 49
Part D – Multiwire Circuits — 51
Introduction To Multiwire Circuits — 51
2–9 Neutral Current Calculations — 51
2–10 Dangers Of Multiwire Circuits — 53
Unit 2 – Electrical Circuits Summary Questions — 56
Challenge Questions — 58
NEC Questions — 61

Unit 3 Understanding Alternating Current — 77
Part A – Alternating Current Fundamentals — 78
3–1 Current Flow — 78
3–2 Alternating Current — 78
3–3 Alternating Current Generator — 78
3–4 Waveform — 78
3–5 Armature Turning Frequency — 79
3–6 Phase – In And Out — 79
3–7 Phase Differences In Degrees — 80
3–8 Values Of Alternating Current — 80
Part B – Induction — 82
Induction Introduction — 82
3–9 Induced Voltage And Applied Current — 82
3–10 Conductor Impedance — 83
3–11 Induction And Conductor Shape — 84
3–12 Induction And Magnetic Cores — 84
Part C – Capacitance — 85
Capacitance Introduction — 85
3–13 Charge, Testing, And Discharging — 85
3–14 Use Of Capacitors — 86
Part D – Power Factor And Efficiency — 86
Power Factor Introduction — 86
3–15 Apparent Power (Volt-Amperes) — 86
3–16 Power Factor — 88
3–17 True Power (Watts) — 89
3–18 Efficiency — 90
Unit 3 – Understanding Alternating Current Summary Questions — 92
Challenge Questions — 96
NEC Questions — 99

Unit 4 Motors And Transformers — 110
Part A – Motors — 110
Motor Introduction — 110
4–1 Motor Speed Control — 111
4–2 Reversing A Direct Current Motor — 111
4–3 Alternating Current Motors — 112
4–4 Reversing Alternating Current Motors — 112
4–5 Motor Volt-Ampere Calculations — 113
4–6 Motor Horsepower/Watts — 114
4–7 Motor Nameplate Amperes — 115
Part B – Transformer Basics — 116
Transformer Introduction — 116
4–8 Transformer Primary Versus Secondary — 116
4–9 Transformer Secondary And Primary Voltage — 116
4–10 Auto Transformers — 117
4–11 Transformer Power Losses — 117
4–12 Transformer Turns Ratio — 118
4–13 Transformer kVA Rating — 118

4–14 Transformer Current	119
Unit 4 – Motors And Transformers Summary Questions	121
Challenge Questions	124
NEC Questions	127

Chapter 2 NEC Calculations And Code Questions — 141

Unit 5 Raceway, Outlet Box, And Junction Boxes Calculations — 143

Part A – Raceway Fill — 143

5–1 Understanding The National Electrical Code Chapter 9	143
5–2 Raceway And Nipple Calculation	148
5–3 Existing Raceway Calculation	150
5–4 Tips For Raceway Calculations	151

Part B – Outlet Box Fill Calculations — 151

Introduction [370-16]	151
5–5 Sizing Box – Conductors All The Same Size [Table 370–16(a)]	151
5–6 Conductor Equivalents [370–16(b)]	152
5–7 Sizing Box – Different Size Conductors [370–16(b)]	154

Part C – Pull, Junction Boxes, And Conduit Bodies — 156

Introduction	156
5–8 Pull And Junction Box Size Calculations	156
5–9 Depth Of Box And Conduit Body Sizing [370–28(a)(2), Exception]	157
5–10 Junction And Pull Box Sizing Tips	158
5–11 Pull Box Examples	159
Unit 5 – Raceway Fill, Box Fill, Junction Boxes, And Conduit Bodies Summary Questions	161
Challenge Questions	164
NEC Questions	165

Unit 6 Conductor Sizing And Protection Calculations — 184

6–1 Conductor Insulation Property [Table 310–13]	184
6–2 Conductor Allowable Ampacity [310–15]	186
6–3 Conductor Sizing [110–6]	187
6–4 Terminal Ratings [110–14(c)]	188
6–5 Conductors In Parallel [310–4]	190
6–6 Conductor Size – Voltage Drop [210–19(a), FPN No. 4, And 215–2(b), FPN No. 2]	191
6–7 Overcurrent Protection [Article 240]	192
6–8 Overcurrent Protection Of Conductors – General Requirements [240–3]	193
6–9 Overcurrent Protection Of Conductors – Specific Requirements	193
6–10 Equipment Conductors Size And Protection Examples	194
6–11 Conductor Ampacity [310–10]	195
6–12 Ambient Temperature Derating Factor [Table 310–16]	196
6–13 Conductor Bunching Derating Factor, Note 8(a) Of Table 310–16	198
6–14 Ambient Temperature And Conductor Bundling Derating Factors	199
6–15 Current-Carrying Conductors	200
6–16 Conductor Sizing Summary	202
Unit 6 – Conductor Sizing And Protection Summary Questions	204
Challenge Questions	207
NEC Questions	208

Unit 7 Motor Calculations — 221

Introduction	221
7–1 Motor Branch Circuit Conductors [430–22(a)]	221
7–2 Motor Overcurrent Protection	222
7–3 Overload Protection [430–32(a)]	223
7–4 Branch Circuit Short-Circuit Ground-Fault Protection [430–52(c)(1)]	225
7–5 Feeder Conductor Size [430–24]	226
7–6 Feeder Protection [430–62(a)]	227
7–7 Highest Rated Motor [430–17]	228
7–8 Motor Calculation Steps	228
7–9 Motor Calculation Review	230
7–10 Motor VA Calculations	232

Unit 7 – Motor Calculations Summary Questions ... 233
Challenge Questions ... 236
NEC Questions ... 237

Unit 8 Voltage Drop Calculations ... 249

8–1 Conductor Resistance ... 249
8–2 Conductor Resistance – Direct Current Circuits, [Chapter 9, Table 8 Of The NEC] ... 250
8–3 Conductor Resistance – Alternating Current Circuits ... 251
8–4 Alternating Current Resistance As Compared To Direct Current ... 251
8–5 Resistance Alternating Current [Chapter 9, Table 9 Of The NEC] ... 252
8–6 Voltage Drop Considerations ... 254
8–7 NEC Voltage Drop Recommendations ... 254
8–8 Determining Circuit Conductors Voltage Drop ... 254
8–9 Sizing Conductors To Prevent Excessive Voltage Drop ... 257
8–10 Limiting Conductor Length To Limit Voltage Drop ... 258
8–11 Limiting Current To Limit Voltage Drop ... 260
8–12 Extending Circuits ... 260
Unit 8 – Voltage Drop Summary Questions ... 263
Challenge Questions ... 266
NEC ... 268

Unit 9 One-family Dwelling-Unit Load Calculations ... 279

Part A - General Requirements ... 279

9–1 General Requirements ... 279
9–2 Voltages [220–2] ... 279
9–3 Fraction Of An Ampere ... 279
9–4 Appliance (Small) Circuits [220–4(b)] ... 280
9–5 Cooking Equipment – Branch Circuit [Table 220–19, Note 4] ... 280
One Wall-Mounted Oven Or One Counter-Mounted Cooking Unit [220–19 Note 4] ... 281
One Counter-Mounted Cooking Unit And Up To Two Ovens [220–19 Note 4] ... 282
9–6 Laundry Receptacle(s) Circuit [220–4(c)] ... 283
9–7 Lighting And Receptacles ... 283
Number Of Circuits Required [Chapter 9, Example No. 1(a)] ... 284

Part B - Standard Method – Feeder/Service Load Calculations ... 285

9–8 Dwelling Unit Feeder/Service Load Calculations (Part B Of Article 240) ... 285
9–9 Dwelling-Unit Feeder/Service Calculations Examples ... 286
Step 1: General Lighting, Small Appliance And Laundry Demand [220–11] ... 286
Step 2: Air Conditioning Versus Heat [220–15] ... 287
Step 3: Appliance Demand Load [220–17] ... 287
Step 4: Dryer Demand Load [220–18] ... 288
Step 5: Cooking Equipment Demand Load [220–19] ... 288
Step 6: Service Conductor Size [Note 3 Of Table 310–16] ... 289

Part C - Optional Method – Feeder/Service Load Calculations ... 290

9–10 Dwelling Unit Optional Feeder/Service Calculations [220–30] ... 290
9–11 Dwelling Unit Optional Calculation Examples ... 290
9–12 Neutral Calculations – General [220–22] ... 292
Cooking Appliance Neutral Load [220–22] ... 292
Dryer Neutral Load [220–22] ... 293
Unit 9 – One-family Dwelling-Unit Load Calculations Summary Questions ... 294
Challenge Questions ... 298
NEC questions ... 300

Chapter 3 Advanced NEC Calculations And Code Questions ... 311

Unit 10 Multifamily Dwelling-Unit Load Calculations ... 315

10–1 Multifamily Dwelling-Unit Calculations – Standard Method ... 315

10–2 Multifamily Dwelling-Units Calculation Examples – Standard Method   316

Step 1. General Lighting, Small Appliance, And Laundry Demand [220–11]   316

Step 2. Air Conditioning Versus Heat [220–15 And 440-34]   317

Step 3. Appliance Demand Load [220–17]   317

Step 4. Dryer Demand Load [220–18]   318

Step 5. Cooking Equipment Demand Load [220–19]   318

Step 6. Service Conductor Size [Note 3 Of Table 310–16]   319

10–3 Multifamily Dwelling-Unit Calculations [220–32] – Standard Method   319

Step 1. General Lighting, Small Appliance, And Laundry Demand [220–11]   320

Step 2. Air Conditioning Versus Heat [220–15 And 440–34]   320

Step 3. Appliance Demand Load [220–17]   320

Step 4. Dryer Demand Load [220–18]   320

Step 5. Cooking Equipment Demand Load [220–19]   320

Step 6. Service Conductor Size [Note 3 Of Table 310–16]   320

10–4 Multifamily Dwelling-Unit Calculations [220–32] – Optional Method   321

10–5 Multifamily Dwelling-Unit Example Questions [220–32]   321

Unit 10 – Multifamily Dwelling-Unit Load Calculations Summary Questions   323

Challenge Questions   325

NEC Questions   327

Unit 11 Commercial Load Calculations   338

Part A – General   338

11–1 General Requirements   338

11–2 Conductor Ampacity [Article 100]   338

11–3 Conductor Overcurrent Protection [240–3]   339

11–4 Voltages [220–2]   339

11–5 Fraction Of An Ampere   339

Part B – Loads   339

11–6 Air Conditioning   339

11–7 Dryers   341

11–8 Electric Heat   341

11–9 Kitchen Equipment   342

11–10 Laundry Equipment   343

11–11 Lighting – Demand Factors [Table 220–3(b) And 220–11]   344

11–12 Lighting Without Demand Factors [Table 220–3(b) And 220–10(b)].   344

11–13 Lighting – Miscellaneous   345

11–14 Multioutlet Receptacle Assembly [220–3(c) Exception No. 1]   345

11–15 Receptacles VA Load [220–3(c)(7) And 220–13]   346

11–16 Banks And Offices General Lighting And Receptacles   347

11–17 Signs [220-3(c)(6) And 600–5]   348

11–18 Neutral Calculations [220–22]   349

Part C – Load Calculations   349

Marina [555–5]   349

Mobile/Manufactured Home Park [550–22]   350

Recreational Vehicle Park [551–73]   350

Restaurant – Optional Method [220–36]   351

School – Optional Method [220–34]   353

Service Demand Load Using The Standard Method   353

Part D – Load Calculation Examples   354

Bank (120/240-Volt Single-Phase).   354

Office Building (480Y/277-Volt Three-Phase)   356

Restaurant (standard) (208Y/120 Volt, Three-Phase System)   357

Unit 11 – Commercial Load Calculations Summary Questions   360

Challenge Questions   363

NEC Questions   365

Unit 12 Delta/Delta And Delta/Wye Transformer Calculations   376

Introduction   376

Definitions   376

12–1 Current Flow	379	*12–17* Wye Transformers Current	390
Part A – Delta/Delta Transformers	380	*12–18* Line Current	390
12–2 Delta Transformer Voltage	380	*12–19* Phase Current	391
12–3 Delta High Leg	381	*12–20* Wye Phase Versus Line	392
12–4 Delta Primary And Secondary Line Currents	381	*12–21* Wye Transformer Loading And Balancing	394
12–5 Delta Primary Or Secondary Phase Currents	382	*12–22* Wye Transformer Sizing	394
12–6 Delta Phase Versus Line	383	*12–23* Wye Panel Schedule In kVA	395
12–7 Delta Current Triangle	384	*12–24* Wye Panelboard And Conductor Sizing	395
12–8 Delta Transformer Balancing	384	*12–25* Wye Neutral Current	395
12–9 Delta Transformer Sizing	385	*12–26* Wye Maximum Unbalanced Load	396
12–10 Delta Panel Schedule In kVA	386	*12–27* Delta/Wye Example	396
12–11 Delta Panelboard And Conductor Sizing	386	*12–28* Delta Versus Wye	398
12–12 Delta Neutral Current	386	Unit 12 – Delta/Delta And Delta/Wye Transformers Summary Questions	399
12–13 Delta Maximum Unbalanced Load	387	Challenge Questions	404
12–14 Delta/Delta Example	387	NEC Questions	407
Part B – Delta/wye Transformers	389		
12–15 Wye Transformer Voltage	389		
12–16 Wye Voltage Triangle	390	**Index**	**417**

Preface

INTRODUCTION

Passing the electrical exam is the dream of every electrician; unfortunately, many electricians don't pass it the first time. The primary reasons that people fail their exam is because they are not prepared on the technical material and/or on how to pass an exam.

Master Electrician's Exam Preparation, is a result of the author preparing thousands of electricians to pass their exam the first time. He has been doing this since 1975. Typically, an electrical exam contains 25 percent Electrical Theory/Basic Calculations, 40 percent National Electrical Code and 35 percent National Electrical Code Calculations. This book contains hundreds of explanations with illustrations, examples, and 2,300 practice questions covering all these subjects.

The writing style of this book is informal and relaxed, and the book contains clear graphics and examples that apply to the electrical exam.

To get the most out of this book, you should answer the two hundred questions at the end of each unit. The twenty-three hundred questions contained in this book are typical questions from electrician exams across the country. After you have reviewed each unit, you need to consider that it will take you about ten hours to complete the two hundred unit practice questions. If you have difficulty with a question, skip it and get it later. You will find that the answer key (Instructors Guide) contains detailed explanations for each question.

HOW TO USE THIS BOOK

Each unit of this book contains objectives, explanations with graphics, examples, steps on calculations, formulas and practice questions. This book is intended to be used with the 1996 National Electrical Code. As you read this book, review the author's comments, graphics, and examples with your Code Book and discuss the subjects with others.

This book contains many cross references to other related Code rules. Take the time to review the cross references.

As you progress through this book, you will find some rules or some comments that you don't understand. Don't get frustrated, highlight the section in the book that you are having a problem with. Discuss it with your boss, inspector, co-worker etc., maybe they'll have some additional feedback. After you have completed this book, review the highlighted sections and see if you now understand those problem areas.

Note. Some words are italicized to bring them to your attention. Be sure that you understand the terms before you continue with each unit.

ABOUT THE AUTHOR

Charles "Mike" Holt Sr. of Coral Springs, Florida, has worked his way up through the electrical trade as an apprentice, journeyman, master electrician, electrical inspector, electrical contractor, electrical designer, and developer of training programs and software for the electrical industry. Formerly contributing editor to Electrical Construction and Maintenance magazine (EC&M), and Construction Editor to Electrical Design and Installation magazine (EDI). Mr. Holt is currently a contributing writer for Electrical Contractor Magazine (EC). With a keen interest in continuing education, Mike Holt attended the University of Miami Masters in Business Administration Program (MBA) for Finance.

The authour has provided custom in-house seminars for: IAEI, NECA, ICBO, IBM, AT&T, Motorola, and the U.S. Navy, to name a few. He has taught over 1,000 classes on over 30 different electrical-related subjects ranging from alarm installations to exam preparation and voltage drop calculations. Many of Mike Holt's seminars are available on video. He continues to develop additional courses, seminars, and workshops to meet today's changing needs for individuals, organizations, vocational, and apprenticeship training programs.

Since 1982 Mike Holt has been helping electrical contractors improve the management of their business by offering business consulting, business management seminars, and computerized estimating and billing software. These software programs are used by hundreds of electrical contractors throughout the United States.

Mike Holt's extensive knowledge of the exam preparation, his hands-on experience, and his unique style of presenting information make this book a must read for those interested in passing the exam the first time.

On the personal side, Mike Holt is a national competitive barefoot water skier and has held several barefoot water ski records. He was the National Barefoot Water-ski Champion for 1988 and is currently training to regain the title by 1998. In addition to barefoot skiing, the author enjoys the outdoors, playing the guitar, reading, working with wood, and spending time with his family (he has seven children).

ACKNOWLEDGMENTS

I would like to say thank you, to all the people in my life who believed in me, even those who didn't. There are many people who played a role in the development and production of this book.

I would like to thank the Culbreath Family. Mike (Master Electrician) for helping me transform my words into graphics. I could not have produced such a fine book without your help. Toni, thanks for those late nights editing the manuscript. Dawn, you're too young to know, but thanks for being patient with your parents while they worked so hard.

Next, Paul Bunchuk (PNB Graphics)—Design, Layout, and Typesetting—thank you. Paul (Master Electrician) for the layout, and editing you did. Your knowledge of computers and the NEC has helped me put my ideas into reality.

Mike Culbreath and Paul Bunchuk, thank you both for not sacrificing quality, and for the extra effort to make sure that this book is the best that it can be.

To my family, thank you for your patience and understanding. I would like to give special thanks to my beautiful wife, Linda and my children, Belynda, Melissa, Autumn, Steven, Michael, Meghan, and Brittney.

I thank all my students; you know how much I care about you.

And thanks to all those who helped me in the electrical industry, Electrical Construction and Maintenance Magazine for my big break, and Joe McPartland, "My Mentor." Joe, you were always there to help and encourage me. I would like thank the following for contributing to my success: James Stallcup, Dick Lloyd, Mark Ode, D. J. Clements, Joe Ross, John Calloggero, Tony Selvestri, and Marvin Weiss.

The final personal thank you goes to Sarina, my friend and office manager. Thank you for covering the office for me the past few years while I spent so much time writing books. Your love and concern for me has helped me through many difficult times.

The author and Delmar Publishers would also like to thank those individuals who reviewed the manuscript and offered invaluable suggestions and feedback. Their assistance is greatly appreciated.

David Figueredo, Electrical Inspector
Metro-Dade County, Florida

John Mills, Master Electrician
Dade County, Florida

Kurt A. Stout, Electrical Inspector
Plantation, Florida

Richard Kurtz, Consultant
Boynton Beach, Florida

Ray Cotten, Electrical Instructor
North Tech Education Center
Palm Beach, Florida

GETTING STARTED

THE EMOTIONAL ASPECT OF LEARNING

To learn effectively, you must develop an attitude that learning is a process that will help you grow both personally and professionally. The learning process has an emotional as well as an intellectual component that we must recognize. To understand what affects our learning, consider the following:

Positive Image. Many feel disturbed by the expectations of being treated like children and we often feel threatened with the learning experience.

Uniqueness. Each of us will understand the subject matter from different perspectives and we all have some unique learning problems and needs.

Resistance To Change. People tend to resist change and resist information that appears to threaten their comfort level of knowledge. However, we often support new ideas that support our existing beliefs.

Dependence And Independence. The dependent person is afraid of disapproval and often will not participate in class discussion and will tend to wrestle alone. The independent person spends too much time asserting differences and too little time trying to understand others' views.

Fearful. Most of us feel insecure and afraid with learning, until we understand what is going to happen and what our role will be. We fear that our performance will not match the standard set by us or others.

Egocentric. Our ego tendency is to prove someone is wrong, with a victorious surge of pride. Learning together without a win/lose attitude can be an exhilarating learning experience.

Emotional. It is difficult to discard our cherished ideas in the face of contrary facts when overpowered by the logic of others.

HOW TO GET THE BEST GRADE ON YOUR EXAM

Studies have concluded that for students to get their best grades, they must learn to get the most from their natural abilities. It's not how long you study or how high your IQ is, it's what you do and how you study that counts. To get your best grade, you must make a decision to do your best and follow as many of the following techniques as possible.

Reality. These instructions are a basic guide to help you get the maximum grade. It is unreasonable to think that all of the instructions can be followed to the letter all of the time. Day-to-day events and unexpected situations must be taken into consideration.

Support. You need encouragement in your studies and you need support from your loved ones and employer. To properly prepare for your exam, you need to study 10 to 15 hours per week for about 3 to 6 months.

Communication With Your Family. Good communication with your family members is very important. Studying every night and on weekends may cause tension. Try to get their support, cooperation, and encouragement during this trying time. Let them know the benefits. Be sure to plan some special time with them during this preparation period; don't go overboard and leave them alone too long.

Stress. Stress can really take the wind out of you. It takes practice, but get into the habit of relaxing before you begin your studies. Stretch; do a few sit-ups and push-ups; take a 20-minute walk or a few slow, deep breaths. Close your eyes for a couple of minutes; deliberately relax the muscle groups that are associated with tension, such as the shoulders, back, neck, and jaw.

Attitude. Maintaining a positive attitude is important. It helps keep you going and helps keep you from getting discouraged.

Training. Preparing for the exam is the same as training for any event. Get plenty of rest and avoid intoxicating drugs, including alcohol. Stretch or exercise each day for at least 10 minutes. Eat light meals such as pasta, chicken, fish, vegetables, fruit, etc. Try to avoid heavy foods, such as red meats, butter, and other high-fat foods. They slow you down and make you tired and sleepy.

Eye Care. It is very important to have your eyes checked! Human beings were not designed to do constant seeing less than arm's length away. Our eyes were designed for survival, spotting food and enemies at a distance. Your eyes will be under tremendous stress because of prolonged, near-vision reading, which can result in headaches, fatigue, nausea, squinting, or eyes that burn, ache, water, or tire easily.

Be sure to tell the eye doctor that you are studying to pass an exam (bring this book and the Code Book), and you expect to do a tremendous amount of reading and writing. Prescribed nearpoint lenses can reduce eye discomfort while making learning more enjoyable and efficient.

Reducing Eye Strain. Be sure to look up occasionally, away from near tasks to distant objects. Your work area should be three times brighter than the rest of the room. Don't read under a single lamp in a dark room. Try to eliminate glare. Mixing of fluorescent and incandescent lighting can be helpful.

Sit straight, chest up, shoulders back, and weight over the seat so both eyes are an equal distance from what is being seen.

Getting Organized. Our lives are so busy that simply making time for homework and exam preparation is almost impossible. You can't waste time looking for a pencil or missing paper. Keep everything you need together. Maintain folders, one for notes, one for exams and answer keys, and one for miscellaneous items.

It is very important that you have a private study area available at all times. Keep your materials there. The dinning room table is not a good spot.

Time Management. Time management and planning is very important. There simply are not enough hours in the day to get everything done. Make a schedule that allows time for work, rest, study, meals, family, and recreation. Establish a schedule that is consistent from day to day.

Have a calendar and immediately plan your homework. Follow it at all costs. Try not to procrastinate (put off something). Try to follow the same routine each week and try not to become overtired. Learn to pace yourself to accomplish as much as you can without the need for cramming.

Learn How To Read. Review the book's contents and graphics. This will help you develop a sense of the material.

Clean Up Your Act. Keep all of your papers neat, clean, and organized. Now is not the time to be sloppy. If you are not neat, now is an excellent time to begin.

Speak Up In Class. If you are in a classroom setting, the most important part of the learning process is class participation. If you don't understand the instructor's point, ask for clarification. Don't try to get attention by asking questions you already know the answer to.

Study With A Friend. Studying with a friend can make learning more enjoyable. You can push and encourage each other. You are more likely to study if someone else is depending on you.

Students who study together perform above average because they try different approaches and explain their solutions to each other. Those who study alone spend most of their time reading and rereading the text and trying the same approach time after time even though it is unsuccessful.

Study Anywhere/Anytime. To make the most of your limited time, always keep a copy of the book(s) with you. Any time you get a minute free, study. Continue to study any chance you get. You can study at the supply house when waiting for your material; you can study during your coffee break, or even while you are at the doctor's office. Become creative!

You need to find your best study time. For some it could be late at night when the house is quiet. For others, it's the first thing in the morning before things get going.

Set Priorities. Once you begin your study, stop all phone calls, TV shows, radio, snacks, and other interruptions. You can always take care of it later.

HOW TO TAKE THE EXAM

Being prepared for an exam means more than just knowing electrical concepts, the Code, and the calculations. Have you felt prepared for an exam, then choke when actually taking it? Many good and knowledgeable electricians couldn't pass their exam because they did not know how to take an exam.

Taking exams is a learned process that takes practice and involves strategies. The following suggestions are designed to help you learn these methods.

Relax. This is easier said than done, but it is one of the most important factors in passing your exam. Stress and tension cause us to choke or forget. Everyone has had experiences where they get tense and couldn't think straight. The first step is becoming aware of the tension and the second step is to make a deliberate effort to relax. Make sure you're comfortable; remove clothes if you are hot, or put on a jacket if you are cold.

There are many ways to relax and you have to find a method that works for you. Two of the easiest methods that work very well for many people follow:

• Breathing Technique: This consists of two or three slow deep breaths every few minutes. Be careful not to confuse this with hyperventilation, which is abnormally fast breathing.

• Single-Muscle Relaxation: When we are tense or stressful, many of us do things like clench our jaw, squint our eyes, or tense our shoulders without even being aware of it. If you find a muscle group that does this, deliberately relax that one group. The rest of the muscles will automatically relax also. Try to repeat this every few minutes, and it will help you stay more relaxed during the exam.

Have The Proper Supplies. First of all, make sure you have everything needed several days before the exam. The night before the exam is not the time to be out buying pencils, calculators, and batteries. The night before the exam, you should have a checklist (prepared in advance) of every-thing you could possibly need. The following is a sample checklist to get you started.

• Six sharpened #2H pencils or two mechanical pens with extra #2H leads. The kind with the larger leads are faster and better for filling in the answer circles.

• Two calculators. Most examining boards require quiet, paperless calculators. Solar calculators are great, but there may not be enough light to operate them.

• Spare batteries. Two sets of extra batteries should be taken. It's very unlikely you'll need them but.

• Extra glasses if you use them.

• A watch for timing questions.

• All your reference materials, even the ones not on the list. Let the proctors tell you which ones are not permitted.

• A thermos of something you like to drink. Coffee is excellent.

• Some fruit, nuts, candy, aspirin, analgesic, etc.

• Know where the exam is going to take place and how long it takes to get there. Arrive at least 30 minutes early.

Note. It is also a good idea to pack a lunch rather than going out. It can give you a little time to review the material for the afternoon portion of the exam, and it reduces the chance of coming back late.

Understand The Question. To answer a question correctly, you must first understand the question. One word in a question can totally change the meaning of it. Carefully read every word of every question. Underlining key words in the question will help you focus.

Skip The Difficult Questions. Contrary to popular belief, you do not have to answer one question before going on to the next one. The irony is that the question you get stuck on is one that you will probably get wrong anyway no matter how much time you spend on it. This will result in not having enough time to answer the easy questions. You will get all stressed-out and a chain reaction is started. More people fail their exams this way than for any other reason.

The following strategy should be used to avoid getting into this situation.

• **First Pass:** Answer the questions you know. Give yourself about 30 seconds for each question. If you can't find the answer in your reference book within the 30 seconds, go on to the next question. Chances are that you'll come across the answers while looking up another question. The total time for the first pass should be 25 percent of the exam time.

• **Second Pass:** This pass is done the same as the first pass except that you allow a little more time for each question, about 60 seconds. If you still can't find the answer, go on to the next one. Don't get stuck. Total time for the second pass should be about 30 percent of the exam time.

• **Third Pass:** See how much time is left and subtract 30 minutes. Spend the remaining time equally on each question. If you still haven't answered the question, it's time to make an educated guess. Never leave a question unanswered.

• **Fourth pass:** Use the last 30 minutes of the exam to transfer your answers from the exam booklet to the answer key. Read each question and verify that you selected the correct answer on the test book. Transfer the answers carefully to the answer key. With the remaining time, see if you can find the answer to those questions you guessed at.

Guessing. When time is running out and you still have questions remaining, GUESS! Never leave a question unanswered.

You can improve your chances of getting a question correct by the process of elimination. When one of the choices is "none of these," or "none of the above," "d" is usually not the correct answer. This improves your chances from one-out-four (25 percent), to one-out-three (33 percent). Guess "All of these" or "All of the Above," and don't select the high or low number.

How do you pick one of the remaining answers? Some people toss a coin, others will count up how many of the answers were A's, B's, C's, and D's and use the one with the most as the basis for their guess.

CHECKING YOUR WORK

The first thing to check (and you should be watching out for this during the whole exam) is to make sure you mark the answer in the correct spot. People have failed the exam by ½ of a point. When they reviewed their exam, they found they correctly answered several questions on the test booklet, but marked the wrong spot on the exam answer sheet. They knew the answer was "(b) False" but marked in "(d)" in error.

Another thing to be very careful of, is marking the answer for, let's say question 7, in the spot reserved for question 8.

CHANGING ANSWERS

When re-reading the question and checking the answers during the fourth pass, resist the urge to change an answer. In most cases, your first choice is best and if you aren't sure, stick with the first choice. Only change answers if you are sure you made a mistake. Multiple choice exams are graded electronically so be sure to thoroughly erase any answer that you changed. Also erase any stray pencil marks from the answer sheet.

ROUNDING OFF

You should always round your answers to the same number of places as the exam's answers. Numbers below "5" are rounded down, while numbers "5" and above are rounded up.

Example: If an exam has multiple choice of:
(a) 2.2   (b) 2.1   (c) 2.3   (d) none of these

And your calculation comes out to 2.16, do not choose the answer (d) none of these. The correct answer is (b) 2.2, because the answers in this case are rounded off to the tenth.

Example: It could be rounded to tens, such as:
(a) 50   (b) 60   (c) 70   (d) none of these.

For this group, an answer such as 67 would be (c) 70, while an answer of 63 would be (b) 60. The general rule is to check the question's choice of answers then round off your answer to match it.

SUMMARY

• Make sure everything is ready and packed the night before the exam.

• Don't try to cram the night before the exam, if you don't know it by then.

• Have a good breakfast. Get the thermos and energy snacks ready.

• Take all your reference books. Let the proctors tell you what you can't use.

• Know where the exam is to be held and be there early.

• Bring ID and your confirmation papers from the license board if there are any.

• Review your NEC while you wait for your exam to begin.

• Try to stay relaxed.

• Determine the time per question for each pass and don't forget to save 30 minutes for transferring your answers to the answer key.

• Remember, in the first pass answer only the easy questions. In the second pass, spend a little more time per question, but don't get stuck. In the third pass, use the remainder of the time minus 30 minutes. In the fourth pass, check your work and transfer the answers to the answer key.

THINGS TO BE CAREFUL OF

- Don't get stuck on any one question.
- Read each question carefully.
- Be sure you are marking the answer in the correct spot on the answer sheet.
- Don't get flustered or extremely tense.

Mike Holt

To request examination copies of the Book, or Instructor's Guide, call or write to:

Delmar Publishers Inc.
3 Columbia Circle
P.O. Box 15015
Albany, NY 12212-5015
Phone: 1-800-347-7707 • 1-518-464-3500 • Fax: 1-518-464-0301

Delmar Publishers' Online Service
To access Delmar on the World Wide Web, point your browser to:
http://www.delmar.com/delmar.html
To access through Gopher: gopher://gopher.delmar.com
(Delmar Online is part of "thomson.com", an Internet site with information on more than 30 publishers of the International Thomson Publishing organization.)
For information on our products and services:
email: info@delmar.com

CHAPTER 1
Electrical Theory And Code Questions

Scope of Chapter 1

UNIT 1 ELECTRICIAN'S MATH AND BASIC ELECTRICAL FORMULAS

UNIT 2 ELECTRICAL CIRCUITS

UNIT 3 UNDERSTANDING ALTERNATING CURRENT

UNIT 4 MOTORS AND TRANSFORMERS

Unit 1

Electrician's Math and Basic Electrical Formulas

OBJECTIVES

After reading this unit, the student should be able to briefly explain the following concepts:

Part A - Electricians Math	Percent increase	Part B - Basic Electrical Formulas	Ohm's law
Fractions	Percentage	Conductance and	PIE circle formula
Kilo	Reciprocals	resistance	Power changes with
Knowing your answer	Square root	Electric meters	the square of the
Multiplier	Squaring	Electrical circuit	voltage
Parentheses	Transposing formulas	Electrical circuit values	Power source
			Power wheel

After reading this unit, the student should be able to briefly explain the following terms:

Part A - Electricians Math	Transposing	Direct current	Ohmmeter
Fractions	Part B - Basic Electrical Formulas	Directly proportional	Ohms
Kilo		E – Electromotive	P – Power
Multiplier	A – Ampere	Force	
Parentheses	Alternating current	Electrical meters	Perpendicular
Percentage	Ammeter	Electromagnetic	Polarity
Ratio	Ampere	Electromagnetic field	Polarized
Reciprocal	Armature	Electron pressure	Power
Rounding off	Clamp-on ammeters	Helically wound	Power source
Square root	Conductance	Intensity	Resistance
Squaring a number	Conductors	Inversely proportional	Shunt bar
	Current	Megohmeter	Shunt meter

PART A – ELECTRICIAN'S MATH

1–1 FRACTIONS

Fractions represent a part of a number. To change a fraction to a decimal form, divide the numerator (top number of the fraction) by the denominator (bottom number of the fraction).

❑ **Examples**

$1/6$ = one divided by six = 0.166

$5/4$ = five divided by four = 1.25

$7/2$ = seven divided by two = 3.5

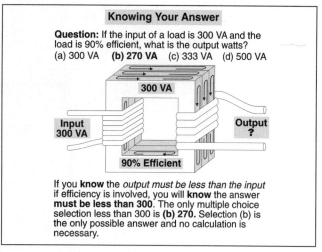

Figure 1-1
Know Your Answer Example

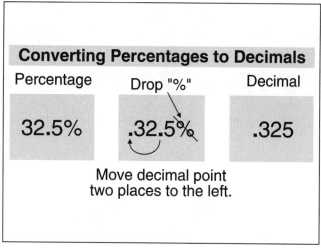

Figure 1-2
Converting Percentages to Decimals

1-2 KILO

The letter *k* is the abbreviation of *kilo*, which represents 1,000.

❑ **Kilo Example No. 1**
What is the wattage for an 8-kW rated range?

(a) 8 watts   (b) 8,000 watts   (c) 4,000 watts   (d) none of these

• Answer: (b) 8,000 watts

Wattage = kW × 1,000. In this case 8 kW × 1,000 = 8,000 watts.

❑ **Kilo Example No. 2**
What is the kVA of a 300-VA load?

(a) 300 kVA   (b) 3,000 kVA   (c) 30 kVA   (d) .3 kVA

• Answer: (d) 0.3 kVA, kW is converted to watts by dividing the watts by 1,000.

In this case; $\frac{300 \text{ VA}}{1,000} = 0.3$ kVA

Note. The use of *k* is not limited to *kW*. It is used for kcmils, such as 250 kcmils.

1-3 KNOWING YOUR ANSWER

When working with mathematical calculations, you should know if the answer is greater than, or less than the values given.

❑ **Knowing-Your-Answer Example**
If the input of a load is 300 watts and the load is 90 percent efficient, what is the output watts? Note. Because of efficiency, the output is always less than the input, Fig. 1–1.

(a) 300 VA   (b) 270 VA   (c) 333 VA   (d) 500 VA

• Answer: (b) 270 VA

Since the question stated that the output had to be less than the 300 watt input, the answer must be less than 300 watts. The only choice that is less than 300 watts, is 270 watts.

1-4 PERCENTAGES

A *percentage* is a ratio of two numbers. When changing a percent to a decimal or whole number, simply move the decimal point two places to the left, Fig. 1–2.

❑ **Percentage Example**
32.5% = .325   100% = 1.00
125% = 1.25   300% = 3.00

1-5 MULTIPLIER

Often a number is required to be increased or decreased by a percentage. When a percentage or fraction is used as a *multiplier*, follow these steps:
- Step 1: → Convert the multiplier to a decimal form, then
- Step 2: → Multiply the number by the decimal value from Step 1.

❏ **Increase By 125 Percent Example**

An overcurrent protection device (breaker or fuse) must be sized no less than 125 percent of the continuous load. If the load is 80 amperes, the overcurrent protection device would have to be sized no less than _____ amperes.

(a) 80 amperes (b) 100 amperes (c) 125 amperes (d) none of these step

- Answer: (b) 100 amperes

Step 1: → Convert 125% to 1.25
Step 2: → Multiply 80 amperes by 1.25 = 100 amperes

❏ **Limit To 80 Percent Example**

The maximum continuous load on an overcurrent protection device is limited to 80 percent of the device rating. If the device is rated 50 amperes, what is the maximum continuous load?

(a) 80 amperes (b) 125 amperes (c) 50 amperes (d) 40 amperes

- Answer: (d) 40 amperes

Step 1: → Convert 80% to a decimal; 0.8
Step 2: → Multiply the number by the decimal; 50 amperes × 0.8 = 40 amperes

1-6 PERCENT INCREASE

Increasing a number by a specific *percentage* is accomplished by:
- Step 1: → Converting the percentage to decimal form.
- Step 2: → Adding one to the decimal value from Step 1.
- Step 3: → Multiplying the number to be increased by the multiplier from Step 2.

❏ **Percent Increase Example**

Increase the whole number 45 by 35%.

(a) 61 (b) 74 (c) 83 (d) 104

- Answer: (a) 61

Step 1: → Convert 35% into .35.
Step 2: → Add one to the decimal value from Step 1; 1 + .35 = 1.35
Step 3: → Multiply the number by the multiplier; 45 × 1.35 = 60.75

1-7 PERCENTAGE RECIPROCALS

A *reciprocal* is a whole number converted into a fraction, with the number one as the top number. This fraction is then converted to a decimal form.
- Step 1: → Convert the number to a decimal.
- Step 2: → Divide the number into one.

❏ **Reciprocal Example No. 1**

What is the reciprocal of 80 percent?

(a) .80 percent (b) 100 percent (c) 125 percent (d) none of these

- Answer: (c) 1.25 or 125 percent

Step 1: → Convert the number to a decimal; 80% = .8
Step 2: → Divide the number into one; $1/0.8$ = 1.25 or 125%

❏ **Reciprocal Example No. 2**

A continuous load requires an overcurrent protection device sized no smaller than 125 percent of the load. What is the maximum continuous load permitted on a 100-ampere overcurrent protection device?

(a) 100 amperes (b) 125 amperes (c) 80 amperes (d) none of these

- Answer: (c) 80 amperes

Step 1: → Convert the number to a decimal; 125% = 1.25

Step 2: → Divide the number into one; $1/1.25 = 0.8$ or 80%

If the overcurrent device is sized no less than 125 percent of the load, the load is limited to 80 percent of the overcurrent protection device (reciprocal). Therefore, the maximum load is limited to 100 amperes × .8 = 80 amperes

1-8 SQUARING

Squaring a number is simply multiplying the number by itself such as; $23^2 = 23 \times 23 = 529$

❏ Squaring Example No. 1

What is the power consumed, in watts, of a No. 12 conductor that is 200 feet long and has a resistance of 0.4 ohm? The current flowing in the circuit is 16 amperes. Formula: Power = $I^2 \times R$

(a) 50 watts (b) 150 watts (c) 100 watts (d) 200 watts

- Answer: (c) 100 watts

  $P = I^2 \times R$, I = 16 amperes, R = 0.4 ohm

  $P = 16 \text{ amperes}^2 \times 0.4 \text{ ohm}$

  P = 102.4 watts; answers are rounded to 50's.

❏ Squaring Example No. 2

What is the area, in square inches, of a one inch raceway whose diameter is 1.049 inches? Use the formula: Area = $\pi \times r^2$, $\pi = 3.14$, r = radius, radius is ½ the diameter.

(a) 1 square inch (b) .86 square inch (c) .34 square inch (d) .5 square inch

- Answer: (b) 0.86 square inch

  Raceway area = $\pi \times r^2 = 3.14 \times (½ \times 1.049)^2 = 3.14 \times .5245^2$

  $3.14 \times (.5245 \times .5245) = 3.14 \times .2751 = 0.86$ square inch

1-9 SQUARE ROOT

The *square root* of a number is the opposite of squaring a number. For all practical purposes, to determine the square root of any number, you must use a calculator with a square root key. For exam preparation purposes, the only number you need to know is the square root of is 3, which is 1.732. To multiply, divide, add, or subtract a number by a square root value, simply determine the square root value first, and then perform the math function. The steps to determine the square root of a number follow.

Step 1: → Enter number in calculator.
Step 2: → Press the √‾ key of the calculation.

❏ Example

What is the $\sqrt{3}$?

(a) 1.55 (b) 1.73 (c) 1.96 (d) none of these

- Answer: (b) 1.732

Step 1: → Type 3:
Step 2: → Press the √‾ key = 1.732.

❏ Square Root Example No. 1

$\dfrac{36,000 \text{ watts}}{(208 \text{ volts} \times \sqrt{3})}$ is equal to _____ amperes?

(a) 120 (b) 208 (c) 360 (d) 100

- Answer: (d) 100

Step 1: → Determine the $\sqrt{3} = 1.732$
Step 2: → Multiply 208 volts × 1.732 = 360 volts
Step 3: → Divide 36,000 watts/360 volts = 100 amperes

❏ Square Root Example No. 2

The phase voltage is equal to $\left(\dfrac{208 \text{ volts}}{\sqrt{3}}\right)$ _____ ?

(a) 120 volts (b) 208 volts (c) 360 volts (d) none of these

- Answer: (a) 120 volts

Step 1: → Determine the $\sqrt{3} = 1.732$.
Step 2: → Divide 208 volts by 1.732 = 120 volts.

1–10 ROUNDING OFF

Numbers below 5 are rounded down, while numbers 5 and above are rounded up. *Rounding* to three significant figures should be sufficient for most calculations, such as

.12459 – the fourth number is 5 or above = .125 rounded up
1.6744 – the fourth number is below 5 = 1.67 rounded down
21.996 – the fourth number is 5 or above = 22 rounded up
367.28 – the fourth number is below 5 = 367 rounded down

Rounding For Exams

You should always round off your answer in the same manner as the answers to an exam question. Do not choose none of these in an exam until you have checked the answers for rounding off. If after rounding off your answer to the exam format there is no answer, choose "none of these."

❑ **Rounding Example**
The sum of 12, 17, 28, and 40 is equal to _____?
(a) 80   (b) 90   (c) 100   (d) none of these
• Answer: (c) 100

The answer is actually 97, but there is no 97 as a choice. Do not choose "none of these" in an exam until you have checked how the choices are rounded off. The choices in this case are all rounded off to the nearest tens.

1–11 PARENTHESES

Whenever numbers are in *parentheses*, we must complete the mathematical function within the parentheses before proceeding with the problem.

❑ **Parenthesis Example**
What is the voltage drop of two No. 14 conductors carrying 16 amperes for a distance of 100 feet? Use the following example for the answer:

$$VD = \frac{2 \times K \times I \times D}{CM} = \frac{(2 \text{ wires} \times 12.9 \text{ ohms} \times 16 \text{ amperes} \times 100 \text{ feet})}{4{,}110 \text{ circular mils}}$$

(a) 3 volts   (b) 3.6 volts   (c) 10.04 volts   (d) none of these
• Answer: (c) 10.04 volts
   Work out the parentheses,
   2 wires × 12. 9-ohms resistance × 16 amperes × 100 feet = 41,280
   41,280/4,110 circular mils = 10.4 volts dropped

1–12 TRANSPOSING FORMULAS

Transposing is an algebraic function used to rearrange formulas. The formula $I = P/E$ can be transposed to $P = I \times E$ or $E = P/I$, Fig. 1–3.

❑ **Transpose Example**
Transform the formula $CM = \frac{(2 \times K \times I \times D)}{VD}$, to find the voltage drop of the circuit, Fig. 1–4.
(a) VD = CM
(b) VD = (2 × K × I × D)
(c) VD = (2 × K × I × D) × CM
(d) none of these
• Answer: (d) none of these

$$VD = \frac{(2 \times K \times I \times D)}{CM}$$

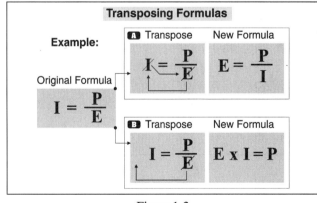

Figure 1-3
Transpose Formulas

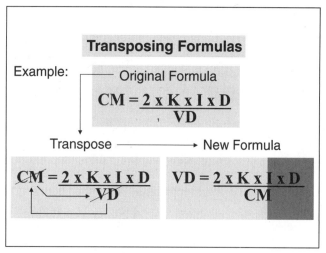

Figure 1-4
Transpose Example

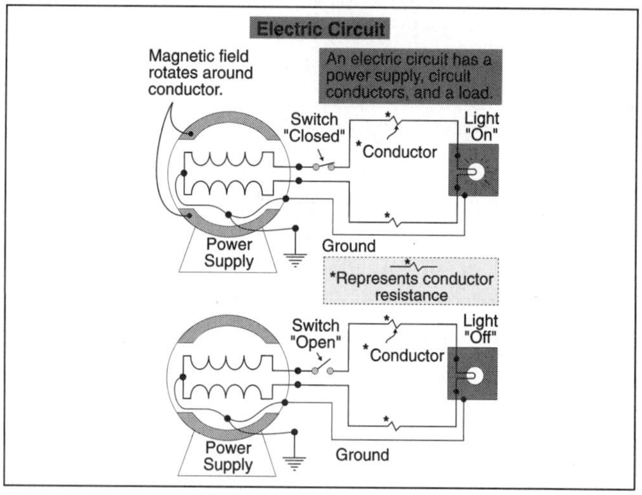

Figure 1-5
The Electrical Circuit

PART B – BASIC ELECTRICAL FORMULAS

1–13 ELECTRICAL CIRCUIT

An electric circuit consists of the power source, the conductors, and the load. For current to travel in the circuit, there must be a complete path from one terminal of the power supply, through the conductors and the load, back to the other terminal of the power supply, Fig. 1–5.

1–14 ELECTRON FLOW

Inside a direct current power source (such as a battery) the electrons travel from the positive terminal to the negative terminal; however, outside of the power source, electrons travel from the negative terminal to the positive terminal, Fig. 1–6.

1–15 POWER SOURCE

In any completed circuit, it takes a force to push the electrons through the power source, conductor, and load. The two most common types of power sources are *direct current* and *alternating current*.

Direct Current

The polarity from direct current power sources never changes; that is, the current flows out of the negative terminal of the power source always in the same direction. When the power supply is a *battery*, the polarity and the magnitude remain the same, Fig. 1–7.

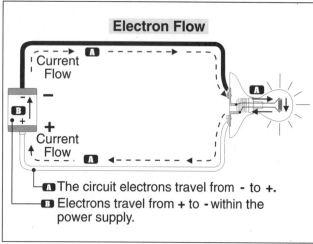

Figure 1-6
Electron Flow

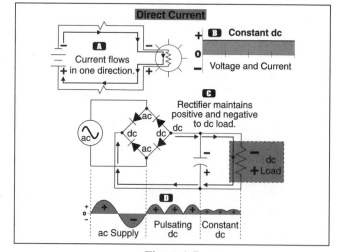

Figure 1-7
Direct Current – Constant Polarity

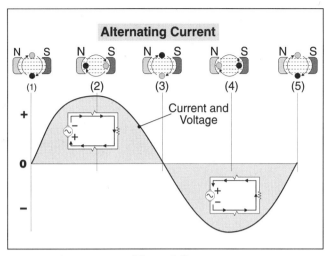

Figure 1-8

Alternating Current – Alternating Polarity

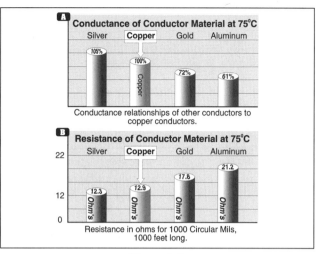

Figure 1-9

Conductance and Resistance

Alternating Current

Alternating current power sources produce a voltage and current that has a constant change in polarity and magnitude at a constant frequency. Alternating current flow is produced by a *generator* or an *alternator*. Fig. 1–8.

1-16 CONDUCTANCE AND RESISTANCE

Conductance

Conductance is the property of metal that permits current to flow. The best conductors, in order of their conductivity are: silver, copper, gold, and aluminum. Although silver is a better conductor of electricity than copper, copper is used more frequently because it is less expensive, Fig. 1–9 Part A.

Resistance

Resistance is the opposite of conductance. It is the property that opposes the flow of electric current. The resistance of a conductor is measured in ohms according to a standard length of 1,000 feet, Fig. 1–9 Part B. This value is listed in the National Electrical Code, Chapter 9, Table 8, for direct current circuits and Chapter 9, Table 9, for alternating current circuits.

1-17 ELECTRICAL CIRCUIT VALUES

In an electrical circuit there are four circuit values that we must understand. They are voltage, resistance, current and power, Fig. 1–10.

Voltage

Electron pressure is called *electromotive force* and is measured by the unit *volt*, abbreviated by the letter E or V. Voltage is also a term used to described the difference of potential between any two points.

Resistance

The friction opposition to the flow of electrons is called *resistance*, and the unit of measurement is the *ohm*, abbreviated by the letter R. Every component of an electric circuit contains resistance including the power supply.

Current

Free electrons moving in the same direction in a conductor produce an electrical *current* sometimes called *intensity*. The rate at which electrons move is measured by the unit called *ampere*, abbreviated by the letter I or A.

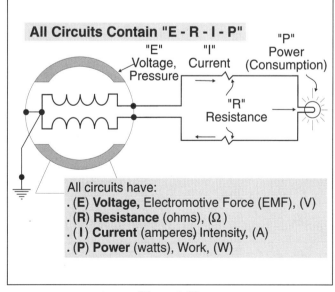

Figure 1-10

Electrical Circuit Values

Power

The rate of work that can be produced by the movement of electrons is called *power*, and the unit is the *watt*. It is very common to see the W symbol for watts used instead of P for power.

Note. A 100 watt lamp will consume 100 watts per hour.

1–18 OHM'S LAW I = E/R

Ohm's law, $I = E/R$, demonstrates the relationship between current, voltage, and resistance in a direct current, or an alternating current circuit that supplies only resistive loads, Fig. 1–11.

$$I = \frac{E}{R} \qquad E = I \times R \qquad R = \frac{E}{I}$$

This law states that:

1. **Current is directly proportional to the voltage.** This means that if the voltage is increased by a given percentage, the current would increase by that same percentage; or if the voltage is decreased by a given percentage, the current will decrease by the same percentage, Fig. 1–12 Part A.
2. **Current is inversely proportional to the resistance.** This means that an increase in resistance will result in a decrease in current, and a decrease in resistance will result in an increase in current, Fig. 1–12 Part B.

Opposition To Current Flow

In a direct current circuit, the physical resistance of the conductor opposes the flow of electrons. In an alternating current circuit, there are three factors that oppose current flow. Those factors are *conductor resistance*, *inductive reactance*, and *capacitive reactance*. The opposition to current flow, due to a combination of resistance and r*eactance*, is called *impedance*, measured in ohms, and abbreviated with the letter Z. Impedance will be covered later, so for now assume that all circuits have very little or no reactance.

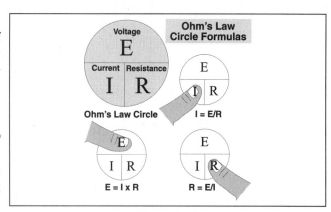

Figure 1-11
Ohms Law Formulas

❑ **Ohms Law Ampere Example**

A 120-volt power source supplies a lamp that has a resistance of 192 ohms. What is the current flow of the circuit, Fig. 1–13.

(a) 0.6 ampere (b) 0.5 ampere (c) 2.5 amperes (d) 1.3 amperes

• Answer: (a) .6 amperes

Step 1: → What is the current, I?
Step 2: → What do you know? E = 120 volts, R = 192 ohms
Step 3: → The formula is $I = E/R$.
Step 4: → The answer is
$$I = \frac{120 \text{ volts}}{192 \text{ ohms}} = 0.625 \text{ ampere}$$

❑ **Ohm's Law Voltage Example**

What is the voltage drop of two No. 12 conductors that supply a 16-ampere load located 50 feet from the power supply? The total resistance of both conductors is 0.2 ohm. Fig. 1–14.

(a) 16 volts (b) 32 volts
(c) 1.6 volts (d) 3.2 volts

• Answer: (d) 3.2 volts

Step 1: → What is the question? It is, what is voltage drop, E?
Step 2: → What do you know about the conductors? I = 16 amperes, R = 0.2 ohm.
Step 3: → The formula is $E = I \times R$.
Step 4: → The answer is E = 16 amperes × .2 ohms, E = 3.2 volts.

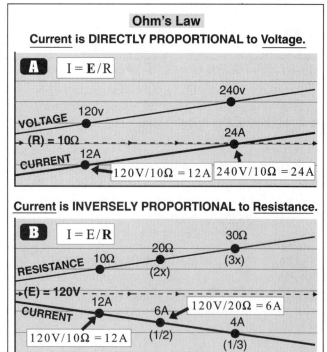

Figure 1-12
Part A – Current Proportional to Voltage
Part B – Current Inversely to Resistance

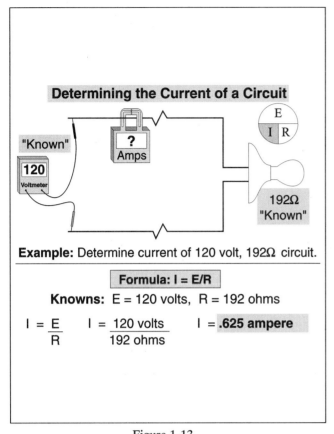

Figure 1-13
Determining the Current of a Circuit

Determining the Current of a Circuit

Example: Determine current of 120 volt, 192Ω circuit.

Formula: $I = E/R$

Knowns: E = 120 volts, R = 192 ohms

$I = \dfrac{E}{R}$ $I = \dfrac{120 \text{ volts}}{192 \text{ ohms}}$ $I = $ **.625 ampere**

Determining Voltage Drop With Ohm's Law

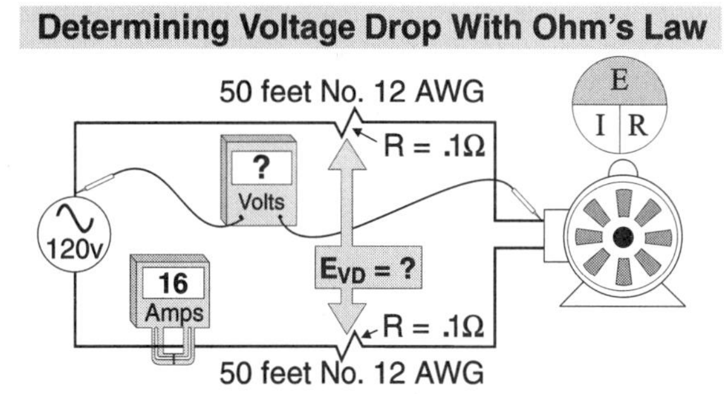

Example: Determine conductor VD on a 120 volt circuit.

Formula: $E_{VD} = I \times R$

To determine the voltage drop of conductors, use resistance of conductors.

Known: I = 16 Amperes (given)
Known: R of each Conductor = .1 ohm

$E_{VD} = I \times R$ $E_{VD} = 16 \text{ amps} \times .1 \text{ ohm}$

$E_{VD} = $ **1.6 volts per conductor**

Note: Voltage drop of both conductors = 16 amperes x .2 ohm = 3.2 volts
Note: Load operates at 120 volts - 3.2 vd = 116.8 volts

Figure 1-14
Voltage Example

❑ Ohms Law Resistance Example

What is the resistance of the circuit conductors when the conductor voltage drop is 3 volts and the current flowing through the conductors is 100 amperes, Fig. 1–15?

(a) .03 ohm
(b) 0.2 ohm
(c) 3 ohms
(d) 30 ohms

• Answer: (a) .03 ohm

Step 1: → What is the question? It is What is resistance, R?

Step 2: → What do you know about the conductors?
E = 3 volts drop,
I = 100 amperes

Step 3: → The formula is $R = E/I$

Step 4: → The answer is

$R = \dfrac{3 \text{ volts}}{100 \text{ amperes}} = R = .03 \text{ ohm}$

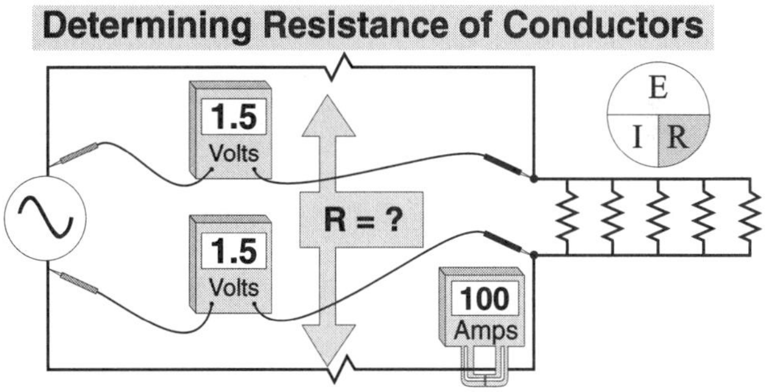

Determining Resistance of Conductors

Example: Determine resistance of conductors.

Formula: $R = E/I$

Known: E_{VD} = 1.5 volts per conductor
Known: I = 100 amperes

$R = \dfrac{E}{I}$ $R = \dfrac{1.5 \text{ volts}}{100 \text{ amp}}$

$R = $ **.015 ohm per conductor**

R = .015Ω x 2 conductors = **.03 ohm for both conductors**

OR... $R = \dfrac{3 \text{ volts}}{100 \text{ amps}} = $ **.03 ohm for both conductors**

Figure 1-15
Resistance Example

1–19 PIE CIRCLE FORMULA

The PIE circle formula, shows the relationships between power, current, and voltage, Fig. 1–16.

$P = E \times I$ $I = P/E$ $E = I \times R$

❏ **Power Example**

What is the power loss, in watts, for two conductors that carry 12 amperes and have a voltage drop of 3.6 volts. Fig. 1–17?

(a) 4.3 watts   (b) 43 watts
(c) 432 watts   (d) none of these

• Answer: (b) 43 watts

Step 1: → What is the question? It is, what is the power, P?
Step 2: → What do you know?
   I = 12 amperes, E = 3.6 volts drop
Step 3: → The formula is $P = I \times E$.
Step 4: → The answer is
   P = 12 amperes × 3.6 volts = 43.2 watts per hour.

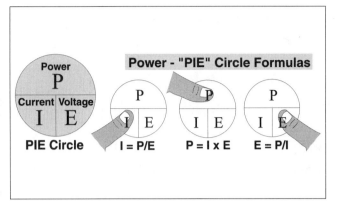

Figure 1-16
PIE Formula Circle

❏ **Current Example**

What is the current flow, in amperes, in the circuit conductors that supply a 7.5-kW heat strip rated 240 volts when connected to a 240-volt power supply, Fig. 1–18?

(a) 25 amperes   (b) 31 amperes   (c) 39 amperes   (d) none of these

• Answer: (b) 31 amperes

Step 1: → What is the question? It is, what is the current, I?

Step 2: → What do you know?
   P = 7,500 watts, E = 240 volts

Step 3: → The formula is $I = P/E$.

Step 4: → The answer is $I = \dfrac{7{,}500 \text{ watts}}{240 \text{ volts}} = 31.25$ amperes.

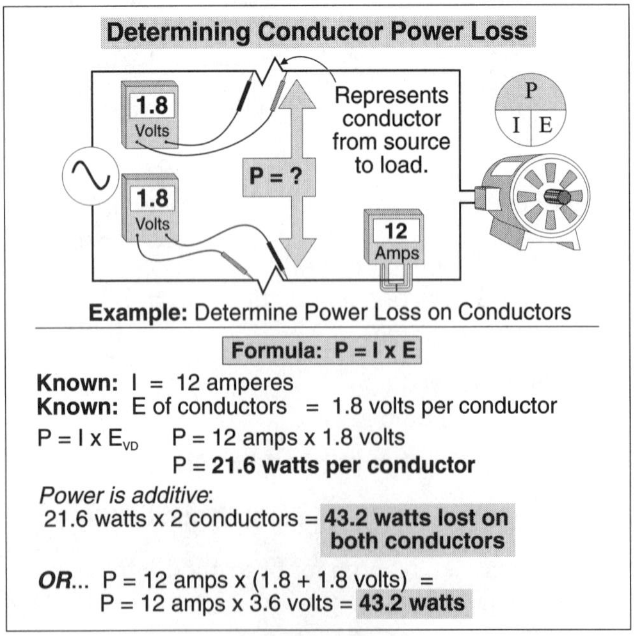

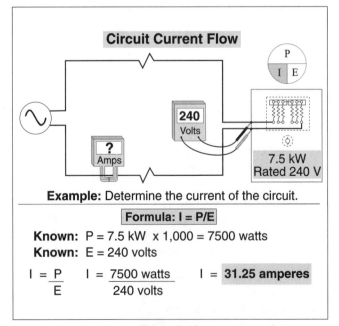

Figure 1-17
Power Example

Figure 1-18
Current Example

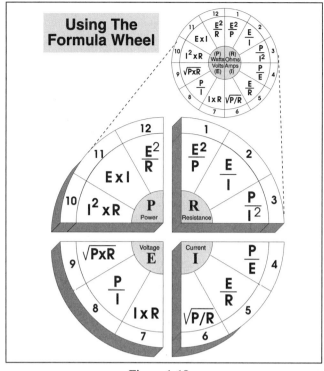

Figure 1-19

Power Wheel

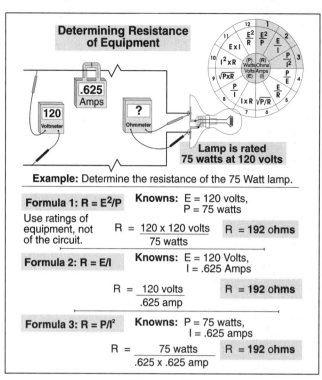

Figure 1-20

Resistance Example

1–20 FORMULA WHEEL

The formula wheel combines the Ohm's law and the PIE formulas. The formula wheel is divided up into four sections with 3 formulas in each section, Fig. 1–19.

❑ **Resistance Example**

What is the resistance of a 75-watt light bulb that is rated 120 volts, Fig. 1–20?

(a) 100 ohms (b) 192 ohms
(c) 225 ohms (d) 417 ohms

- Answer: (b) 192 ohms
  $R = E^2/P$
  E = 120-volt rating
  P = 75 watt rating
  $R = 120 \text{ volts}^2/75 \text{ watts}$
  R = 192 ohms

❑ **Current Example**

What is the current flow of a 10-kW heat strip connected to a 230-volt power supply, Fig. 1–21?

(a) 13 amperes (b) 26 amperes
(c) 43 amperes (d) 52 amperes

- Answer: (c) 43 amperes
  I = P/E
  P = 10,000 watts
  E = 230 volts
  I = 10,000 watts/230 volts
  I = 43 amperes

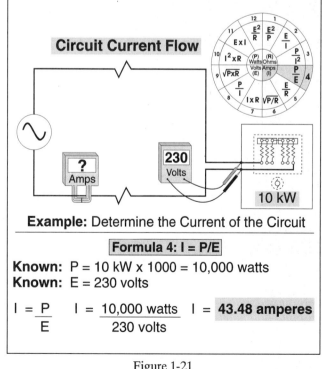

Figure 1-21

Current Example

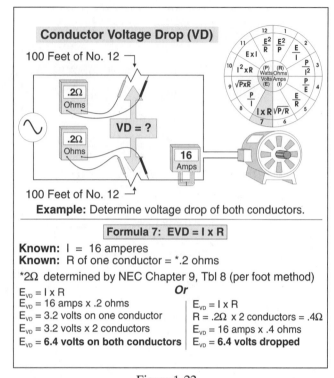

Figure 1-22
Voltage Example

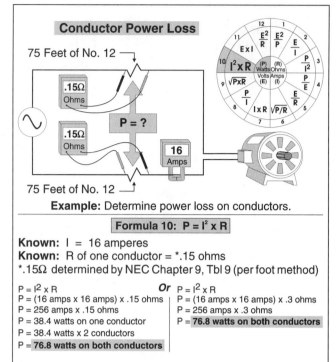

Figure 1-23
Power Example

❏ **Voltage Example**

What is the voltage drop of 200 feet of No. 12 conductor that carries 16 amperes, Fig. 1–22?

Note. The resistance of No. 12 conductor is 2-ohms per 1,000 feet.

(a) 1.6 volts drop (b) 2.9 volts drop (c) 3.2 volts drop (d) 6.4 volts drop

- Answer: (d) 6.4 volts drop

  E = I × R, I = 16 amperes, R = 0.2 ohm, 2 ohms/1,000 = .002-ohm per foot × 200 = .4 ohm

  E = 16 amperes × 0.4 ohm

  E = 6.4 volts drop

❏ **Power Example**

The total resistance of two No. 12 conductors, 75 feet long, is 0.3 ohm (.15 ohm for each conductor). The current of the circuit is 16 amperes. What is the power loss of the conductors in watts per hour. Fig. 1–23?

(a) 19 watts (b) 77 watts (c) 172.8 watts (d) none of these

- Answer: (b) 77 watts

  $P = I^2R$, I = 16 amperes, R = 0.3 ohm

  P = 16 amperes^2 × 0.3 ohm

  P = 76.8 watts per hour

1–21 POWER CHANGES WITH THE SQUARE OF THE VOLTAGE

The *power* consumed by a resistor is dramatically affected by the voltage applied. The power is affected by the square of the voltage and directly to the resistance, Fig. 1–24.

$$P = \frac{E^2}{R}$$

❏ **Power Changes With Square Of Voltage Example**

What is the power consumed of a 9.6-kW heat strip rated 230 volts connected to a 115-, 230- and 460-volt power supplies? The resistance of the heat strip is 5.51 ohms, Fig. 1–25.

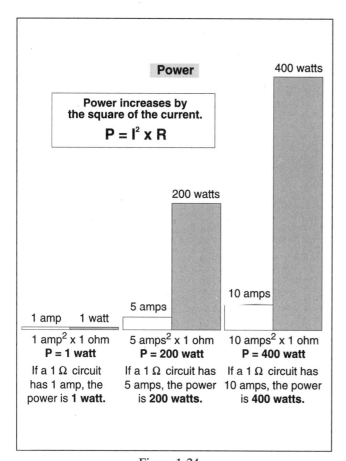

Figure 1-24
Power Calculations

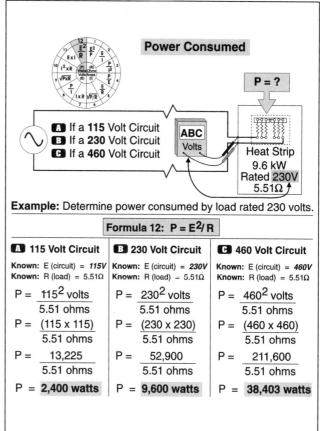

Figure 1-25
Power Changes with the Square of the Voltage

Step 1: → What is the question? It is, what is the power consumed, P?
Step 2: → What do you know about the heat strip?
E = 115 volts, 230 volts and 460 volts, R = 5.51 ohms
Step 3: → The formula to determine power is, $P = E^2/R$.
E = 115 volts, 230 volts, and 460 volts, R = 5.51 ohms
P = 115 volts^2/5.51 ohms, P = 2,400 watts
P = 230 volts^2/5.51 ohms, P = 9,600 watts (2 times volts = 4 times power)
P = 460 volts^2/5.51 ohms, P = 38,403 watts (4 times volts = 16 times power)

1–22 ELECTRIC METERS

Basic electrical meters use a helically wound coil of conductor, called a *solenoid*, to produce a strong electromagnetic field to attract an iron bar inside the coil. The iron bar that moves inside the coil is called an *armature*. When a meter has positive (+) and negative (–) shown for the meter leads, the meter is said to be *polarized*. The negative (–) lead must be connected to the negative terminal, and the positive (+) lead must be connected to the positive terminal of the power source.

Ammeter

An ammeter is a meter that has a *helically* wound coil, and it directly utilizes the circuit energy to measure direct current. As current flows through the meter's helically wound coil, the combined electromagnetic field of the coil draws in the iron bar. The greater the current flow through the meter's coil, the greater the electromagnetic field, and the further the armature is drawn into the coil. Ammeters are connected in series with the circuit and are used to measure only direct current, Fig. 1–26.

Ammeters are connected in *series* with the power supply and the load. If the ammeter is accidentally connected in *parallel* to the power supply, the current flow through the meter will be extremely high. The excessive high current through the meter will destroy the meter due to excessive heat.

If the ammeter is not connected to the proper *polarity* when measuring direct current, the meter's needle will quickly move in the reverse direction and possibly damage the meter's calibration.

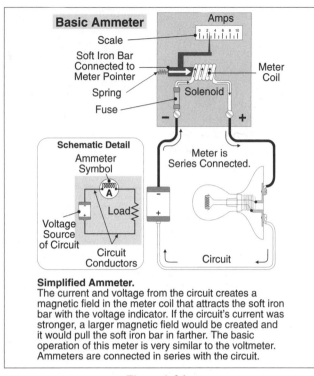

Figure 1-26
Basic Ammeter

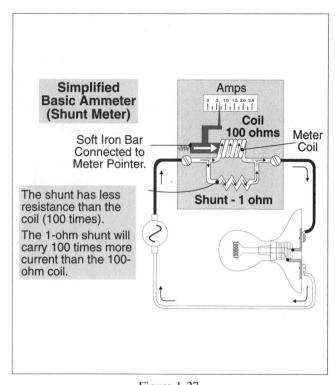

Figure 1-27
Shunt Ammeter Current Example

Ammeters that measure currents larger than 10/1,000 ampere (10 milliamperes, 10 mA) often contain a device called a *shunt*. The shunt bar is placed in parallel with the meter coil. This permits the current flow to divide between the meter's coil and the shunt bar. The current through the meter coil depends on the resistance of the shunt bar.

❑ **Shunt Ammeter Current Example**
What is the current flow through the meter if the shunt bar is 1 ohm and the coil is 100 ohms. Fig. 1–27?

(a) the same as the shunt
(b) 10 times less than the shunt
(c) 100 times less than the shunt
(d) 1,000 times less than the shunt

• Answer: (c) 100 times less

Since the shunt is 100 times less resistant than the meter's coil, the shunt bar will carry 100 times more current than the meter's coil, or the meter's coil will carry 100 times less current than the shunt bar.

Clamp-on Ammeter

Clamp-on ammeters are used to measure alternating current. They are connected *perpendicular* around the conductor (90 degrees) without breaking the circuit. A clamp-on ammeter indirectly utilizes the circuit energy by *induction* of the electromagnetic field. The clamp on ammeter is actually a transformer (sometimes called a *current transformer*). The primary winding is the phase conductor, and the secondary winding is the meter's coil, Fig. 1–28.

The electromagnetic field around the phase conductor expands and collapses, which causes electrons to flow in the meter's circuit. As current flows through the meter's coil, the electromagnetic field of the meter draws in the armature. Since the phase conductor serves as the primary (one turn), the current to be measured must be high enough to produce an electromagnetic field that is strong enough to cause the meter to operate.

Ohmmeter and Megohmeter

Ohmmeters are used to measure the resistance of a circuit or component and can be used to locate open circuits or shorts. An ohmmeter has an armature, a coil and it's own power supply, generally a battery. Ohm meters are always connected to deenergized circuits and polarity is not required to be observed. When an ohm meter is used, current flows through the meter's coil causing an electromagnetic field around the coil and drawing in the armature. The greater the current flow, as a result of lower resistance ($I = E/R$), the greater the magnetic field and the further the armature is drawn into the coil. A short circuit will be indicated by a reading of zero and a circuit that is opened will be indicated by infinity (∞), Fig. 1–29.

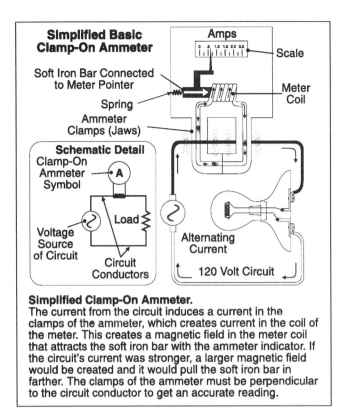

Figure 1-28

Clamp–on Ammeters

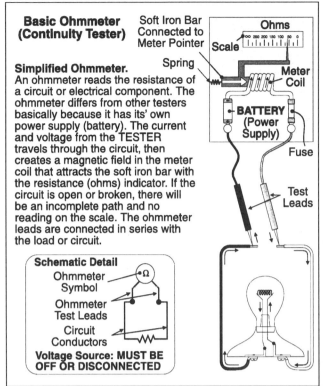

Figure 1-29

Ohm Meter

The *megger*, also called *megohmmeter* or *megohmer* is an instrument designed to measure very high resistances, such as those found in cable insulation between motor or transformer windings.

Voltmeter

Voltmeters are used to measure both direct current and alternating current voltage. A voltmeter contains a *resistor* in *series* with the coil and utilizes the circuit energy for its operation. The purpose of the resistor in the meter is to reduce the current flow through the meter. As current flows through the meter's helically wound coil, the combined electromagnetic field of the coil draws in the iron bar. The greater the circuit voltage, the greater the current flow through the meter's coil ($I=E/R$). The greater the current flow through the meter's coil, the greater the electromagnetic field and the further the armature is drawn into the coil, Fig. 1–30.

Polarity must be observed when connecting voltmeters to direct current circuits. If the meter is not connected to the proper polarity, the meter's needle will quickly move in the reverse direction and damage the meter's calibration. Polarity is not required when connecting voltmeters to alternating current circuits.

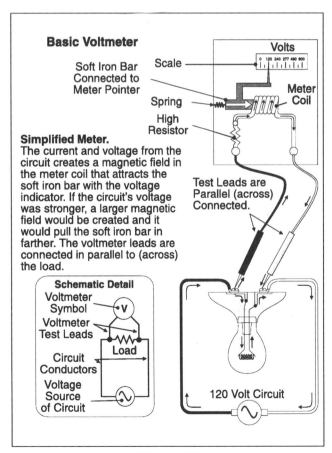

Figure 1-30

Voltmeter

Unit 1 – Electrician's Math and Basic Electrical Formulas

Part A – Electrician's Math (• Indicates that 75% or less get the question correct)

1–1 Fractions

1. The decimal equivalent for the fraction ½ is _____.
   (a) .5   (b) 5   (c) 2   (d) .2

2. The decimal equivalent for the fraction ⁴⁄₁₈ is _____.
   (a) 4.5   (b) 1.5   (c) 2.5   (d) .2

1–2 Kilo

3. What is the kW of a 75-watt load?
   (a) 75 kW   (b) 7.5 kW   (c) .75 kW   (d) .075 kW

1–3 Knowing Your Answer

4. • The output VA of a transformer is 100 VA. The transformer efficiency is 90 percent. What is the transformer input? Note: Because of efficiency, input is always greater than output.
   (a) 90 watts   (b) 110 watts   (c) 100 watts   (d) 125 watts

1–4 Percentages

5. When changing a percent value to a decimal or whole number, simply move the decimal point two places to the _____.
   (a) right   (b) left   (c) depends   (d) none of these

6. The decimal equivalent for 75 percent is _____.
   (a) .075   (b) .75   (c) 7.5   (d) 75

7. The decimal equivalent for 225 percent is _____.
   (a) 225   (b) 22.5   (c) 2.25   (d) .225

8. The decimal equivalent for 300 percent is _____.
   (a) .03   (b) .3   (c) 3   (d) 30.0

1–5 Multiplier

9. The method of increasing a number by another number is call the _____.
   (a) percentage   (b) decimal
   (c) fraction   (d) multiplier

10. An overcurrent protection device (breaker or fuse) is required to be sized no less than 115 percent of the load. If the load is 20 amperes, the overcurrent protection device would have to be sized at no less than _____.
    (a) 20 amperes   (b) 23 amperes   (c) 17 amperes   (d) 30 amperes

11. The maximum continuous load on an overcurrent protection device is limited to 80 percent of the device rating. If the device is rated 90 amperes, the maximum continuous load is _____ amperes.
    (a) 72   (b) 90   (c) 110s   (d) 125

12. A 50-ampere rated wire is required to be adjusted for temperature. If the correct multiplier is .80, which of the following statements is/are correct?
    (a) the answer will be less than 50 amperes
    (b) 80 percent of the ampacity (50) can be used
    (c) the formula is 50 amperes × .8
    (d) all the above

1–6 Percent Increase

13. The feeder demand load for an 8-kW load, increased by 20 percent is _____ kVA.
    (a) 8 (b) 9.6 (c) 6.4 (d) 10

1–7 Percentage Reciprocals

14. What is the reciprocal of 125 percent?
    (a) .8
    (b) 100 percent
    (c) 125 percent
    (d) none of these

15. A continuous load requires an overcurrent protection device sized no smaller than 125 percent of the load. What is the maximum continuous load permitted on a 100-ampere overcurrent protection device?
    (a) 100 amperes
    (b) 125 amperes
    (c) 80 amperes
    (d) 110 amperes

1–8 Squaring

16. What is the power consumed in watts of a No. 12 conductor that is 100 feet long and has a resistance of (R) 0.2 ohm, the current (I) in the circuit is 16 amperes? Formula: Power = $I^2 \times R$
    (a) 75 watts
    (b) 50 watts
    (c) 100 watts
    (d) 200 watts

17. • What is the area in square inches of a 2-inch raceway? Formula: Area = πr^2, $\pi = 3.14$, r = radius, which equals ½ the diameter.
    (a) 1 square inch
    (b) 2 square inches
    (c) 3 square inches
    (d) 4 square inches

18. The numeric equivalent of 4^2 is _____.
    (a) 2 (b) 8 (c) 16 (d) 32

19. The numeric equivalent of 12^2 is _____.
    (a) 3.46 (b) 24 (c) 144 (d) 1,728

1–9 Square Root

20. What is the square root of 1,000 ($\sqrt{1000}$)?
    (a) 3 (b) 32 (c) 100 (d) 500

21. The square root of the number 3 is _____.
    (a) 1.732 (b) 9 (c) 729 (d) 1.5

1–10 Rounding Off

22. • The sum of 5, 7, 8 and 9 is _____ approximately.
    (a) 20 (b) 25 (c) 30 (d) 35

1–11 Parentheses

23. What is the distance of two No. 14 conductors carrying 16 amperes with a voltage drop of 10 volts? Formula: $D = \dfrac{(4{,}100 \text{ circularmils} \times 10 \text{ volts drop})}{(2 \text{ wires} \times 12.9 \text{ amperes} \times 16 \text{ ohms})}$
    (a) 50 feet (b) 75 feet (c) 100 feet (d) 150 feet

24. What is the current in amperes of a three-phase, 18-kW, 208-volt load? Formula: $I = \dfrac{W}{E \times \sqrt{3}}$
    (a) 25 amperes
    (b) 50 amperes
    (c) 100 amperes
    (d) 150 amperes

1–12 Transposing Formulas

25. • Transform the formula $CM = \dfrac{(2 \times K \times I \times D)}{VD}$ to find the distance of the circuit.

    (a) $D = VD \times CM$
    (b) $D = \dfrac{(2 \times K \times I \times C \times M)}{VD}$
    (c) $D = (2 \times K \times I) \times CM$
    (d) $D = \dfrac{CM \times VD}{2 \times K \times I}$

26. If $I = P/E$, which of the following statements contain the correct transposed formula?
    (a) $P = E/I$
    (b) $P = 1/E$
    (c) $P = I \times E$
    (d) $P = I^2 E$

Part B – Basic Electrical Formulas

1–13 Electrical Circuits

27. An electric circuit consists of the _____.
    (a) power source
    (b) conductors
    (c) load
    (d) all of these

1–14 Electron Flow

28. Inside the power source, electrons travel from the positive terminal to the negative terminal.
    (a) True
    (b) False

1–15 Power Source

29. The polarity of a(n) _____ current power sources never change. One terminal is always negative and the other is always positive. _____ current flows out of the negative terminal of the power source at the same polarity.
    (a) static, Static
    (b) direct, Direct
    (c) alternating, Alternating
    (d) all of the above

30. _____ current power sources produce a voltage that has a constant change in polarity and magnitude in one direction exactly the same as it does in the other.
    (a) Static
    (b) Direct
    (c) Alternating
    (d) All of the above

1–16 Conductance And Resistance

31. Conductance is the property of metal that permits current to flow. The best conductor, in order of conductivity, are: _____.
    (a) gold, silver, copper, aluminum
    (b) copper, gold, copper, aluminum
    (c) gold copper, silver, aluminum
    (d) silver, copper, gold, aluminum

1–17 Electrical Circuit Values

32. The _____ is the pressure required to force one ampere of electrons through a one ohm resistor.
    (a) ohm
    (b) watt
    (c) volt
    (d) ampere

33. All conductors have resistance that opposes the flow of electrons. Some materials have more resistance than others. _____ has the lowest resistance and _____ is more resistant than copper.
    (a) Silver, gold
    (b) Gold, aluminum
    (c) Gold, silver
    (d) None of these

34. Resistance is represented by the letter R, and it is expressed in _____.
    (a) volts
    (b) impedance
    (c) capacitance
    (d) ohms

35. The opposition to the flow of current can be thought of as restricting the flow of electrons in the circuit. Every component of an electric circuit contains resistance, except the generator or transformer.
    (a) True   (b) False

36. In electrical systems, the volume of electrons that moves through a conductor is called the _____ of the circuit.
    (a) intensity   (b) voltage   (c) power   (d) resistance

37. The rate of work that can be produced by the movement of electrons is called _____.
    (a) voltage   (b) current   (c) power   (d) none of these

1–18 Ohm's Law (I = E/R)

38. The Ohms's law formula, $I = E/R$ demonstrates that current is _____ proportional to the voltage, and _____ proportional to the resistance.
    (a) indirectly, inversely   (b) inversely, directly
    (c) inversely, indirectly   (d) directly, inversely

39. In an alternating current circuit, which factors oppose current flow?
    (a) resistance   (b) capacitance reactance
    (c) induction reactance   (d) all of these

40. The opposition to current flow due in an alternating current circuit is called _____ and is often represented by the letter Z.
    (a) resistance   (b) capacitance   (c) induction   (d) impedance

41. • What is the voltage drop of two No. 12 conductors supplying a 16-ampere load, located 100 feet from the power supply?
    Formula: $E_{VD} = I \times R$, I = 16 amperes, R = 200 feet of No. 12 wire = 0.4 ohm
    (a) 6.4 volts   (b) 12.8 volts   (c) 1.6 volts   (d) 3.2 volts

42. What is the resistance of the circuit conductors when the conductor voltage drop is 7.2 volts and the current flow is 50 amperes?
    (a) 0.14 ohm   (b) 0.3 ohm   (c) 3 ohms   (d) 14 ohms

1–19 Pie Circle Formula

43. What is the power loss in watts for a conductor that carries 24 amperes and has a voltage drop of 7.2 volts?
    (a) 175 watts   (b) 350 watts   (c) 700 watts   (d) 2,400 watts

44. • What is the power of a 10-kW heat strip rated 240 volts, when connected to a 208-volt circuit?
    (a) 8 kW   (b) 9 kW   (c) 12 kW   (d) 15 kW

1–20 Formula Wheel

45. • The formulas listed in the formula wheel apply to _____.
    (a) direct current circuits only
    (b) alternating current circuits with unity power factor
    (c) a and b
    (d) none of these

46. When working any formula, the key to getting the correct answer, is following these four simple steps:
    Step 1: → Know what the question is asking.
    Step 2: → Determine the knowns of the circuit or resistor.
    Step 3: → Select the formula.
    Step 4: → Work out the formula calculation.
    (a) True   (b) False

47. The total resistance of two No. 12 conductors, 150 feet long is 0.6 ohm, and the current of the circuit is 16 ampere. What is the power loss of the conductors in watts per hour?
    (a) 75 watts   (b) 150 watts   (c) 300 watts   (d) 600 watts

48. • What is the conductor power loss in watts for a 120-volt circuit that has a 3 percent voltage drop and carries a current flow of 12 amperes? The load operates 24 hours per day, 365 days each year.
    (a) 43 watts (b) 86 watts (c) 172 watts (d) 722 watts

49. • What does it cost per year (8.6 cents per kW) for the power loss of a conductor? The No. 12 circuit conductor resistance is 0.3 ohm and the current flow is 12 amperes.
    (a) $33.00 (b) $13.00 (c) $130.00 (d) $1,300.00

1–21 Power Changes With The Square Of The Voltage

50. • What is the power consumed of a 10-kW heat strip rated 230 volts, connected to a 115-volt circuit?
    (a) 10 kW (b) 2.5 kW (c) 5 kW (d) 20 kW

51. A(n) _____ is connected in series with the load.
    (a) watt-hour meter (b) voltmeter (c) power meter (d) ammeter

52. • Ammeters are used to measure _____ current and are connected in series with the circuit and are said to shunt the circuit.
    (a) direct current (b) alternating current
    (c) a and b (d) none of these

53. Clamp on ammeters have one coil connected _____ around the circuit conductor.
    (a) series (b) parallel
    (c) series-parallel (d) at right angles (perpendicular)

54. An ohmmeter has a _____ connected in series with the resistor. As current flows through the meter coil, the magnetic field around the coil draws in the soft iron bar. The greater the current flow through the circuit, the greater the magnetic field and the further the iron bar is drawn into the coil.
    (a) coil and resistor (b) coil and power supply
    (c) two coils (d) none of these

Challenge Questions

1–2 Kilo

55. • kVA is equal to _____.
    (a) 100 VA (b) 1,000 volts (c) 1,000 watts (d) 1,000 VA

1–16 Conductance And Resistance

56. • _____ is not an insulator.
    (a) Bakelite (b) Oil (c) Air (d) Salt water

1–17 Electrical Circuit Values

57. • _____ is not the force that moves electrons.
    (a) EMF (b) Voltage (c) Potential (d) Current

58. Conductor resistance varies with _____.
    (a) material (b) voltage (c) current (d) power

1–18 Ohm's Law (I = E/R)

59. • If the contact resistance of a connection increases and the current remains the same, the voltage drop across the connection will _____.
    (a) increase (b) decrease (c) remain the same (d) cannot be determined

60. • To double the current of a circuit when the voltage remains constant, the R (resistance) must be _____.
    (a) doubled (b) reduced by half
    (c) increased (d) none of these

61. • An ohmmeter is being used to test a relay coil. The equipment instructions indicate that the resistance of the coil should be between 30 and 33 ohms. The ohmmeter indicates that the actual resistance is less than 22 ohms. This reading would most likely indicate _____.
    (a) the coil is okay   (b) an open coil
    (c) a shorted coil   (d) a meter problem

1–20 Formula Wheel

62. • To calculate the power consumed by a resistive appliance, one needs to know _____.
    (a) voltage and current   (b) current and resistance
    (c) voltage and resistance   (d) any of these

63. • The number of watts of heat given off by a resistor is expressed by the formula $I^2 \times R$. If 10 volts is applied to a 5 ohm resistor, then ____ watts of heat will be given off.
    (a) 500   (b) 250   (c) 50   (d) 20

64. • Power loss in a circuit because of heat can be determined by the formula _____.
    (a) $P = R \times I$ (b) $P = I \times R$ (c) $P = I^2 \times R$ (d) none of these

65. • If current remains the same and resistance increases, the circuit will consume _____ power.
    (a) higher   (b) lower

66. When a lamp that is rated 500-watts at 115 volts is connected to a 120-volt power supply, the current of the circuit will be _____ amperes. *Tip:* Does power remain the same when voltage is changed?
    (a) 3.8   (b) 4.5   (c) 2.7   (d) 5.5

1–21 Power Changes With The Square Of The Voltage

67. A toaster will produce less heat at low voltage. As a result of the low voltage, _____.
    (a) its total watt output will decrease   (b) the current flow will decrease
    (c) the resistance is not changed   (d) all of these

68. • When a resistive load is operated at a voltage 10 percent higher than the nameplate rating of the appliance, the appliance will _____.
    (a) have a longer life   (b) draw a lower current
    (c) use more power   (d) none of these

69. • A 1,500-watt heater rated 230 volts is connected to a 208-volt supply. The power consumed for this load is _____ watts. *Tip:* When the voltage is reduced, will the power be greater or less!
    (a) 1,625   (b) 1,750   (c) 1,850   (d) 1,225

70. • The total resistance of a circuit is 12 ohms; the load is 10 ohms, and the wire 2 ohms. If the current of the circuit is 3 amperes, then the power consumed by the circuit conductors is _____ watts.
    (a) 28   (b) 18   (c) 90   (d) 75

1–22 Electric Meters

71. • The best instrument for detecting an electric current is a(n) _____.
    (a) ohmmeter   (b) voltmeter   (c) ammeter   (d) wattmeter

72. The polarity of a circuit being tested must be observed when connecting an ohmmeter to _____.
    (a) an alternating current circuit
    (b) a direct current circuit
    (c) any circuit
    (d) polarity doesn't matter because the circuit is not energized

73. • When the test leads of an ohmmeter are shorted together, the meter will read _____ on the scale.
    (a) 0 (zero ohms)   (b) high
    (c) either a or b   (d) both a and b

74. • A short circuit is indicated by a reading of _____ when tested with an ohmmeter.
    (a) zero   (b) ohms   (c) infinity   (d) R

75. Voltmeters are used to measure _____.
    (a) voltages to ground   (b) voltage differences
    (c) AC voltages only   (d) DC voltages only

76. Voltmeters must be connected in _____ with the circuit component being tested.
    (a) series   (b) parallel   (c) series-parallel   (d) multiwire

77. To measure the voltage across a load, you would connect a(n) _____.
    (a) voltmeter across the load
    (b) ammeter across the load
    (c) voltmeter in series with the load
    (d) ammeter in series with the load

78. A voltmeter is connected in _____ to the load.
    (a) series   (b) parallel   (c) series-parallel   (d) none of these

79. • In the course of normal operation, the least effective instrument in indicating that a generator may overheat because it is overloaded is a(n) _____.
    (a) ammeter   (b) voltmeter   (c) wattmeter   (d) none of these

80. • A direct-current voltmeter (not a digital meter) can be used to measure _____.
    (a) power   (b) frequency   (c) polarity   (d) power factor

81. • Polarity must be observed when connecting a voltmeter to _____ current circuit.
    (a) an alternating   (b) a direct
    (c) any   (d) polarity doesn't matter

82. The minimum number of wattmeters necessary to measure the power in the load of a balanced three-phase, four-wire system is _____.
    (a) 1   (b) 2   (c) 3   (d) 4

Introduction to the National Electrical Code

The National Electrical Code (NEC) is Volume 70 (NFPA–70) of the National Fire Protection Association © (NFPA). This Code has been published, edited, and revised since 1897 and sponsored by the NFPA since 1911. It is used not only throughout most of the United States but in many countries throughout the world.

A comprehensive knowledge of the NEC takes years for most people. Whether you need to learn the Code for exam preparation or for everyday use with your job, learning the NEC is the same. Learn to use the NEC to find answers to questions and information needed on the exam as well as for the job.

This chapter is designed to help you learn the National Electrical Code through understanding its structure with study questions. For most people, the NEC is an intimidating and complicated book. It is about 875 pages long, and it is full of technical information. On the other hand, the NEC is generally well organized and structured. Understanding how the NEC is structured and organized will help you use it effectively. The study questions are for practice in using the NEC. As with most learning experiences, practice is the best way to learn. All the study questions in this workbook are in order of appearance in the NEC. This will help you learn the structure better and use your study time more efficiently.

Suggestions

You should read this textbook and highlight those areas in your Code Book that are important to you. If there is any area of this textbook you don't understand, don't worry about it now. Simply highlight the textbook and later you can review these highlighted areas. Having a general understanding of the Code will make it easier to understand those difficult areas when you review them later.

Caution

Before you get into the Code, remember the following tips. Most of the articles are broken down into parts. The information following a part heading applies to that part only. ALWAYS remember which part you are in. You wouldn't want to be reading a rule in the over-600-volt systems for an installation under 600 volts!

Warning
Understanding the Code can turn good people into mouthy know-it-alls, who go around showing their co-workers how brilliant they are. If you are going to attempt to show how much you know, do it in a positive way. Be constructive with your comments. If you know more about the Code than your supervisor and inspector (and everyone thinks he or she does), be careful how you tell them they are wrong.

Getting The Most
If your are using this workbook for exam preparation, complete the following steps. Adjust as necessary to fit your schedule.
Step 1: → Read all of the text in this workbook.
Step 2: → Tab your copy of the NEC according to the instructions in this Chapter.
Step 3: → Highlight the NEC Section as you answer each question.
After you have completed each exam, find the Code sections for the questions you missed and highlight those sections in a different color. This helps to indicate particularly difficult Code section.
Tip: When an item in the NEC index helps you find an answer to a study question, highlight that item in the index.
Step 4: → Answer every question in the workbook. Read each question very carefully. One word can change the entire meaning of the question.

Exceptions: Generally NEC rules (sections) take priority over NEC exceptions. Don't use exceptions unless the wording of the question is the same or very similar to the wording of the exception. The following example questions show when an exception should be used.

❑ **Example Question:** Service-entrance conductors shall not be spliced.
    (a) True   (b) False

To answer this question, go to section 230–46 of your Code Book, which says service-entrance conductors shall not be spliced. There are several exceptions that permit splicing of service-entrance conductors under specific conditions. The answer is **(a) True** Service conductors cannot be spliced.

❑ **Example of using the Exception:** Service-entrance conductors can be spliced where an underground wiring method is changed to another type of wiring method.
    (a) True   (b) False

The general rule is service conductors cannot be spliced but the question contains wording similar to 230–46 Exception No. 3. The answer to this question is *(a) True*. Service conductors underground can be spliced under certain conditions.

Understanding The Terms And Theories

Terms
There are many technical words and phrases used in the NEC. It is crucial that you understand the meanings of words like *ground, grounded, grounding,* and *neutral*. If you do not clearly understand the terms used in the Code, you will not understand the rule itself.

It isn't always the technical words that require close attention. In the NEC, even the simplest words can make a big difference. The word *or* can mean: alternate choices for equipment, wiring methods, or other requirements. Sometimes, the word *or* can mean any item in a group. The word *and* can mean; an additional requirement or any item in a group. Be aware of how simple words are used.

Theory
Theory is simply understanding how and why things work the way they do. Why can a bird sit on an energized power line without getting shocked? Why does installing a lot of wires close together reduce the amount of current they can individually carry? Why can't a single conductor be installed in a metal raceway? Why can the circuit breaker to a motor circuit be 40 amperes, when the conductors are only rated 20 amperes? Why can't a 20-ampere receptacle be installed on a 15-ampere circuit? If you understand why or how things work you will probably understand the Code rules.

This book does not cover all theoretical aspects of the NEC. As much as we would like to, I can't put everything into one book.

Layout Of The Nec
Contrary to popular belief, the NEC is a fairly well-organized document. The NEC does have some areas that are somewhat vague. The slang term in the trade for a vague Code rule is **gray areas**, but overall, the NEC is very good. Understanding the NEC structure and writing style is extremely important for understanding and using the Code Book.

The main components of the NEC are:
1. Chapters (categories)
2. Articles (subjects)
3. Parts (sub-headings)
4. Sections and Tables (Code)

5. Exceptions (Code)
6. Fine Print Notes (explanatory material, *not* Code)
7. Definitions (Code)
8. Superscript Letter X
9. 1996 Code Changes and Deletions

A short description of each component follows:

1. Chapters

There are nine chapters in the NEC. Each chapter is a group of articles, parts, sections, and tables. The nine chapters fall into four categories:

Chapters 1 through 4: General Rules (the scope of this book)

Chapters 5 through 7: Specific Rules (Motion Picture Projectors, Recreational Vehicles, Cranes and Hoists, X-Ray Equipment, etc.)

Chapter 8: Communication Systems (Radio and Television Equipment and Cable TV Systems)

Chapter 9: Physical Properties and NEC Examples (Tables for conductor and raceways)

2. Articles

The NEC contains approximately 125 articles. An article is a specific subject, such as grounding, services, feeders, branch circuits, fixtures, motors, appliances, air conditioning, etc.

3. Parts

When an article is sufficiently large, the article is subdivided into parts. The parts break down the main subject of the article into organized groups of information. For example, Article 230 contains eight parts, such as Part A–General, Part B–Overhead Service Conductors, and Part H–Services Exceeding 600 Volts, Nominal.

4. Sections and Tables

The actual NEC Code rule is called a section and is identified with numbers and letters. Many in the electrical industry use the slang term article when referring to a Code section. A Code section may be broken down into subsections (by letters in parentheses) and each subsection may be broken down further by numbers in parentheses. For example, the section that requires all receptacles in a dwelling unit bathroom to be GFCI protected is Section 210–8(a)(1).

Many Code sections contain tables, which are a systematic list of Code rules in an orderly arrangement. Example, Table 310-16 contains the ampacity of conductors.

5. Exceptions

Exceptions are in *italic* type and provide an alternate choice to a specific rule. There are two types of exceptions: One exception is mandatory and the other is permissible. When a rule has several exceptions, those exceptions with mandatory requirements are listed before those with permissible exceptions.

A mandatory exception uses the words *shall* or *shall not* in the wording. The word *shall* in an exception means that if you are using the exception, you shall do it in a particular way. The term *shall not* in an exception means that you shall not do something. A permissible exception uses such words as *shall be permitted*, this means that it's okay to do it in this way.

6. Fine Print Notes

Fine Print Notes (FPN) are explanatory material, not Code rules. Fine print notes attempt to clarify a rule or give assistance, but they are not a Code requirement. For example, FPN No. 4 of Section 210–19(a) states that the voltage drop for branch circuits should not exceed 3% of the circuit voltage. This is not a Code requirement but only a suggestion; therefore, there is no NEC requirement for conductor voltage drop.

7. Definitions

Definitions are listed in Article 100 and throughout the NEC. In general, the definitions listed in Article 100 apply to more than one Code article. Definitions at the beginning of a specific article apply to that article only. Definitions in a part of an article apply only to that part, and definitions in a specific Code section only apply to that section.

8. Superscript Letter ^{x}

This superscript letter is used only in Chapter 5. The superscript letter x means the material was extracted from other NFPA documents. Appendix A, at the rear of the Code Book, identifies the NFPA document and the section(s) that the material was extracted from.

9. Changes and Deletions

Changes and deletions to the 1993 NEC are identified in the margins of the 1996 NEC in the following manner: Changes are marked with a vertical line (|) and deletions of a Code rule are identified by a bullet (•). For those of you that haven't had a lot of experience using the National Electrical Code Book, it is important that you understand the structure of the NEC. The four main components of the NEC are Chapters, Articles, Parts, and Sections /Tables. A short definition for each component is provided below:

Understanding A Code Section

What does Section 210–8(b)(1) mean? The first number indicates the chapter (Chapter 2). The numbers before the hyphen indicate the article (Article 210). The numbers after the hyphen indicate the Code section (Section 8), and the letter and numbers in parentheses after the section indicate the subsection. In this case, Section 210-8(b)(1) is the rule that requires all receptacles in commercial and industrial bathrooms to be GFCI protected.

CAUTION: The "Parts" of a Code article are not included in the section numbers. Because of this, we have a tendency to forget what "Part" the Code rule is relating to. For example, Table 110–34 gives the dimensions of working space clearances in front of electrical equipment. If we are not careful, we might think that this table applies to all electrical installations. But, Section 110–34 is located in Part B Over-600-volt Systems of Article 110! The working clearance rule for under-600-volt systems is located in Part A of Article 110, Table 110–16.

How To Use The NEC

Everybody wants to be able to use the NEC quickly, but unfortunately it doesn't work that way. How you use the NEC depends on your experience. An experienced Code user rarely uses the index. He or she knows which Code articles contain which rules and uses the Table of Contents instead of the index.

Let me give you an example of how handy the Table of Contents can be. What Code rule indicates the maximum number of disconnects permitted for a service? Answer: You need to know that there is an Article for Services - 230, and this article has a Part for Disconnection Means. Use the Table of Contents to find that the answer is around page 90. I'm sorry that I can't tell you the exact page number, because this book was written about six months before the actual Code Book was printed.

Once you are familiar with the NEC (completed this book) you should use the index to find your way around the Code. The index lists subjects in alphabetical order, and is usually the best place to start for specific information. Unlike most books, the NEC index does not list page numbers, it lists sections, tables, articles, parts, or appendices by numbers.

Finding Information In The NEC

Finding information in the NEC is very difficult if you don't understand the layout of its contents. In this exercise, we will explore some of the problems and challenges you will face while using the Code Book. Please use your Code Book for this exercise.

❏ **Example 1**

Find the ampacity of a No. 10 THHN. The key word is *ampacity*. Look up ampacities in the NEC index and you should see different sub-headings under the boldface word *ampacity*. The first sub heading is Conductors 310–15, Tables 310–16 through 310–19, B-310–1 through B-310–10, 310–61 through 310–84, and 374–6.

The first (310–15) and last (374–6) listings are section numbers. The B-310–1 through B-310–10 are tables, but the B indicates that they are found in Appendix B at the rear of the NEC. If you turn to the beginning of Appendix B in the rear of your Code Book, the very first line tells you that the appendix is not part of the Code and that it is included only for informational purposes.

Back to the index. If we turn to Section 310–15 of the NEC, it tells us that the ampacity can be determined by one of two methods. In almost all cases we will use the method listed in subsection (a), which refers to Table 310–16. Find Table 310–16 in your Code Book, the first column (down) is the conductor size in American Wire Gauge or kcmil. Go down until you find the number 10. Now move across to the right to the THHN column (third column), you will find that No. 10 THHN has an ampacity of 40 amperes.

This ampacity is used only when there are no more than three conductors in the raceway or earth, based on an ambient temperature of 86°F (conditions listed at the top of Table 310–16). What is the No. 10 THHN ampacity if there are 6 conductors in the raceway? What if the ambient temperature is 106ºF? What is ambient temperature anyway? It seems we're getting more questions than answers!

❏ **Example 2**

What is the maximum size circuit breaker (overcurrent protection device) that can be used to protect a No. 10 THHN conductor?

The Index has no reference to size, and the sub-headings for circuit breakers refers you to sections that will not give you the answer. Try looking up the word conductors, nothing there about breakers. Looks as if we are at a dead end.

The problem is not the NEC or it's index; it's our lack of knowledge of the NEC terms. If you don't know that a circuit breaker is an overcurrent protection device (Article 240) you will go in circles trying to use the index. The key words are conductor overcurrent protection. Go to the sub-headings overcurrent protection under conductors. The 240–3 and 240–4 listings mean Sections 240–3 and 240–4.

Go to Section 240–3 in the NEC, now we seem to be getting somewhere. Section 240–3 contains the requirement of conductor overcurrent protection. But, all this section says is that conductors must be protected at their ampacity as specified in Section 310–15. If you go to Section 310–15, you will find a FPN (Fine Print Note) that refers you to Table 310–16.

The ampacity of No. 10 THHN is 40 amperes. Can we really put a 40-ampere breaker on No. 10 THHN rated 40 amperes? No, the 40 amperes listed in the table has an obelisk † symbol next to 40. The obelisk refers to the note at the bottom of Table 310–16, which indicates that the maximum overcurrent protection for No. 10 conductors is 30 amperes. The No. 10 THHN conductor has an ampacity of 40 amperes, but the maximum overcurrent protection device (circuit breaker) permitted on the conductor is 30 amperes.

❏ **Example 3**

You are wiring a motor control circuit with No. 18 fixture wire (TF, TFF, TFFN). How do you know the ampacity of No. 18 TF? Why isn't the ampacity of No. 18 TF, TFF or TFFN listed in Table 310–16? Again, it seems that we're getting more questions than answers.

If you don't know that TF stands for fixture wire, you would not be able to find the answer. Now, go back to ampacities in the index. There is a sub-headings fixture wires, and it lists Table 402–3. If you look at Section 402–3, you will find all kinds of information; but, it doesn't mention anything about ampacity. So, where is the ampacity? Look at Section 402–5, which is immediately after Table 402–3. Why didn't the index simply say Table 402–5 instead of Table 402–3? I don't know. There are some things about the NEC that you just have to know, and the only way to learn is through practice and experience.

Many people say the Code takes you in circles; it sometimes does, but most often, you take yourself in circles. The problem generally isn't the Code, but your inexperience in using and understanding the NEC. Becoming proficient with the NEC is a matter of practice.

This book should help you understand the NEC terms, the structure of the NEC, how to use the NEC, and what the NEC means. This can get you off to a great start, but you need to practice what you are learning to become truly proficient, so be sure to complete each of the NEC practice questions at the end of each unit.

Customizing Your Code Book

You will want to customize your Code Book to increase your speed in finding exam answers in the NEC. It is very important that your Code Book be set up properly with tabs and highlighted code sections. The NEC Tab list that follows is to help you set up your Code Book for quick reference.

The following is a list of articles and sections of the National Electrical Code Book most commonly referred to in test questions. I recommended that you place colored tabs on the edge of the pages of your Code Book as specified. Put the Article title on the front of the tab and the article number on the back. Make a color code that is logical to you. The articles listed are to be used as a guide only. If you feel more articles are necessary you may add additional tabs as needed. However, please be aware that tabbing is a convenience to you and too many tabs defeat the purpose.

Place tabs on the following selections to the side of your Code Book. Abbreviate as necessary:

Table of Contents
Article Title
90 .... Introduction

Chapter 1. Tabs. General
100 .... Definitions
110 .... Requirements for Electrical Installations

Chapter 2 Tabs. Wiring and Protection
210 .... Branch Circuits
215 .... Feeders
220 .... Branch Circuit, Feeder, & Service Calculations
225 .... Outside Branch Circuits & Feeders
230 .... Services
240 .... Overcurrent Protection
250 .... Grounding

Table 250–94 Grounding Electrode Conductor
Chapter 3 Tabs. Wiring Methods and Material
300 .... Wiring Methods
Table 310–16 Ampacities of Copper/Aluminum
336 .... Nonmetallic Sheathed Cable, NM, NMC
346 .... Rigid Metal Conduit
364 .... Busways
370 .... Pull and Junction Boxes and Box Fill

Chapter 4 Tabs. Equipment for General Use
410	Lighting Fixtures
422	Appliances
430–22	Motor Branch Circuit Wire Size
430–32	Motor Heater Size Standard
430–52	Motor Branch Circuit Breaker or Fuse Size
430–62	Motor Feeder Breaker of Fuse Size
430–148	Motor Full Load Current
430–152	Branch Circuit Short Circuit Protection
450	Transformers and Transformer Vault

Chapter 5 Tabs. Special Occupancies
500	Hazardous Locations (Classified)
550–22	Mobile Home Parks
551–44	Recreational Vehicles Parks
555–5	Marinas and Boatyards

Chapter 6 Tabs. Special Equipment
600	Electric Signs
620	Elevators
680	Swimming Pools

Chapter 7 Tabs. Special Conditions
700	Emergency Systems
701	Legally Required Standby Systems
720	Circuits and Equipment Operating at Less than 50 volts
725	Class 1, 2, and 3 Circuits
760	Fire Alarm Circuits
770	Optical Fiber Cable

Chapter 8 Tabs. Communications Systems
800	Communication

Chapter 9 Tabs. Tables and Examples
Chapter 9	Chapter 9
Chapter 9 Table 4	Chapter 9, Table 5 (Side 2 of tab)
Chapter 9, Table 8	Circular Mils, Chapter 9, Table 8
Chapter 9 Table 9	Chapter 9, Table 9
Examples	Examples
Index	Index
Appendix "C"	Raceway Fill Tables

NEC Introduction And Practice Questions

Article 90 – Introduction. Article 90 is the introduction of the NEC. This is a short article. Read all of it.

Section 90–3. Code Arrangement is very important for understanding the structure of the Code. Understanding the structure of the Code is the key to finding answers and information quickly. It tells you that this Code is divided into the Introduction (Article 90) and nine chapters.

Chapters 1, 2, 3, and 4 are the general rules of the Code. Chapters 1, 2, and 3 are very general while Chapter 4 is more specific in that it deals with equipment such as flexible cords, appliances, motors, etc. These chapters apply to everything in the Code except as modified in other chapters.

Chapters 5, 6, and 7 apply to special (specific) occupancies, equipment, or conditions. These chapters modify the general rules of chapters 1 through 4. The basic rule of thumb is that Chapters 1, 2, 3, and 4 apply except as modified by Chapters 5, 6, and 7.

Chapter 8 covers communications systems and is independent of the other chapters except where they are specifically referenced.

Chapter 9 consists of Tables and Examples. This chapter is very helpful for everyone that uses the Code on a regular basis. The tables have a variety of applications and have a great deal of specific information about conductors.

NEC Questions From Section 90–1 Through Sections 225–19

The following National Electrical Code questions are in consecutive order. Starting in Unit 6 the code questions become mixed and you will gain practice on how to use the index. Questions with • indicate that 75 percent or less get the question correct.

Article 90 – Introduction

83. Compliance with the provisions of the Code will result in _____.
    (a) good electrical service
    (b) an efficient electrical system
    (c) an electrical system freed from hazard
    (d) all of these

84. • The Code applies to the installation of _____.
    (a) Installations of electric conductors and equipment within or on public and private buildings
    (b) Installations of optical fiber cable.
    (c) Installations of other outside conductors and equipment on the premises.
    (d) all these

85. Chapters 1 through 4 of the NEC apply _____.
    (a) generally to the entire NEC
    (b) to special installations and conditions
    (c) to special equipment and material
    (d) all of these

86. Equipment listed by a qualified electrical testing laboratory is not required to have the _____ wiring to be reinspected at the time of installation.
    (a) external (b) associated (c) internal (d) all of these

NEC Chapter 1 – General Requirements

As the name implies, this chapter has very general information about the NEC. If the average person using the NEC read Article 90 and Chapter 1, he or she would gain a good understanding of the Code Book structure, terminology (definitions), and general assumptions (such as copper unless otherwise specified), which would make the other chapters much easier to use and understand.

Article 100 – Definition

Article 100 – Definitions contains the definitions of many terms that apply to the NEC. It is very important to understand the meanings of these terms as they apply to the different Code rules.

87. • Admitting close approach; not guarded by locked doors, elevation, or other effective means is commonly referred to as _____.
    (a) accessible (equipment)
    (b) accessible (wiring methods)
    (c) accessible, readily
    (d) all of these

88. Capable of being removed or exposed without damaging the building structure or finish, or not permanently closed in by the structure or finish of the building defines _____.
    (a) accessible (equipment)
    (b) accessible (wiring methods)
    (c) accessible, readily
    (d) all of these

89. • A device that, by insertion in a receptacle, establishes connection between the conductors of the attached flexible cord and the conductors connected permanently to the receptacle is called a(n) _____.
    (a) attachment plug
    (b) plug cap
    (c) cap
    (d) all of these

90. The conductors between the final overcurrent device and the outlet(s) are known as the _____ conductors.
    (a) feeder
    (b) branch circuit
    (c) home run
    (d) none of these

Chapter 1 Electrical Theory And Code Questions   Unit 1 Math and Basic Electrical Formulas 31

91. • For a circuit to be considered a multiwire branch circuit, it must have _____.
(a) two or more ungrounded conductors with a potential difference between them
(b) a grounded conductor having equal potential difference between it and each ungrounded conductor of the circuit
(c) a grounded conductor connected to the neutral (grounded) conductor of the system
(d) all of these

92. A circuit breaker is a device designed to _____ a circuit by nonautomatic means and to open the circuit automatically on a predetermined overcurrent without damage to itself when properly applied within its rating.
(a) blow   (b) disconnect   (c) connect   (d) open and close

93. A separate portion of a conduit or tubing system that provides access through a removable cover(s) to the interior of the system, at a junction of two or more sections of the system, or at a terminal point of the system is defined as a(n) _____.
(a) junction box   (b) accessible raceway
(c) conduit body   (d) pressure connector

94. A device or group of devices that serves to govern in some predetermined manner the electric power delivered to the apparatus to which it is connected is a _____.
(a) relay   (b) breaker   (c) transformer   (d) controller

95. Which of the following does the Code recognize as a device?
(a) switch   (b) switch and light bulb
(c) lock nut and switch   (d) lock nut, and bushing

96. So constructed that dust will not enter the enclosing case under specified test conditions is known as _____.
(a) dusttight   (b) dust proof   (c) dust rated   (d) all of these

97. Continuous duty is defined as _____.
(a) when the load is expected to continue for three hours or more
(b) operation at a substantially constant load for an indefinite length of time
(c) operation at loads and for intervals of time, both of which may be subject to wide variations
(d) operation at which the load may be subject to maximum current for 6 hours or more

98. _____ is defined as intentionally connected to earth through a ground connection or connections, of sufficiently low impedance and having sufficient current-carrying capacity, to prevent the build-up of voltages that may result in undue hazards to connected equipment or to persons.
(a) Effectively grounded   (b) A proper wiring system
(c) A lighting rod   (d) A grounded conductor

99. A _____ is the circuit conductors between the service equipment or source of a separately derived system, and the final branch circuit overcurrent device.
(a) branch circuit   (b) service entrance conductor
(c) feeder   (d) circuit wiring

100. A building or portion of a building in which self-propelled vehicles carrying volatile flammable liquids for fuel or power are kept for use, sale, storage, and exhibition is called _____.
(a) a garage   (b) a service station   (c) a service garage   (d) all of these

101. • In a grounded system, the conductor that connects the circuit grounded (neutral) conductor at the service to the grounding electrode is called the _____ conductor.
(a) main grounding   (b) common main
(c) equipment grounding   (d) grounding electrode

102. Recognized as suitable for the specific purpose, function, use, environment, and application is the definition of _____.
(a) labeled   (b) identified   (c) listed   (d) approved

103. The environment of a wiring method under the eave of a house or in an outside patio would be considered a _____ location.
(a) dry   (b) damp   (c) wet   (d) moist

104. An area classified as a _____ location may be temporarily subject to dampness and wetness.
    (a) dry (b) damp (c) moist (d) wet

105. A circuit in which any arc or thermal effect produced, under intended operating conditions of the equipment or due to opening, shorting, or grounding of field wiring, is not capable, under specified test conditions, of igniting the flammable gas, vapor, or dust-air mixture is a _____ circuit.
    (a) nonconductive (b) branch (c) nonincendive (d) closed

106. Any current in excess of the rated current of equipment or the ampacity of a conductor is called _____.
    (a) overload (b) smoked
    (c) overcurrent (d) current in excess of rating

107. • Premises wiring system extends from the _____.
    (a) service point of the utility conductors
    (b) source of the separately derived system
    (c) main service disconnect
    (d) a and b

108. Something that is so constructed, protected, or treated as to prevent rain from interfering with the successful operation of the apparatus under specified test conditions is defined as _____.
    (a) raintight (b) waterproof (c) weathertight (d) rainproof

109. A multiple receptacle is a _____ device containing two or more receptacles.
    (a) dual (b) single (c) duplex (d) live

110. Equipment enclosed in a case or cabinet that is provided with a means of sealing or locking so that live parts cannot be made accessible without opening the enclosure is said to be _____.
    (a) guarded
    (b) protected
    (c) sealable equipment
    (d) lockable equipment

111. Service _____ are the conductors from the service point or other source of power to the service disconnecting means.
    (a) conductors (b) drops (c) cables (d) equipment

112. The service conductor between the terminals of the service equipment and a point usually outside the building, clear of building walls, where joined by tap or splice to the service drop are service entrance conductors, _____ system.
    (a) underground (b) complete (c) overhead (d) grounded

113. The _____ is the point of connection between the facilities of the serving utility and the premises wiring.
    (a) service entrance
    (b) service point
    (c) overcurrent protection
    (d) beginning of the wiring system

114. The total components and subsystems that, in combination, convert solar energy into electrical energy, are called a _____ by the Code.
    (a) solar system (b) solar voltaic system
    (c) separately derived source (d) solar photovoltaic system

115. A thermal protector may consist of one or more heat-sensing elements integral with the motor or motor-compressor and an external control device.
    (a) True (b) False

116. The voltage of a circuit is defined by the Code as the _____ root-mean-square (effective) difference of potential between any two conductors of the circuit.
    (a) lowest (b) greatest
    (c) average (d) nominal

117. Something so constructed so that moisture will not enter the enclosure under specific test conditions is called _____.
    (a) watertight (b) moisture proof
    (c) waterproof (d) rainproof

Article 110 – General Requirements

Part A. General

118. In judging equipment, considerations such as the following should be evaluated:
    I. Mechanical strength
    II. Cost.
    III. Arcing effects
    IV. Guarantee
    (a) I only
    (b) I and II
    (c) II and IV
    (d) I and III

119. All wiring shall be so installed, that when completed, the system will be free from _____.
    I. short circuits
    II. grounds
    (a) I only
    (b) II only
    (c) I and II
    (d) none of these

120. The _____ of the circuit shall be so selected and coordinated as to permit the circuit protective devices used, to clear a fault without extensive damage to the electrical components of the circuit.
    (a) overcurrent protective devices
    (b) total circuit impedance
    (c) component short-circuit withstand ratings
    (d) all of these

121. Unless identified for use in the operating environment, no conductors or equipment shall be _____ having a deteriorating effect on the conductors or equipment.
    (a) located in damp or wet locations
    (b) exposed to fumes, vapors, and gases
    (c) exposed to liquids and excessive temperatures
    (d) all of these

122. The NEC requires that electrical work must be _____.
    (a) done in a neat and worker manner
    (b) done under the supervision of qualified personnel
    (c) completed before being inspected
    (d) all of these

123. Conductors shall be _____ to provide ready and safe access in underground and subsurface enclosures, into which persons enter for installation and maintenance.
    (a) bundled
    (b) tied together
    (c) color coded
    (d) racked

124. Many equipment terminations are marked with _____.
    (a) electrical tape
    (b) a nameplate
    (c) a tightening torque
    (d) ink

125. Connection of conductors to terminal parts shall ensure a thoroughly good connection without damaging the conductors and shall be made by means of _____.
    (a) solder lugs
    (b) pressure connectors
    (c) splices to flexible leads
    (d) all of these

126. • What size THHN conductor is required for a 50 ampere circuit if the equipment terminals are listed for 75° conductor sizing? *Tip.* Table 310–16 contains conductor ampacities.
    (a) No. 10
    (b) No. 8
    (c) No. 6
    (d) none of these

127. • Working space distances shall be measured from the _____ if such are enclosed.
    (a) front
    (b) opening
    (c) a or b
    (d) none of these

128. • The dimension of working clearance for access to live parts operating at 300 volts, nominal to ground where there are exposed live parts on one side and grounded parts on the other side is _____ feet according to Table 110–16(a).
    (a) 3
    (b) $3\frac{1}{2}$
    (c) 4
    (d) $4\frac{1}{2}$

129. The minimum headroom of working spaces about service equipment, switchboards, panelboards, or motor control centers shall be 6½ feet except service equipment or panelboards, in existing dwelling units, that do not exceed 200 amperes.
(a) True (b) False

Part B. Over 600 Volts, Nominal

130. Electrical installations over 600 volts located in _____, where access is controlled by lock and key or other approved means, shall be considered to be accessible to qualified persons only.
(a) a room or closet
(b) a vault
(c) an area surrounded by a wall, screen, or fence
(d) any of these

131. Openings in ventilated dry-type _____ or similar openings in other equipment over 600 volts, shall be designed so that foreign objects inserted through these openings will be deflected from energized parts.
(a) lamp holders (b) motors (c) fuse holders (d) transformers

132. • An outdoor installation (over 600 volts nominal) that has metal-enclosed equipment accessible to unqualified persons shall _____.
(a) not be permitted
(b) be designed so that exposed nuts or bolts cannot be readily removed
(c) have suitable guards where exposed to vehicular traffic
(d) b and c

133. • A 6-foot wide control panel supplied by over 600 volts, requires a minimum of _____ entrances to the required working space about electrical equipment.
(a) one (b) two (c) three (d) four

134. • The minimum depth of clear working space in front of electric equipment for 5,000 volts, nominal, to ground is _____ feet when there are exposed live parts on both sides of the work space with the operator between.
(a) 4 (b) 5 (c) 6 (d) 9

135. • Warning signs for over 600 volts, shall read "Warning – High Voltage – Keep Out".
(a) True (b) False

136. Unguarded live parts operating at 30,000 volts above a working space shall be maintained at an elevation of _____ feet.
(a) 24 (b) 18 (c) 12 (d) 9

Chapter 2 – Wiring and Protection

Chapter 2 of the NEC is a general rules chapter as applied to Wiring and Protection of conductors and equipment. The rules in this chapter apply everywhere in the NEC except as modified (such as in Chapters 5, 6, and 7). Along with Chapter 3, it can be considered the heart of the Code. Many of the everyday applications of the NEC can be found in this chapter.

Article 200 – Use/Identification of Grounded Conductor

Article 200 covers the requirements for the identification of terminals, grounded conductors in premises wiring systems, and identification of grounded conductors.

The general rule is that – All premises wiring systems shall have an identified grounded conductor. A *grounded conductor* is usually the system neutral or white wire. Be careful not to confuse grounded conductors (neutral wire) with grounding conductors (ground wire).

137. The grounded conductors of _____ metal-sheathed cable shall be identified by distinctive marking at the terminals during the process of installation.
(a) armored
(b) mineral-insulated
(c) copper
(d) aluminum

138. If two different wiring systems are present in the same raceway or enclosure, the grounded conductor(s) of _____.
    (a) one system must be white and the grounded conductor of the other system must be gray
    (b) both systems can be both white or both gray
    (c) one system must be white or gray and the other system can have any color with white or gray tape
    (d) one system must be white or gray, and the grounded conductor of the other system must be white with a color tracer

139. • An insulated conductor with a white or natural gray finish shall be permitted as a(n) _____ conductor where permanently reidentified to indicate its use, by painting or other effective means at its termination, and at each location where the conductor is visible and accessible.
    (a) terminating   (b) grounded   (c) ungrounded   (d) grounding

140. • A terminal for a grounded conductor on a polarized plug when not visible the conductor entrance hole shall be *marked* with _____.
    (a) a white color   (b) a green color
    (c) the word white   (d) an identifiable metal coating in a white color

141. • Two wire attachment plugs _____.
    (a) need not have their terminals marked for identification
    (b) need not have their terminals marked for identification unless polarized
    (c) never need to have their terminals identified
    (d) always need to have their terminals identified

Article 210 – Branch Circuit

Article 210 covers the requirements for branch circuits. A branch circuit is the conductors between the final overcurrent device and the outlet. Article 210 applies generally to most branch circuits, but not to all!

Part A. General Provisions

142. • Multiwire branch-circuits shall _____.
    (a) supply only line to neutral loads
    (b) provide a means to disconnect simultaneously all ungrounded conductors if the multiwire branch circuit supplies more than one device or equipment on the same yoke for commercial installations
    (c) have its conductors originate from different panelboards
    (d) none of these

143. 208Y/120- or 480Y/277-volt, three-phase, 4-wire wye power systems used to supply nonlinear loads such as, personal computers, energy efficient electronic ballast, electronic dimming, etc. cause a distortion of the phase and neutral currents which produces high, unwanted and potentially hazardous harmonic neutral currents. The Code cautions us that the system design should allow for the possibility of high harmonic neutral currents.
    (a) True   (b) False

144. • A multiwire branch circuit (3-phase, 4-wire) used to supply power to air conditioning units or other similar cooling loads may necessitate that the power system design allow for the possibility of high harmonic neutral currents.
    (a) True   (b) False

145. The means of identification of multiwire ungrounded conductors shall be permanently posted at each _____.
    (a) accessible opening
    (b) terminal
    (c) branch-circuit panelboard
    (d) none of these

146. The use of conductor insulation having a continuous green color or a continuous green color with one or more yellow stripes shall be permitted for ungrounded conductor _____ wiring of equipment if such wiring does not serve as the lead wires for connection to branch-circuit conductors.
    (a) internal   (b) external
    (c) grounding   (d) power

147. • Grounding conductors (ground wires) must be insulated when run in the same raceway with other insulated conductors.
    (a) True   (b) False

148. A branch-circuit voltage that exceeds 277 volts to ground and does not exceed 600 volts between conductor, is used to wire the auxiliary equipment of electric discharge lamps mounted on poles. The minimum height of these fixtures shall not be less than _____ feet.
    (a) 31   (b) 15   (c) 18   (d) 22

149. • If you are replacing an ungrounded receptacle in a bedroom of a dwelling unit, and a grounding means does not exist in the receptacle enclosure, you must use a _____ .
    I. nongrounding receptacle
    II. grounding receptacle
    III. GFCI-type of receptacle
    (a) I only   (b) II only   (c) I or II   (d) I or III

150. All 125-volt, single-phase 15-and 20-ampere receptacles installed in bathrooms of _____ shall have ground fault circuit interrupter protection for personnel.
    I. guest rooms in hotels/motels
    II. hotel common areas
    III. dwelling units
    IV. office buildings
    (a) I, II, and IV   (b) I, II, and III
    (c) I and III   (d) I, II, III, and IV

151. Ground fault circuit protection for personnel is required for all 125-volt, single phase, 15-and 20-ampere receptacles that are installed in a dwelling unit _____.
    (a) attic   (b) garage   (c) laundry   (d) All of These

152. All 125-volt, single-phase 15- and 20-ampere receptacles installed in crawl spaces at or below grade level and in _____ of dwelling units shall have ground-fault, circuit-interrupter protection for personnel.
    (a) unfinished attics   (b) finished attics
    (c) unfinished basements   (d) finished basements

153. • Dwelling unit GFCI protection is required for all 15- and 20- ampere, 125-volt receptacles installed to supply counter top appliances in kitchens.
    (a) True   (b) False

154. GFCI protection is required for all 15- and 20-ampere, 125-volt single-phase receptacles installed _____ of commercial, industrial, and all other nondwelling occupancies.
    (a) outside   (b) in wet locations
    (c) within 6 feet of sinks   (d) in bathrooms

Part B. Branch Circuit Ratings

155. The recommended maximum total voltage drop on both the feeders and branch circuits is _____ percent.
    (a) 3   (b) 2   (c) 5   (d) 3.6

156. • A single receptacle installed on an individual branch circuit must be rated at least _____ percent of the rating of the circuit.
    (a) 50   (b) 60   (c) 90   (d) 100

157. • A branch circuit rated 20 amperes serves four receptacles. The rating of the receptacles must not be less than _____ amperes.
    (a) 10   (b) 15   (c) 25   (d) 30

158. Continuous loads such as store lighting shall not exceed _____ percent of the branch circuit rating.
    (a) 80   (b) 90   (c) 70   (d) 60

159. • If a 277-volt fluorescent lighting fixture (in an office, continuously loaded) draws a 2 ampere load current. The number of such fixtures that may be connected on a 2-wire 277-volt circuit protected by a 20-ampere circuit breaker is _____.
    (a) 8   (b) 9   (c) 10   (d) 12

Chapter 1 Electrical Theory And Code Questions   Unit 1 Math and Basic Electrical Formulas 37

160. Multioutlet circuits rated 15- or 20-amperes are permitted to supply fixed appliances (utilization equipment fastened in place) as long as the fixed appliances do not exceed _____ percent of the circuit rating.
 (a) 125 (b) 100 (c) 75 (d) 50

161. • The rating of one cord-and-plug connected appliance with other lighting loads shall not exceed _____ percent of a 15- or 20-ampere rated branch circuit.
 (a) 50 (b) 60 (c) 80 (d) 100

162. A permanently installed cord connector on a cord pendant is considered a receptacle outlet.
 (a) True (b) False

163. In a dwelling unit, each wall space _____ feet or more requires a receptacle.
 (a) 2 (b) 3 (c) 4 (d) 5

164. In dwelling units, when determining the spacing of general use receptacles, _____ on exterior walls are not considered wall space.
 (a) fixed panels (b) fixed glass (c) sliding panels (d) all of these

165. No point along the floor line in any wall space may be more than _____ feet from an outlet.
 (a) 12 (b) 6½ (c) 8 (d) 6

166. • In dwelling units outdoor receptacles are permitted to be connected to the 20-ampere small appliance branch circuit.
 (a) True (b) False

167. • A receptacle outlet must be installed in dwelling-units for every kitchen and dining area counter top space _____, and no point along the wall line shall be more than 2 feet, measured horizontally from a receptacle outlet.
 (a) wider than 12 inches (b) wider than 2 feet
 (c) 2 feet and wider (d) 12 inches and wider

168. • The intent of Section 210–52(c) is that there be a receptacle outlet for every 4 linear feet or fraction thereof of counter length, measured along the wall or centerline of dwelling-unit kitchen and dining counter tops.
 (a) True (b) False

169. • One wall receptacle shall be installed in every dwelling unit bathroom within 3 feet of a mirror and higher than the rim of the lavatory.
 (a) True (b) False

170. • Which rooms in a dwelling unit must have switch-controlled lighting outlet?
 (a) every habitable room (b) outdoor entrances or exits
 (c) hallways and stairways (d) all of these

Article 215 – Feeders

Article 215 covers feeders, which are the conductors between the service point to the service equipment.

171. Dwelling unit or mobile home feeder conductors (not over 400 amperes) can be sized according to Note _____ of Table 310–16.
 (a) 3 (b) 8 (c) 10 (d) 11

172. The feeder conductor ampacity shall not be less than that of the service-entrance conductors where the feeder conductors carry the total load supplied by service-entrance conductors with an ampacity of _____ amperes or less.
 (a) 100 (b) 60 (c) 55 (d) 30

173. Where required, drawings for feeder installations must be submitted before _____.
 (a) completion of the installation (b) beginning the installation
 (c) the feeders are energized or tested (d) drawings are not required

174. • Ground fault protection is required for the feeder disconnect if _____.
 (a) the feeder is rated 1,000 amperes or more (b) it is a solidly grounded wye system
 (c) the system voltage is 480Y/277 volts (d) all of these

Article 220 – Branch-Circuit and Feeder Calculations

Article 220 covers the requirements for most feeders and service conductor sizes in residential, commercial, and industrial occupancies.

Part A. General

175. • The branch-circuit conductor ampacity shall be rated not less than _____ percent of the continuous load plus _____ percent of the noncontinuous load.
    (a) 125, 100
    (b) 80, 125
    (c) 125, 80
    (d) 100, 100

176. • When determining the load for recessed lighting fixtures for branch circuits, the load shall be based on the _____.
    (a) wattage rating of the fixture   (b) VA of the equipment and lamps
    (c) wattage rating of the lamps   (d) none of these

177. The number of small appliance branch circuits required in a dwelling unit is _____.
    (a) two or more 20 ampere
    (b) two or more 15 or 20 ampere
    (c) two (maximum) 20 ampere
    (d) not restricted

178. Where the load is computed on a volt-amperes-per-square-foot basis, the wiring system up to and including the branch-circuit panelboard(s) shall be provided to serve not less than the _____ load.
    (a) service entrance   (b) calculated
    (c) branch-circuit   (d) maximum

Part B. Feeders

179. • The demand factors of Table 220–11 shall apply to the computed load of feeders to areas in hospitals, hotels, and motels where the entire lighting is likely to be used at one time, as in operating rooms, ballrooms, or dining rooms.
    (a) True   (b) False

180. • Receptacle loads for non-dwelling units computed at not more than 180 VA per outlet in accordance with Section 220–3(c)(7) shall be permitted to be _____.
    (a) added to the lighting loads and made subject to demand factors of Table 220–11
    (b) made subject to the demand factors of Table 220–13
    (c) made subject to the lighting demand loads of Table 220–3(b)
    (d) a or b

181. The small appliance branch circuit in dwelling units can use _____ -ampere rated receptacles.
    (a) 20   (b) 15
    (c) 15 or 20   (d) none of these

182. • To determine the feeder demand load for ten 3-kW household cooking appliances, _____ of Table 220–19 would be used.
    (a) Column A   (b) Column B
    (c) Column C   (d) Column B or C

183. • The maximum unbalanced load for household electric dryers shall be calculated at _____ percent of the demand load as determined by Table 220–18.
    (a) 50   (b) 70   (c) 100   (d) 125

Part C. Optional Calculations for Computing Feeder and Service Loads

184. Feeder and service-entrance conductors with demand load determined by the use of Table 220–30 shall be permitted to have the _____ load determined by Section 220–22.
    (a) feeder (b) circuit (c) neutral (d) none of these

185. A demand factor of _____ percent is applicable for a multifamily dwelling with ten units, if the optional calculation method is used.
    (a) 75 (b) 60 (c) 50 (d) 43

186. The connected load to which the demand factors of Table 220–32 apply shall include 3 volt-amperes per _____ for general lighting and general-use receptacles.
    (a) inch (b) foot (c) square inch (d) square foot

187. To use the optional method for calculating a service to a school, the school must be equipped with _____.
    (a) cooking facilities (b) electric space heating
    (c) air conditioning (d) b or c

188. • Service-entrance or feeder conductors whose demand load is determined by the optional calculation as permitted in Section 220–36 shall not be permitted to have the neutral load determined by Section 220–22.
    (a) True (b) False

Part D. Method for Computing Farm Loads

189. Where the farm dwelling has electric heat and the farm has electric grain drying systems, Part _____ of Article 220 shall not be used to compute the dwelling load.
    (a) A (b) B (c) C (d) none of these

190. The FPN for Section 220–40(b) refers to Section 230–21 for overhead conductors from a pole to a _____.
    (a) grain elevator (b) barn
    (c) dwelling unit (d) building or other structure

Article 225 – Outside Branch Circuits and Feeders

191. • Branch circuit and feeder conductors for outside wiring must be sized according to the calculated load to Article _____.
    (a) 220 (b) 210 (c) 230 (d) 225

192. Where within _____ feet of any building or other structure, open wiring on insulators shall be insulated or covered.
    (a) 4 (b) 3 (c) 6 (d) 10

193. • The minimum size conductor that may be used for an overhead feeder from a residence to a remote garage is No. _____.
    (a) 10 copper (b) 12 copper
    (c) 6 al (d) 10 al

194. Where more than one building, structure, or pole is on the same property and under single management, each building, structure, or pole must have a disconnect (at the separate building, structure, or pole) that disconnects all ungrounded conductors. The disconnect may be located _____ of a building or structure.
    (a) outside (b) inside
    (c) readily accessible location (d) all of these

195. • Disconnects for separate buildings, structures, and poles must be grouped and each disconnect must be _____ to indicated the load served.
    (a) listed (b) labeled (c) identified (d) marked

196. • The disconnecting means for a remote building or structure shall be suitable for use as _____.
    (a) service equipment (b) ground-fault protection
    (c) feeders (d) branch circuits

197. Open conductors installed outside shall be separated from open conductors of other circuits by not less than _____ inches.
    (a) 4 (b) 6

    (c) 8   (d) 10

198. Conductors on poles shall have a separation of not less than 1 foot (305 mm) where not placed on racks or brackets. Power conductors 300 volts or less installed above communication conductors supported on poles shall provide a horizontal climbing space not less than _____ inches.
    (a) 12   (b) 24
    (c) 6   (d) none of these

199. If the voltage between conductors does not exceed 300 volts, the outside conductor clearance over the roof overhang can be reduced from 8 feet to 18 inches. But, no more than _____ feet of conductors can pass over the roof overhang.
    (a) 6   (b) 8
    (c) 10   (d) none of these

200. Outside branch circuits and feeder conductors shall have a vertical clearance of _____ feet from the surface of a roof.
    (a) 8   (b) 10
    (c) 12   (d) 18

Unit 2

Electrical Circuits

OBJECTIVES

After reading this unit, the student should be able to briefly explain the following concepts:

Series circuit calculations	Series-parallel circuit	Multiwire circuit calculations
Parallel circuit calculations	Calculations	Neutral current calculations
		Dangers of multiwire circuits

After reading this unit, the student should be able to briefly explain the following terms:

Amp-hour	Parallel circuits	Series-parallel circuit
Equal resistor method	Phases	Unbalanced current
Kirchoff's law	Pigtailing	Ungrounded conductors
Multiwire circuits	Product of the sum method	(hot wires)
Neutral conductor	Reciprocal method	
Nonlinear loads	Series circuit	

PART A – SERIES CIRCUITS

INTRODUCTION TO SERIES CIRCUITS

A *series circuit* is a circuit in which the current leaves the voltage source and flows through every electrical device with the same intensity before it returns to the voltage source. If any part of a series circuit is opened, the current will stop flowing in the entire circuit, Fig. 2–1.

For most practical purposes, series (*closed loop*) circuits are not used for building wiring, but they are important for the operation of many *control* and *signal circuits*, Fig. 2–2. Motor control circuit stop switches are generally wired in series with the starter's coil and the line conductors. Dual-rated motors, such as 460/230 volts, will have their winding connected in series when supplied by the higher voltage and connected in parallel when supplied by the lower voltage.

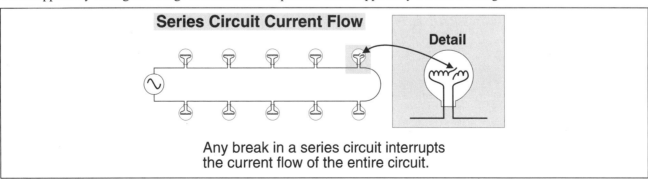

Figure 2-1

Series Circuit Current Flow

2–1 UNDERSTANDING SERIES CALCULATIONS

It is important that you understand the relationship between resistance, current, voltage, and power of series circuits. Fig. 2–3.

Calculating Resistance Total

In a series circuit, the total resistance of the circuit is equal to the sum of all the series resistor's resistance, according to the formula, $R_T = R_1 + R_2 + R_3 + R_4$

❏ **Total Resistance Example**
What is the total resistance of the loads in Figure 2–4?

(a) 2.5 ohms   (b) 5.5 ohms

(c) 7.5 ohms   (d) 10 ohms

• Answer: (c) 7.5 ohms

R_1 – Power Supply .... 0.05 ohm
R_2 – Conductor No. 1 .... 0.15 ohm
R_3 – Appliance .... 7.15 ohms
R_4 – Conductor No. 2 .... 0.15 ohm
Total Resistance: .... 7.50 ohms

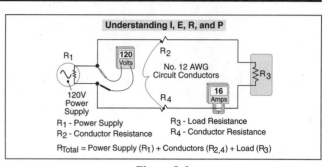

Figure 2-3

Understanding I, E, R, and P

Calculating Voltage Drop

The result of current flowing through a resistor is voltage lost across the resistor which is called *voltage drop*. In a closed loop circuit, the sum of the voltage drops of all the loads is equal to the voltage source, Fig. 2–5. This is known as *Kirchoff's First Law*, of the *voltage law*: The voltage drop (E_{VD}) of each resistor can be determined by the formula:

$E_{VD} = I \times R$ I = Current of the circuit   R = Resistance of the resistor

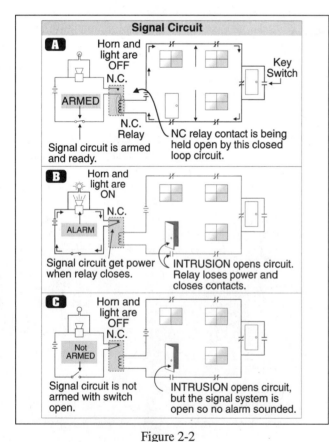

Figure 2-2

Control and Signal Circuits

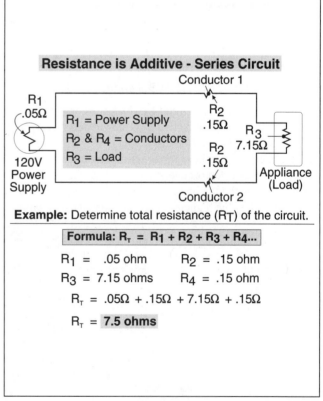

Figure 2-4

Total Resistance Example

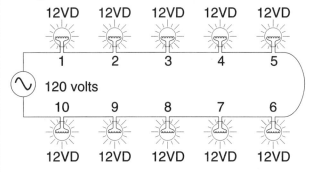

Figure 2-5

Voltage Drop Distribution

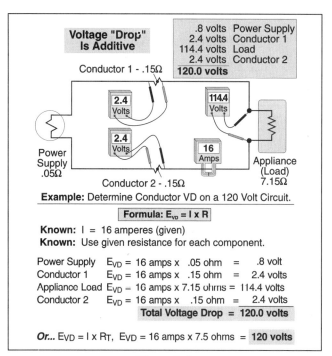

Figure 2-6

Series – Voltage Drop Example

☐ Voltage Drop Example

What is the voltage drop across each resistor in Figure 2–6 using the formula $E_{VD} = I \times R$?

- Answer: $E_{VD} = I \times R$

R_{1vd} – Power Supply . . . . 16 amperes × 0.05 ohm . . . . . . = ... 0.8 volt drop
R_{2vd} – Conductor No. 1 . . . . 16 amperes × 0.15 ohm . . . . . . = ... 2.4 volts drop
R_{3vd} – Appliance . . . . 16 amperes × 7.15 ohms . . . . . = ... 144.4 volts drop
R_{4vd} – Conductor No. 2 . . . . 16 amperes × 0.15 ohm . . . . . . = ... 2.4 volts drop
Total Voltage Drop . . . . 16 amperes × 7.50 ohms . . . . . = ... 120 volts drop

Note. Due to rounding off, the sum of the voltage drops might be slightly different than the voltage source.

Voltage of Series Connected Power Supplies

When *power supplies* are connected in series, the voltage of each power supply will add together, providing that all the polarities are connected properly.

☐ Series Connected Power Supplies Example

What is the total voltage output of four 1.5-volt batteries connected in series, Fig. 2–7?

(a) 1.5 volts   (b) 3.0 volts
(c) 4.5 volts   (d) 6.0 volts

- Answer: (d) 6 volts

Current of Resistor or Circuit

In a series circuit, the *current* throughout the circuit is constant and does not change. The current through each resistor of the circuit can be determined by the formula:

I = E/R
E = Voltage drop of the load or circuit
R = Resistance of the load or circuit

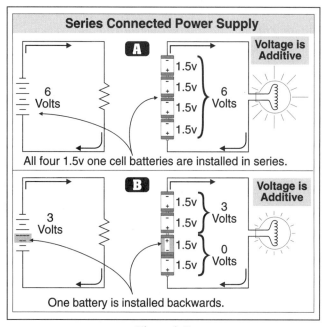

Figure 2-7

Series Power Supplies Example

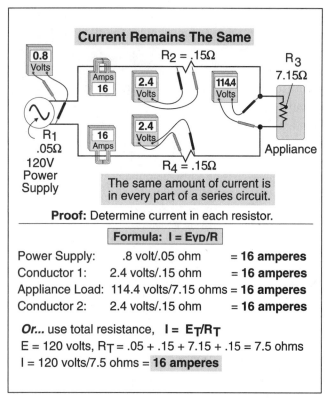

Figure 2-8
Series Current Example

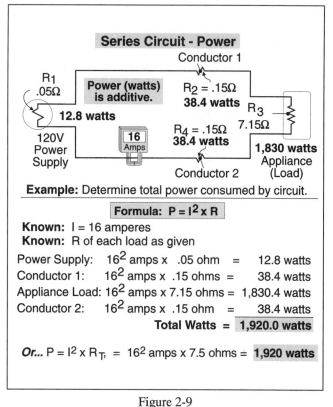

Figure 2-9
Series Power Example

Note. If the resistance is not given, you can determine the resistance of a resistor (if you know the nameplate voltage and power rating of the load) by the formula: $R = E^2/P$, E = Nameplate voltage rating (squared), P = Nameplate power rating

☐ **Current Example**

What is the current flow through the series circuit in Figure 2–8?

(a) 4 amperes   (b) 8 amperes   (c) 12 amperes   (d) 16 amperes

• Answer: (d) 16 amperes

$I = E/R$

R_1 – Power Supply .... 0.8 volt drop/0.05 ohm ...... = ... 16 amperes
R_2 – Conductor No. 1 .... 2.4 volts drop /0.15 ohm .... = ... 16 amperes
R_3 – Appliance .... 114.4 volts drop/7.15 ohms .. = ... 16 amperes
R_4 – Conductor No. 2 .... 2.4 volts drop/0.15 ohm ..... = ... 16 amperes
Circuit Current .... 120 volts/7.5 ohms .... = ... 16 amperes

Power of Resistor or Circuit

The *power* consumed in a series circuit is equal to the sum of the power of all of the resistors in the series circuit. The resistor with the highest resistance will consume the most power, and the resistor with the smallest resistance will consume the least power. You can calculate the power consumed (watts) of each resistor or of the circuit by the formula:

$P = I^2R$ I^2 = **Current of the circuit (squared)**   R = **Resistance of circuit or resistor**

☐ **Power Example**

What is the power consumed of each resistor in Figure 2–9?

• Answer: $P = I^2R$

R_1 – Power Supply .... 16 amperes^2 × 0.05 ohm ..... = ... 12.8 watts
R_2 – Conductor No. 1 .... 16 amperes^2 × 0.15 ohm ..... = ... 38.4 watts
R_3 – Appliance.. .... 16 amperes^2 × 7.15 ohms .... = ... 1,830.4 watts
R_4 – Conductor No. 2 .... 16 amperes^2 × 0.15 ohm ..... = ... 38.4 watts

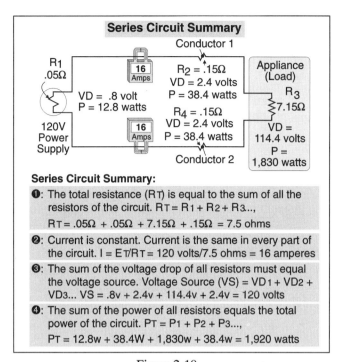

Figure 2-10

Series Circuit Summary

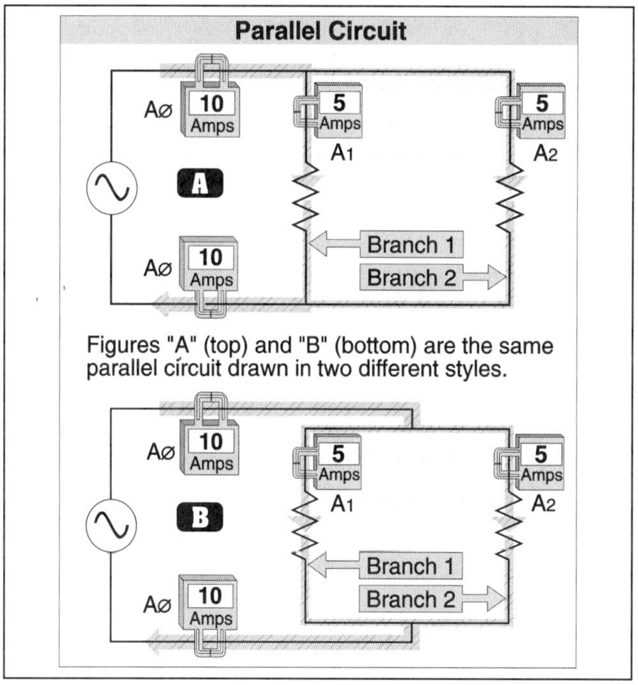

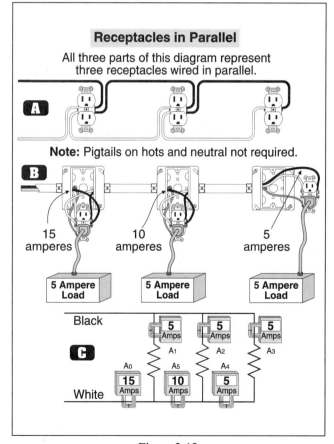

Figure 2-11

Parallel Circuit – Current Flow

2–2 SERIES CIRCUIT SUMMARY

Figure 2–10, Series Circuit Summary.
- Note 1: → The total resistance of the series circuit is equal to the sum of all the resistors of the circuit.
- Note 2: → Current is constant.
- Note 3: → The sum of the voltage drop of all resistors must equal the voltage source.
- Note 4: → The sum of the power of all resistors equals the total power of the circuit.

PART B – *PARALLEL CIRCUIT*

INTRODUCTION TO PARALLEL CIRCUITS

A *parallel circuit* is a circuit in which current leaves the voltage source, branches through different parts of the circuit, and then returns to the voltage source, Fig. 2–11. Parallel is a term used to describe a method of connecting electrical components so that the current can flow through two or more different branches of the circuit.

2–3 *PRACTICAL USES OF PARALLEL CIRCUITS*

Parallel circuits are commonly used for building wiring; in addition, parallel (*open loop*) circuits are often used for fire alarm systems and the internal wiring of many types of electrical equipment, such as motors and transformers. Dual-rated motors, such as 460/230 volts, have their winding connected in parallel when supplied by the lower voltage, and in series when supplied by the higher voltage, Fig. 2–12.

Figure 2-12

Parallel Wiring of Equipment

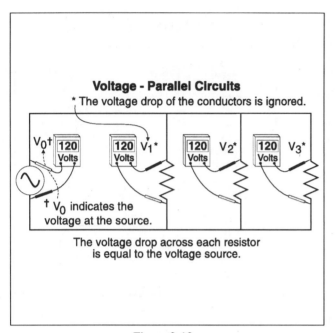

Figure 2-13
Parallel Voltage of Each Branch

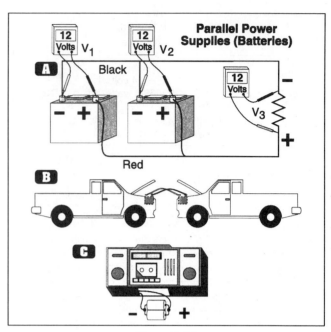

Figure 2-14
Parallel Power Supply Example

2–4 UNDERSTANDING PARALLEL CALCULATIONS

It is important that you understand the relationship between voltage, current, power, and resistance of a parallel circuit.

Voltage of Each Branch

The *voltage drop* across loads connected in parallel is equal to the voltage that supplies each parallel branch, Fig. 2–13.

Power Supplies Connected in Parallel

When *power supplies* are connected in parallel, voltage remains the same but the current, or amp-hour, capacity is increased. When connecting batteries in parallel, always connect batteries of the same voltage with the proper polarity.

❑ **Parallel Connected Power Supplies Example**

If two 12-volt batteries are connected in parallel, what is the output voltage, Fig. 2–14?

(a) 3 volts (b) 6 volts (c) 12 volts (d) 24 volts

• Answer: (c) 12 volts, voltage remains the same when connected in parallel

Note. Two 12-volt batteries connected in parallel will result in an output voltage of 12 volts, but the amp-hour capacity will be increased resulting in longer service.

Current Through Each Branch

The *current* from the power supply is equal to the sum of the branch circuits. The current in each branch depends on the branch voltage and branch resistance and can be calculated by the formula:

$I = E/R$ E = Voltage of Branch   R = Resistance of Branch

❑ **Current Through Each Branch Example**

What is the current of each appliance in Figure 2–15?

• Answer: $I = E/R$.

R_1 – Coffee Pot . . . . 120 volts/16 ohms . . . . =. . 7.50 amperes
R_2 – Skillet . . . . 120 volts/13 ohms . . . . =. . 9.17 amperes
R_3 – Blender . . . . 120 volts/36 ohms . . . . =. . <u>3.33 amperes</u>
Total Current = 7.5 amperes + 9.17 amperes + 3.33 amperes   =. 20.00 amperes

Power Consumed Of Each Branch

The total *power* consumed of any circuit is equal to the sum of the branch powers. Each branch power depends on the branch current and resistance. The power can be found by the formula:

$P = I^2R$ I = Current of Each Branch   R = Resistance of Each Branch

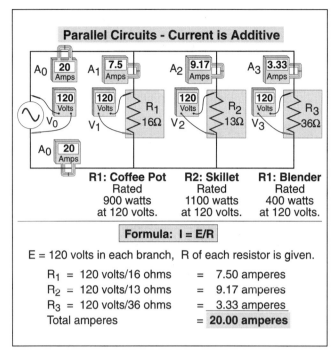

Figure 2-15

Parallel Circuits – Current Is Additive

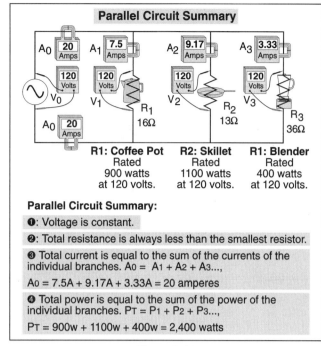

Figure 2-16

Parallel Circuits Summary

❑ **Power of Each Branch Example**
What is the power consumed of each appliance in Figure 2–16?
- Answer: $P = I^2R$

R_1 – Coffee Pot . . . . 7.5 amperes^2 × 16 ohms . . . = ... 900 watts
R_2 – Skillet . . . . 9.17 amperes^2 × 13 ohms . . = ... 1,093 watts, round to 1,100 watts
R_3 – Blender . . . . 3.33 amperes^2 × 36 ohms . . = ... <u>399 watts, round to 400 watts</u>
Total Power = 900 watts + 1,100 watts + 400 watts . . . = ... 2,400 watts

2–5 PARALLEL CIRCUIT RESISTANCE CALCULATIONS

In a parallel circuit, the total circuit *resistance* is always less than the smallest resistor and can be determined by one of three methods:

Equal Resistor Method

The *equal resistor method* can be used when all the resistors of the parallel circuit have the same resistance. Simply divide the resistance of one resistor by the number of resistors in parallel.

$R_T = R/N$ R = Resistance of One Resistor   N = Number of Resistors

❑ **Equal Resistors Method Example**
The total resistance of three, 10-ohm resistors is _____, Fig. 2–17.

(a) 10 ohms   (b) 20 ohms   (c) 30 ohms   (d) none of these

- Answer: (d) none of these

$$R_T = \frac{\text{Resistance of One Resistor}}{\text{Number of Resistors}} = \frac{10 \text{ ohms}}{3 \text{ resistors}} = 3.33 \text{ ohms}$$

Product Of The Sum Method

The product of the sum method can be used to calculate the resistance of two resistors.

$$R_T = \frac{R_1 \times R_2 \text{ (Product)}}{R_1 + R_2 \text{ (Sum)}}$$

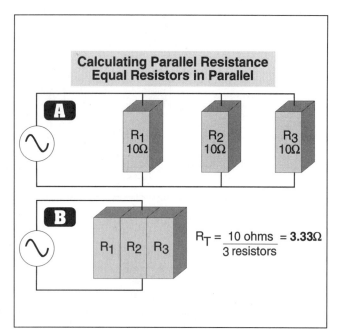

Figure 2-17
Parallel Resistance – Equal Resistors Example

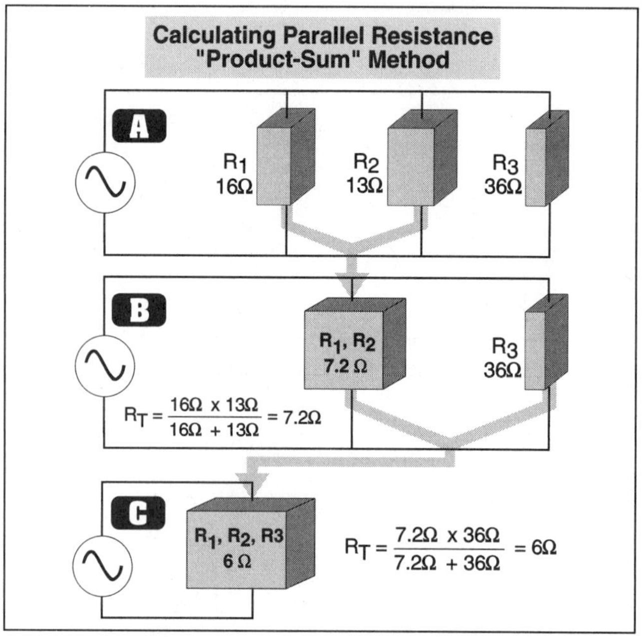

Figure 2-18
Parallel Resistance – Product/Sum Method Example

The term *product* means the answer of numbers that are multiplied together. The term *sum* is the answer to numbers that are added together. The product of the sum method can be used for more than two resistors, but only two can be calculated at a time.

❑ **Product Of The Sum Method Example**

What is the total resistance of a 16 ohm coffee pot and a 13 ohm skillet connected in parallel, Fig. 2–18?

(a) 16 ohms (b) 13.09 ohms (c) 29.09 ohms (d) 7.2 ohms

• Answer: 7.2 ohms

The total resistance of parallel circuit is always less than the smallest resistor (13 ohms).

$$R_T = \frac{R_1 \times R_2}{R_1 + R_2} = \frac{16 \text{ ohms} \times 13 \text{ ohms}}{16 \text{ ohms} + 13 \text{ ohms}} = 7.20 \text{ ohms}$$

Reciprocal Method

The advantage of the *reciprocal method* is that this formula can be used for an unlimited number of parallel resistors.

$$R_T = \frac{1}{1/R_1 + 1/R_2 + 1/R_3 \ldots}$$

❑ **Reciprocal Method Example**

What is the resistance total of a 16-ohm, 13-ohm, and 36-ohm resistor connected in parallel, Fig. 2–19?

(a) 13 ohms (b) 16 ohms (c) 36 ohms (d) 6 ohms

• Answer: (d) 6 ohms

$$R_T = \frac{1}{1/16 \text{ ohm} + 1/13 \text{ ohm} + 1/36 \text{ ohm}} = R_T = \frac{1}{0.0625 \text{ ohm} + 0.0769 \text{ ohm} + 0.0278 \text{ ohm}}$$

$$R_T = \frac{1}{0.1672 \text{ ohm}} = R_T = 6 \text{ ohms}$$

2–6 PARALLEL CIRCUIT SUMMARY

Note 1: → The total resistance of a parallel circuit is always less than the smallest resistor, Fig. 2–20.

Note 2: → Current total of a parallel circuit is equal to the sum of the currents of the individual branches.

Note 3: → Power total is equal to the sum of the power in all the individual branches.

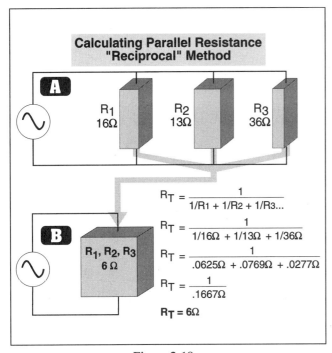

Figure 2-19
Parallel Resistance Reciprocal Method Example

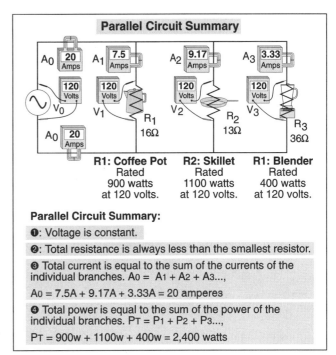

Figure 2-20
Parallel Summary

PART C – *SERIES–PARALLEL AND MULTIWIRE CIRCUITS*

INTRODUCTION TO SERIES-PARALLEL CIRCUITS

A *series-parallel* circuit is a circuit that contains some resistors in series and some resistors in parallel to each other, Fig. 2–21.

2–7 *REVIEW OF SERIES AND PARALLEL CIRCUITS*

To have a better understanding of series-parallel circuits, let's review the rules for series and parallel circuits. That portion of the circuit that contains resistors in series must comply with the rules of series circuits, and that portion of the circuit that is connected in parallel must comply with the rules of parallel circuits, Fig. 2–22.

Series Circuit Rules:
Note 1: → Resistance is additive.
Note 2: → Current is constant.
Note 3: → Voltage is additive.
Note 4: → Power is additive.

Parallel Circuit Rules
Note 1: → Resistance is less than the smallest resistor.
Note 2: → Current is additive.
Note 3: → Voltage is constant.
Note 4: → Power is additive.

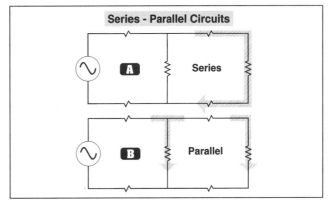

Figure 2-21
Series-Parallel Circuit

2–8 *SERIES-PARALLEL CIRCUIT RESISTANCE CALCULATIONS*

When determining the *resistance total* of a series-parallel circuit, it is best to redraw the circuit so you can see the series components and the parallel branches. Determine the resistance of the series components first or the parallel components depending on the circuit, then determine the resistance total of all the branches. Keep breaking the circuit down until you have determined the total effective resistance of the circuit as one resistor.

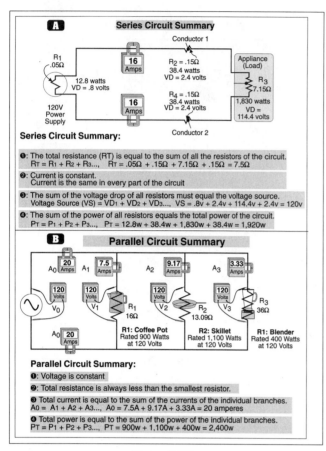

Figure 2-22

Series and Parallel Review

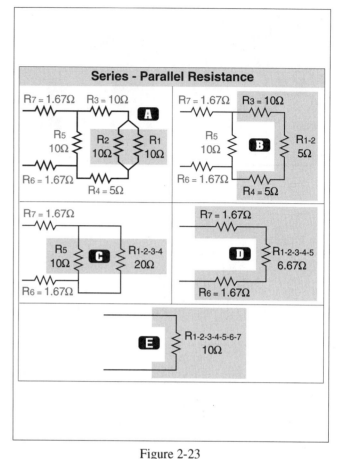

Figure 2-23

Series – Parallel Resistance

❑ **Series-Parallel Circuit Resistance Example**

What is the resistance total of the circuit shown in Figure 2–23, Part A?

- Answer:

Step 1: → Determine the resistance of the equal parallel resistors - R_1 and R_2 – Figure 2–23 Part A:
Using the *Equal Resistor Method*
The *equal resistor method* can be used when all the resistors of the parallel circuit have the same resistance. Simply divide the resistance of one resistor by the number of resistors in parallel.
$R_T = R/N$, R = Resistance of one resistor, N = Number of resistors
R_T = 10 ohms/2 resistors, R_T = 5 ohms

Step 2: → Redraw the circuit – Figure 2–23 Part B, now determine the series resistance of $R_{1/2}$ (5 ohms), plus R_3 (10 ohms), plus R_4 (5 ohms).
Series resistance total is equal to $R_{1/2} + R_3 + R_4$
R_T = 5 ohms + 10 ohms + 5 ohms, R_T = 20 ohms

Step 3: → Redraw the circuit – Figure 2–23 Part C, now determine the parallel resistance of R_5, plus the resistance of $R_{1,2,3,4}$. Remember the resistance total of a parallel circuit is always less than the smallest parallel branch (10 ohms). Since we have only two parallel branches, the resistance can be determined by the product of the sum method.

$$R_T = \frac{R_1 \times R_2 \;(Product)}{R_1 + R_2 \;(Sum)} \qquad R_T = \frac{(10 \text{ ohms} \times 20 \text{ ohms})}{(10 \text{ ohms} + 20 \text{ ohms})} \qquad R_T = 6.67 \text{ ohms}$$

Step 4: → Redraw the circuit – Figure 2–23 Part D, now determine the series resistance total of R_7, plus R_6, plus $R_{1,2,3,4,5}$
Series resistance total is equal to $R_7 + R_6 + R_{1,2,3,4,5}$
R_T = 1.67 ohms + 1.67 ohms + 6.67 ohms, R_T = 10 ohms

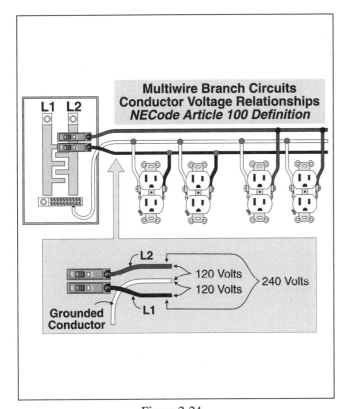

Figure 2-24
Multiwire Branch Circuit

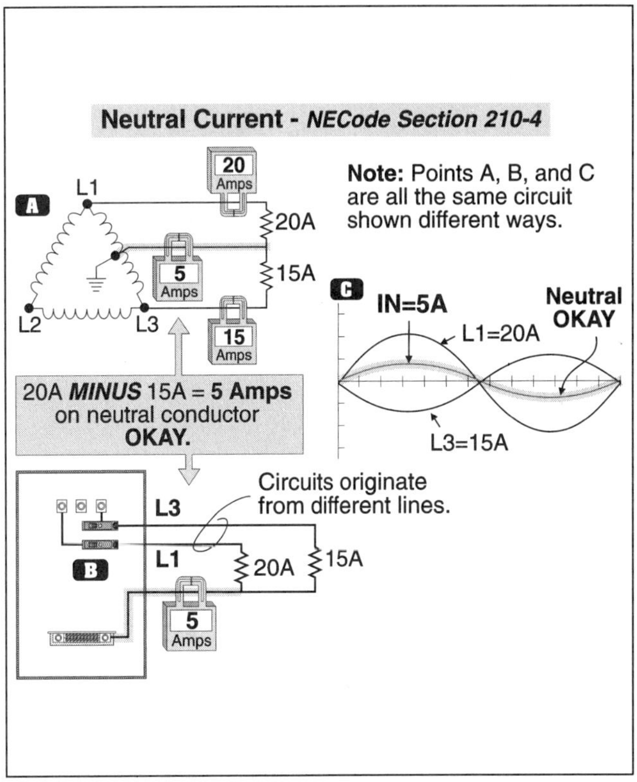

Figure 2-25
Neutral Current – 120/240 Volt, 3-Wire Example

PART D – *MULTIWIRE CIRCUITS*

INTRODUCTION TO MULTIWIRE CIRCUITS

A multiwire circuit is a circuit consisting of two or more *ungrounded* conductors (hot conductors) having a *potential difference* between them, and an equal difference of potential between each hot wire and the *neutral* or *grounded conductor*. Figure 2–24.

Note. The National Electrical Code contains specific requirements on multiwire circuits, see Article 100 definition of multiwire circuit, Section 210–4 branch circuit requirements, and Section 310–13(b) requirements on pigtailing of the neutral conductor.

2–9 *NEUTRAL CURRENT CALCULATIONS*

When current flows on the *neutral* conductor of a multiwire circuit, the current is called *unbalanced current*. This current can be determined according to the following:

120/240-Volt, 3-Wire Circuits

A 120/240-volt, 3-wire circuit, consisting of two hot wires and a neutral, will carry no current when the circuit is balanced. However, the neutral (or *grounded*) conductor will carry unbalanced current when the circuit is not balanced. The neutral current can be calculated as:

Line 1 Current less Line 2 Current

☐ **120/240-Volt, 3-Wire Neutral Current Example**

What is the neutral current if $line_1$ current is 20 amperes and $line_2$ current is 15 amperes, Fig. 2–25?

(a) 0 amperes (b) 5 amperes (c) 10 amperes (d) 35 amperes

• Answer: (b) 5 amperes
  $Line_1$ current = 20 amperes
  $Line_2$ current = 15 amperes
  Unbalance = 5 amperes

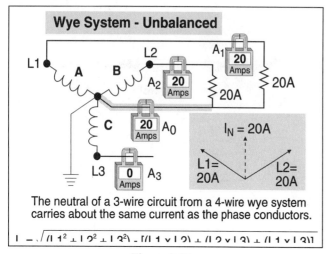

Figure 2-26
Neutral Current – 120/208 Volt, 3-Wire Example

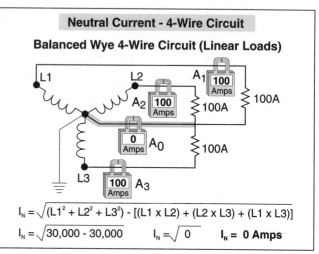

Figure 2-27
Neutral Current – Wye 4-Wire Balanced Example

Wye 3-Wire Circuit Neutral Current

A 3-wire wye, 208Y/120- or 480Y/277-volt circuit, consisting of two phases and a neutral always carries unbalanced current. The current on the neutral or grounded conductor is determined by the formula:

$$I_N = \sqrt{(\text{Line 1}^2 + \text{Line 2}^2) - (\text{Line 1} \times \text{Line 2})}$$

Line 1 = Current of one phase
Line 2 = Current of other phase

❑ **Three-wire Wye Circuit Neutral Current Example**

What is the neutral current for a 20 ampere, 3-wire circuit (two hots and a neutral)? Power is supplied from a 208Y/120-volt feeder, Fig. 2–26?

(a) 40 amperes (b) 20 amperes (c) 60 amperes (d) 0 amperes

• Answer: (b) 20 amperes

$$I_N = \sqrt{(20 \text{ amperes}^2 + 20 \text{ amperes}^2) - (20 \text{ amperes} \times 20 \text{ amperes})} = \sqrt{400}$$

I_N = 20 amperes

Wye 4-wire Circuit Neutral Current

A 4-wire wye, 120/208Y- or 277/480Y-volt circuit will carry no current when the circuit is balanced but will carry unbalanced current when the circuit is not balanced.

$$I_N = \sqrt{(\text{Line 1}^2 + \text{Line 2}^2 + \text{Line 3}^2) - [(\text{Line 1} \times \text{Line 2}) + (\text{Line 2} \times \text{Line 3}) + (\text{Line 1} \times \text{Line 3})]}$$

❑ **Example Balanced Circuits**

What is the neutral current for a 4-wire, 208Y/120-volt feeder where L_1 = 100 amperes, L_2 = 100 amperes, and L_3 = 100 amperes, Fig. 2–27?

(a) 50 amperes (b) 100 amperes (c) 125 amperes (d) 0 amperes

• Answer: (d) 0 amperes

$$I_N = \sqrt{(100 \text{ Amps}^2 + 100 \text{ Amps}^2 + 100 \text{ Amps}^2) - [(100A \times 100A) + (100A \times 100A) + (100A \times 100A)]} = \sqrt{0}$$

I_N = 0 amperes

❑ **Unbalanced Circuits – Example**

What is the neutral current for a 4-wire, 208Y/120-volt feeder where L_1 = 100 amperes, L_2 = 100 amperes, and L_3 = 50 amperes, Fig. 2–28?

(a) 50 amperes (b) 100 amperes (c) 125 amperes (d) 0 amperes

• Answer: (a) 50 amperes

$$I_N = \sqrt{(100A^2 + 100A^2 + 50A^2) - [(100A \times 100A) + (100A \times 50A) + (100A \times 50A)]} = \sqrt{2{,}500}$$

I_N = 50 amperes

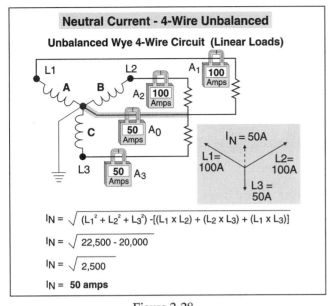

Figure 2-28
Neutral Current – Wye 4-Wire Unbalanced

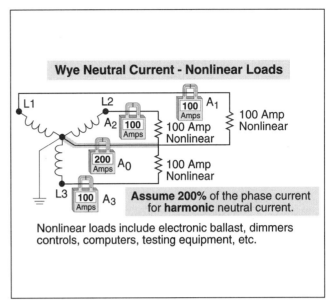

Figure 2-29
Neutral Current – Nonlinear Loads

Nonlinear Load Neutral Current

The neutral conductor of a balanced wye 4-wire circuit will carry current when supplying power to *nonlinear* loads, such as computers, copy machines, laser printers, fluorescent lighting, etc. The current can be as much as two times the phase current, depending on the harmonic content of the loads.

❑ **Nonlinear Loads – Example**

What is the neutral current for a 4-wire, 208Y/120-volt feeder supplying power to a nonlinear load, where $L_1 = 100$ amperes, $L_2 = 100$ amperes, and $L_3 = 100$ amperes? Assume that the harmonic content results in neutral current equal to 200 percent of the phase current, Fig. 2–29.

(a) 80 amperes (b) 100 amperes (c) 125 amperes (d) 200 amperes

• Answer: 200 amperes

2–10 DANGERS OF MULTIWIRE CIRCUITS

Improper wiring, or mishandling of multiwire circuits, can cause excessive neutral current (*overload*) or destruction of electrical equipment because of *overvoltage* if the neutral conductor is opened.

Overloading of the Neutral Conductor

If the ungrounded conductors (hot wires) of a multiwire circuit are connected to different phases, the current on the grounded conductor (neutral) will cancel. If the ungrounded conductors are not connected to different phases, the current from each phase will add on the neutral conductor. This can result in an *overload* of the neutral conductor, Fig. 2–30.

Note. Overloading of the neutral conductor will cause the insulation to look discolored. Now you know why the neutral wires sometimes look like they were burned.

Overvoltage To Electrical Equipment

If the neutral conductor of a multiwire circuit is opened, the multiwire circuit changes from a parallel circuit into a series circuit. Instead of two 120-volt circuits, we now have one 240-volt circuit, which can result in fires and the destruction of the electric equipment because of *overvoltage*, Fig. 2–31.

To determine the operating voltage of each load in an open multiwire circuit, use the following steps:

Step 1: → Determine the resistance of each appliance,
$R = E^2/P$
E = Appliance voltage nameplate rating
P = Appliance power nameplate rating

Step 2: → Determine the circuit resistance, $R_T = R_1 + R_2$

Step 3: → Determine the current of the circuit,
$I_T = E_S/R_T$,
E_S = Voltage, phase to phase
R_T = Resistance total, from Step 2

Step 4: → Determine the voltage for each appliance,
$E = I_T \times R_X$.
I_T = Current of the circuit
R_X = Resistance of each resistor

Step 5: → Determine the power consumed of each appliance; $P = E^2/R_X$
E = Voltage the appliance operates at (squared)
Rx = Resistance of the appliance

❑ **Example – At what voltage does each of the loads operate at if the neutral is opened,** Fig. 2–31.?
- Answer: Hair dryer = 77 volts, T.V. = 163 volts

Step 1: → Determine the resistance of each appliance, $R = E^2/P$
E = Appliance voltage nameplate rating
P = Appliance power nameplate rating
Hair dryer rated 1,275 watts at 120 volts.
$R = E^2/P$, R = 120 volts^2/1,275 watts = 11.3 ohms
Television rated 600 watts at 120 volts.
$R = E^2/P$, R = 120 volts^2/600 watts = 24 ohms

Step 2: → Determine the circuit resistance, $R_t = R_1 + R_2$
R_T = 11.3 ohms + 24 ohms = R_T = 35.3 ohms

Step 3: → Determine the current of the circuit, $I_T = E_S/R_T$.
I_T = Current of the Circuit, R_T = ResistanceTotal
$I_T = \dfrac{240 \text{ volts}}{35.3 \text{ ohms}} = 6.8$ amperes

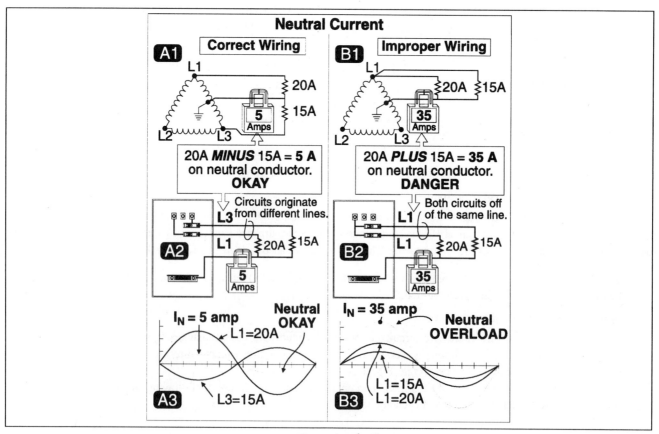

Figure 2-30

Neutral Overload

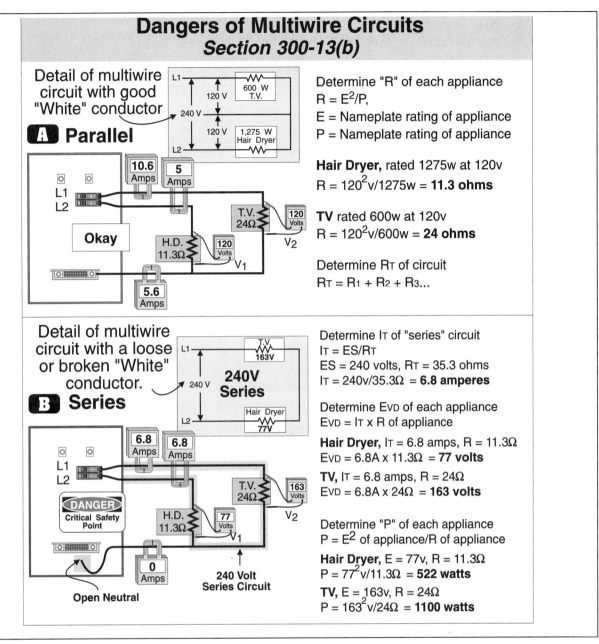

Figure 2-31

Overvoltage – Open Neutral

Step 4: → Determine the voltage for each appliance, $E = I_T \times R_x$

I_T = Current of the circuit, R = Resistance of each resistor

Hair dryer: 6.8 amperes × 11.3 ohms = 76.84 volts

Television: 6.8 amperes × 24 ohms = 163.2 volts

The 120-volt rated TV in the split second before it burns up or explodes is operating at 163.2 volts.

Step 5: → Determine the power consumed of each appliance, $P = E^2/R$

E^2 = Voltage the appliance operates at (squared)

R = Resistance of the appliance

Hair Dryer: P = 76.8 volts^2/11.3 ohms = 522 watts

Television: P = 163.2 volts^2/24 ohms = 1,110 watts

Note. The 600-watt, 120-volt rated TV will operate at 163 volts and consume 1,110 watts. You can kiss this TV good-by. Because of the dangers associated with multiwire branch-circuits, don't use them for sensitive or expensive equipment, such as computers, stereos, research equipment, etc.

Unit 2 – Electrical Circuits Summary Questions

Part A – Series Circuits

1. A closed-loop circuit is a circuit in which a specific amount of current leaves the voltage source and flows through every electrical device in a single path before it returns to the voltage source.
   (a) True	(b) False

2. For most practical purposes, closed-loop system circuits are used for signal and control circuits.
   (a) True	(b) False

2–1 Understanding Series Calculations

3. Resistance opposes the flow of electrons. In a series circuit, the total circuit resistance is equal to the sum of all the series resistor.
   (a) True	(b) False

4. The opposition to current flow results in _____.
   (a) current	(b) voltage	(c) voltage drop	(d) none of these

5. In a series circuit, the current is _____ through the transformer, the conductors, and the appliance.
   (a) proportional	(b) distributed	(c) additive	(d) constant

6. • When power supplies are connected in series, the voltage remains the same, provided that all the polarities are connected properly.
   (a) True	(b) False

7. The power consumed in a series circuit is equal to the power of the largest resistor in the series circuit.
   (a) True	(b) False

Part B – Parallel Circuits

Introduction To Parallel Circuits

8. A _____ circuit is a circuit in which current leaves the voltage source, branches through different parts of the circuit in different magnitudes, and then branches back to the voltage source.
   (a) series	(b) parallel	(c) series-parallel	(d) multiwire

2–4 Understanding Parallel Calculations

9. • The power supply provides the pressure needed to move the electrons; however, the _____ oppose(s) the current flow.
   (a) power supply	(b) conductors
   (c) appliances	(d) all of these

10. When power supplies are connected in parallel, the amp-hour capacity remains the same.
    (a) True	(b) False

11. The total current of a parallel circuit is equal to the sum of the branch currents. The current in each branch can be calculated by the formula, $I = E/R$.
    (a) True	(b) False

12. When current flows through a resistor, power is consumed. The power consumed of each branch can be determined by the formula, $P = I^2 \times R$. The total power consumed in a parallel circuit is equal to the largest branch power.
    (a) True	(b) False

2-5 Parallel Circuit Resistance Calculations

13. The basic method(s) of calculating total resistance, R_T of a parallel circuit is/are _____.
    (a) equal resistor method
    (b) product of the sum method
    (c) reciprocal method
    (d) all of these

14. The total resistance of three 6-ohm resistors in parallel is _____.
    (a) 6 ohms
    (b) 12 ohms
    (c) 18 ohms
    (d) none of these

15. The circuit resistance of a 600-watt coffee pot and a 1,000-watt skillet is _____ ohms when connected to a 120 volt parallel circuit.
    (a) 24
    (b) 14.4
    (c) 38.4
    (d) 9

16. The resistance total of a 20 ohms, 20 ohms, and 10 ohms resistor in parallel is _____ ohms.
    (a) 5
    (b) 20
    (c) 30
    (d) 50

2-6 Parallel Circuit Summary

17. • Which of the following statements is/are true about parallel circuits?
    I. The total resistance of a parallel circuit is less than the smallest resistor of the circuit.
    II. Current total is equal to the sum of the branch currents.
    III. The power of all resistors is equal to the sum of the branch powers.
    (a) I and II
    (b) I, II, and III
    (c) II and III
    (d) I and III

Part C – Series–Parallel And Multiwire Circuits

Introduction To Series-Parallel Circuits

18. A _____ is a circuit that contains some resistors in series and some resistors in parallel to each other.
    (a) parallel circuit
    (b) series circuit
    (c) series-parallel circuit
    (d) none of these

Part D – Multiwire Branch Circuits

Introduction To Multiwire Circuits

19. A multiwire circuit has two or more ungrounded conductors having a potential difference between them and an equal difference of potential between each ungrounded conductor and the grounded conductor.
    (a) True
    (b) False

2-9 Neutral Current Calculations

20. • The current on the grounded conductor of a 2-wire circuit will be _____ percent of the current on the ungrounded conductor.
    (a) 50
    (b) 70
    (c) 80
    (d) 100

Three-Wire Circuits

21. • A balanced 3-wire circuit is connected so the ungrounded conductors are from different transformer phases (Line$_1$ and Line$_2$). The current on the grounded conductor will be _____ percent of the ungrounded conductor current.
    (a) 0
    (b) 70
    (c) 80
    (d) 100

22. A 3-wire, 120/240-volt circuit will carry 10 amperes unbalanced neutral current if Line$_1$ = 20 amperes and Line$_2$ = 10 amperes.
    (a) True
    (b) False

23. • What is the neutral current for a 20 ampere, 3-wire, 208Y/120-volt circuit?
    (a) 0 amperes
    (b) 10 amperes
    (c) 20 amperes
    (d) 40 amperes

Four-Wire Circuits

24. Wye 4-wire Circuits: The neutral of a 4-wire, 208Y/120- or 480Y/277-volt circuit will carry the unbalanced current when the circuit is balanced.
    (a) True
    (b) False

25. What is the neutral current for a 4-wire, 208Y/120-volt circuit, L_1 = 20 amperes, L_2 = 20 amperes, L_3 = 20 amperes?
    (a) 0 amperes
    (b) 10 amperes
    (c) 20 amperes
    (d) 40 amperes

26. • The grounded conductor (neutral) of a balanced wye 4-wire circuit will carry no current when supplying power to balanced *nonlinear loads*.
    (a) True
    (b) False

27. • A three-phase, 4-wire, 208Y/120-volt circuit supplying nonlinear load, L_1 = 20 amperes, L_2 = 20 amperes, L_3 = 20 amperes. The neutral conductor will carry as much current as _____ amperes.
    (a) 0
    (b) 10
    (c) 15
    (d) 40

2–10 Dangers Of Multiwire Circuits

28. Improper wiring or mishandling of multiwire branch circuits can cause _____ connected to the circuit.
    (a) overloading of the ungrounded conductors
    (b) overloading of the grounded conductors
    (c) destruction of equipment because of over voltage
    (d) b and c

29. • Because of the dangers associated with the open neutral (grounded conductor), the continuity of the _____ conductor cannot be dependent on the receptacle. (Code Rule in Article 300)
    (a) ungrounded
    (b) grounded
    (c) a and b
    (d) none of these

Challenge Questions

Part A – Series Circuits

30. • Two resistors, one 4 ohms, and one 8 ohms, are connected in series. If the voltage dropped across both resistors is 12 volts, then the current that would pass through the 4-ohm resistor is _____ amperes.
    (a) 1
    (b) 2
    (c) 4
    (d) 8

31. • A series circuit has four 40-ohm resistors and the power supply is 120 volts. The voltage drop of each resistor would be _____.
    (a) one-quarter of the source voltage
    (b) 30 volts
    (c) the same across each resistor
    (d) all of these

32. • The power consumed in a series circuit is _____.
    (a) the sum of the power consumed of each load
    (b) determined by the formula $P_T = I^2 \times R_T$
    (c) determined by the formula $P_T = E \times I$
    (d) all of these

33. • The reading on voltmeter 2 (V_2) is _____, Fig. 2–32.
    (a) 5 volts
    (b) 7 volts
    (c) 10 volts
    (d) 6 volts

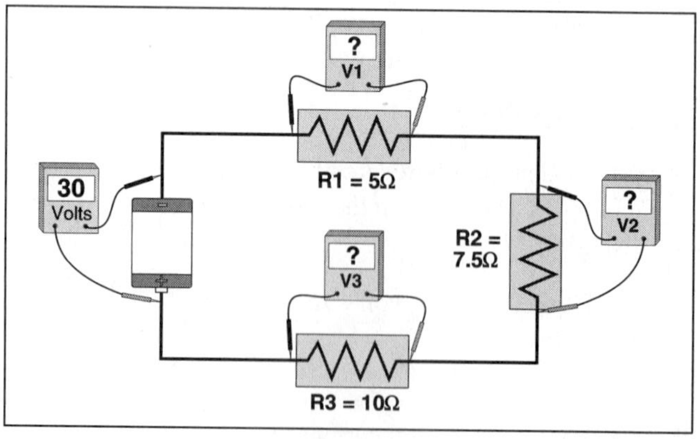

Figure 2-32

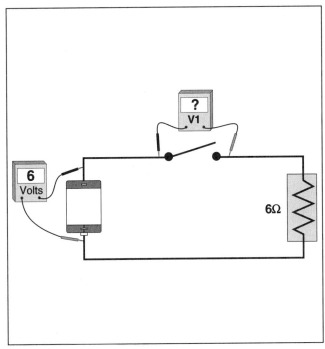

Figure 2-33

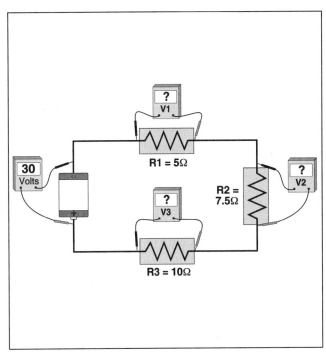

Figure 2-34

34. The voltmeter connected to the circuit on the right would read _____, Fig. 2–33.
    (a) 3 volts (b) 12 volts (c) 6 volts (d) 18 volts

Part B – Parallel Circuits

35. • In general, when multiple light bulbs are wired in a single fixture, they are connected in _____ to each other.
    (a) series
    (b) series-parallel
    (c) parallel
    (d) order of wattage

36. • A single-phase, dual-rated, 120/240-volt motor will have it winding connected in _____ when supplied by 120 volts.
    (a) series
    (b) parallel
    (c) series-parallel
    (d) parallel-series

37. • The voltmeters shown in Figure 2–34 are connected _____ each of the loads.
    (a) in series to
    (b) across from
    (c) in parallel to
    (d) b and c

38. • If the supply voltage remains constant, four resistors will consume the most power when they are connected _____, Fig. 2–34.
    (a) all in series
    (b) all in parallel
    (c) with two parallel pairs in series
    (d) with one pair in parallel and the other two in series

Circuit Resistance

39. A parallel circuit has three resistors. One resistor is 2 ohms, one is 3 ohms, and the other is 7 ohms. The total resistance of the parallel circuit would be _____ ohm(s).
    Remember the total resistance of any parallel circuit is always less than the smallest resistor.
    (a) 12
    (b) 1
    (c) 42
    (d) 1.35

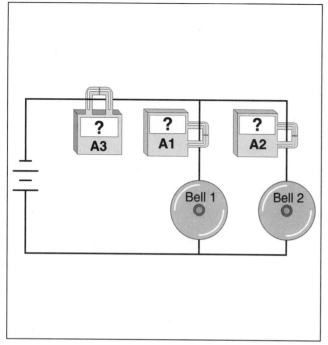

Figure 2-35

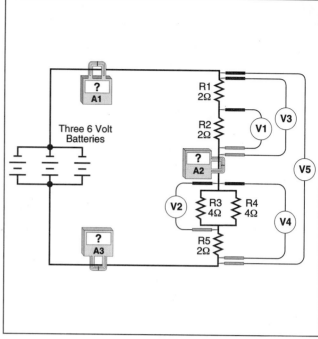

Figure 2-36

Figure 2–35 (4. applies to the next three questions.

40. • The total current of the circuit can be measured by ammeter _____ only,.. Fig. 2–35.
    (a) 1 (b) 2 (c) 3 (d) none of these

41. • If bell 2 consumed 12 watts of power when supplied by two 12-volt batteries (connected in series), the resistance of this bell would be ____ ohms, Fig. 2–35.
    (a) 9.6 (b) 44 (c) 576 (d) 48

42. Determine the total circuit resistance of the parallel circuit based on the following facts: Figure 2–35.
    1. The current on ammeter 1 reads .75 ampere.
    2. The voltage of the circuit is 30 volts.
    3. Bell 2 has a resistance of 48 ohms.
    Tip: Resistance total of a parallel circuit is always.
    (a) 22 ohms (b) 48 ohms (c) 1920 ohms (d) 60 ohms

Part C – Series-Parallel Circuits

Figure 2–36 (5. applies to the next three questions:

43. • The total current of this circuit can be read on _____, Fig. 2–36.
    I. Ammeter 1
    II. Ammeter 2
    III. Ammeter 3
    (a) I only (b) II only (c) III only (d) I, II and III

44. • The voltage reading of V_2 is _____ volts, Fig. 2–36.
    (a) 1.5 (b) 4 (c) 4 (d) 8

45. • The voltage reading of V_4 is _____ volts, Fig. 2–36.
    (a) 1.5 (b) 3 (c) 5 (d) 8

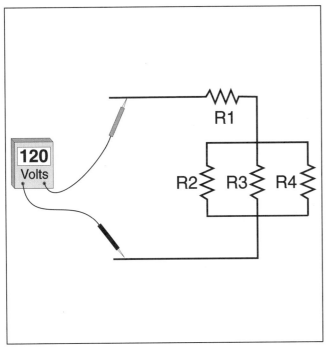
Figure 2-37

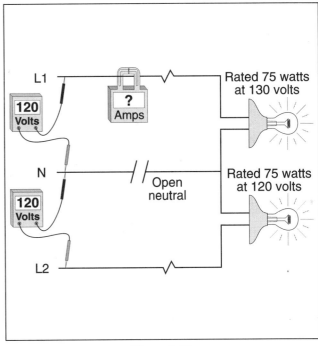
Figure 2-38

Figure 2–37 (6. applies to the next two questions:

46. Resistor R_1 has a resistance of 5 ohms, resistors R_2, R_3, and R_4 have a resistance of 15 ohms each. The total resistance of this series-parallel circuit is _____ ohms, Fig. 2–37.
    (a) 50 (b) 35
    (c) 25 (d) 10

47. • What is the voltage drop accross R_1? R_1 is 5 ohms and total resistance of R_2, R_3, and R_4 is 5 ohms, Fig. 2–37.
    (a) 60 volts (b) 33 volts
    (c) 40 volts (d) 120 volts

Part D – Multiwire Circuits

48. • If the neutral of the circuit in the diagram is opened, the circuit becomes one series circuit of 240 volts. Under this condition, the current of the circuit is _____ ampere(s). *Tip:* Determine the resistance total, Fig. 2–38.
    (a) .67
    (b) .58
    (c) 2.25
    (d) .25

NEC Questions From Section 225–19 Through Section 310–13

Article 225 – Outside Branch Circuits And Feeders

49. If the voltage between conductors exceeds 300 volts, outside conductor clearance over the roof overhang can be reduced from 8 feet to 18 inches. However, no more than _____ feet of conductors can pass over the roof overhang, and the point of attachment must be located not more than 4 feet measured horizontally from the nearest edge of the roof.
    (a) 6 (b) 8
    (c) 10 (d) none of these

50. Final spans of feeders to a building they supply shall be kept _____ feet from windows that are designed to be opened, doors, balconies, ladders, stairs, and similar locations.
    (a) 3
    (b) 5
    (c) 6
    (d) 10

51. Raceways on the exterior surfaces of buildings shall be _____.
    (a) weatherproof and covered
    (b) watertight and arranged to drain
    (c) raintight and arranged to drain
    (d) rainproof and guarded

52. Which of the following cannot be attached to vegetation?
    (a) overhead conductor spans
    (b) overhead conductor spans for temporary wiring
    (c) lighting fixtures
    (d) all of these

Article 230 – Services

Article 230 covers service conductors and equipment for control and protection of services and their installation requirements.

Part A. General

53. A building or structure shall be supplied by a maximum of _____ service(s).
    (a) one   (b) two   (c) three   (d) as many as desired

54. Service conductors supplying one building shall not _____ of another building.
    (a) be installed on the exterior walls   (b) pass through the interior
    (c) both a and b   (d) none of these

55. • Branch circuit and feeder conductors shall not be installed in the same _____ with service conductors.
    (a) raceway
    (b) cable
    (c) enclosure
    (d) a or b

56. Service conductors installed as unjacketed multiconductor cable shall have a minimum clearance of _____ feet from windows designed to be opened, doors, porches, fire escapes, and similar equipment.
    (a) 3   (b) 4   (c) 6   (d) 10

Part B. Overhead Service – Drop Conductors

57. • Open conductors run individually as service drops shall be _____.
    I. insulated   II. bare   III. covered
    (a) I only   (b) II only
    (c) III only   (d) either I or III

58. Service drop conductors shall have _____.
    (a) sufficient ampacity to carry the load
    (b) adequate mechanical strength
    (c) a or b
    (d) a and b

59. Two hundred forty-volt service conductors terminate at a through-the-roof raceway, and less than 6 feet of the conductors pass over the roof overhang. The minimum clearance above the roof for these service conductors is _____.
    (a) 12 inches   (b) 18 inches   (c) 2 feet   (d) 5 feet

60. The minimum clearance for service drops, not exceeding 600 volts, over commercial areas subject to truck traffic is _____ feet.
    (a) 10 (b) 12 (c) 15 (d) 18

61. Service-drop conductors where not in excess of 600 volts, nominal, shall have a minimum of _____ feet vertical clearance from final grade over residential property and driveways, and those commercial areas not subject to truck traffic where the voltage is limited to 300 volts to ground.
    (a) 10 (b) 12 (c) 15 (d) 18

62. • Overhead service drop conductors shall have a horizontal clearance of _____ feet from a pool.
    (a) 6 (b) 10 (c) 8 (d) 4

Part C. Underground Service-Lateral Conductors

63. Service lateral conductors must have _____.
    (a) adequate mechanical strength
    (b) sufficient ampacity for the loads computed
    (c) a and b
    (d) none of these

64. To supply power to a single-phase water heater, the size of the underground service lateral conductors shall not be smaller than _____.
    (a) No. 4 copper (b) No. 8 aluminum (c) No. 12 copper (d) No. 10 aluminum

Part D. Service-Entrance Conductors

65. • Service-entrance conductors are required to be insulated except the grounded conductor when _____.
    (a) it is bare copper and installed in a raceway
    (b) the cable assembly identified for underground use in a raceway or for direct burial
    (c) it is copper-clad aluminum
    (d) a and b

66. The service rating for service entrance conductors to a single-family dwelling shall not be less than 100 amperes, if the initial load is _____ kW or more.
    (a) 7.5 (b) 10 (c) 12.5 (d) 15

67. Wiring methods permitted for service conductors include _____.
    (a) metal-clad cable
    (b) electrical metallic tubing
    (c) liquidtight flexible nonmetallic conduit
    (d) all of these

68. • Splices on service-entrance conductors shall _____.
    (a) be permitted in metering equipment enclosures if clamped or bolted connections are made
    (b) be permitted at a properly enclosed junction box when an underground method is changed to another type of wiring method
    (c) a and b
    (d) never be permitted

69. Service cables mounted in contact with a building shall be supported at intervals not exceeding _____.
    (a) 4½ feet (b) 3 feet (c) 30 inches (d) 24 inches

70. Where individual open conductors enter a building or other structure through tubes, _____ shall be formed on the conductors before they enter the tubes.
    (a) drop loops (b) knots
    (c) drip loops (d) none of these

71. Service heads for service raceways shall be _____.
    (a) raintight (b) weatherproof (c) rainproof (d) watertight

72. Service heads must be located _____.
    (a) above the point of attachment
    (b) below the point of attachment
    (c) within 36 inches from the point of attachment
    (d) none of these

73. Service-drop conductors and service-entrance conductors shall be arranged so that _____ will not enter service raceway or equipment.
    (a) moisture (b) condensation (c) water (d) air

74. • The high leg service entrance conductor of a four-wire, delta-connected service is required to be permanently marked _____.
    (a) in orange
    (b) with orange marking tape
    (c) by other effective means
    (d) any of these

Part E. Service Equipment - General

75. Sufficient working space and illumination is required to permit safe operation, inspection, maintenance, and repairs of service equipment. The working space shall not be less than _____ inches wide by _____ inches deep.
    (a) 30, 36 (b) 24, 30 (c) 25, 30 (d) 30, 42

Part F. Service Equipment - Disconnecting Means

76. Each service disconnect must be permanently marked to identify it as a service disconnecting means.
    (a) True (b) False

77. • When the service disconnection means consists of two to six service disconnects, they are required to be _____.
    (a) the same size
    (b) grouped at one location
    (c) in the same enclosure
    (d) none of these

78. A _____ service disconnecting means shall be externally operable without exposing the operator to contact with energized parts.
    (a) open (b) building (c) safe (d) enclosed

79. For installations consisting of not more than two 2-wire branch circuits, the service disconnecting means shall have a rating of not less than _____ amperes.
    (a) 15 (b) 20 (c) 25 (d) 30

Part F. Service Equipment – Disconnecting Means

80. _____ shall operate such that all ungrounded conductors of one source of supply are disconnected before any ungrounded conductors of the second source are connected.
    (a) Service disconnecting means
    (b) Transfer equipment
    (c) Ground-fault protection
    (d) Transformer equipment

Part G. Service Equipment – Overcurrent Protection

81. Each _____ service conductor shall have overload protection.
    (a) overhead
    (b) underground
    (c) ungrounded
    (d) none of these

Part G. Service Equipment – Overcurrent Protection

82. • The provisions of Section 230–90 (overcurrent protection) do not apply where the service to the fire pump room is judged to be inside of a building.
    (a) True (b) False

83. In a service, overcurrent devices shall never be placed in the grounded service conductor except circuit breakers that simultaneously open all conductors of the circuit.
    (a) True   (b) False

84. Where the service overcurrent devices are locked or sealed, or otherwise are not readily accessible, _____ overcurrent devices shall be installed on the load side, shall be mounted in a readily accessible location, and shall be of lower ampere rating than the service overcurrent device.
    (a) sub-service   (b) branch-circuit   (c) additional   (d) protected

85. • The maximum setting of the ground fault protection in a disconnecting means shall be _____ amperes.
    (a) 800   (b) 1000   (c) 1200   (d) 2000

86. As used in Section 230–95, the rating of the service disconnecting means is considered to be the rating of the largest _____ that can be installed or the highest continuous current trip setting for which the actual overcurrent device installed in a circuit breaker is rated or can be adjusted.
    (a) fuse   (b) circuit
    (c) wire   (d) all of these

87. Ground-fault protection that functions to open the service disconnecting means _____ protect(s) service conductors or the service disconnecting means.
    (a) will   (b) will not   (c) adequately   (d) totally

88. The ground-fault protection system shall be _____ when first installed on site.
    (a) inspected   (b) identified
    (c) turned on   (d) performance tested

Part H. Services Exceeding 600 Volts, Nominal

89. For services exceeding 600 volts, nominal, the _____ switch shall be accessible to qualified persons only.
    (a) power   (b) isolating   (c) emergency shut-off   (d) all of these

Article 240 – Overcurrent Protection

Overcurrent protection is the first line of defense against the potential of electricity to cause damage to life and property. Proper sizing and application of overcurrent protection devices is critical for every system.

Part A. General

90. • When protecting conductors against overcurrent, a No. 12 THHN conductor can be protected by an overcurrent protection device that is greater than 20 amperes. This is permitted when the conductors are used for _____.
    (a) tap conductors   (b) motor circuit conductors
    (c) Class 1 control and signaling circuit conductors   (d) all of these

91. • A feeder conductor to a panelboard is run with No. 6 THHN. The maximum size overcurrent protection for this conductor is _____ amperes. *Note:* Terminal rating of 60ºC.
    (a) 60   (b) 40   (c) 90   (d) 100

92. Flexible cords approved for use with listed appliances or portable lamps shall be considered as protected by a 20-ampere circuit breaker if it is _____.
    (a) not more than 6 feet in length   (b) No. 20 and larger
    (c) No. 18 and larger   (d) No. 16 and larger

93. Which of the following are *not* standard size fuses or inverse time circuit breakers?
    I. 45   II. 70   III. 75   IV. 80
    (a) I only   (b) III only   (c) I or III   (d) II or IV

94. When can breakers or fuses be used in parallel?
    (a) When factory assembled in parallel
    (b) When listed as a unit
    (c) both a and b
    (d) either a or b

95. Which of the following statements about supplementary overcurrent protection is correct?
    (a) It shall not be used in lighting fixtures.
    (b) It may be used as a substitute for a branch circuit overcurrent device.
    (c) It may be used to protect internal circuits of equipment.
    (d) It shall be readily accessible.

96. Where an orderly shutdown is required to minimize hazard(s) to personnel and equipment, a system of coordination based on two conditions shall be permitted. Those two conditions are (1) _____ short-circuit protection, and (2) _____ indication based on monitoring systems or devices.
    (a) uncoordinated, overcurrent
    (b) coordinated, overcurrent
    (c) coordinated, overload
    (d) none of these

Part B. Location

97. A(n) _____ shall be considered equivalent to an overcurrent trip unit.
    (a) current transformer   (b) overcurrent relay
    (c) both a and b   (d) either a or b

98. • Circuit breaker handle tie-bars are required for multiwire branch-circuits in commercial and industrial buildings if they supply line to neutral loads.
    (a) True   (b) False

99. A tap from a feeder that is over 10 feet but less than 25 feet is permitted without overcurrent protection at the tap point, providing the _____.
    (a) ampacity of the tap conductors is not less than 1/3 the ampacity of the overcurrent device protecting the feeders being tapped
    (b) tap conductors terminate in a single circuit breaker or set of fuses that limit the load to the ampacity of the tap conductors
    (c) tap conductors are protected from physical damage
    (d) all of these

100. In the case of transformer feeder taps with primary plus secondary not over 25 feet long, conductors supplying a transformer shall be permitted to be tapped, without overcurrent protection at the tap, from a feeder where five conditions are met. One of those conditions is that the total length of one primary plus one secondary conductor, excluding any portion of the _____ conductor that is protected at its ampacity, is not over 25 feet.
    (a) primary   (b) secondary   (c) tertiary   (d) none of these

101. The maximum length of an unprotected feeder tap conductor (indoors) is _____ feet.
    (a) 15   (b) 20   (c) 50   (d) 100

102. Outside conductors tapped to a feeder or connected to a transformer secondary shall be permitted to be protected by complying with four conditions. One of those conditions is that the overcurrent device for the _____ is an integral part of a disconnecting means or shall be located immediately adjacent thereto.
    (a) feeder tap(s)   (b) conductors
    (c) branch-circuit   (d) none of these

103. No overcurrent device shall be connected in series with any conductor that is intentionally grounded except when _____.
    (a) less than 50 volts
    (b) the device opens all conductors of the circuit, including the grounded conductor
    (c) it opens all ungrounded conductors simultaneously
    (d) all of these

104. Overcurrent devices shall be _____.
    (a) accessible (as applied to wiring methods)
    (b) accessible (as applied to equipment)
    (c) readily accessible
    (d) inaccessible to unauthorized personnel

Part C. Enclosures

105. Enclosures for overcurrent devices in damp or wet locations shall comply with Section _____.
 (a) 372–1(a) (b) 372–2(a) (c) 373–1(a) (d) 373–2(a)

Part D. Disconnecting and Guarding

106. Handles or levers of circuit breakers and similar parts that may move suddenly in such a way that persons in the vicinity are likely to be injured by being struck by them shall be _____.
 (a) guarded (b) isolated (c) remote (d) a or b

Part E. Plug Fuses, Fuseholders, And Adapters

107. Plug fuses of the Edison-base type shall be used _____.
 (a) where overfusing is necessary
 (b) only as replacement items in existing installations
 (c) as a replacement for type S fuses
 (d) only for 50 amperes and above

108. • Which of the following statements about type S fuses are not true?
 (a) Adapters will fit Edison-base fuse holders.
 (b) Adapters are designed to be easily removed.
 (c) Type S fuses shall be classified as not over 125 volts and 30 amperes.
 (d) All of These

109. Type S fuses, fuse holders, and adapters are required to be designed so that _____ would be difficult.
 I. installation II. tampering III. shunting
 (a) I only (b) II only (c) III only (d) II and III

110. Dimensions of Type S fuses, fuseholders, and adapters shall be standardized to permit interchange ability regardless of the _____.
 (a) model (b) manufacturer (c) amperage (d) voltage

Part F. Cartridge Fuses And Fuseholders

111. • An 800-ampere fuse rated 600 volts _____ on a 250-volt system.
 (a) shall not be used (b) are required to be used
 (c) can be installed (d) none of these

112. Cartridge fuses and fuseholders shall be classified according to _____ ranges.
 (a) voltage (b) amperage
 (c) voltage or amperage (d) voltage and amperage

Part G. Circuit Breakers

113. Circuit breakers shall be marked with their _____ rating in a manner that will be durable and visible after installation.
 (a) voltage (b) ampere (c) type (d) all of these

114. Markings on circuit breakers required by the Code shall be permitted to be made visible by removal of a _____ or cover.
 (a) plate (b) panel (c) trim (d) all of these

115. Every circuit breaker having an interrupting (IC) rating of other than _____ amperes shall have its interrupting rating shown on the circuit breaker.
 (a) 50,000 (b) 10,000 (c) 15,000 (d) 5,000

Article 250 – Grounding and Bonding

Grounding is one of the largest, most important, and least understood articles in the NEC. As specified in Section 90–1(a), safety is the key element and purpose of the NEC. Proper grounding and bonding is essential for maximum protection of life and property. If overcurrent protection is considered the first line of defense, grounding could be considered the last line of defense.

Most of the articles in NEC Chapter 2 are very detailed and difficult to understand, especially for the novice. It would be impossible to explain these subjects properly in this type of book.

Part A. General

116. Conductive materials enclosing electrical conductors or equipment are grounded to limit the voltage to ground and are bonded to facilitate the operation of overcurrent devices under ground-fault conditions.
    (a) True (b) False

Part C. Location of System Grounding Connections

117. • Currents that introduce noise or data errors in electronic equipment shall be considered the objectionable currents addressed in Section 250–21.
    (a) True (b) False

118. The grounding electrode conductor at the service is permitted to terminate on an equipment grounding terminal bar, if a (main) bonding jumper is installed between the neutral bus and the equipment grounding terminal.
    (a) True (b) False

119. • When grounding service-supplied alternating-current systems, the grounding electrode conductor shall be connected (bonded) to the grounded service conductor (neutral) at _____.
    (a) the load end of the service drop
    (b) the meter equipment
    (c) the service disconnect
    (d) any of these

120. Where an ac system operating at less than 1,000 volts is grounded at any point, the _____ conductor shall be run to each service disconnecting means and shall be bonded to each disconnect enclosure.
    (a) ungrounded (b) grounded (c) grounding (d) none of these

Part D. Enclosure Grounding

121. Isolated metal elbows installed underground are permitted if located so that no part is less than _____ inches below finish grade.
    (a) 6 (b) 12
    (c) 18 (d) according to Table 300–5

122. Metal enclosures used to provide support or protection of _____ from physical damage shall not be required to be grounded.
    (a) conductors (b) feeders (c) cables (d) none of these

Part E. Equipment Grounding

123. An electrically operated pipe organ shall have both the generator and motor frame grounded or _____.
    (a) the generator and motor shall be effectively insulated from ground
    (b) the generator and motor shall be effectively insulated from ground and from each other
    (c) the generator shall be effectively insulated from ground and from the motor driving it
    (d) both shall have double insulation

124. Which of the following appliances installed in residential occupancies need not be grounded?
    (a) a toaster (b) an aquarium
    (c) a dishwasher (d) a refrigerator

125. Metal raceways and other noncurrent-carrying metal parts of electric equipment shall be kept at least _____ feet away from lightning rod conductors, or they shall be bonded to the lightning rod conductors.
    (a) 3 (b) 6 (c) 8 (d) 10

Part F. Methods of Grounding

126. The path to ground from circuits, equipment, and metal conductor enclosures shall _____.
    (a) be permanent and continuous
    (b) have the capacity to conduct safely any fault current likely to be imposed on it
    (c) have sufficiently low impedance to limit the voltage to ground and to facilitate the operation of the circuit protective devices
    (d) all of these

127. When a building has two or more electrodes for different services, the separate electrodes must be _____. This shall be considered as a single electrode in this sense.
    (a) effectively bonded together
    (b) spaced no more than 6 feet apart
    (c) both a and b
    (d) neither a nor b

128. When considering whether equipment is effectively grounded, the structural metal frame of a building shall be permitted to be used as the required equipment grounding conductor for ac equipment.
    (a) True
    (b) False

129. Noncurrent-carrying metal parts of cord and plug-connected equipment, where grounding is required, can be accomplished by means of the metal enclosure of the conductors supplying such equipment if _____.
    (a) a grounding-type attachment plug with one fixed grounding contact is used for grounding the metal enclosure
    (b) the metal enclosure of the conductors is secured to the attachment plug and equipment by approved means
    (c) both a and b
    (d) neither a nor b

130. On the load side of the service disconnecting means, the _____ circuit conductor is permitted to ground meter enclosures if all meter enclosures are located near the service disconnecting means and no service ground-fault protection is installed.
    (a) grounding
    (b) bonding
    (c) grounded
    (d) phase

Part G. Bonding

131. The noncurrent carrying metal parts of equipment, such as _____, shall be effectively bonded together.
    (a) service raceways, cable trays, or service cable armor
    (b) service equipment enclosures containing service conductors, including meter fittings, boxes, or the like, interposed in the service raceway or armor
    (c) the metallic raceway or armor enclosing a grounding electrode conductor
    (d) all of these

132. Service raceways threaded into metal service equipment such as bosses (hubs) are considered to be effectively _____ to the service metal enclosure.
    (a) attached
    (b) bonded
    (c) grounded
    (d) attached

133. Equipment bonding jumper shall be used to connect the grounding terminal of a grounding-type receptacle to a grounded box. Where the box is surface mounted, direct metal-to-metal contact between the device yoke and the box shall be permitted to ground the receptacle to the box.
    (a) True
    (b) False

134. For circuits over 250-volts to ground (480Y/277-volt), electrical continuity can be maintained between a box or enclosure (with no ringed knockouts) and a metal conduit by _____.
    (a) threadless fitting for cables with metal sheath
    (b) double locknuts on threaded conduit (one inside and one outside the box or enclosure)
    (c) set screw conduit fittings that have shoulders that seat firmly against the box or enclosure with a single locknut on the inside.
    (d) all of these

135. • All metal raceways, boxes, and enclosures from a hazardous (classified) location to the service or source of a separately derived system are required to be _____ according to the requirements of Section 250–72(b) through (e).
    (a) grounded
    (b) continuous
    (c) rigid metal conduit
    (d) bonded

70   Unit 2   Electrical Circuits   Chapter 1   Electrical Theory And Code Questions

136. A service is supplied by three metal raceways. Each raceway contains 600-kcmil phase conductors. Determine the size of the service bonding jumper for each raceway.
(a) 1/0   (b) 2/0   (c) 225 kcmil   (d) 500 kcmil

137. The bonding jumper for service raceways must be sized according to the _____.
(a) calculated load
(b) service-entrance conductor size
(c) service drop size
(d) load to be served

138. A 200-ampere residential service consists of 2/0 THHN copper conductors. What is the minimum size service bonding jumper for the service raceway?
(a) 6 aluminum
(b) 3 copper
(c) 4 aluminum
(d) 4 copper

139. What is the minimum size equipment grounding conductor required for equipment connected to a 40-ampere rated circuit?
(a) 12   (b) 14   (c) 8   (d) 10

140. Interior metal water piping systems are required to be bonded to the _____.
(a) grounded conductor at the service
(b) service equipment enclosure
(c) equipment grounding bar or bus at any panelboard within the building
(d) a and b only

Part H. Grounding Electrode System

141. Intentionally connected to earth through a ground connection or connection of sufficiently low impedance is defined as _____.
(a) effectively grounded
(b) effectively bonded
(c) a and b
(d) none of these

142. If none of the electrodes specified in Section 250–81 are available, then _____ can be used as the required grounding electrode.
(a) local metal underground systems or structures (not gas)
(b) ground rods or pipes
(c) plate electrodes
(d) any of these

143. • The smallest diameter rod that may be used for a made electrode is _____ inch.
(a) ½   (b) ¾   (c) 1   (d) ¼

144. The upper end of the made electrode shall be _____ ground level unless the aboveground end and the grounding electrode conductor attachment are protected against physical damage.
(a) above the ground level
(b) flush with ground level
(c) below the ground level
(d) b or c

145. • Where the resistance to ground of a single made electrode exceeds 25 ohms, _____.
(a) additional electrodes must be added until the resistance to ground is less than 25 ohms
(b) one additional electrode must be added
(c) no additional electrodes are required
(d) the electrode can be omitted

Part J. Grounding Conductors

146. The grounding electrode conductor shall be made of which of the following materials?
(a) copper
(b) aluminum
(c) copper-clad aluminum
(d) any of these

147. Flexible metal conduit shall be permitted for grounding if the length in any ground return path does not exceed 6 feet and the circuit conductors are protected by an overcurrent protection device that does not exceed _____ amperes.
    (a) 15 (b) 20 (c) 30 (d) 60

148. When aluminum grounding electrode conductors are used outdoors to connect to the grounding electrode, the aluminum shall not be installed within _____ inches of the earth.
    (a) 6 (b) 12 (c) 15 (d) 18

149. • A service that contains No. 12 service-entrance conductors, shall require a No. _____ grounding electrode conductor.
    (a) 6 (b) 12 (c) 8 (d) 10

150. • What size copper grounding electrode conductor is required for a service that has three sets of 500-kcmil copper per phase?
    (a) No. 2/0 (b) No. 3/0 (c) No. 0 (d) No. 2

151. The size of the grounding electrode conductor of a grounded or ungrounded ac system to a concrete-encased electrode as in Section 250–81(c) shall not be required to be larger than No. _____ copper wire.
    (a) 4
    (b) 6
    (c) 8
    (d) 10

152. • What size equipment grounding conductor is required for a nonmetallic raceway that contains the following three circuits?
    No. 12 circuit protected by a 20-ampere device
    No. 10 circuit protected by a 30-ampere device
    No. 8 circuit protected by a 40-ampere device
    (a) No. 10
    (b) No 6
    (c) No. 8
    (d) No. 12

Part K. Grounding Conductor Connections

153. The connection (attachment) of the grounding electrode conductor to a grounding electrode shall _____.
    (a) be readily accessible
    (b) be made in a manner that will ensure a permanent and effective ground
    (c) not require bonding around insulated joints of a metal piping system
    (d) none of these

154. When an underground metal water piping system is used as a grounding electrode, effective bonding shall be provided around insulated joints and sections around any equipment that is likely to be disconnected for repairs. _____ conductors shall be of sufficient length to permit removal of such equipment while retaining the integrity of the bond.
    (a) Main bonding jumper
    (b) Equipment grounding
    (c) Bonding
    (d) None of these

155. • Where one or more equipment grounding conductors enter a box, all equipment grounding conductors shall be spliced together (in the enclosure). This does not apply to insulated equipment grounding conductors for isolated ground receptacles for electronic equipment.
    (a) True (b) False

156. The connection of the grounding electrode conductor to a buried grounding electrode (driven ground rod) shall be made with a listed terminal device that is _____.
    (a) suitable for direct burial
    (b) accessible
    (c) readily accessible
    (d) concrete proof

157. _____ on equipment to be grounded shall be removed from contact surfaces to ensure good electrical continuity.
    (a) Conductive coatings
    (b) Nonconductive coatings
    (c) Manufacturer's instructions
    (d) all of these

Part L. Instrument Transformers, Relays, Etc.

158. Secondary circuits of current and potential instrument transformers shall be grounded where the primary windings are connected to circuits of _____ -volts or more to ground and, where on switchboards, shall be grounded irrespective of voltage.
    (a) 300 (b) 600 (c) 1000 (d) 150

159. • Cases or frames of instrument transformers are not required to be grounded _____.
    (a) when accessible to qualified persons only
    (b) for current transformers where the primary is not over 150-volts to ground and which are used exclusively to supply current to meters
    (c) for potential transformers where the primary is less than 150-volts to ground
    (d) a or b

Article 280 – Surge Arresters

Part A. General

160. A surge arrestor is a protective device for limiting surge voltages by _____ or by passing surge current.
    (a) decreasing (b) discharging (c) limiting (d) derating

Part B. Installation

161. The conductor used to connect the surge arrester to _____ shall not be any longer than necessary.
    I. line II. ground III. bus
    (a) I only (b) II only (c) I or II (d) I, II or III

Part C. Connecting Surge Arresters

162. The conductor between a lightning arrester and the line and surge arrestor and the grounding connection shall not be smaller than (installations operating at 1000 volts or more) _____ copper.
    (a) No. 4 (b) No. 6 (c) No. 8 (d) No. 2

Chapter 3 – Wiring Methods And Material

NEC Chapter 3 is a general rules chapter as applied to wiring methods and material and is generally related to wiring raceways, cables, junction boxes, cabinets, etc. The rules in this chapter apply everywhere in the NEC except as modified (such as in Chapters 5, 6, and 7). Along with Chapter 2, Chapter 3 can be considered the heart of the Code.

If you turn to the NEC Table of Contents, and review the Chapter 3 listings, the articles are generally grouped as follows. Articles 318 through 344 (except for 331) are related to specific types of conductors and cables.

345 through 362 are specific types of raceways. These are followed by Articles 364 and 365 (Busways and Cablebus) which are raceways with conductors for assembly.

The rest, Articles 370 through 384, are equipment associated with wiring, such as boxes and panelboards.

Article 300 – Wiring Methods

Section 300–1. Scope tells us that Article 300 applies to all wiring installations. The exceptions that follow list several articles that have their own wiring methods requirements. Throughout the rest of the Code, the rules (sections) of Article 300 will apply except where specified, which is usually in Chapters 5, 6, 7, and 8. If you flip through Article 300 and read the bolded section and subsection titles, you will see very general types of requirements.

Part A. General Requirements

163. Unless specified elsewhere in the Code, Chapter 3 shall be used for voltages of _____.
    (a) 600-volts to ground or less
    (b) 300-volts between conductors or less
    (c) 600-volts, nominal, or less

(d) 600-volts RMS

164. • Circuit conductors that operate at 277 volts (with 600-volt insulation) may occupy the same enclosure or raceway with other conductors that have an insulation rating of 300 volts.
(a) True (b) False

165. Where nonmetallic sheath cable is installed through punched or factory-made holes in metal studs or metal framing members, _____ must be installed before the cable is installed.
(a) grommets (b) bushings (c) plates (d) a or b

166. Wiring methods installed behind panels that allow access, such as the space above a drop ceiling, are required to be _____ according to their applicable articles.
(a) supported (b) painted (c) in a metal raceway (d) all of these

167. UF cable is used to supply power for a 120-volt, 30-ampere circuit. If the cable is installed outdoors underground, the minimum cover requirement is _____ inches.
(a) 12 (b) 24 (c) 16 (d) 6

168. • UF cable is used to supply power of a 120-volt, 15-ampere GFCI circuit. If the cable is installed outdoors under a one-family drive way, the minimum cover requirement is _____ inches.
(a) 12 (b) 24 (c) 16 (d) 6

169. • When listed liquidtight flexible metal conduit that is approved for direct burial is installed underground, it must have no less than _____ inches of cover.
(a) 8 (b) 12 (c) 18 (d) 22

170. • When installing raceways underground rigid nonmetallic conduit and other approved raceways must have a minimum of _____ inches of cover.
(a) 6 (b) 12 (c) 18 (d) 22

171. A 24-volt landscape lighting system installed with UF cable is permitted with a minimum cover of _____ inches.
(a) 6 (b) 12
(c) 18 (d) 24

172. Direct buried conductors or cables shall be _____, providing those splices are made with approved methods and identified materials.
(a) permitted to be spliced without a junction box
(b) spliced only in an approved junction box
(c) spliced only in a weatherproof splice box
(d) none of these

173. Metal raceways, cable armor, boxes, cable sheathing, cabinets, elbows, couplings, fittings, supports, and support hardware shall be of materials suitable for _____.
(a) corrosive locations
(b) wet locations
(c) the environment in which they are to be installed
(d) all of these

174. Which of the following metal parts shall be protected from corrosion inside and out?
(a) ferrous raceways
(b) metal elbows
(c) boxes
(d) all of these

175. • Wiring systems installed in outdoor wet locations are required to be _____.
(a) placed so a permanent ½ air space separates them from the supporting surface
(b) separated by insulated bushings
(c) separated by noncombustible tubing
(d) none of these

176. Circulation of warm and cold air in _____ must be prevented.
 (a) raceways (b) boxes
 (c) cabinets (d) panels

177. Suspended ceiling support wires, can be used to support branch circuit wiring when _____.
 (a) associated with equipment that is located with the drop ceiling
 (b) in a fire-rated drop ceiling
 (c) in a nonfire-rated drop ceiling
 (d) none of these

178. In multiwire circuits, the continuity of the _____ conductor shall not be dependent upon the device connections.
 I. ungrounded II. grounded III. grounding
 (a) I (b) II (c) III (d) I and II

179. The number of conductors permitted in a raceway shall be limited to _____.
 (a) permit heat to dissipate
 (b) prevent damage to insulation during installation
 (c) prevent damage to insulation during removal of conductors
 (d) all of these

180. • A 100-foot vertical run of 4/0 copper requires the conductors to be supported at _____ locations.
 (a) 4 (b) 5 (c) 2 (d) 3

181. Three conductors (phases A, B, and neutral) are run from a busway to a panel, these conductors should not be run in three separate raceways because _____.
 (a) it's cheaper to run three in one raceway
 (b) it's easier to install
 (c) of inductive heating effect
 (d) none of these

182. Electrical installations in hollow spaces, vertical shafts, and ventilation or air-handling ducts shall be so made that the possible spread of fire or products of combustion will not be _____.
 (a) substantially increased (b) allowed (c) increased (d) possible

183. No wiring of any type shall be installed in ducts used to transport _____.
 (a) dust (b) flammable vapors (c) loose stock (d) all of these

184. Type AC cable can be used in ducts or plenums used for environmental air.
 (a) True (b) False

185. Wiring methods permitted in ducts or plenums used for environmental air are _____.
 (a) flexible metal conduit of any length
 (b) electrical metallic tubing
 (c) armor cable (BX)
 (d) nonmetallic sheath cable

186. The space above a drop ceiling used for environmental air is considered as _____ and the wiring limitations of section _____ apply.
 (a) a plenum, 300–22(b)
 (b) other spaces, 300–22(c)
 (c) a duct, 300–22(b)
 (d) none of these

187. • The air handling area beneath raised floors for data processing systems is not a plenum and is not required to comply with the requirements of Section 300–22 but shall be permitted in accordance with Article 645.
 (a) True (b) False

Article 305 – Temporary Wiring

Article 305 contains specific rules for temporary wiring methods.

188. Temporary electrical power and lighting is permitted for emergencies and _____.
(a) tests
(b) experiments
(c) developmental work
(d) all of these

189. Receptacles for construction sites shall not be installed on the _____ or connected to _____ that supply temporary lighting.
(a) same branch circuit, feeders
(b) feeders, multiwire branch circuits
(c) same branch circuit, multiwire branch circuits
(d) all of these

190. Multiwire branch circuits for temporary wiring shall be provided with a means to disconnect simultaneously all _____ conductors at the power outlet or panelboard where the branch circuit originated.
(a) underground
(b) overhead
(c) ungrounded
(d) grounded

191. At construction sites, boxes are not required for temporary wiring splices of _____.
(a) multiconductor cords
(b) multiconductor cables
(c) a or b
(d) none of these

192. All 125-volt, single-phase, 15- and 20-ampere receptacle outlets that are not a part of the permanent wiring of the building or structure and that are in use by personnel shall have ground-fault, circuit-interrupter protection for personnel, except receptacles on a two-wire, single-phase portable or vehicle-mounted generator rated not more than _____ kW, where the circuit conductors of the generator are insulated from the generator frame and all other grounded surfaces.
(a) 3
(b) 4
(c) 5
(d) 6

Article 310 – Conductors For General Wiring

Article 310 – Conductors for General Wiring, along with Article 300, are the two general-use articles in Chapter 3 that most electricians use on a daily basis. Article 310 covers the general requirements for conductors and their type designations, insulation, markings, mechanical strengths, ampacity ratings, and uses. These requirements do not apply to conductors that form an integral part of equipment, such as motors and appliances.

Table 310–13 is a very informative table. The second column Type letter alphabetically lists the common use conductor insulation types most of us are familiar with. On the fourth page of this table is the conductor type *THW*. The double and triple obelisk (††) and (†††) listed with THW tell you notes could apply; these notes should always be read when looking up information. For example, the double obelisk note (††) says that if you see THW with the suffix /LS (THW/LS), the insulation meets the requirements for flame-retardant, limited smoke. The fourth column, Application Provisions, lists information for each insulation type. Using THW as an example, this column tells us that even though THW is 75ºC in wet and dry locations, it is rated 90ºC for use in special applications within electric discharge lighting equipment.

In Section 310–15(a), ampacities for conductors rated 0 to 2,000 volts are listed in Tables 310–16 through 310–19. There are the tables that apply to Article 220 calculations and the tables used by electricians, architects, and engineers. Tables 310–69 through 310–84 are for 2,001 through 35,000 volts and do not apply to the average person using the NEC.

Table 310–16 is one of the most used tables in the NEC. It lists the ampacity ratings (current carrying capacity) of the most commonly used conductors. These are continuous load ratings for conductors. See definition of ampacity in Article 100. As with most tables in the NEC, the notes associated with the table should always be considered when using that table. In this case, the Notes to Ampacity Tables of 0 to 2000 volts, located after Table 310–19, are very important for understanding how to use Tables 310–16 through 310–19.

193. Where installed in raceways, conductors No. _____ and larger shall be stranded, except as permitted in Section 680–20(b)(1).
(a) 10
(b) 6
(c) 8
(d) 4

194. The parallel conductors in each phase or neutral shall _____.
(a) be the same length and conductor material
(b) have the same circular mil area and insulation type
(c) be terminated in the same manner

(d) all of these

195. The minimum size conductor permitted in a commercial building for branch circuits under 600 volts is No. _____ AWG.
 (a) 14 (b) 12 (c) 10 (d) 8

196. • Where conductors of different insulations are associated together, the limiting temperature of any conductor shall not be exceeded.
 (a) True (b) False

197. The _____ rating of a conductor is the maximum temperature, at any location along its length, that the conductor can withstand over a prolonged period of time without serious degradation.
 (a) ambient
 (b) temperature
 (c) maximum withstand
 (d) none of these

198. Suffixes to designate the number of conductors within a cable are _____.
 (a) D – Two insulated conductors laid parallel
 (b) M – Two or more insulated conductors twisted spirally
 (c) T – Two or more insulated conductors twisted in parallel
 (d) a and b

199. • Type THW insulation has a _____ °C rating when installed within electric discharge lighting equipment, such as through fluorescent fixtures.
 (a) 60 (b) 75

200. Which conductor has a temperature rating of 90°C?
 (a) RH (b) RHW (c) THHN (d) TW

Unit 3

Understanding Alternating Current

OBJECTIVES

After reading this unit, the student should be able to briefly explain the following concepts:

Part A – Alternating Current Fundamentals
Alternating current
Armature turning frequency
Current flow
Generator
Magnetic cores
Phase differences in degrees
Phase– In and Out
Values of alternating current
Waveform

Part B - Induction And Capacitance
Charge, discharging and testing of capacitors
Conductor impedance
Conductor shape
Induced voltage and applied current
Magnetic cores
Use of capacitors

Part C - Power Factor And Efficiency
Apparent power (volt-amperes)
Efficiency
Power factor
True power (watts)

After reading this unit, the student should be able to briefly explain the following terms:

Part A – Alternating Current Fundamentals
Ampere-turns
Armature speed
Coil
Conductor cross-sectional area
Effective (RMS)
Effective to peak
Electromagnetic field
Frequency
Impedance
Induced voltage
Magnetic field
Magnetic flux lines
Peak
Peak to effective
Phase relationship

RMS
Root-mean-square
Self-inductance
Skin effect
Waveform

Part B And C - Induction And Capacitance
Back-EMF
Capacitance
Capacitive reactance
Capacitor
Coil
Counterelectromotive force
Eddy currents
Farads
Frequency
Henrys

Impedance
Induced voltage
Induction
Inductive reactance
Phase relationship
Self-inductance
Skin effect

Part D - Power Factor And Efficiency
Apparent power
Efficiency
Input watts
Output watts
Power factor
True power
Volts-amperes
Wattmeter

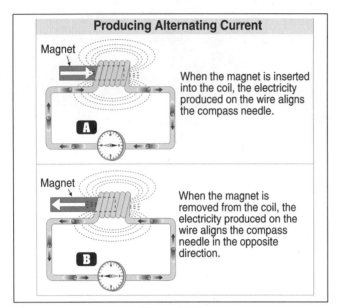

Figure 3-1
Alternating Current

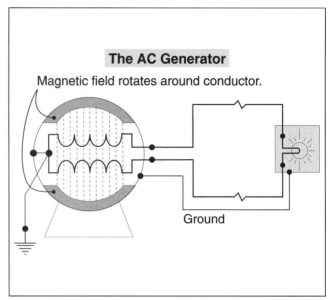

Figure 3-2
Alternating Current Generator

PART A – *ALTERNATING CURRENT FUNDAMENTALS*

3–1 CURRENT FLOW

For current to flow in a circuit, the circuit must have a *closed loop* and the power supply must push the electrons through the completed circuit. The transfer of electrical energy can be accomplished by electrons flowing in one constant direction (*direct current*), or by electrons flowing in one direction and then reversing in the other direction (*alternating current*).

3–2 ALTERNATING CURRENT

Alternating current is generally used instead of direct current for electrical systems and building wiring. This is because alternating current can easily have voltage variations (*transformers*). In addition alternating current can be transmitted inexpensively at *high voltage* over long distances resulting in reduced *voltage drop* of the power distribution system as well as smaller power distribution wire and equipment. Alternating current can be used for certain applications for which direct current is not suitable.

Alternating current is produced when electrons in a conductor are forced to move because there is a moving magnetic field. The lines of force from the magnetic field cause the electrons in the wire to flow in a specific direction. When the lines of force of the magnetic field move in the opposite direction, the electrons in the wire will be forced to flow in the opposite direction, Fig. 3–1. We must remember that electrons will flow only when there is relative motion between the conductors and the magnetic field.

3–3 ALTERNATING CURRENT GENERATOR

An alternating current *generator* consists of many loops of wire that rotates between the *flux lines* of a magnetic field. Each conductor loop travels through the magnetic lines of force in opposite directions, causing the electrons within the conductor to move in a specific direction, Fig. 3–2. The force on the electrons caused by the *magnetic flux lines* is called *voltage*, or *electromotive force*, and is abbreviated as EMF, E or V.

The magnitude of the electromotive force is dependent on the number of turns (wraps) of wire, the strength of the magnetic field, and the speed at which the coil rotates. The rotating conductor loop mounted on a shaft is called a *rotor*, or *armature*. Slip, or collector rings, and carbon brushes are used to connect the output voltage from the generator to an external circuit.

3–4 WAVEFORM

A *waveform* is a pictorial view of the shape and magnitude of the level and direction of current or voltage over a period of time. The waveform represents the magnitude and direction of the current or voltage overtime in relationship with the generators armature.

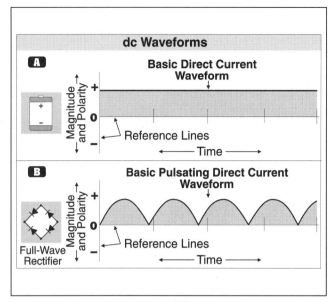

Figure 3-3
Direct Current Waveform

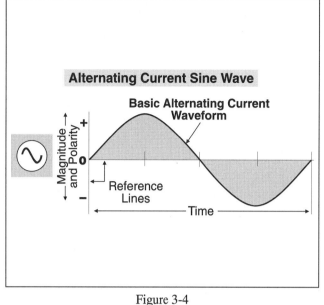

Figure 3-4
Alternating Current Sine Wave

Direct Current Waveform

The *polarity* of the voltage of direct current is positive, and the current flows through the circuit in the same direction at all times. In general, direct current voltage and current will remain the same magnitude, particularly when supplied from batteries or capacitors, Fig. 3–3.

Alternating Current Waveform

The waveform for alternating current displays the level and direction of the current and voltage for every instant of time for one full revolution of the *armature*. When the waveform of an alternating current circuit is symmetrical with positive above, and negative below the zero reference level the waveform is called a *sine wave*, Fig. 3–4.

3–5 ARMATURE TURNING FREQUENCY

Frequency is a term used to indicate the number of times the generator's armature turns one full revolution (360 degrees) in one second, Fig. 3–5. This is expressed as *hertz*, or *cycles per second*, and is abbreviated as *Hz* or cycles per second (*cps*). In the Unites States, frequency is not a problem for most electrical work because it remains constant at 60 hertz.

Armature Speed

A current or voltage that is produced when the generator's armature makes one cycle (360 degrees) in 1/60th of a second is called 60-Hz. This is because the armature travels at a speed of one cycle every 1/60th of a second. A 60-Hz circuit will take 1/60th of a second for the armature to complete one full cycle (360 degrees), and it will take 1/120th of a second to complete 1/2 cycle (180 degrees), Fig. 3–6.

3–6 PHASE – IN AND OUT

Phase is a term used to indicate the time relationship between two waveforms, such as voltage to current, or voltage to voltage. *In phase* means that two waveforms are in step with each other, that is they cross the horizontal zero axis at the same time, Fig. 3–7. The terms lead and lags are used to describe the relative position, in degrees, between two waveforms. The waveform that is ahead, *leads*, the waveform behind, *lags*, Fig. 3–8.

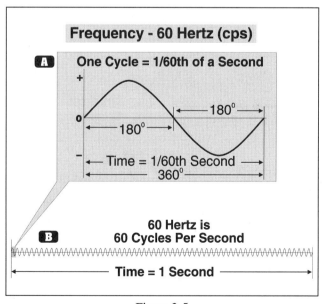

Figure 3-5
Armature Turn Frequency

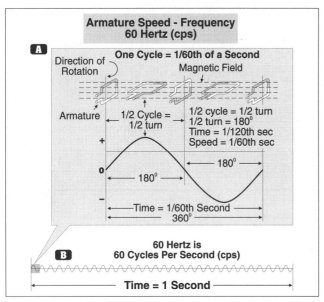

Figure 3-6

Armature Speed

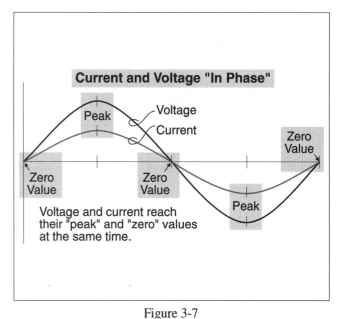

Figure 3-7

In Phase

3–7 PHASE DIFFERENCES IN DEGREES

Phase differences are expressed in *degrees*, one full revolution of the armature is equal to 360 degrees. Single-phase, 120/240-volt power is out of phase by 180 degrees, and three-phase power is out of phase by 120 degrees. Figure 3–9.

3–8 VALUES OF ALTERNATING CURRENT

There are many different types of values in alternating current, the most important are *peak* (maximum) and *effective*. In alternating current systems, effective is the value that will cause the same amount of heat to be produced as in a direct current circuit containing only resistance. Another term for effective is *RMS*, which is *root-mean-square*. In the field, whenever you measure voltage, you are always measuring effective voltage. Newer types of clamp-on ammeters have the ability to measure peak as well as effective current, Fig. 3–10.

Note. The values of direct current generally remains constant and are equal to the effective values of alternating current.

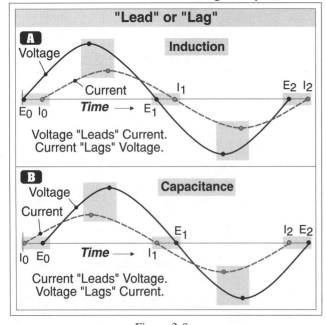

Figure 3-8

Out of Phase – Lead or Lag

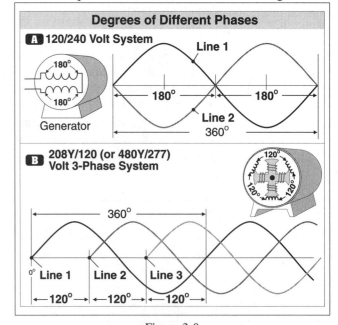

Figure 3-9

Phase Differences in Degrees

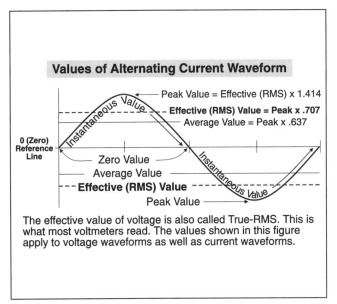

Figure 3-10
Values of Alternating Current

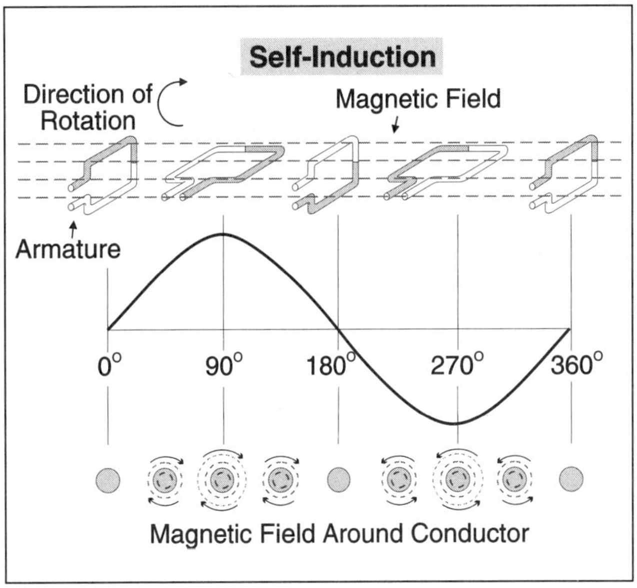

Figure 3-11
Electromagnetic Field Within Conductor

❏ **Alternating Current Versus Direct Current Heating Example**
Twelve amperes effective alternating current will have the same heating effect as _____-amperes direct current.
(a) 7 (b) 9s (c) 11 (d) 12

• Answer: (d) 12

❏ **Voltage Example**
One Hundred and twenty volt effective alternating current will produce the same heating effect as _____ volts direct current.
(a) 70 (b) 90 (c) 120 (d) 170

• Answer: = 120

Effective to Peak
To convert effective to peak, or vise versa, the following formulas should be helpful:
Peak = Effective (RMS) × 1.414 times or Peak = Effective (RMS)/0.707

❏ **Effective to Peak Example**
What is the peak voltage of a 120-volt effective circuit?
(a) 120 volts (b) 170 volts (c) 190 volts (d) 208 volts

• Answer: (b) 170 volts

   Peak Volts = Effective (RMS) volts × 1.411   Effective (RMS) = 120 volts
   Peak Volts = 120 volts × 1.414   Peak Volts = 170 volts

Peak to Effective
Peak to effective can be determine by the formula:

Effective (RMS) = Peak × .707

❏ **Peak to Effective Example**
What is the effective current of 60 amperes peak?
(a) 42 amperes (b) 60 amperes (c) 72 amperes (d) none of these

• Answer: (a) 42 amperes

   Effective Current = Peak Current × 0.707   Peak Current = 60 amperes
   Effective Current = 60 amperes × 0.707   Effective Current = 42 amperes

PART B – INDUCTION

INDUCTION INTRODUCTION

When a magnetic field is moved through a conductor, the electrons in the conductor will move. The movement of electrons caused by electromagnetism is called *induction*. In an alternating current circuit, the current movement of the electrons increases and decreases with the rotation of the *armature* through the magnetic field. As the current flows through the conductor increases or decreases, it causes an expanding and collapsing electromagnetic field within the conductor, Fig. 3–11. This varying electromagnetic field within the conductor causes the electrons in the conductor to move at 90 degrees to the flowing electrons within the conductor. This movement of electrons because of the conductor's electromagnetic field is called *self-induction*. Self-induction (induced voltage) is commonly called *counterelectromotive-force, CEMF,* or *back-EMF*.

3–9 INDUCED VOLTAGE AND APPLIED CURRENT

The *induced voltage* in the conductor because of the conductor's electromagnetism (CEMF) is always 90 degrees *out of phase* with the applied current, Fig. 3–12. When the current in the conductor increases, the induced voltage tries to prevent the current from increasing by storing some of the electrical energy in the *magnetic field* around the conductor. When the current in the conductor decreases, the induced voltage attempts to keep the current from decreasing by releasing the the energy from the magnetic field back into the conductor. The opposition to any change in current because of induced voltage is called *inductive reactance*, which is measured in ohms, abbreviated as X_L, and is known as *Lentz's Law*.

Inductive reactance changes proportionally with frequency and can be found by the formula:

$X_L = 2\pi fL$ $\pi = 3.14,$ f = frequency   L= Inductance in henrys

❑ **Inductive Reactance Example**

What is the inductive reactance of a conductor in a transformer that has an inductance of 0.001 henrys? Calculate at 60-Hz, 180-Hz, and 300-Hz, Fig. 3–13.

- Answers:

$X_L = 2\pi fL$

$\pi = 3.14$, f = 60-Hz, 180-Hz and 300-Hz, L= Inductance measured in henrys, 0.001

X_L at 60-Hz = 2 × 3.14 × 60-Hz × 0.001 henrys, X_L = 0.377 ohms

X_L at 180-Hz = 2 × 3.14 × 180-Hz × 0.001 henrys, X_L = 1.13 ohms

X_L at 300-Hz = 2 × 3.14 × 300-Hz × 0.001 henrys, X_L = 1.884 ohms

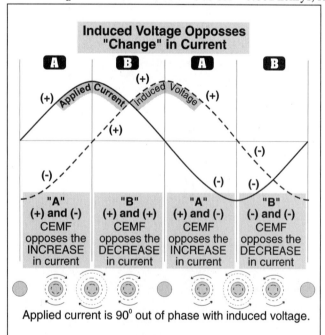

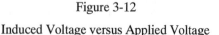

Figure 3-12

Induced Voltage versus Applied Voltage

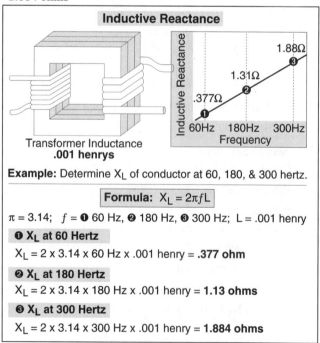

Figure 3-13

Inductive Reactance Example

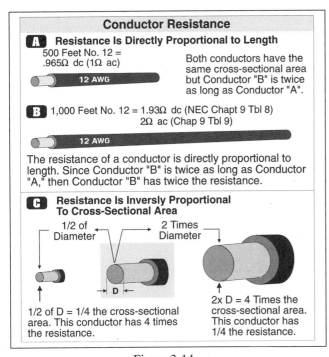

Figure 3-14
Conductor Resistance

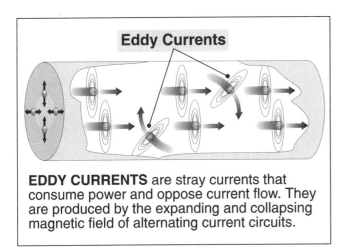

Figure 3-16
Skin Effect

3–10 CONDUCTOR IMPEDANCE

The *resistance* of a conductor is a physical property of the conductors that oppose the flow of electrons. This resistance is directly proportional to the conductor length, and is inversely proportional to the conductor cross-sectional area. This means that the longer the wire, the greater the conductor resistance; the smaller the wire, the greater the conductor resistance, Fig. 3–14.

The opposition to alternating current flow due to *inductive reactance* and conductor resistance is called *impedance (Z)*. Impedance is only present in alternating current circuits and is measured in ohms. The opposition to alternating current (impedance - Z) is greater than the resistance (R) of a direct current circuit, because of counterelectromotive-force, *eddy currents*, and *skin effect*.

Eddy Currents and Skin Effect

Eddy currents are small independent currents that are produced as a result of the expanding and collapsing magnetic field of alternating current flow, Fig. 3–15. The expanding and collapsing magnetic field of an alternating current circuit also induces a counterelectromotive-force in the conductors that repel the flowing electrons towards the surface of the conductor, Fig. 3–16. The counterelectromotive-force causes the circuit current to flow near the surface of the conductor rather than at the conductors center. The flow of electrons near the surface is known as *skin effect*. Eddy currents and skin effect decrease the effective conductor cross-sectional area for current flow, which results in an increase in the the conductor's *impedance* which is measured in ohms.

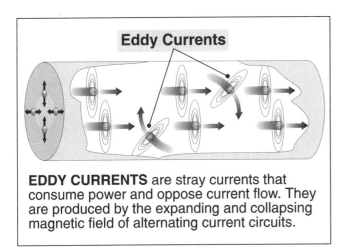

Figure 3-15
Eddy Currents

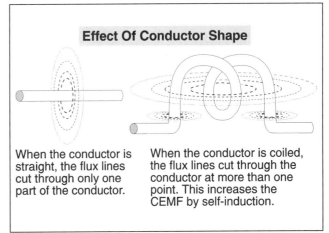

Figure 3-17
Induction and Conductor Loops

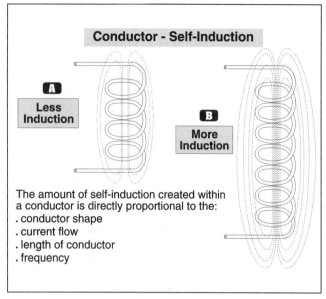

Figure 3-18
Conductor Self-Induction Magnitude

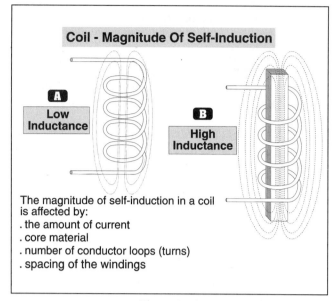

Figure 3-19
Coil Self-Induction Magnitude

3–11 INDUCTION AND CONDUCTOR SHAPE

The physical shape of a conductor affects the amount of *self-induced voltage* in the conductor. If a conductor is coiled into adjacent loops *(winding)*, the electromagnetic field *(flux lines)* of each conductor loop adds together to create a stronger overall magnetic field, Fig. 3–17. The amount of self-induced voltage created within a conductor is directly proportional to the current flow, the length of the conductor, and the frequency at which the magnetic fields cut through the conductors. Fig. 3–18.

3–12 INDUCTION AND MAGNETIC CORES

Self-inductance in a *coil* of conductors is affected by the winding current magnitude, the core material, the number of conductor loops *(turns)*, and the spacing of the winding, Fig. 3–19. *Iron cores* within the conductor winding permits an easy path for the electromagnetic flux lines which produce a greater counterelectromotive-force than air core windings. In addition, conductor coils are measured in *ampere-turns*, that is, the current in amperes times the number of conductor loops or turns.

Ampere-Turns = Coil Amperes × Number of Coil Turns

❏ **Ampere-Turns Example**

A magnetic coil in a transformer has 1,000 turns and carries 5 amperes. What is the ampere-turns rating of this transformer, Fig. 3–20?

(a) 5,000 ampere-turns

(b) 7,500 ampere-turns

(c) 10,000 ampere-turns

(d) none of these

• Answer: (a) 5,000 ampere-turns

Ampere-Turns = Amperes × Number of Turns

Ampere-turns = 5 amperes × 1,000 turns

Ampere-turns = 5,000

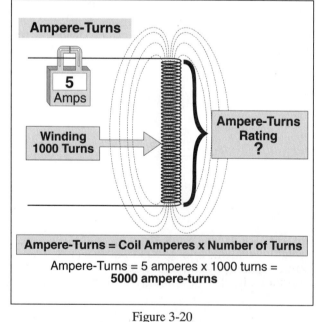

Figure 3-20
Ampere Turn Example

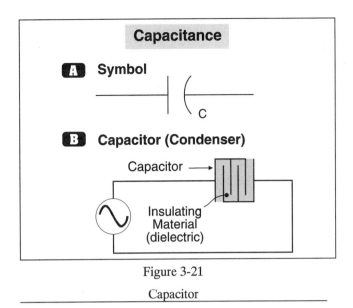

Figure 3-21
Capacitor

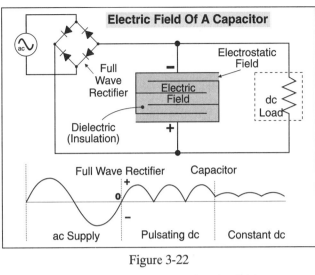

Figure 3-22
Capacitor – Charge and Electric Field

PART C – CAPACITANCE

CAPACITANCE INTRODUCTION

Capacitance is the property of an electric circuit that enables the storage of electric energy by means of an *electrostatic field*, much like a spring stores mechanical energy, Fig. 3–21. Capacitance exists whenever an insulating material separates two objects that have a difference of potential. A capacitor is simply two metal foils separated by wax paper rolled together. Devices that intentionally introduce capacitance into circuits are called *capacitors* or *condensers*.

3–13 CHARGE, TESTING AND DISCHARGING

When a capacitor has a potential difference between the plates, it is said to be *charged*. One plate has an excess of free electrons (–), and the other plate has a lack of electrons (+). Because of the insulation (*dielectric*) between the capacitor foils, electrons cannot flow from one plate to the other. However, there are electric lines of force between the plates, and this force is called the *electric field*, Fig. 3–22.

Factors that determine capacitance are; the surface area of the plates, the distance between the plates, and the *dielectric* insulating material between the plates. Capacitance is measured in *farads*, but the opposition offered to the flow of current by a capacitor is called *capacitive reactance*, which is measured in ohms. Capacitive reactance is abbreviated X_C, is directly proportional to frequency and can be calculated by the equation:

$X_C = \dfrac{1}{2\pi fC}$ $\pi = 3.14$, f = frequency, C = Capacitance measure in farads

❏ **Capacitive Reactance Example**

What is the capacitive reactance of a 0.000001 farad capacitor supplied by 60-Hz, 180-Hz, and 300-Hz, Fig. 3–23?

• Answer: $X_C = \dfrac{1}{2\pi fC}$.

$\pi = 3.14$

f = frequency; 60-Hz, 180-Hz, and 300-Hz

C = Capacitance measure in farads, 0.0000001

X_C at 60–Hz $= \dfrac{1}{2 \times 3.14 \times 60\text{–hertz} \times 0.000001 \text{ farads}}$, $X_C = 2{,}654$ ohms

X_C at 180–Hz $= \dfrac{1}{2 \times 3.14 \times 180\text{–hertz} \times 0.000001 \text{ farads}}$, $X_C = 885$ ohms

X_C at 300–Hz $= \dfrac{1}{2 \times 3.14 \times 300\text{–hertz} \times 0.000001 \text{ farads}}$, $X_C = 530.46$ ohms

Capacitive Reactance

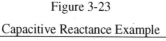

Example: Determine X_C of conductor at 60, 180, and 300 hertz.

Formula: $X_C \text{ (ohms)} = \dfrac{1}{2\pi f C}$

$\pi = 3.14$; $f =$ ❶ 60 Hz, ❷ 180 Hz, ❸ 300 Hz; C = .000001 farad

❶ X_C at 60 Hertz
$X_C = \dfrac{1}{2 \times 3.14 \times 60 \text{ Hz} \times .000001 \text{ farad}} = $ **2,654 ohms**

❷ X_C at 180 Hertz
$X_C = \dfrac{1}{2 \times 3.14 \times 180 \text{ Hz} \times .000001 \text{ farad}} = $ **885 ohms**

❸ X_C at 300 Hertz
$X_C = \dfrac{1}{2 \times 3.14 \times 300 \text{ Hz} \times .000001 \text{ farad}} = $ **530.46 ohms**

Figure 3-23
Capacitive Reactance Example

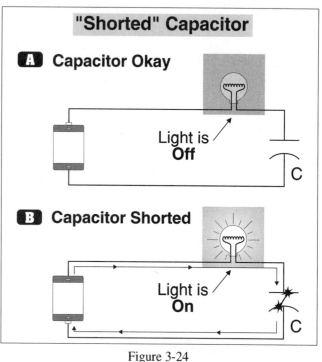

Figure 3-24
Capacitor Testing

Capacitor Short and Discharge

The more the capacitor is charged, the stronger the electric field. If the capacitor is overcharged, the electrons from the negative plate could be pulled through the dielectric insulation to the positive plate, resulting in a short of the capacitor. To discharge a capacitor, all that is required is a conducting path between the capacitor plates.

Testing a Capacitor

If a capacitor is connected in series with a direct current power supply and a test lamp, the lamp will be continuously illuminated if the capacitor is shorted, but will remain dark if the capacitor is good, Fig. 3–24.

3–14 USE OF CAPACITORS

Capacitors (condensers) are used to prevent arcing across the *contacts* of low ampere switches and direct current power supplies for electronic loads such as computer, Fig. 3–25. In addition, capacitors cause the current to lead the voltage by as much as 90 degrees and can be used to correct poor power factor due to inductive reactive loads such as motors. Poor power factor, because of induction, causes the current to lag the voltage by as much as 90 degrees. Correction of the power factor to *unity* (100%) is known as *resonance*, and occurs when inductive reactance is equal to capacitive reactance in a series circuit; $X_L = X_C$.

PART D – POWER FACTOR AND EFFICIENCY

POWER FACTOR INTRODUCTION

Alternating current circuits develop inductive and capacitive reactance which causes some power to be stored temporarily in the electromagnetic field of the inductor, or in the electrostatic field of the capacitor, Fig. 3–26. Because of the temporary storage of energy, the voltage and current of inductive and capacitive circuits are not in phase. This out of phase relationship between voltage and current is called *power factor*. Figure 3–27. Power factor affects the calculation of power in alternating current circuits, but not direct current circuits.

3–15 APPARENT POWER (VOLT-AMPERES)

In alternating current circuits, the value from multiplying volts times amperes equals *apparent power* and is commonly referred to as *volt-amperes*, *VA*, or *kVA*. Apparent power (VA) is equal to or greater than *true power* (*watts*) depending on the *power factor*. When sizing circuits or equipment, always size the circuit equipment according to the apparent power (volt-ampere), not the true power (watt).

Apparent Power = Volts × Amperes, or Apparent Power = True Power/Power Factor

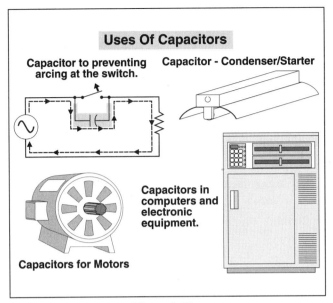

Figure 3-25
Capacitor Uses

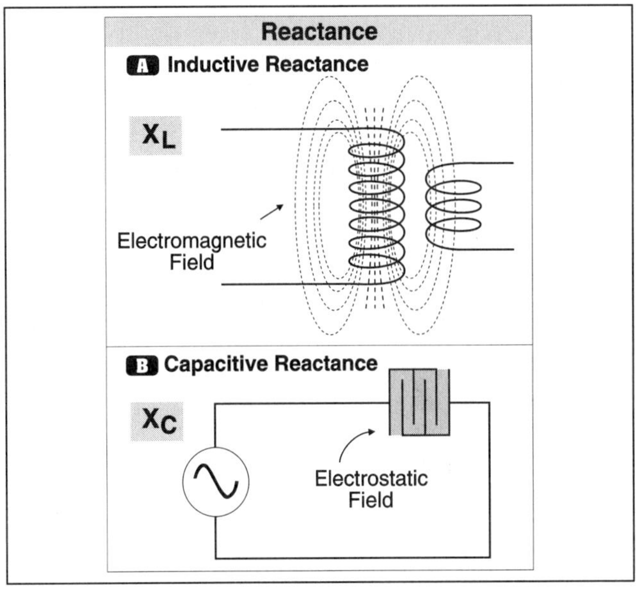

Figure 3-26
Reactance

❑ **Apparent Power (VA) Example No. 1**
What is the apparent power of a 16-ampere load operating at 120 volts with a power factor of 80 percent, Fig. 3–28?
(a) 2,400 VA (b) 1,920 VA (c) 1,632 VA (d) none of these

• Answer: (b) 1,920 VA

Apparent Power = Volt × Amperes, VA = 120 volts × 16 amperes = 1,920 VA

❑ **Apparent Power Example No. 2**
What is the apparent power of a 250-watt fixture that has a power factor of 80 percent, Fig. 3–29?
(a) 200 VA (b) 250 VA (c) 313 VA (d) 375 VA

• Answer: (c) 313 VA

Apparent Power = True Power/Power Factor

True Power = 250 watts, Power Factor = 80% or 0.8

Apparent Power = 250 watts/.8

Apparent Power = 313 volt-amperes

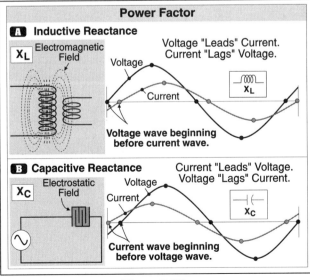

Figure 3-27
Power Factor

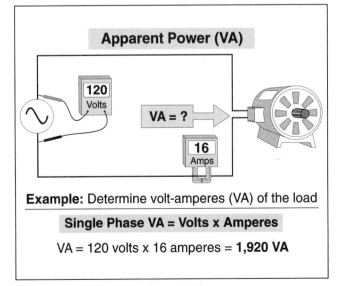

Figure 3-28
Apparent Power Example

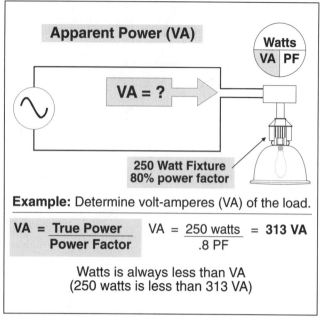

Figure 3-29
Apparent Power Example

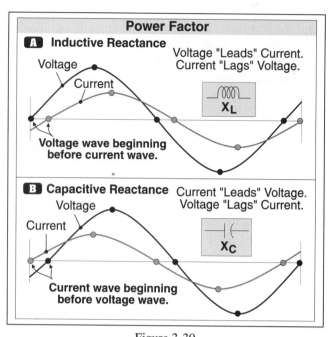

Figure 3-30
Power Factor Example

3–16 POWER FACTOR

Power Factor is the ratio of active power (R) to apparent power (Z) express as a percent that does not exceed 100 percent. Power Factor is a form of measurement of how far the voltage and current are out of phase with each other. In an alternating current circuit that supplies power to resistive loads, such as incandescent lighting, heating elements, etc., the circuit voltage and current will be *in phase*. The term in phase means that the voltage and current reach their zero and peak values at the same time, resulting in a power factor of 100 percent or *unity*. Unity can occur if the circuit only supplies resistive loads or capacitive reactance (X_C) is equal to inductive reactance (X_L).

The formulas for determining power factor are:

Power Factor = True Power/Apparent Power, or
Power Factor = Watts/Volt-Amperes, or
Power Factor = Resistance/Impedance
Power Factor = R/Z, Resistance (Watts)/Impedance (VA)

The formulas for current flow are:

I = Watts/(Volts Line to Line times Power Factor) Single-Phase
I = Watts/(Volts Line to Line times 1.732 times Power Factor) Three-Phase

❏ **Power Factor Example**

What is the power factor of a fluorescent lighting fixture that produces 160 watts of light and has a ballast rated for 200 volt-amperes, Fig. 3–30?

(a) 80 percent   (b) 90 percent   (c) 100 percent   (d) none of these

• Answer: (a) 80 percent

$$\text{Power Factor} = \frac{\text{True Power}}{\text{VA}} = \frac{160 \text{ watts}}{200 \text{ volt-amperes}} = 0.80 \text{ or } 80\%$$

❏ **Current Example**

What is the current flow for three 5-kW, 115-volt, single-phase loads that have a power factor of 90 percent?

(a) 72 amperes   (b) 90 amperes   (c) 100 amperes   (d) none of these

• Answer: (a) 72 amperes

I = Watts/(Volts Line to Line times Power Factor) Single-Phase
I = 15,000 Watts/(230 volts × 0.9)
I = 72 amperes

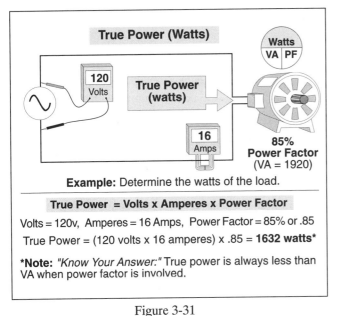

Figure 3-31
Alternating Current – True Power Example

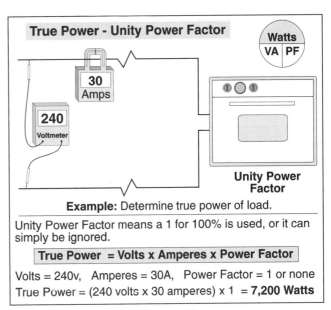

Figure 3-32
Direct Current – True Power Example

3–17 TRUE POWER (WATTS)

Alternating Current
True power is the energy consumed for work, expressed by the unit called the *watt*. Power is measured by a wattmeter which is connected series-parallel [series (measures amperes) and parallel (measures volts)] to the circuit conductors. The true power of a circuit that contains inductive or capacitance reactance can be calculated by the use of one of the following formulas:

True Power (alternating current) = Volts × Amperes × Power Factor, or

True power (alternating current) = Volts × Amperes × θ

θ (Cosine) is the symbol for power factor

Note. True power for a direct current circuit is calculated as Volts × Amperes.

☐ **Alternating Current – True Power Example**
What is the true power of a 16-ampere load operating at 120 volts with a power factor of 85 percent, Fig. 3–31?

(a) 2,400 watts　　　(b) 1,920 watts　　　(c) 1,632 watts　　　(d) none of these

• Answer: (c) 1,632 watts

True Power = Volts-Amperes × Power Factor

Watts = (120 volts × 16 amperes) × 0.85, Watts = 1,632 watts per hour

Note. True Power is always equal to or less than apparent power.

Direct Current
In direct current or alternating current circuits at unity power factor, the true power is simply:

True Power Direct Current = Volts × Amperes

Note. True power for alternating current with reactance = Volts × Amperes × θ Cosine

☐ **Direct Current – True Power Example**
What is true power of a 30-ampere, 240-volt resistive load that has unity power factor, Fig. 3–32?

(a) 4,300 watts　　　(b) 7,200 watts　　　(c) 7,200 VA　　　(d) none of these

• Answer: (b) 7,200 watts, not VA

True Power = Volts × Amperes

Volts = 240 volts, Amperes = 30 amperes

True Power = 240 volts × 30 amperes

True Power = 7,200 watts

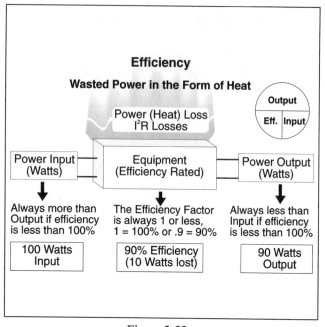

Figure 3-33
Efficiency

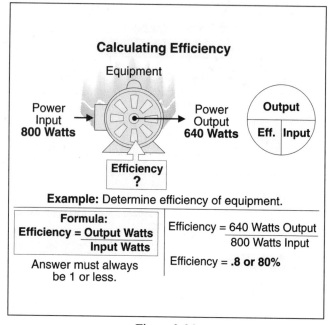

Figure 3-34
Efficiency Example

3–18 EFFICIENCY

Efficiency has nothing to do with power factor. Efficiency is the ratio of input power to output power, where power factor is the ratio of true power (watts) to apparent power (volt-amperes). Energy that is not used for its intended purpose is called power loss. Power losses can be caused by conductor resistance, friction, mechanical loading, etc. Power losses of equipment are expressed by the term efficiency, which is the *ratio* of the input power to the output power, Fig. 3–33. The formulas for efficiency are:

Efficiency = Output watts/Input watts
Input watts = Output watts/Efficiency
Output watts = Input watts × Efficiency

☐ **Efficiency Example**

If the input of a load is 800 watts and the output is 640 watts, what is the efficiency of the equipment, Fig. 3–34?

(a) 60 percent   (b) 70 percent   (c) 80 percent   (d) 100 percent

- Answer: (c) 80 percent

Efficiency is always less than 100 percent,

$$\text{Efficiency} = \frac{\text{Output}}{\text{Input}}$$

$$\text{Efficiency} = \frac{640 \text{ watts}}{800 \text{ watts}}$$

Efficiency = 0.8 or 80%

☐ **Input Example**

If the output of a load is 250 watts and the equipment is 88 percent efficient, what are the input watts, Fig. 3–35?

(a) 200 watts   (b) 250 watts   (c) 285 watts   (d) 325 watts

- Answer: (c) 285 watts

Input is always greater than the output,

$$\text{Input} = \frac{\text{Output}}{\text{Efficiency}}$$

$$\text{Input} = \frac{250 \text{ watts}}{0.88 \text{ efficiency}}$$

Input = 284 watts

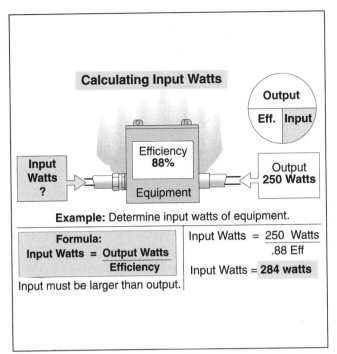

Figure 3-35
Input Example

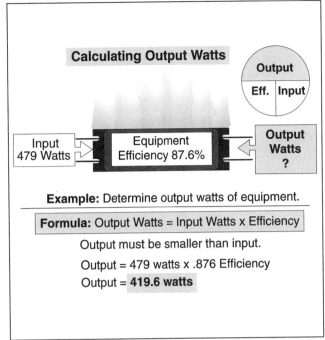

Figure 3-36
Output Example

❑ **Output Example**

If we have a load that is 87.6 percent efficient, for every 479 watts of input we will have _____ watts of output, Fig. 3–36.

(a) 440 (b) 480 (c) 390 (d) 420

- Answer: (d) 420 watts
  Output is always less than input.
  Output = Input × Efficiency
  Output = 479 watts × 0.876
  Output = 419.6 watts

Unit 3 – Understanding Alternating Current Summary Questions

Part A – Alternating Current Fundamentals

3–1 Current Flow

1. The effects of electron movement is the same regardless of the direction of the current flow.
   (a) True   (b) False

3–2 Alternating Current

2. Alternating current is primarily used because it can be transmitted inexpensively and because it can be used for certain applications for which direct current is not suitable.
   (a) True   (b) False

3. Faradays experiments revealed that when a magnetic field moves through a coil of wire, the lines of force of the magnetic field makes the electrons in the wire to flow in a specific direction. When the magnetic field moves in the opposite direction, electrons in the wire flow continue to flow in the same direction.
   (a) True   (b) False

3–3 Alternating Current Generator

4. A simple alternating current generator consists of a rotating loop of wire between the lines of force of opposite poles of a magnet. The magnitude of the voltage is dependent on _____.
   (a) number of turns of wire
   (b) strength of the magnetic field
   (c) speed at which the coil rotates
   (d) all of these

3–4 Waveform

5. • A waveform is used to display the level and direction of the current and voltage. The waveform for _____ circuits displays the level and direction of the current and voltage for every instant of time for one full revolution of the armature.
   (a) direct current   (b) alternating current
   (c) a and b   (d) none of these

6. For alternating current circuits, the waveform is called the _____.
   (a) frequency   (b) cycle
   (c) degree   (d) none of these

3–5 Armature Turns Frequency

7. The number of times the armature turns in one second is called the frequency. Frequency is expressed as _____ or cycles per second.
   (a) degrees   (b) sign wave
   (c) phase   (d) hertz

3–6 Phase

8. When two waveforms are in step with each other, they are said to be in phase. In a purely resistive alternating current circuit, the current and voltage are in phase. This means that they both reach their zero and _____ values at the same time.
   (a) peak   (b) effective
   (c) average   (d) none of these

3–7 Degrees

9. Phase differences are expressed in _____.
   (a) sign   (b) phase   (c) hertz   (d) degrees

10. The terms _____ and _____ are used to describe the relative positions in time of two waveforms (voltage or current).
    (a) hertz, phase
    (b) frequency, phase
    (c) sine, degrees
    (d) lead and lag

3–8 Values Of Alternating Current

11. • _____ is the value of the voltage or current at any one particular moment of time.
    (a) Peak
    (b) Root mean square
    (c) Effective
    (d) Instantaneous

12. _____ is the maximum value that alternating current current or voltage reaches, both for positive and negative.
    (a) Peak
    (b) Root mean square
    (c) Instantaneous
    (d) None of these

13. • Effective is the alternating current voltage or current value that produces the same amount of heat in a resistor that would be produced by the same amount of direct current voltage or current. _____ is the same as effective.
    (a) Peak
    (b) Root mean square
    (c) Instantaneous
    (d) None of these

Part B – Induction

Induction Introduction

14. The movement of electrons because of electromagnetism is called _____.
    (a) flux lines   (b) voltage   (c) induction   (d) magnetic field

15. The induction of voltage in a conductor because of expanding and collapsing magnetic fields is know as _____.
    (a) flux lines
    (b) power
    (c) self-induced voltage
    (d) magnetic field

3–9 Induced Voltage And Applied Current

16. A change in current produces a magnetic field through the conductor which produces an induced voltage that always opposes the change in current. The induced voltage that opposes the change in current is called _____.
    (a) cemf
    (b) counter electromotive force
    (c) back-emf
    (d) all of these

17. When the conductor current increases, the direction of the induced voltage (cemf) in the conductor is opposite to the direction of the conductor current and tries to prevent the current from _____.
    (a) decreasing   (b) increasing

18. When the current in the conductor decreases, the direction of the induced voltage in the conductor attempts to keep the current from decreasing by releasing the energy from the magnetic field back into the conductor.
    (a) True   (b) False

3–10 Conductor Impedance

19. In direct current circuits, the only property that effects current and voltage flow is _____ which is a physical property of the conductors that oppose current flow.
    (a) voltage   (b) cemf   (c) back-emf   (d) none of these

20. Conductor resistance is directly proportional to the conductor length and inversely proportional to conductor cross sectional area.
    (a) True   (b) False

21. Alternating currents produce a cemf that is set up inside the conductor which increases the effective resistance of the conductor because of eddy currents and skin effect.
    (a) True   (b) False

22. The opposition to current flow in a conductor because of resistance and induction is called _____.
    (a) resistance   (b) capacitance
    (c) induction   (d) impedance

23. • Eddy currents are small independent currents that are produced as a result of direct current. Eddy currents flow erratic through a conductor, consume power and increase the effective conductor resistance by opposing the current flow.
    (a) True   (b) False

24. The expanding and collapsing magnetic field of the conductors current induces a voltage in the conductors which repels the flowing electrons towards the surface of the conductor. This has the effect of decreasing the effective conductor cross-sectional area which causes an increased in the conductor impedance. The flow of electrons near the surface is know as _____.
    (a) eddy currents   (b) induced voltage
    (c) impedance   (d) skin effect

3–11 Conductor Shape

25. The amount of the self-induced voltage created within the conductor is directly proportional to the current flow, the length of the conductor, and frequency at which the magnetic fields cut through the conductors.
    (a) True   (b) False

3–12 Magnetic Cores

26. • Self-inductance (cemf) in a coil is effected by the _____.
    (a) winding length and shape   (b) core material
    (c) frequency   (d) all of these

Part C – Capacitance

Capacitance Introduction

27. _____ is a property of an electric circuit that enables it to store electric energy by means of an electrostatic field and to release this energy at a latter time.
    (a) Capacitance   (b) Induction
    (c) Self-induction   (d) None of these

3–13 Charged, Testing, And Discharging

28. When a capacitor has a potential difference between the plates, it is said to be _____. One plate has an excess of free electrons, and the other plate has a lack of them.
    (a) induced   (b) charged   (c) discharged   (d) shorted

29. If the capacitor is overcharged, the electrons from the negative plate could be pulled through the insulation to the positive plate. The capacitor is said to have _____.
    (a) charged   (b) discharged
    (c) induced   (d) shorted

30. To discharge a capacitor, all that is required is a _____ path between the capacitor plates.
    (a) conducting   (b) insulating
    (c) isolating   (d) none of these

3–14 Uses Of Capacitors

31. What helps prevent arcing across the contacts of electric switches?
    (a) springs   (b) condenser
    (c) inductor   (d) resistor

32. The current in a purely capacitive circuit _____.
    (a) leads the applied voltage by 90 degrees
    (b) lags the applied voltage by 90 degrees
    (c) leads the applied voltage by 180 degrees
    (d) lags the applied voltage by 180 degrees

33. Circuits containing inductive or capacitive reactance temporarily store power in the electromagnetic field of induction and the electrostatic field of capacitors.
    (a) True   (b) False

Part D – Power Factor And Efficiency

3–15 Apparent Power

34. If you measure voltage and current in an inductive or capacitive circuit and then multiply them together, you would obtain the circuits _____.
    (a) true power   (b) power factor   (c) apparent power   (d) power loss

35. Apparent power is equal to or greater than true power depending on the power factor.
    (a) True   (b) False

36. When sizing circuits or equipment, always size the circuit components and transformers according to the apparent power, not the True power.
    (a) True   (b) False

3–16 Power Factor

37. Power factor is a measurement of how far the voltage and current are out of phase with each other. Power factor is the ratio between True Power (resistive load) to Apparent Power (reactive load). Power factor can be express by the formula _____.
    (a) P/E   (b) R/Z   (c) I^2R (d) Z/R

38. • When current and voltage are in phase, the power factor is _____.
    (a) 100 percent   (b) unity   (c) 90 degrees   (d) a and b

39. In an alternating current circuit that supplies power to resistive loads such as incandescent lighting, heating elements, etc., the circuit voltage and current are said to be _____, resulting in a power factor of unity.
    (a) out of phase   (b) leading by 90 degrees
    (c) 90 degrees out of phase   (d) none of these

3–17 True Power (Watts)

40. True power is the energy consumed for work expressed by the term watts. To determine the True power of a circuit that contains inductive or capacitance reactance, we must multiply the volts times the current times the _____.
    (a) efficiency   (b) sine wave   (c) power factor   (d) none of these

41. In direct current circuits, the voltage and current are constant and the true power is simply, volts times amperes.
    (a) True   (b) False

42. What does it cost per year (9 cents per kWH) for 10 - 150 watt recessed fixture to operate if they are on 6 hours per day?
    (a) $150   (b) $300   (c) $500   (d) $800

Power Factor Examples

43. A 2 x 4 recessed fixture contains four 34 watt lamps, ballast is rated 1.5 amperes at 120 volts. What is the power factor of the ballast assuming 100 percent efficiency?
    (a) 55 percent   (b) 65 percent   (c) 70 percent   (d) 75 percent

44. • What is the apparent power of a 20 ampere load operating at 120 volts, power factor 85 percent?
    (a) 2,400 watts   (b) 1,920 watts   (c) 1,632 watts   (d) 2,400 VA

45. • What is the true power of a 20 ampere load operating at 120 volts, power factor 85 percent?
    (a) 2,400 watts   (b) 1,920 watts   (c) 1,632 watts   (d) none of these

46. • Since power factor cannot be greater than 100 percent, True power is equal to or less than the apparent power. Because of power factor, the VA of the load is greater than the watts, which results in less loads per circuit, greater number of circuits, and larger transformers.
    (a) True   (b) False

47. • What is the true power of a 10-ampere circuit operating at 120 volts with unity power factor?
    (a) 1,200 VA   (b) 2,400 VA   (c) 1,200 watts   (d) 2,400 watts

48. What size transformer is required for a 125-ampere, 240 volt single-phase load?
    (a) 3 kVA   (b) 30 kVA   (c) 12.5 kVA   (d) 15 kVA

49. What size transformer is required for a 30 kW-load that has a power factor of 85 percent?
    (a) 12.5 kVA   (b) 35 kVA   (c) 7.5 kVA   (d) 15 kVA

50. • How many 20-ampere, 120-volt circuits are require for 42 - 300 watt recessed fixtures (noncontinuous load)? Tip: Only so many fixtures are permitted on a circuit.
    (a) 3 circuits   (b) 4 circuits   (c) 5 circuits   (d) 6 circuits

51. • How many 20-ampere, 120-volt circuits are require for 42 - 300 watt recessed fixtures (noncontinuous load) with a power factor of 85 percent?
    (a) 5 circuits   (b) 6 circuits   (c) 7 circuits   (d) 8 circuits

3–18 Efficiency

52. Efficiency is the ratio of the input power to output power.
    (a) True   (b) False

53. If the output is 1,320 watts and the input is 1,800 watts, what is the efficiency of the equipment?
    (a) 62 percent   (b) 73 percent   (c) 0 percent   (d) 100 percent

54. If the output is 160 watts and the equipment is 88 percent efficient, what is the input amperes at 120 volts?
    (a) .75 amperes   (b) 1.500 amperes   (c) 2.275 amperes   (d) 3.250 amperes

55. If we have a transformer that is 97 percent efficient. For every 1 kW input, we have _____ watts output.
    (a) 970 watts   (b) 1,000 watts   (c) 1,030 watts   (d) 1,300 watts

Challenge Questions

Part A – Alternating Current Fundamentals

3–2 Alternating Current

56. The primary reason(s) for high voltage transmission lines is/are _____.
    I. reduced voltage drop   II. smaller wire   III. smaller equipment
    (a) I only   (b) II only
    (c) III only   (d) all of these

57. One of the advantages of a higher voltage system as compared to a lower voltage system (for the same wattage loads) is _____.
    (a) reduced voltage drop   (b) reduced power use
    (c) large currents   (d) lower electrical pressure

58. The advantage of alternating current over direct current is that alternating current provides for _____.
    (a) better speed control   (b) ease of voltage variation
    (c) lower resistance at high currents   (d) none of these

3–4 Waveforms

59. A waveform represents _____.
    (a) the magnitude and direction of current or voltage
    (b) how current or voltage can vary with time
    (c) how output voltage can vary with the generator armature
    (d) all of these

3–5 Armature Turning Frequency

60. Frequency, of an alternating current waveform, is the number of times the current or voltage goes through 360º in _____.
    (a) 1/10th of a second   (b) 5 seconds   (c) 1 second   (d) 60 seconds

61. How much time does it take for 60 Hz alternating current to travel through 180 degrees?
    (a) 1/120 of a second   (b) 1/40 of a second
    (c) 1/180 of a second   (d) none of these

3–8 Values Of Alternating Current

62. • The heating effects of 10 amperes alternating current as compared with 10 amperes direct current is _____.
    (a) the same   (b) less   (c) greater   (d) none of these

63. If the maximum value of an alternating current system is 50 amperes, the RMS value would be approximately _____.
    (a) 25 amperes   (b) 30 amperes   (c) 35 amperes   (d) 40 amperes

64. • The maximum value of 120 volt direct current is equal to the maximum value of an equivalent alternating current.
    (a) True   (b) False

65. • You are getting a 120 volt reading on your voltmeter, this is an indication of the _____ value of the voltage source.
    (a) average   (b) peak   (c) effective   (d) instantaneous

Part B – Induction

3–9 Voltage And Applied Current

66. A change in current that produces a counter electromotive force whose direction is such that it opposes the change in current is known as _____ Law.
    (a) Kirchoff's 2nd   (b) Kirchoff's 1st   (c) Lenz's   (d) Hertz

67. Inductive reactance is abbreviated as _____.
    (a) IR   (b) L_X (c) X_L (d) Z

68. Inductive reactance changes proportionally with frequency.
    (a) True   (b) False

69. Inductive reactance is measured in _____.
    (a) farads   (b) watts   (c) ohms   (d) coulombs

70. • If the frequency is constant, the inductive reactance of a circuit will _____.
    (a) remain constant regardless of the current and voltage changes
    (b) vary directly with the voltage
    (c) vary directly with the current
    (d) not affect the impedance

3–10 Conductor Impedance

71. The total opposition to current flow in an alternating current circuit is expressed in ohms and is called _____.
    (a) impedance   (b) conductance   (c) reluctance   (d) none of these

72. Impedance is present in _____ type circuit(s).
    (a) resistance (b) direct current (c) alternating current (d) none of these

73. Conductor resistance to alternating current flow is _____ the resistance to direct current.
    (a) higher than (b) lower than

Part C – Capacitance

3–13 Charge, Testing And Discharging

74. If a test lamp is placed in series with a capacitor with a direct current voltage source and the lamp is continuously illuminated, it is an indication that the capacitor is _____.
    (a) fully charged (b) shorted (c) fully discharged (d) open-circuited

75. The insulating material between the surface plates of a capacitor is called the _____.
    (a) inhibitor (b) electrolyte (c) dielectric (d) regulator

76. Capacitors are measured in _____.
    (a) watts (b) volts (c) farads (d) henrys

77. Capacitive reactance is measured in _____.
    (a) ohms (b) volts (c) watts (d) henrys

3–14 Uses Of Capacitors

78. • In a circuit that has only capacitive reactance (X_C), the voltage and current are said to be out of phase to each other because the voltage _____.
    (a) leads the current by 90º (b) lags the current by 90º
    (c) leads the current (d) none of these

79. Resonance occurs when _____.
    (a) only resistance occurs in the system
    (b) the power factor is equal to 0
    (c) $X_L = X_C$ are in a series circuit
    (d) when R = Z

Part D – Power Factor And Efficiency

3–15 Apparent Power

80. If you multiply the voltage times the current in an inductive or capacitive circuit, the answer you obtain will be the _____ of the circuit.
    (a) watts (b) true power (c) apparent power (d) all of these

81. • The apparent power of a 19.2 ampere, 120 volt load is _____.
    (a) 2,304 kVA (b) 2.3 kVA (c) 2.3 VA (d) 230 kVA

3–16 Power Factor

82. Power factor in an alternating current circuit will be unity (100%), if the circuit contains only _____.
    (a) induction motors (b) transformers
    (c) reactance coils (d) resistive load such as electric space heater coils

83. • Three 8 kW electric discharge lighting bank circuits, which have a 92 percent power factor, are connected to a 230-volt, three-phase source. The current flow of these lights is _____.
    (a) 37 amperes (b) 50 amperes (c) 65 amperes (d) 75 amperes

3–17 True Power

84. • A wattmeter is connected in _____ in the circuit.
    (a) series (b) parallel (c) series-parallel (d) none of these

85. • True power is always voltage times current for _____ .
    (a) all alternating current circuits
    (b) direct current circuits
    (c) alternating current circuits at unity power factor
    (d) b and c

86. Power consumed in either a single-phase alternating current or direct current system is always equal to _____ .
    (a) E × I cos   (b) E × R   (c) $I^2 \times R$ (d) E/(I × R)

87. • The power consumed on a 76 ampere, 208 volt three-phase circuit that has a power factor of 89 percent is _____ .
    (a) 27,379 watts   (b) 35,808 watts   (c) 24,367 watts   (d) 12,456 watts

88. • The true power of a single phase 2.1 kVA load with a power factor of 91 percent is _____ .
    (a) 2.1 kW   (b) 1.91 kW   (c) 1.75 kW   (d) 1,911 kW

3–18 Efficiency

89. • Motor efficiency can be determined by which of the following formulas?
    (a) HP × 746   (b) HP × 746/VA Input
    (c) HP × 746/Watts Input   (δ) HP × 746/kVA Input

90. The efficiency ratio of a 4,000 VA transformer with a secondary VA of 3,600 VA is _____ .
    (a) 80   (b) 9   (c) .9   (d) 1.1

NEC Questions From Section 310–13 Through Section 354–8

Article 310 – Conductors For General Wiring

91. TFE-insulated conductors are manufactured in sizes from No. 14 through No. _____ .
    (a) 2   (b) 1   (c) 2/0   (d) 4/0

92. • When bare grounded conductors are allowed, their ampacities are limited to _____ .
    (a) 60ºC   (b) 75ºC
    (d) 90ºC   (d) that permitted for the insulated conductors of the same size

93. • Aluminum and copper-clad aluminum of the same circular mil size and insulation have _____ .
    (a) the same physical characteristics
    (b) the same termination
    (c) the same ampacity
    (d) different ampacities

94. Where six current carrying-conductors are run in the same conduit or cable, the ampacity of each conductor shall be adjusted by a factor of _____ percent.
    (a) 90   (b) 60   (c) 40   (d) 80

95. The ampacity of a single insulated No. 1/0 THHN copper conductor in free air is _____ amperes.
    (a) 260   (b) 300   (c) 185   (d) 215

96. • The ampacity of a No. 6 bare copper conductor in free air is _____ amperes.
    (a) 60   (b) 95   (c) 75   (d) 106

97. Table 310–72 provides ampacities of an insulated three-conductor aluminum cable isolated in air based on conductor temperature of 90°C (194°F) and ambient air temperature of 40°C (104°F). If the conductor size is 4 AWG, and the voltage range is 5,001 to 35,000, then the ampa0(b) 72(c) 81(d) 95

98. Table 310–75 provides ampacities of an insulated three-conductor copper cable in isolated conduit in air based on conductor temperature of 90°C (194°F) and ambient air temperature of 40°C (104°F). If the conductor size is 2/0 AWG, and the voltage range is 5,001 to 35,000, then the ampacity is _____ amperes.
    (a) 165 (b) 190 (c) 195 (d) 220

99. Thermal resistivity is the reciprocal of thermal conductivity and is designated Rho and expressed in the units of _____.
    (a) °F–cm/volt (b) °F–cm/watt (c) °C–cm/volt (d) °C–cm/watt

100. Thermal resistivity, as used in the Code, refers to the heat _____ capability through a substance by conduction.
    (a) assimilation (b) generation (c) transfer (d) dissipation

Article 318 – Cable Trays

101. • It is the intent of Article 318 to limit the use of cable trays to industrial establishments only.
    (a) True (b) False

102. Nonmetallic cable trays shall be made of _____ material.
    (a) fire-resistant (b) waterproof (c) corrosive (d) flame-retardant

103. Each run of cable tray shall be _____ before the installation of cables.
    (a) inspected (b) tested (c) completed (d) all of these

104. Where approved and designed to support the load, cable trays shall be permitted as a means of incidental support for _____ that are supported in accordance with Articles 345, 346, 347, and 348.
    (a) busways (b) plenums
    (c) wiring systems (d) raceways

105. Steel cable trays shall not be used as equipment grounding conductors for circuits with ground fault protected above _____ amperes.
    (a) 200 (b) 60 (c) 600 (d) 1200

106. • Steel or aluminum cable tray systems shall be permitted to be used as equipment grounding conductors provided four requirements are met. One of those requirements is that cable tray sections, fittings, and connected raceways shall be _____ in accordance with Section 250–75 using bolted mechanical connectors or bonding jumpers sized and installed in accordance with Section 250–79.
    (a) grounded (b) bonded (c) either a or b (d) both a and b

107. Where single conductor cables comprising each phase or neutral of a circuit are connected in parallel in a cable tray, the conductors shall be installed _____ to prevent current unbalance in the paralleled conductors due to inductive reactance.
    (a) in groups consisting of not more than three conductors per phase or neutral
    (b) in groups consisting of not more than one conductor per phase or neutral
    (c) as individual conductors securely bound to the cable tray
    (d) in separate groups

108. Where ladder or ventilated trough cable trays contain multiconductor power or lighting cables, or any mixture of multiconductor power, lighting, control, and signal cables, the maximum number of cables shall conform to three requirements. One of those requirements is where all of the cables are No. _____ or larger, the sum of the diameters of all cables shall not exceed the cable tray width, and the cables shall be installed in a single layer.
    (a) 1/0 (b) 2/0 (c) 3/0 (d) 4/0

109. Where a solid bottom cable tray, having a usable inside depth of 6 inches or less, contains multiconductor control and/or signal cables only, the sum of the cross-sectional areas of all cables at any cross section shall not exceed _____ percent of the interior cross-sectional area of the cable tray.
    (a) 25 (b) 30 (c) 35 (d) 40

110. Where ladder or ventilated trough cable trays contain single conductor cables (rated 2000 volts or less), the maximum number of single conductors shall conform to four requirements. One of those requirements is where all of the cables are from _____ kcmil up to _____ kcmil, the sum of the cross-sectional areas of all single conductor cables shall not exceed the maximum allowable cable fill area in Column 1 of Table 318–10, for the appropriate cable tray width.
(a) 250, 1000 (b) 8, 250 (c) 8, 1000 (d) 350, 1000

Article 320 – Open Wiring on Insulatorss

111. Open wiring on insulators shall comply with its own article and also with the applicable provisions of other articles in the Code, especially Article(s) _____.
I. 220 II. 225 III. 300 IV. 321
(a) I and II (b) II and III
(c) III and IV (d) IV and II

112. Conductors for open wiring on insulators shall be supported within _____ inches of a tap or splice.
(a) 6 (b) 8
(c) 10 (d) 12

113. Conductor No. _____ or larger supported on solid knobs shall be securely tied thereto by tie wires having an insulation equivalent to that of the open wire.
(a) 14 (b) 12
(c) 10 (d) 8

114. Open conductors shall be separated at least _____ inches from metal raceways, piping, or other conducting material, and from any exposed lighting, power, or signaling conductor, or shall be separated by a continuous and firmly fixed nonconductor in addition to the insulation of the conductor.
(a) 2 (b) 2½
(c) 3 (d) 3½

Article 321 – Messenger Supported Wiring

115. _____ shall be permitted to be installed in messenger supported wiring.
(a) Multiconductor service entrance cable
(b) Type MI cable
(c) Multiconductor underground feeder cable
(d) all of these

116. The conductors supported by messenger shall be permitted to come into contact with the messenger supports or any structural members, walls, or pipes.
(a) True (b) False

Article 324 – Concealed Knob-And-Tube Wiring

117. Concealed knob-and-tube wiring shall be permitted to be used in commercial garages, theaters and similar locations, motion picture studios, hazardous (classified) locations, or in the hollow spaces of walls, ceilings, and attics where such spaces are insulated by loose, rolled, or foamed-in-place insulating material that envelops the conductors.
(a) True (b) False

118. Where concealed solid knobs are used, conductors shall be securely tied thereto by _____ equivalent to that of the conductor.
(a) wires having insulation
(b) wires having an AWG
(c) nonconductive material
(d) none of these

Article 325 – Integrated Gas Spacer Cable (Type IGS)

Part A. General

119. Type IGS cable is a factory assembly of one or more conductors, each individually insulated and enclosed in a loose fit nonmetallic flexible conduit as an integrated gas spacer cable rated _____ volts.
    (a) 0 – 6000 (b) 600 – 6000 (c) 0 – 600 (d) 3000 – 6000

Article 328 – Flat Conductor Cable (Type FCC)

Part A. General

120. A field installed wiring system for branch circuits designed for installation under carpet squares is defined as _____.
    (a) underfloor wiring
    (b) undercarpet wiring
    (c) flat conductor cable
    (d) underfloor conductor cable

121. Use of FCC systems shall be permitted both for general-purpose and appliance branch circuits; it shall not be permitted for individual branch circuits.
    (a) True (b) False

122. Flat conductor cable cannot be installed in _____.
    (a) residential units (b) schools (c) hospitals (d) any of these

Part B. Installation

123. Floor-mounted-type flat conductor cable and fittings shall be covered with carpet squares no larger than _____.
    (a) 36 square inches
    (b) 36 inches square
    (c) 30 square inches
    (d) 24 inches square

124. All bare FCC cable ends shall be _____.
    I. sealed II. insulated III. listed
    (a) I only (b) II only (c) III only (d) I, II and III

125. Metal shields for flat conductor cable (FCC) must be electrically continuous to the _____.
    (a) floor
    (b) cable
    (c) equipment grounding conductor
    (d) none of these

126. Receptacles, receptacle housings, and self-contained devices used with flat conductor cable systems shall be _____.
    (a) rated a minimum of 20 amperes
    (b) rated a minimum of 15 amperes
    (c) identified for this use
    (d) none of these

Part C. Construction

127. Type FCC cable shall be clearly and durably marked _____.
    (a) on the top side at intervals not exceeding 30 inches
    (b) on both sides at intervals not exceeding 24 inches
    (c) with conductor material, maximum temperature, and ampacity
    (d) b and c only

128. Each FCC transition assembly shall incorporate means for _____.
    (a) facilitating the entry of the Type FCC cable into the assembly
    (b) connecting the Type FCC cable to grounded conductors
    (c) electrically connecting the assembly to the metal cable shields and grounding conductors
    (d) all of these

Article 330 – Mineral-Insulated Cable (Type MI)

Part A. General

129. • Which of the following statements about MI cable is correct?
    (a) It may be used in any hazardous location.
    (b) A single run of cable shall not contain more than the equivalent of four quarter bends.
    (c) It shall be securely supported at intervals not exceeding 10 feet.
    (d) All of these ·

Part B. Installation

130. The radius of the inner edge of any bend shall not be less than _____ times the MI cable diameter for cables that have a diameter of over ¾ inch.
    (a) 6   (b) 3   (c) 8   (d) 10

131. Where MI cable terminates, a _____ shall be provided immediately after stripping to prevent the entrance of moisture into the insulation.
    (a) bushing   (b) connector   (c) fitting   (d) seal

Part C. Construction Specifications

132. The conductor insulation in Type MI cable shall be a highly compressed refractory mineral that will provide proper _____ for the conductors.
    (a) covering   (b) spacing   (c) resistance   (d) none of these

Article 331 – Electrical Nonmetallic Tubing

Part A. General

133. The use of electrical nonmetallic tubing and fittings shall be permitted: In any building not exceeding three floors above grade, for exposed work, where not subject to physical damage, or concealed within walls, floors, and ceilings.
    (a) True   (b) False

134. The use of electrical nonmetallic tubing (ENT) shall be permitted _____.
    (a) concealed within walls, floors, and ceilings
    (b) embedded in poured concrete
    (c) in locations subject to severe corrosive influences
    (d) all of these

135. Electrical nonmetallic tubing is not permitted in hazardous (classified) locations.
    (a) True   (b) False

136. • Electrical nonmetallic tubing is permitted for direct earth burial, when used with fittings listed for this purpose.
    (a) True   (b) False

Part B. Installation

137. All cut ends of electrical nonmetallic tubing shall be trimmed inside and _____ to remove rough edges.
    (a) outside
    (b) outside, smoothing all surfaces,
    (c) the surface smoothed
    (d) none of these

138. The maximum number of bends between pull points cannot exceed _____ degrees, including any offsets.
    (a) 320   (b) 270   (c) 360   (d) unlimited

Article 333 – Armored Cable (Type AC)

Part B. Installation

139. Armor cable used for the connection of recessed fixtures or equipment within an accessible ceiling is not required to be secured for lengths up to _____ feet.
    (a) 2 (b) 3 (c) 4 (d) 6

140. Where AC cable is run across the top of a floor joist in an attic without permanent ladders or stairs, the cable shall be protected by substantial guard strips within _____ feet of the scuttle hole.
    (a) 7 (b) 6 (c) 5 (d) 3

141. Where armor cable is run on the side of rafters, studs or floor joists in an accessible attic, protection is required for the cable with running boards.
    (a) True (b) False

Article 334 – Metal-Clad Cable (Type MC)

Part A. General

142. MC cables can always be installed _____.
    (a) direct burial (b) in concrete (c) in cinder fill (d) none of these

Part B. Installation

143. • Smooth sheath MC cables that have external diameters of not greater than $1\frac{1}{2}$ inches, shall have a bending radius of not more than _____ times the cable external diameter.
    (a) 5 (b) 10 (c) 12 (d) 13

Article 336 – Nonmetallic-Sheathed Cable (Romex)

Part A. General

144. NM cable shall not be used _____.
    (a) in commercial buildings
    (b) in the air void of masonry block not subject to excessive moisture
    (c) for exposed work
    (d) embedded in poured cement, concrete, or aggregate

Part B. Installation

145. Where NMC cable is run at angles with joists in unfinished basements it shall be permissible to secure cables not smaller than _____ conductors directly to the lower edges of the joist.
    I. two No. 6 II. three No. 8 III. three No. 10
    (a) I only (b) II only (c) I and II (d) I, II, and III

146. Nonmetallic sheath cable shall be secured in place within _____ inches of every cabinet, box, or fitting.
    (a) 6 (b) 10 (c) 12 (d) 18

147. Nonmetallic sheath cables run through wood framing members are considered supported.
    (a) True (b) False

Part C. Construction Specifications

148. • The difference in the overall covering between NM cable and NMC cable is that it is _____.
    I. corrosion-resistant II. flame-retardant
    III. fungus-resistant IV. moisture-resistant
    (a) I and III (b) II and IV
    (c) I and II (d) III and IV

Article 338 – Service-Entrance Cable (Types SE and USE)

149. _____ is a type of multiconductor cable permitted for use as an underground service entrance cable.
 (a) SE (b) NMC (c) UF (d) USE

150. Service-entrance cable may be used in interior wiring systems where the circuit conductors of the cable are of the _____ type.
 I. rubber-covered II. thermoplastic III. linked polymer
 (a) I only (b) II only (c) III only (d) I and II

Article 339 – Underground Cable (Type UF)

151. • The maximum size underground feeder cable is No. _____ copper.
 (a) 14 (b) 10 (c) 1/0 (d) 4/0

152. The overall covering of UF cable shall be _____.
 I. suitable for direct burial in the earth
 II. flame-retardant
 III. moisture, fungus, and corrosion resistant
 (a) III and II (b) I only (c) I and III (d) I, II and III

153. Type UF cable is permitted for _____.
 (a) interior use
 (b) wet and corrosive locations
 (c) direct burial
 (d) all of these

154. Type UF cable shall not be used in _____.
 (a) motion picture studios
 (b) storage battery rooms
 (c) hoistways
 (d) all of these

Article 340 – Power and Control Tray Cable (Type TC)

155. Type TC tray cable shall not be installed _____.
 (a) where they will be exposed to physical damage
 (b) as open cable on brackets or cleats
 (c) direct buried, unless identified for such use
 (d) all of these

Article 342 – Nonmetallic Extensions

156. Nonmetallic surface extension are permitted in _____.
 I. Buildings not over three stories   II. Buildings over four stories
 III. All buildings
 (a) I only (b) I and II
 (c) III only (d) none of these

157. Nonmetallic surface extensions with one or more extensions shall be permitted to be run in any direction from an existing outlet, but not on the floor or within _____ inches from the floor.
 (a) 6 (b) 4
 (c) 3 (d) 2

158. Aerial cable suspended over work benches, not accessible to pedestrians, shall have a clearance of not less than _____ feet above the floor.
 (a) 12 (b) 15
 (c) 14 (d) 8

Article 343 – Preassembled Cable in Nonmetallic Conduit

Part A. General

159. The use of preassembled cable in nonmetallic conduit and fittings shall be permitted _____.
    (a) for direct burial underground installation
    (b) encased or embedded in concrete
    (c) in cinder fill
    (d) all of these

Part B. Installation

160. For _____, preassembled cable conduit shall be trimmed away from the conductors or cables using an approved method that will not damage the conductor or cable insulation or jacket.
    (a) convenience
    (b) appearances
    (c) termination
    (d) safety to persons

Article 345 – Intermediate Metal Conduit

Part A. General

161. Intermediate metal conduit cannot serve as the circuit equipment grounding conductor.
    (a) True (b) False

162. Intermediate metal conduit shall be permitted to be installed in or under cinder fill where subject to permanent moisture where protected by _____ and judged suitable for the condition.
    (a) not less than 18 inches under the fill
    (b) 2 inches of concrete
    (c) corrosion protection judged suitable
    (d) any of these

Part B. Installation

163. The area of a 1-inch IMC raceway permitted for conductor fill, assuming three or more conductors, is _____ percent.
    (a) 53 (b) 31 (c) 40 (d) 60

164. Running threads of intermediate metal conduit shall not be used on conduit for connection at couplings.
    (a) True (b) False

165. For industrial machinery, straight exposed vertical risers of intermediate metal conduit with threaded couplings are permitted, if supported at the top and bottom no more than _____ feet apart.
    (a) 10 (b) 12 (c) 15 (d) 20

166. When IMC installed through bored or punched holes in framing members, additional support requirements are not necessary. This applies to both wood and metal framing members.
    (a) True (b) False

Article 346 – Rigid Metal Conduit

167. • Aluminum fittings and enclosures shall be permitted to be used with _____ conduit.
    (a) steel rigid metal
    (b) aluminum rigid metal
    (c) rigid nonmetallic
    (d) a and b

Part A. Installation

168. Materials such as straps, bolts, etc., associated with the installation of rigid metal conduit in a wet location are required to be _____.
    (a) weatherproof
    (b) weathertight
    (c) corrosion resistant
    (d) none of these

169. Where threadless couplings and connectors used with rigid metal conduit are buried in masonry or concrete, they shall be of the _____ type.
    (a) raintight   (b) wet and damp location
    (c) nonabsorbent   (d) concrete-tight

170. The minimum radius for a bend of 1-inch rigid conduit with three No. 10 TW conductors is _____ inches. (one shot bender).
    (a) 13   (b) 11   (c) 5 ¾   (d) 9

171. • The minimum radius of a field bend on 1¼ rigid metallic conduit is _____ inches.
    (a) 7   (b) 8   (c) 14   (d) 10

172. • Two-inch rigid metal conduit shall be supported every _____ feet.
    (a) 10   (b) 12   (c) 14   (d) 15

Part B. Construction Specifications

173. Rigid conduit as shipped shall _____.
    (a) be in standard lengths of 10 feet   (b) include a coupling on each length
    (c) be reamed and threaded on each end   (d) all of these

174. Rigid conduit shall be _____ every 10 feet as required by Section 110–21.
    (a) stamped   (b) clearly and durably identified
    (c) marked   (d) none of these

Article 347 – Rigid Nonmetallic Conduit

175. The use of listed rigid nonmetallic conduit and fittings shall be permitted in portions of dairies, laundries, canneries, or other wet locations and in locations where walls are frequently washed; however, the entire conduit system including boxes and _____ used therewith shall be so installed and equipped as to prevent water from entering the conduit.
    (a) couplings   (b) fittings   (c) supports   (d) all of these

176. Rigid nonmetallic conduit can be installed, exposed, in buildings _____.
    (a) three floors   (b) twelve floors   (c) six floors   (d) of any height.

Part A. Installations

177. When installing rigid nonmetallic conduit _____.
    I. all joints shall be made by an approved method
    II. there shall be a support within 2 feet of each box, cabinet
    III. all cut ends shall be trimmed inside and outside to remove rough edges
    (a) I, II and III   (b) I and III   (c) I and II   (d) II and III

178. Rigid nonmetallic conduit in the size of 1½ inches is required to be supported every _____ feet.
    (a) 3   (b) 5   (c) 6   (d) 7

179. • Rigid nonmetallic conduit requires bushings or adapters to protect conductors No. _____ and larger from abrasion.
    (a) 8   (b) 6   (c) 3   (d) 4

180. • When installing conductors in PVC, splices and taps shall be made only in _____.
    I. junction boxes   II. conduit bodies   III. outlet boxes
    (a) I only   (b) III only   (c) I or III   (d) I, II, or III

Article 348 – Electrical Metallic Tubing

181. Ferrous electrical metallic tubing and fittings are permitted for use in _____ where protected by corrosion protection and judged suitable for the condition.
   (a) concrete
   (b) in direct contact with the earth
   (c) areas subject to severe corrosive influences
   (d) all of these

182. • Electrical metallic tubing cannot be used to support _____.
   (a) lighting fixtures   (b) outlet boxes   (c) appliances   (d) all of these

Part A. Installation

183. Couplings and connectors used with electrical metallic tubing shall be made up _____.
   (a) of metal   (b) in accordance with industry standards
   (c) tight   (d) none of these

184. • The Code requires that the support of EMT be within 3 feet of each termination and each *coupling*.
   (a) True   (b) False

Part B. Construction Specifications

185. The _____ for use with electrical metallic tubing shall have a circular cross section.
   I. tubing   II. bends   III. elbows
   (a) I only   (b) II only   (c) III only   (d) I, II, and III

Article 349 – Flexible Metallic Tubing

Part B. Construction and Installation

186. Where 3/8 flexible metallic tubing has a fixed bend for installation purposes and not flexed for service, the minimum radius of the bend shall not be less than _____ inches.
   (a) 8   (b) 12½   (c) 3½   (d) 4

Article 350 – Flexible Metal Conduit

187. The maximum run length of ½-inch flexible metal conduit for any circuit is _____ feet.
   (a) 3   (b) 8   (c) 12   (d) no limit

188. The largest size conductor permitted in ⅜-inch flexible conduit is No. _____.
   (a) 12   (b) 16   (c) 14   (d) 10

189. • Three-eights-inch flexible metal conduit used for the supply to recessed fixtures must have the raceway secured every _____ inches.
   (a) 18   (b) 36
   (c) 48   (d) not required to be secured

Article 351 – Liquidtight Flexible Conduit

Part A. Liquidtight Flexible Metal Conduit

190. The use of listed and marked liquidtight flexible metal conduit shall be permitted for _____.
   I. direct burial where listed and marked for the purpose
   II. exposed work
   III. concealed work
   (a) I   (b) II and III   (c) II   (d) I, II, and III

191. • Liquidtight flexible metal conduit smaller than _____ inch(es) shall not be used, except as permitted in the Exception.
    (a) ⅜   (b) ½   (c) 11   (d) 1¼

192. Where flexibility is necessary, securing of the raceway is not required for lengths not exceeding _____ feet at terminals.
    (a) two   (b) three   (c) four   (d) six

193. • Three-quarter-inch listed liquidtight flexible metal conduit with fittings listed for grounding can be used as a grounding path _____.
    I. Circuit conductors protected by a 20-ampere overcurrent device and the total length of the liquidtight is 6 feet or less
    II. Conductors protected by a 60-ampere overcurrent device and the total length of the liquidtight is 6 feet or less
    (a) I   (b) I and II
    (c) II   (d) None of these

Part B. Liquidtight Flexible Nonmetallic Conduit

194. The Code permits liquidtight flexible nonmetallic conduit to be installed exposed or concealed in lengths over _____ feet if it is essential for required flexibility.
    (a) 2   (b) 3   (c) 6   (d) 10

Article 352 – Surface Raceways

Part A. Surface Metal Raceways

195. The derating factors of Article 310 Note 8(a) of Notes to Ampacity Tables of 0 to 2,000 volts shall not apply to conductors installed in surface metal raceways where _____.
    (a) the cross-sectional area exceeds 4 square inches
    (b) the current carrying conductors do not exceed thirty
    (c) the total cross-sectional area of all conductors does not exceed 20 percent of the interior cross-sectional area of the raceway
    (d) all of these conditions combined

196. Where combination surface metal raceways are used for both signaling and for lighting and power circuits, the different systems shall be run in separate compartments identified by sharply _____ of the interior finish, and the same relative position of compartments shall be maintained throughout the premises.
    (a) brilliant colors   (b) etching
    (c) identifying   (d) contrasting colors

197. When an extension is made from a surface metal enclosure, the enclosure must contain a means for terminating a(n) _____.
    (a) grounded conductor
    (b) ungrounded conductor
    (c) equipment grounding conductor
    (d) all of these

Article 354 – Underfloor Raceways

198. Underfloor raceways spaced less than 1 inch apart shall be covered with concrete to a depth of _____ inch(es).
    (a) ½   (b) 4   (c) 1½   (d) 2

199. The combined area of all conductors installed at any point of an underfloor raceway shall not exceed _____ percent of the internal cross-sectional area.
    (a) 75   (b) 60   (c) 40   (d) 30

200. Underfloor raceways shall be laid so that a straight line from the center of one _____ to the center of the next _____ will coincide with the centerline of the raceway system.
    (a) termination point   (b) junction box   (c) receptacle   (d) panelboard

Unit 4

Motors and Transformers

OBJECTIVES

After reading this unit, the student should be able to briefly explain the following concepts:

Part A - Motors	Reversing DC motors	Transformer current
Alternating current motors	Reversing AC motors	Transformer kVA rating
Dual voltage motors	Volt-ampere calculations	Transformer power losses
Horsepower/watts	**Part B - Transformers**	Transformer turns ratio
Motor speed control	Transformer primary vs.	
Nameplate amperes	secondary	

After reading this unit, the student should be able to briefly explain the following terms:

Part A - Motors	Watts rating	Flux leakage loss
Armature winding	**Part B - Transformers**	Hysteresis losses
Commutator	Auto transformers	Kilo volt-amperes
Dual voltage	Circuit Impedance	Line current
Field winding	Conductor losses	Ratio
Horsepower ratings	Core losses	Self-excited
Magnetic field	Counter-electromotive-force	Step-down transformer
Motor full load current	Current transformers (ct)	Step-up transformers
Nameplate amperes	Eddy current losses	
Torque	Excitation current	

PART A – MOTORS

MOTOR INTRODUCTION

The *electric motor* operates on the principle of the attracting and repelling forces of magnetic fields. One magnetic field (permanent or electromagnetic) is *stationary* and the other magnetic field (called the *armature* or *rotor*) rotates between the stationary magnetic field, Fig. 4–1. The turning or repelling forces between the magnetic fields is called *torque*. Torque is dependent on the strength of the stationary magnetic field, the strength of the armatures magnetic field, and the physical construction of the motor.

The repelling force of like magnetic polarities, and the attraction force of unlike polarities, causes the armature to rotate in the electric motor. For a direct current motor, a device called a *commutator* is placed on the end of the conductor loop. The purpose of the commutator is to maintain the proper polarity of the loop, so as to keep the armature or rotor turning.

The *armature winding* of a motor will carry short-circuit current when voltage is first applied to the motor windings. As the armature turns, it cuts the lines of force of the *field winding* resulting in an increase in counter-electromotive-force, Fig. 4–2. The increased counter-electromotive-force results in an increase in inductive reactance, which will increase the circuit impedance. The increased circuit impedance causes a decrease in the motor armature running current, Fig. 4–3.

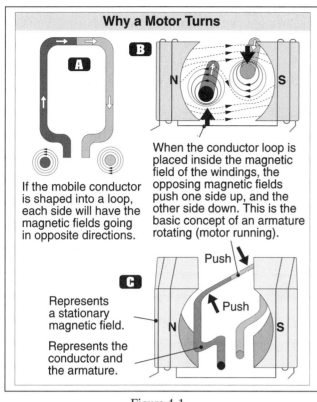

Figure 4-1

Why a Motor Turns

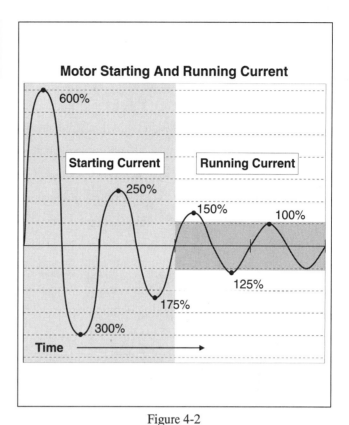

Figure 4-2

Motor Starting Current

4–1 MOTOR SPEED CONTROL

One of the advantages of the *direct current motor* over the alternating current motor is the motor's ability to maintain a constant speed. But, series direct current motors are susceptible to run away when not connected to a load.

If the speed of a direct current motor is increased, the armature winding will be cut by the magnetic field at an increasing rate, resulting in an increase of the armatures counter-electromotive-force. The increased counter-electromotive-force in the armature acts to cut down on the armature current, resulting in the motor slowing down. Placing a load on a direct current motor causes the motor to slow down, which reduces the rate at which the armature winding is cut by the magnetic field flux lines. A reduction of the armature flux lines results in a decrease in the armatures counter-electromotive-force and an increase in motor speed.

4–2 REVERSING A DIRECT CURRENT MOTOR

To reverse a direct current motor, you must reverse either the *magnetic field* of the field winding or the magnetic field of the armature. This is accomplished by reversing either the field or armature current flow, Fig. 4–4. Because most direct current motors have the field and armature winding connected to the same direct current power supply, reversing the polarity of the power supply changes both the field and armature simultaneously. To reverse the rotation of a direct current motor, you must reverse either the field or armature leads, but not both.

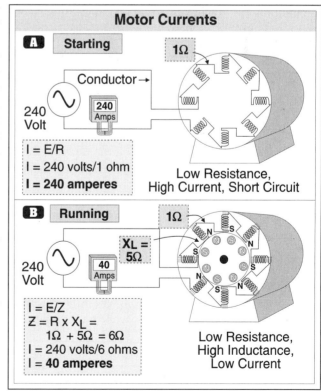

Figure 4-3

Motor Current and Impedance

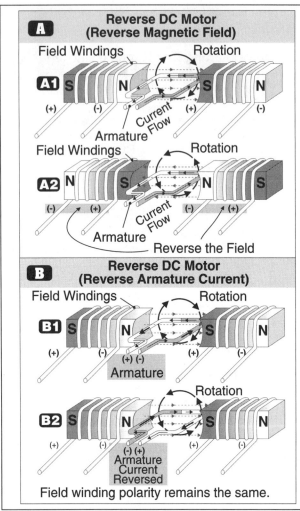

Figure 4-4
Reversing Direct Current Motors

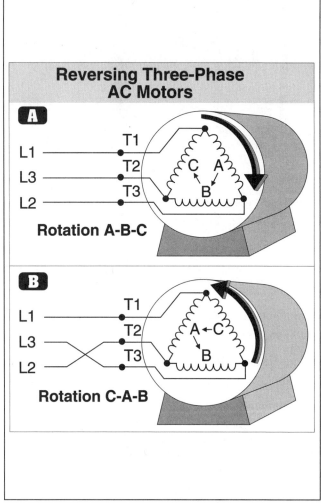

Figure 4-5
Reversing Three-Phase Alternating Current Motors

4–3 ALTERNATING CURRENT MOTORS

Fractional horsepower motors that can operate on either alternating or direct current are called *universal motors*. A motor that will not operate on direct current is called an induction motor.

The purest form of an alternating current motor is the induction motor, which has no physical connection between its rotating member (*rotor*), and stationary member (*stator*). Two common types of alternating current motors are synchronous and wound rotor motors.

Synchronous Motors

The synchronous motor's rotor is locked in step with the rotating stator field. Synchronous motors maintain their speed with a high degree of accuracy and are used for electric clocks and other timing devices.

Wound Rotor Motors

Because of their high starting torque requirements, wound rotor motors are used only in special applications. In addition, wound rotor motors only operate on three-phase alternating current power.

4–4 REVERSING ALTERNATING CURRENT MOTORS

Three-phase alternating current motors can be reversed by reversing any two of the three line conductors that supply the motor, Fig. 4–5.

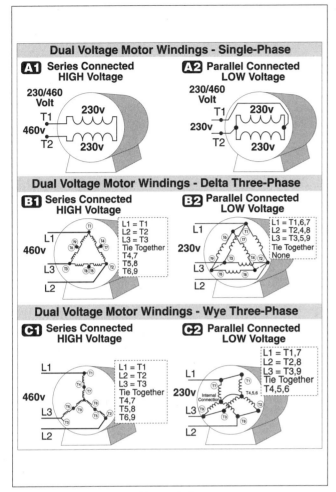

Figure 4-6
Dual Voltage Motors

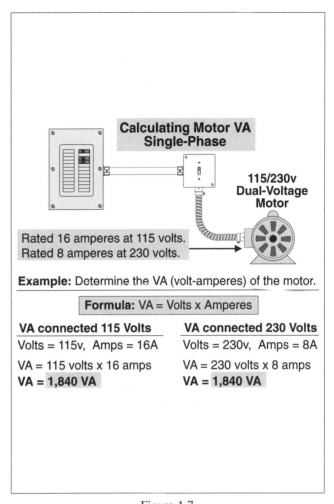

Figure 4-7
Motor VA – Single-Phase Example

4–5 MOTOR VOLT-AMPERE CALCULATIONS

Dual voltage motors are made with two field windings, each rated for the lower voltage marked on the nameplate of the motor. When the motor is wired for the lower voltage, the field windings are connected in parallel; when wired for the higher voltage, the motor windings are connected in series, Fig. 4–6.

Motor Input VA

Regardless of the voltage connection, the power consumed for a motor is the same at either voltage. To determine the *motor input* apparent power (VA), use the following formulas:

Motor VA (single-phase) = Volts × Amperes   **Motor VA (three-phase) = Volts × Amperes × $\sqrt{3}$**

❑ **Motor VA Single-Phase Example**

What is the motor VA of a single-phase 115/230-volt, 1-horsepower motor that has a current rating of 16 amperes at 115-volts and 8 amperes at 230-volts, Fig. 4–7?

    (a) 1,450 VA   (b) 1,600 VA   (c) 1,840 VA   (d) 1,920 VA

- Answer: (c) 1,840 VA

    Motor VA = Volts × Amperes

    Volts = 115 or 230, Amperes = 16 or 8 amperes

    Motor VA = 115 volts × 16 amperes, or

    Motor VA = 230 volts × 8 amperes

    Motor VA = 1,840 VA

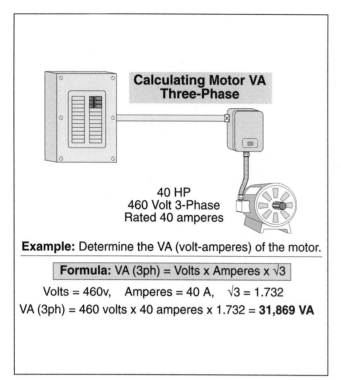

Figure. 4-8
Motor VA – Three-Phase Example

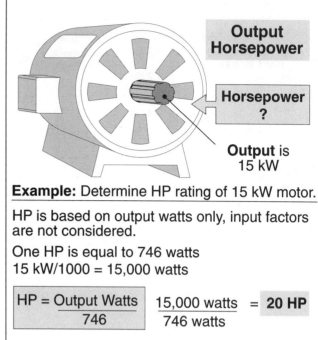

Figure. 4-9
Motor Horsepower Example

❏ **Motor VA Three-Phase Example**

What is the motor VA of a three-phase, 230/460-volt, 30-horsepower motor that has a current rating of 40 amperes at 460-volts, Fig. 4–8?

(a) 41,450 VA (b) 31,600 VA (c) 21,840 VA (d) 31,869 VA

- Answer: (d) 31,869 VA

  Motor VA = Volts × Amperes × $\sqrt{3}$

  Volts = 460 volts, Amperes = 40 amperes, $\sqrt{3}$ = 1.732

  Motor VA = 460 volts × 40 amperes × 1.732

  Motor VA = 31,869 VA

4–6 MOTOR HORSEPOWER/WATTS

The mechanical work (output) of a motor is rated in *horsepower* and can be converted to electrical energy as 746-watts per horsepower.

Horsepower Size = Output Watts/746 Watts   **Motor Output Watts = Horsepower × 746**

❏ **Motor Horsepower Example**

What size horsepower motor is required to produce 15-kW output watts, Fig. 4–9?

(a) 5 HP (b) 10 HP (c) 20 HP (d) 30 HP

- Answer: (c) 20 horsepower

  Horsepower = $\dfrac{\text{Output Watts}}{746}$

  Horsepower = $\dfrac{15{,}000 \text{ watts}}{746 \text{ watts}}$

  Horsepower = 20 horsepower

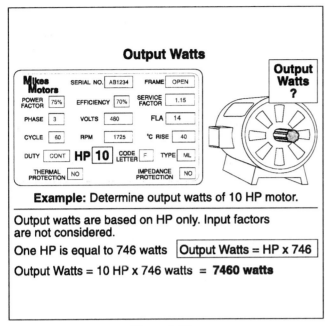

Figure 4-10
Motor Output Watts Example

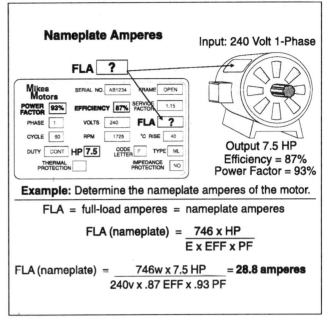

Figure 4-11
Nameplate Amperes – Single-Phase Example

❑ **Motor Output Watt Example**

What is the output watt rating of a 10-HP motor, Fig. 4–10?

(a) 3 kW (b) 2.2 kVA (c) 3.2 kW (d) 7.5 kW

Answer: (d) 7.5 kW

Output Watts = Horespower × 746 Watts

Output watts = 10 horsepower × 746 watts = 7,460 watts

Output watts = $\frac{7{,}460 \text{ watts}}{1{,}000}$ = 7.46 kW

4–7 MOTOR NAMEPLATE AMPERES

The motor nameplate indicates the motor operating voltage and current. The actual current drawn by the motor depends on how much the motor is loaded. It is important not to overload a motor above its rated horsepower, because the current of the motor will increase to a point that the motor winding will be destroyed from excess heat. The nameplate current can be calculated by the following formulas:

Single-Phase Motor Nameplate Amperes = $\frac{\text{Motor horsepower} \times 746 \text{ Watts}}{(\text{Volts} \times \text{Efficiency} \times \text{Power Factor})}$

Three-Phase Motor Nameplate Amperes = $\frac{\text{Motor Horsepower} \times 746 \text{ Watts}}{(\text{Volts} \times \sqrt{3} \times \text{Efficiency} \times \text{Power Factor})}$

❑ **Nameplate Amperes – Single-Phase Example**

What is the nameplate amperes for a 7.5-horsepower motor rated 240-volts, single-phase? The efficiency is 87 percent and the power factor is 93 percent, Fig. 4–11.

(a) 19 amperes (b) 24 amperes (c) 19 amperes (d) 29 amperes

• Answer: (d) 29 amperes

Nameplate amperes = $\frac{\text{Horsepower} \times 746 \text{ Watts}}{(\text{Volts} \times \text{Efficiency} \times \text{Power Factor})}$

Nameplate amperes = $\frac{7.5 \text{ horsepower} \times 746 \text{ watts}}{(240 \text{ volts} \times 0.87 \text{ Efficiency} \times 0.93 \text{ Power Factor})}$

Nameplate amperes = 28.81 amperes

❏ **Nameplate Amperes Three-Phase Example**

What it the nameplate amperes of a 40-horsepower motor rated 208-volts, three-phase? The efficiency is 80 percent and the power factor is 90 percent, Fig. 4–12.

(a) 85 amperes   (b) 95 amperes   (c) 105 amperes   (d) 115 amperes

• Answer: (d) 115 amperes

$$\text{Nameplate amperes} = \frac{\text{Horsepower} \times 746 \text{ Watts}}{(\text{Volts} \times \sqrt{3} \times \text{Efficiency} \times \text{Power Factor})}$$

$$\text{Nameplate amperes} = \frac{40 \text{ horsepower} \times 746 \text{ Watts}}{(208 \text{ volts} \times 1.732 \times 0.8 \text{ Efficiency} \times 0.9 \text{ Power Factor})}$$

Nameplate amperes = 29,840/259.4

Nameplate amperes = 115 amperes

PART B – *TRANSFORMER BASICS*

TRANSFORMER INTRODUCTION

A *transformer* is a stationary device used to raise or lower voltage. Transformers have the ability to transfer electrical energy from one circuit to another by mutual induction between two conductor coils. *Mutual induction* occurs between two conductor coils *(windings)* when electromagnetic lines of force within one winding induces a voltage into a second winding, Fig. 4–13.

Current transformers use the circuit conductors as the primary winding and steps the current down for metering. Often, the *ratio* of the current transformer is 1,000 to 1. This means that if there are 400 amperes on the phase conductors, the current transformer would step the current down to 0.4 ampere for the meter.

The *magnetic coupling* (*magnetomotive force, mmf*) between the primary and secondary winding can be increased by increasing the winding *ampere-turns*. Ampere-turns can be increased by increasing the number of coils and/or the current through each coil. When the current in the core of a transformer is raised to a point where there is high *flux density*, additional increases in current will produce few additional flux lines. The transformer metal iron core is said to be *saturated*.

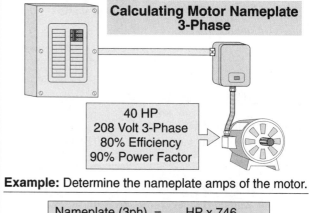

Figure 4-12

Nameplate Amperes – Three-Phase Example

4–8 TRANSFORMER PRIMARY versus SECONDARY

The transformer *winding* connected to the source is called the *primary winding*. The transformer winding connected to the load is called the *secondary winding*. Transformers are reversible, that is, either winding can be used as the primary or secondary.

4–9 TRANSFORMER SECONDARY AND PRIMARY VOLTAGE

Voltage induced in the secondary winding of a transformer is equal to the sum of the voltages induced in each *loop* of the secondary winding. The voltage induced in the secondary of a transformer depends on the number of secondary conductor turns cut by the primary magnetic flux lines. The greater the number of secondary conductor loops, the greater the secondary voltage, Fig. 4–14.

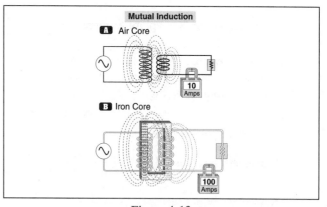

Figure 4-13

Transformer – Mutual Induction

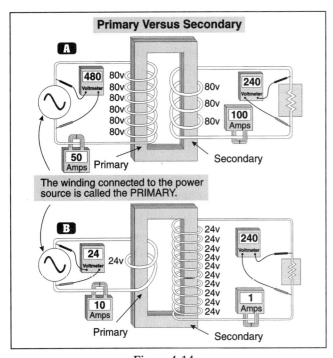

Figure 4-14
Secondary and Primary Voltage

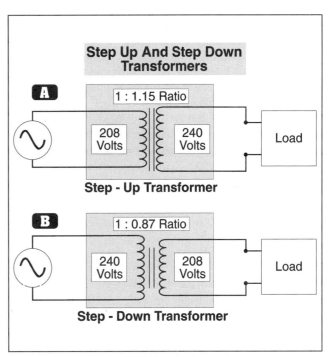

Figure 4-15
Step-Up and Step-Down Transformers

Step-Up and Step-Down Transformers

The secondary winding of a *step-down transformer* has fewer turns than the primary winding, resulting in a lower secondary voltage. The secondary winding of a *step-up transformer* has more turns than the primary winding, resulting in a higher secondary voltage, Fig. 4–15.

4–10 AUTO TRANSFORMERS

Auto transformers are transformers that use the same common winding for both the primary and the secondary. The disadvantage of an auto transformer is the lack of isolation between the primary and secondary conductors, but they are often used because they are less expensive, Fig. 4–16.

4–11 TRANSFORMER POWER LOSSES

When current flows through the winding of a transformer, power is dissipated in the form of heat. This loss is referred to as conductor I^2R loss. In addition, losses include flux leakage, core loss from eddy currents, and hysteresis heating losses, Fig. 4–17.

Conductor Resistance Loss

Transformer windings are made of many turns of wire. The resistance of the conductors is directly proportional to the length of the conductors, and inversely proportional to the cross-sectional area of the conductor. The more turns there are, the longer the conductor is, and the greater the conductor resistance. Conductor losses can be determined by the formula: $P = I^2R$.

Flux Leakage Loss

The *flux leakage loss* represents the electromagnetic flux lines between the primary and secondary winding that are not used to convert electrical energy from the primary to the secondary. They represent wasted energy.

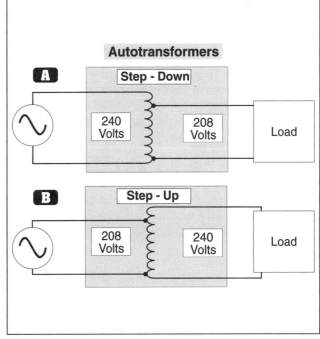

Figure 4-16
Auto Transformer

Core Losses

Iron is the only metal used for transformer cores because it offers low resistance to magnetic flux lines (*low reluctance*). Iron cores permit more flux lines between the primary and secondary winding, thereby increasing the magnetic coupling between the primary and secondary windings. However, alternating currents produce electromagnetic fields within the windings that induce a circulating current in the iron core. These circulating currents (*eddy currents*) flow within the iron core producing power losses that cannot be transferred to the secondary winding. Transformer iron cores are laminated to have a small cross-sectional area to reducing the eddy currents and their associated losses.

Hysteresis Losses

Each time the primary magnetic field expands and collapses, the transformer's iron core molecules realign themselves to the changing polarity of the electromagnetic field. The energy required to realign the iron core molecules to the changing electromagnetic field is called *hysteresis losses*.

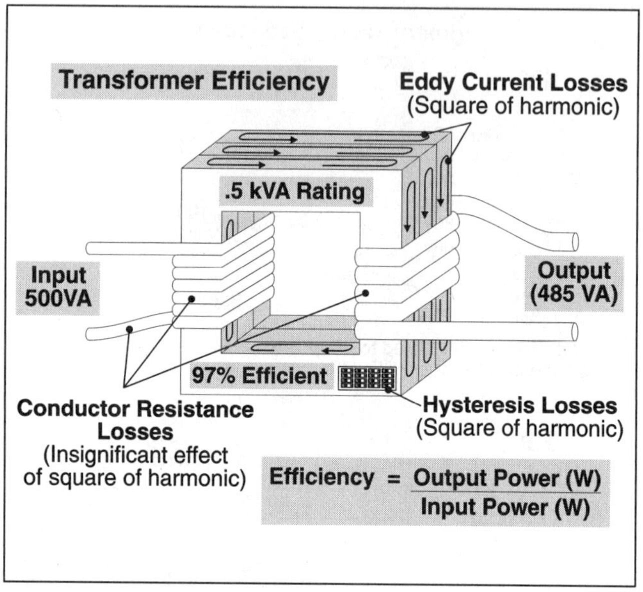

Heating By The Square Of The Frequency

Hysteresis and eddy current losses are affected by the square of the alternating current frequency. For this reason, care must be taken when iron core transformers are used in applications involving high frequencies and nonlinear loads. If the transformer operates at the third harmonic (180-Hz), the losses will by the square of the frequency. At 180-Hz, the frequency is three times the fundamental frequency (60-Hz). Therefore, the heating effect will be $3^2 = 3 \times 3 = 9$.

Figure 4-17
Transformer Losses

❑ **Transformer Losses By Square Of Frequency Example**

What is the effect on 60-Hz rated transfer for third harmonic loads (180 Hz)?

(a) heating increases three times   (b) heating increases six times

(c) heating increases nine times   (d) no significant heating

• Answer: (c) heating increases nine times

4–12 TRANSFORMER TURNS RATIO

The relationship of the primary winding voltage to the secondary winding voltage is the same as the relation between the number of primary turns as compared to the number of secondary turns. This relationship is called turns ratio or *voltage ratio*.

❑ **Delta Winding Turns Ratio Example**

What is the turns ratio of a delta/delta transformer? The primary winding is 480-volts, and the secondary winding voltage is 240-volts, Fig. 4–18?

(a) 4:1   (b) 1:4   (c) 2:1   (d) 1:2

• Answer: (c) 2:1

The primary winding voltage is 480 and the secondary winding voltage is 240. This results in a ratio of 480:240 or 2:1.

❑ **Wye Winding Ratio Example**

What is the turns ratio of a delta/wye transformer? The primary winding is 480-volts, and the secondary winding is 120-volts, Fig. 4–19?

(a) 4:1   (b) 1:4   (c) 2.3:1   (d) 1:2.3

• Answer: (a) 4:1

The primary winding voltage is 480, and the secondary winding voltage is 120. This results in a ratio of 480:120 or 4:1.

4–13 TRANSFORMER kVA RATING

Transformers are rated in kilo volt-amperes, abbreviated as kVA.

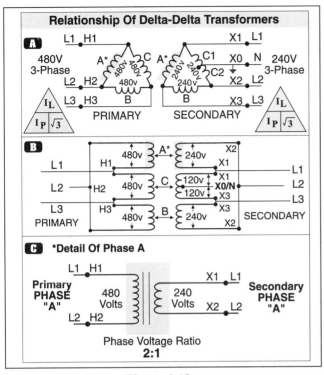

Figure. 4-18

Delta Winding Ratio Example

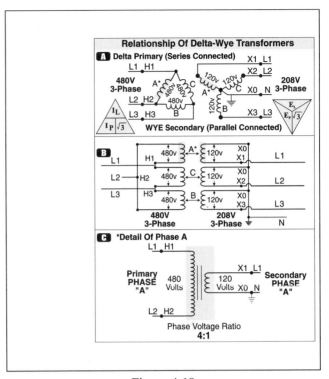

Figure. 4-19

Wye Winding Ratio Example

4–14 TRANSFORMER CURRENT

Whenever the number of primary turns is greater than the number of secondary turns, the secondary voltage will be less than the primary voltage. This will result in secondary current being greater than the primary current because the power remains the same, but the voltage changes. Since the secondary voltage of most transformers is less than the primary voltage, the secondary conductors will carry more current that the primary, Fig. 4–20. Primary and secondary line current can be calculated by:

Current (Single-Phase) = Volt-Amperes/Volts

Current (Three-Phase) = Volt-Amperes/(Volts × 1.732)

❑ **Single-Phase Transformer Current Example**

What is the primary and secondary line current for a single-phase 25-kVA transformer, rated 480-volts primary and 240-volts secondary? Which conductor is larger. Fig. 4–21?

(a) 52/104 amperes   (b) 104/52 amperes

(c) 104/208 amperes   (d) 208/104 amperes

- Answer: (a) 52/104 amperes. The secondary conductor is larger.

  Primary current =

  $VA/E = \dfrac{25{,}000\ VA}{480\ volts} = 52\ amperes$

  Secondary current =

  $VA/E = \dfrac{25{,}000\ VA}{240\ volts} = 104\ amperes$

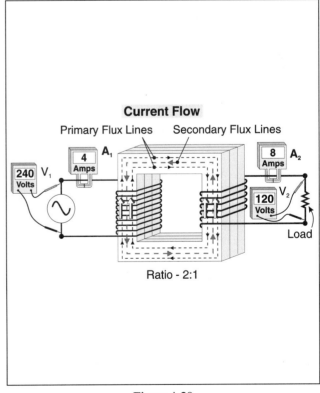

Figure 4-20

Secondary Current Greater Than Primary

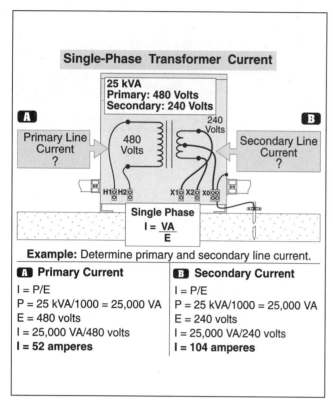

Figure 4-21
Single Phase Current Example

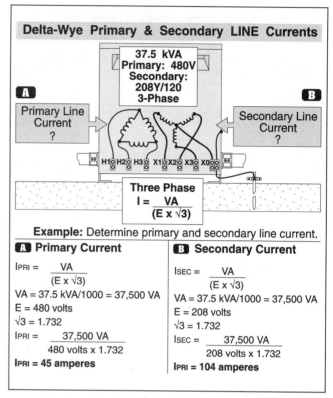

Figure 4-22
Three Phase Transformer Current Example

❏ **Three-Phase Transformer Current Example**

What is the primary and secondary line current for a three-phase 37.5-kVA transformer, rated 480-volts primary and 208-volts secondary, Fig. 4–22?

(a) 45/104 amps (b) 104/40 amps (c) 208/140 amps (d) 140/120 amps

• Answer: (a) 45/104 amperes

The secondary conductor is larger

$$\text{Primary current} = \frac{VA}{E \times \sqrt{3}} = \frac{37{,}500 \text{ VA}}{(480 \text{ volts} \times 1.732)} = 45 \text{ amperes}$$

$$\text{Secondary current} = \frac{VA}{E \times \sqrt{3}} = \frac{37{,}500 \text{ VA}}{(208 \text{ volts} \times 1.732)} = 104 \text{ amperes}$$

Unit 4 – Motors and Transformers Summary Questions

Part A – Motors

Motor Introduction

1. When voltage is applied to the motors armature, short-circuit current will flow and the armature will begin to turn. As the armature starts turning, it cuts the lines of force of the field winding, resulting in induced CEMF in the armature conductor, which cuts down on the short-circuit current.
   (a) True   (b) False

2. An electric motor works because of the effects a _____ has against a wire carrying an electric current.
   (a) magnetic field   (b) commutator   (c) voltage source   (d) none of these

3. The rotating part of a direct current motor or generator is called the _____.
   (a) shaft   (b) rotor   (c) capacitor   (d) field

4. For a direct current motor, a device called a _____ is placed on the end of the conductor loop. The polarity of the loop is maintained with the proper magnetic field to keep the opposing magnetic fields pushing each other. This in turn keeps the loop turning.
   (a) coil   (b) resistor   (c) commutator   (d) none of these

4–1 Motor Speed Control

5. • One of the great advantages of the direct current motor is the motors ability to maintain a constant speed. If the speed of a direct current motor is increased, the armature will be cut through the field winding at an increasing rate resulting in a lower CEMF that acts to cut down on the increased armature current, which slows the motor back down.
   (a) True   (b) False

6. • Placing a load on a direct current motor causes the motor to slow down, which increases the rate at which the field flux lines are being cut by the armature. As a result, the armature CEMF increases, resulting in an increase in the applied armature voltage and current. The increase in current results in increase in motor speed.
   (a) True   (b) False

4–2 Reversing A Direct Current Motor

7. To reverse a direct current motor, you must reverse the direction of the _____.
   (a) field current   (b) armature current   (c) a or b   (d) a and b

4–3 Alternating Current Motors

8. In a(n) _____ motor, the rotor is actually locked in step with the rotating stator field and is dragged along at the synchronous speed of the rotating magnetic field. _____ motors maintain their speed with a high degree of accuracy and are used for electric clocks and other timing devices.
   (a) alternating current   (b) universal   (c) wound rotor   (d) synchronous

9. _____ rotor motors are used only as special applications because of their high starting torque requirements and only operate on three-phase alternating current power.
   (a) Alternating current   (b) Universal
   (c) Wound rotor   (d) Synchronous

10. _____ motors are fractional horsepower motors that operate equally well on alternating current and direct current and are used for vacuum cleaners, electric drills, mixers, and light household appliances.
    (a) Alternating current   (b) Universal
    (c) Wound rotor   (d) Synchronous

4–4 Reversing Alternating Current Motors

11. • Three-phase alternating current motors can be reversed by changing the wiring from the line wiring from ABC phase configuration to _____.
    (a) BCA (b) CAB (c) CBA (d) ABC

4–5 Motor Volt-Ampere Calculations

12. Dual voltage 277/480-volt motors are made with two field windings, each rated at 277 volts. The field windings are connected in parallel for _____-volt operation and in series for _____-volt operation.
    (a) 277, 480 (b) 480, 277 (c) 277, 277 (d) 480, 480

4–6 Motor Horsepower/watts

13. What size horsepower motor is required to produce 30-kW output?
    (a) 20 horsepower (b) 30 horsepower (c) 40 horsepower (d) 50 horsepower

14. What are the output watts of a 15 horsepower?
    (a) 11 kW (b) 15 kVA (c) 22 kW (d) 31 kW

15. What are the output watts of a 5-horsepower, alternating current, three-phase 480 volt motor, efficiency 75 percent and power factor 70 percent?
    (a) 3.75 kW (b) 7.5 kVA (c) 7.5 kW (d) 10 kW

4–7 Motor Nameplate Amperes

16. • For all practical purposes you will not need to calculate the motor nameplate current; but you should understand how it is calculated.
    (a) True (b) False

17. What are the nameplate amperes for a 5-horsepower motor, 240volt, single-phase, efficiency 93 percent, power factor 87 percent?
    (a) 9 amperes (b) 19 amperes (c) 28 amperes (d) 31 amperes

18. What are the nameplate amperes of a 20-horsepower, 208volt, three-phase motor, power factor 90 percent and efficiency 80 percent?
    (a) 50 amperes (b) 58 amperes (c) 65 amperes (d) 80 amperes

Part B – Transformer Basics

Transformer Introduction

19. A _____ is a stationary device used to raise or lower the voltage and has the ability to transfer electrical energy from one circuit to another, with no physical connection between the two.
    (a) capacitor (b) motor (c) relay (d) transformer

20. Transformers operate on the principle of _____.
    (a) magnetoelectricity (b) triboelectric effect
    (c) thermocouple (d) mutual induction

4–8 Primary Versus Secondary

21. The transformer winding that is connected to the source is called the _____ winding and the transformer winding that is connected to the load is called the _____. Transformers are reversible; that is, either winding can be used as the primary or secondary.
    (a) secondary, primary (b) primary, secondary
    (c) depends on the wiring (d) none of these

22. Voltage induced in the secondary winding of a transformer is dependent on the number of secondary turns as compared to the number of primary turns.
    (a) True (b) False

23. The secondary winding of a step-down transformer has _____ turns than the primary, resulting in a _____ secondary voltage as compared to the primary.
    (a) less, higher (b) more, lower (c) less, lower (d) more, higher

24. The secondary winding of a step-up transformer has _____ turns than the primary, resulting in a _____ secondary voltage as compared to the primary.
    (a) less, higher (b) more, lower (c) less, lower (d) more, higher

4–10 Auto Transformers

25. Auto transformers use the same winding for both the primary and secondary. The disadvantage of an autotransformer is the lack of _____ between the primary and secondary conductors.
    (a) power (b) voltage (c) isolation (d) grounding

4–11 Transformer Power Losses

26. The most common causes of power losses for transformers are _____.
    (a) conductor resistance (b) eddy currents (c) hysteresis (d) all of these

27. The leakage of the electromagnetic flux lines between the primary and secondary winding represents wasted energy.
    (a) True (b) False

28. • The expanding and collapsing electromagnetic field of the transformer also induces a voltage in the transformer core. The induced voltage causes _____ to flow within the core, which removes energy from the transformer winding and represents wasted power.
    (a) eddy currents (b) flux (c) inductive (d) hysteresis

29. Eddy currents can be reduced by dividing the core into many flat sections or laminations. Because the laminations have a _____ cross-sectional area, the resistance offered to the eddy currents is greatly increased.
    (a) round (b) porous (c) large (d) small

30. As current flows through the transformer, the iron-core is temporarily magnetized by the electromagnetic field created by the alternating current. Each time the primary magnetic field expands and collapses, the core molecules realign themselves to the changing polarity of the electromagnetic field. The energy required to realign the core molecules to the changing electromagnetic field is called the _____ loss of the core.
    (a) eddy current (b) flux (c) inductive (d) hysteresis

4–12 Transformer Turns Ratio

31. The relationship of the primary winding voltage to the secondary winding voltage is the same as the relation between the number of conductor turns on the primary as compared to the secondary. This relationship is called _____.
    (a) ratio (b) efficiency (c) power factor (d) none of these

32. • The primary phase voltage is 240 and the secondary phase is 480. This results in a ratio of _____.
    (a) 1:2 (b) 2:1 (c) 4:1 (d) 1:4

4–13 Transformer kVA Rating

33. Transformers are rated in _____.
    (a) VA (b) kW (c) watts (d) kVA

4–14 Transformer Current

34. • The current flow in the secondary transformer winding creates a electromagnetic field that opposes the primary electromagnetic field resulting in less primary CEMF. The primary current automatically increases in direct proportion to the secondary current.
    (a) True (b) False

35. • The transformer winding with the _____ number of turns will have the lower current and the winding with the _____ number of turns will have the higher current.
    (a) least, most (b) most, most (c) least, least (d) most, least

Challenge Questions

Part A – Motors

4–1 Motor Speed Control

36. A(n) _____ type of electric motor tends to run away if it is not always connected to its load.
    (a) direct current series
    (b) direct current shunt
    (c) alternating current induction
    (d) alternating current synchronous

37. • A(n) _____ motor has a wide speed range.
    (a) alternating current (b) direct current (c) synchronous (d) induction

4–2 Reversing A Direct Current Motor

38. • If the two line (supply) leads of a direct current series motor are reversed, the motor will _____.
    (a) not run (b) run backwards
    (c) run the same as before (d) become a generator

39. • To reverse a direct current motor, one may simply reverse the supply wires.
    (a) True (b) False

4–3 Alternating Current Motors

40. The _____ induction motor is used only in special applications and is always operated on three-phase alternating current power.
    (a) compound (b) synchronous (c) split phase (d) wound rotor

41. • The rotating part of a direct current motor or generator is called the _____.
    (a) shaft (b) rotor (c) armature (d) b or c

4–5 Motor Volt-Ampere Calculations

42. The input volt-amperes of a 5-horsepower (15.2-amperes), 230-volt, three-phase motor is closest to _____ VA.
    (a) 7,500 (b) 6,100 (c) 5,300 (d) 4,600

43. The VA of a 1-horsepower (16-amperes), 115-volt motor is _____ VA.
    (a) 2,960 (b) 1,840 (c) 3,190 (d) 1,650

Part B – Transformer Basics

4–11 Transformer Power Losses

44. • When the current in the core of a transformer has risen to a point where high flux density has been reached, and additional increases in current produce few additional flux lines, the metal core is said to be _____.
    (a) maximum (b) saturated (c) full (d) none of these

45. • Magnetomotive force (mmf) can be increased by increasing the _____.
    (a) number of ampere-turns (b) current in the coils
    (c) number of coils (d) all of these

4–12 Transformer Turns Ratio

46. • The secondary current of a transformer that has a turns ratio of 2:1 is _____ than the primary current.
    (a) higher (b) lower (c) the same (d) none of these

4–13 Transformer kVA Rating

47. The primary kVA rating for a transformer that is 100 percent efficient, with a secondary of 12 volts (E_s) and with secondary current (I_s) of 5 amperes is _____ kVA.
    (a) 600 (b) 30 (c) 6 (d) 0.06

4–14 Transformer Current

48. • The primary a transformer has 100 turns ($N_p = 100$) and the secondary has 10 turns ($N_s = 10$). What is the primary current of the transformer if the secondary current (I_s) of 5 amperes is _____ amperes?
    (a) 25 (b) 10 (c) 5 (d) 0.5

49. • Which winding of a current transformer will carry more current? *Note.* A clamp-on ammeter is a current transformer, the meter is the secondary.
    (a) primary (b) secondary (c) interwinding (d) tertiary

50. If a transformer primary winding has 900 turns ($N_p = 900$) and the secondary winding has 90 turns ($N_s = 90$ turns). Which winding of the transformer has a larger a larger conductor?
    (a) primary (b) secondary (c) interwinding (d) none of these

51. If the primary of a transformer is 480 volts, and the secondary is 240 volts. Which winding of this transformer has a larger conductor?
    (a) tertiary (b) secondary (c) primary (d) windings are equal

52. If the transformer voltage turns ratio is 5:1, and the secondary has 10 ampere-turns, the primary current would be _____ if the secondary current was 10 amperes.
    (a) 25 amperes (b) 10 amperes (c) 2 amperes (d) cannot be determined

53. The transformer primary is 240 volts, the secondary is 12 volts, and the load is two 100-watt lamps. The transformer is 92 percent efficient. The secondary current of this transformer is _____.
    (a) 1 amperes (b) 17 amperes (c) 28 amperes (d) none of these

54. The transformer primary is 240 volts, the secondary is 120 volts, and the load is 1,500-watt lamps. The transformer is 92 percent efficient. The primary current of this transformer is _____ amperes.
    (a) 6.8 (b) 8.6 (c) 9.9 (d) 7.8

Figure 4–23 applies to the next two questions:

55. • The primary kVA of this transformer is _____ kVA, Fig. 4–23?
    (a) 72.5
    (b) 42
    (c) 30
    (d) 21

56. • The primary current of this transformer is _____ amperes, Fig. 4–23?
    (a) 90
    (b) 50
    (c) 25
    (d) 12

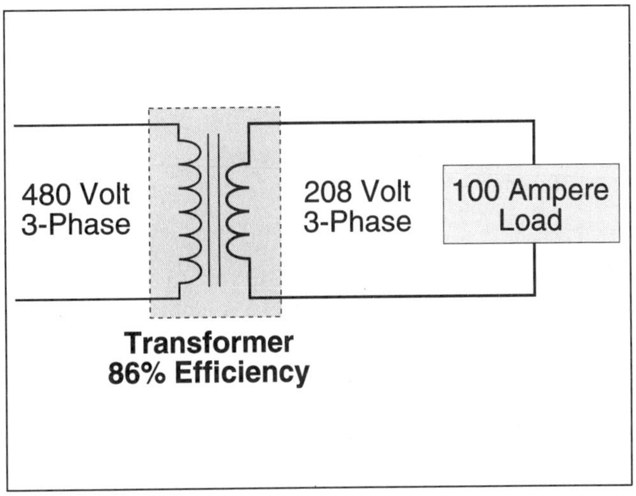

Figure 4-23

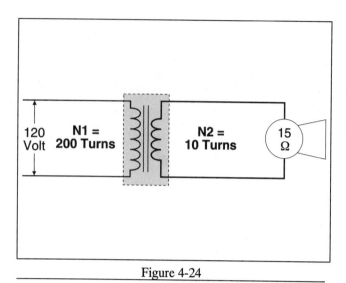

Figure 4-24

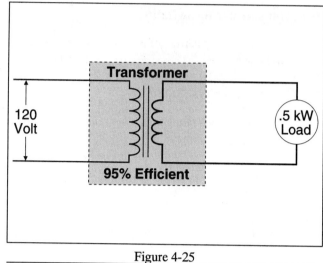

Figure 4-25

57. • The secondary voltage of this transformer is _____ volts,.. Fig. 4–24?
(a) 6 (b) 12
(c) 24 (d) 30

Figure 4–35 applies to the next two questions:

58. The primary current of the transformer is _____ ampere(s), Fig. 4–25?
(a) 0.416 (b) 4.38
(c) 3.56 (d) 41.6

59. The primary power for the transformer is _____ VA, Fig. 4–25?
(a) 526 (b) 400
(c) 475 (d) 550

Transformer Miscellaneous

60. The transformer primary is 240 volts, the secondary is 12 volts, and the load is two 100 watt lamps. The transformer is 92 percent efficient. The secondary VA of this transformer _____.
(a) is 200 VA
(b) is the same as the primary VA
(c) cannot be calculated
(d) none of these

61. • The secondary of a transformer is 24 volts with a load of 5 amperes. The primary is 120 volts and the transformer is 100 percent efficient. The secondary VA of this transformer _____.
(a) 120 VA
(b) is the same as the primary VA
(c) cannot be calculated
(d) a and b

62. The transformer primary is 240 volts, the secondary is 12 volts, and the load is two 100 watt lamps. The transformer is 92 percent efficient. The primary VA of this transformer _____.
(a) is 185 VA
(b) is 217 VA
(c) is 0.217 VA
(d) cannot be determined

63. The output of a generator is 20 kW. The input kVA is _____ kVA if the efficiency rating is 65 percent.
(a) 11 (b) 20 (c) 31 (d) 33

NEC Questions From Section 356–1 Through 430–24

Article 356 – Cellular Metal Floor Raceways

64. A _____ shall be defined as a single, enclosed tubular space in a cellular metal floor member.
    (a) cellular metal floor raceway
    (b) cell
    (c) header
    (d) none of these

Part A. Installation

65. • Loop wiring shall _____ in a cellular metal raceway.
    (a) not be permitted
    (b) not be considered a splice or tap
    (c) be considered a splice or tap when used
    (d) be permitted on conductor sizes No. 10 or less

66. A junction box used with a cellular metal floor raceway shall be _____.
    I. level to the floor grade
    II. sealed against the entrance of water or concrete
    III. metal and electrically continuous
    (a) I only   (b) II and III
    (c) III only   (d) I, II, III

Article 358 – Cellular Concrete Floor Raceways

67. A transverse metal raceway for electrical conductors, furnishing access to predetermined cells of a precast cellular concrete floors, which permits installation of electrical conductors from a distribution center to the floor cells is usually known as a(n) _____.
    (a) cell   (b) header   (c) open-bottom raceway   (d) none of these

68. Connections from a cellular concrete floor raceway headers to cabinets shall be made by means of _____.
    I. Listed metal raceways   II. PVC raceways   III. Listed fittings
    (a) I only   (b) II only
    (c) III only   (d) I and III

69. In cellular concrete floor raceways, a grounding conductor shall connect the insert receptacle to a _____.
    (a) negative ground connection provided in the raceway
    (b) negative ground connection provided on the header
    (c) positive ground connection provided on the header
    (d) grounded terminal located within the insert

70. When an outlet is _____ from a cellular concrete floor raceway, the sections of circuit conductors supplying the outlet shall be removed from the raceway.
    I. discontinued   II. abandoned   III. removed
    (a) I only   (b) II only   (c) III only   (d) I, II or III

Article 362 – Metal Wireways And Nonmetallic Wireways

71. • Wireways shall be permitted for _____.
    (a) exposed work   (b) concealed work
    (c) wet locations if of raintight construction   (d) a and c

72. Wireways longer than _____ feet shall be supported at each end and the distance between supports shall not exceed _____ feet, unless listed for other supports.
    (a) 5, 12   (b) 5, 10   (c) 3, 10   (d) 3, 6

73. Unbroken lengths of wireway shall be permitted to pass transversely through walls _____.
    (a) if in unbroken lengths where passing through
    (b) if in broken lengths where passing through
    (c) when in hazardous locations
    (d) if the wall is fire rated

Article 363 – Flat Cable Assemblies (Type FC)

74. Flat cable assemblies shall not be installed outdoors or in wet or damp locations unless _____ for use in wet locations.
    (a) special permission is granted   (b) approved
    (c) identified   (d) none of these

75. Flat cable assemblies shall have conductors of No. _____ special stranded copper.
    I. 14   II. 12   III. 10
    (a) I only   (b) II only   (c) III only   (d) I, II, III

76. Tap devices used in FC assemblies shall be rated at not less than _____ amperes, or more than 300 volts, and they shall be color-coded in accordance with the requirements of 363–20.
    (a) 20   (b) 15   (c) 30   (d) 40

77. The maximum branch circuit rating of a flat cable circuit is _____ amperes.
    (a) 15   (b) 20   (c) 30   (d) 50

Article 364 – Busways

Part A. General Requirements

78. Busways shall not be installed _____.
    (a) where subject to severe physical damage
    (b) outdoors or in wet or damp locations unless identified for such use
    (c) in hoistways
    (d) all of these

79. Busways shall be securely supported, unless otherwise designed and marked at intervals not to exceed _____ feet.
    (a) 10   (b) 5   (c) 3   (d) 8

80. When a busway is used as a branch circuit, such as the types used in Article 210, the rating of the overcurrent protection device, and the circuit, in all respects shall _____.
    (a) conform with the requirements of Article 365
    (b) conform with the requirements of Article 210 that apply to branch circuits of that rating
    (c) meet the Code requirements of Article 215 that apply to this type of circuit
    (d) meet the requirements listed in a, b, and c above

Part B. Requirements for Over 600 Volts, Nominal

81. Bus runs of over 600 volts having sections located both inside and outside of buildings shall have a _____ at the building wall to prevent interchange of air between indoor and outdoor sections.
    (a) waterproof rating   (b) vapor seal   (c) fire seal   (d) water seal

82. Where bus enclosures terminate at machines cooled by flammable gas _____ shall be provided to prevent accumulation of flammable gas in the bus enclosures.
    (a) seal-off bushings   (b) baffles   (c) a and b   (d) none of these

Article 365 – Cablebus

83. Cablebus shall not be permitted for _____.
    (a) a service   (b) branch circuits
    (c) exposed work   (d) concealed work

84. The size and number of conductors shall be that for which the cablebus is designed, and in no case smaller than No. _____.
    (a) 1/0 (b) 2/0 (c) 3/0 (d) 4/0

85. • Each section of cablebus shall be marked with the manufacturer's name or trade designation and the *minimum* diameter, number, voltage rating, and ampacity of the conductors to be installed. Markings shall be so located as to be visible after installation.
    (a) True (b) False

Article 370 – Outlet, Device, Pull, And Junction Boxes

Article 370 is another article that has numerous applications for the average person using the NEC. It contains information about the number of conductors permitted in boxes as well as how to size junction boxes. It also contains the requirements for supports and mounting equipment as well as many other requirements.

Part A. Scope and General

86. • Nonmetallic boxes are permitted only for _____.
    (a) flexible nonmetallic conduit (b) liquidtight nonmetallic conduit
    (c) nonmetallic sheath ccable (d) all of these

87. Metallic boxes are required to be _____.
    (a) bonded (b) installed (c) grounded (d) all of these

Part B. Installation

88. • An outlet box contains two cable clamps, six No. 12 THHN conductors, and one single-pole switch. The minimum size box for this combination would be _____ cubic inches.
    (a) 12 (b) 13.5 (c) 14.5 (d) 20.25

89. • In combustible walls or ceilings, the front edge of an outlet box or fitting may set back of the finished surface _____ inch.
    (a) 3/8 (b) 1/8 (c) 1/2 (d) not at all

90. • Boxes mounted in walls and ceilings constructed of combustible material shall be _____ the finished surface.
    (a) within 1/4 inch of (b) within 1/8 inch of (c) within 1/2 inch of (d) flush with

91. • Only a _____ wiring method can be used for a surface extension from a cover, and the wiring method must include an equipment grounding conductor.
    (a) solid (b) flexible (c) rigid (d) cord

92. • Boxes must be fastened to ceiling framing members with _____.
    (a) screws (b) clips identified for the purposes
    (c) rivets or bolts (d) any of these

93. • The minimum size box that does not enclose a flush device shall not be less than _____ inch(es) deep.
    (a) 1/2 (b) 1 5/8 (c) 1 (d) 1 1/4

94. • The internal depth of outlet boxes intended to enclose flush devices shall be at least _____ inch.
    (a) 1/2 (b) 7/8
    (c) 5/8 (d) none of these

95. Section 370–28 provides requirements for distances between each raceway entry inside a box and the opposite wall of a box where angle or U pulls are made. The exception to those requirements is where a raceway or cable entry is in the wall of a box or conduit body opposite to a removable cover and where the distance from that wall to the cover is in conformance with the column for one wire per terminal in Table _____.
    (a) 370–28 (b) 373–6(a)
    (c) 300–5 (d) none of these

96. Which of the following are required to be accessible?
    I. outlet boxes   II. junction boxes   III. pull boxes
    (a) I and II   (b) I and III   (c) I only   (d) I, II and III

Part C. Construction Specifications

97. A standard sheet metal outlet box shall be made from steel not less than _____.
    (a) 0.0625 inch   (b) 0.0757 inch   (c) 15 MSG   (d) 16 MSG

98. If metal boxes are over 100 cubic inches in size, and made of sheet steel, the metal thickness shall not be less than _____ inch uncoated.
    (a) 1/8   (b) 1/4   (c) 3/8   (d) none of these

Part D. Pull and Junction Boxes for Use on Systems Over 600 Volts, Nominal

99. For straight pulls, the length of the box shall be not less than _____ times the outside diameter, over sheath, of the largest conductor or cable entering the box on systems over 600 volts.
    (a) 18   (b) 16   (c) 36   (d) 48

100. The distance between a cable or conductor entry and its exit from the box shall be not less than _____ times the outside diameter, over sheath, of that cable or conductor on a 1000 volt system.
    (a) 16   (b) 18   (c) 36   (d) 40

Article 373 – Cabinets, Cutout Boxes, And Meter Enclosures

Part A. Installation

101. In walls constructed of wood or other _____ material, cabinets shall be flush with the finished surface or project therefrom.
    (a) wood-like   (b) thermoplastic   (c) corrosive   (d) combustible

102. Unused openings in cabinets and cutout boxes must be closed with a protection fitting _____ that of the wall of the enclosure.
    (a) equivalent to   (b) the same as   (c) larger than   (d) not required

103. • Cables entering a cutout box shall _____.
    (a) be secured independently to the cutout box
    (b) be permitted to be sleeved through a chase
    (c) have a maximum of two cable per connector
    (d) all of these

104. A switch or circuit breaker enclosure shall not be used as a junction box, except where adequate space is provided so that the conductors do not fill the wiring space at any cross section to more than 40 percent of the cross-sectional area of the space, and so that _____ do not fill the wiring space at any cross section to more than 75 percent of the cross-sectional area of the space.
    (a) splices   (b) taps   (c) conductors   (d) all of these

Article 374 – Auxiliary Gutters

105. An auxiliary gutter is permitted to contain _____.
    I. conductors   II. overcurrent devices   III. busbars
    (a) I only   (b) III only   (c) I or III   (d) I, II, III

106. • The maximum amperage rating of a 4 inch x 1/2 inch busbar that is 4 feet long and installed in an auxiliary gutter is _____ amperes.
    (a) 500   (b) 750   (c) 650   (d) 2000

107. • The continuous current-carrying capacity of a 1½ square inch copper busbar mounted in an unventilated enclosure is _____ amperes.
    (a) 500 (b) 750 (c) 650 (d) 1500

108. • Auxiliary gutters shall be constructed and installed so as to maintain _____ continuity.
    (a) mechanical (b) electrical (c) a or b (d) a and b

Article 380 – Switches

Part A. Installation

109. Switches or circuit breakers shall not disconnect the grounded conductor of a circuit unless the switch or circuit breaker ____.
    (a) can be opened and closed by hand levers only
    (b) simultaneously disconnects all conductors of the circuit
    (c) open the grounded conductor before it disconnects the ungrounded conductors
    (d) none of these

110. All switches and circuit breakers used as switches, must be installed so they can be operated from a readily accessible place.
    (a) True (b) False

111. A faceplate for a flush-mounted snap switch shall not be less than _____ inches thick when made of a nonferrous metal.
    (a) .03 (b) .04 (c) .003 (d) .004

112. Snap switches installed in recessed boxes must have the _____ of the strap seated against the finished wall surface.
    (a) plaster ears (b) body (c) toggle (d) all of these

113. • AC–DC general-use snap switches may be used for control of inductive loads not exceeding _____ percent of the ampere rating of the switch at the applied voltage.
    (a) 75 (b) 90 (c) 100 (d) 50

114. Snap switches rated _____ and _____ amperes must be marked CO/ALR when connected to aluminum wire.
    (a) 15, 20 (b) 15, 25 (c) 20, 30 (d) 30, 40

Article 384 – Switchboards And Panelboards

Part A. General

115. Barriers shall be placed in all service switchboards that will isolate the service _____ and terminals from the remainder of the switchboard.
    (a) busbars (b) conductors (c) cables (d) none of these

116. Each switchboard, or panelboard if used as service equipment, shall be provided with a main bonding jumper sized in accordance with Section 250–79(d) or the equivalent placed within the panelboard or one of the sections of the switchboard for connecting the grounded service conductor on its _____ side to the switchboard or panelboard frame.
    (a) load (b) supply (c) phase (d) high-leg

117. • Ventilated heating or cooling equipment (including ducts) that service the electrical room or space, cannot be installed in the dedicated space above a panelboard or switchboard.
    (a) True (b) False

118. The dedicated space above a panelboard extends from the floor to the structural ceiling. A suspended ceiling is considered the structural ceiling.
    (a) True (b) False

Part B. Switchboards

119. A space of _____ or more shall be provided between the top of any switchboard and any combustible ceiling.
 (a) 12 inches (b) 18 inches (c) 2 feet (d) 3 feet

120. Noninsulated busbars shall have a minimum space of _____ inches between the bottom of enclosure and busbar.
 (a) 6 (b) 8 (c) 10 (d) 12

121. Switchboard frames and structures supporting switching equipment shall be grounded.
 (a) True (b) False

Part C. Panelboards

122. To qualify as a lighting and appliance branch circuit panelboard, the number of circuits rated 30 ampere or less with neutrals must be _____.
 (a) more than 10 percent (b) 10 or more (c) 24 or more (d) 42 or more

123. A lighting and appliance branch circuit panelboard is considered protected by the _____ conductor protection device, if the protection device rating is not greater than the panelboard rating.
 (a) grounded (b) feeder
 (c) branch circuit (d) none of these

124. The total load on any overcurrent device located in a panelboard shall not exceed _____ where in normal operation the load will continue for 3 hours or more.
 (a) 80 percent of the conductor ampacity (b) 80 percent of its rating
 (c) 50 percent of its rating (d) a and b only

125. When equipment grounding conductors are installed in panelboards, a _____ is required for the proper termination of the equipment grounding conductors.
 (a) neutral (b) grounded terminal bar
 (c) grounding terminal bar (d) none of these

Part C. Construction Specifications

126. The minimum spacing of busbars of opposite polarity, held in free air in a panel is _____ inch(es) when not over 125 volts, nominal.
 (a) ½ (b) 1 (c) ¾ (d) ⅝

Chapter 4 – Equipment For General Use

NEC Chapter 4 is a general rules chapter as applied to general-use-equipment. Although the term General is used in the name of the chapter, it deals with specific equipment in a general application. The rules in this chapter apply everywhere in the NEC except as modified (such as in Chapters 5, 6, and 7).

Article 400 – Flexible Cords And Cables

Part A. General

127. Types TPT, TS and TST shall be permitted in lengths not exceeding _____ feet when attached directly, or by means of a special type of plug, to a portable appliance rated at 50 watts or less.
 (a) 8 (b) 10 (c) 15 (d) can't be used at all

128. A three conductor No. 16, SJE cable (one conductor is used for grounding) shall have a maximum ampacity of _____ for each conductor.
 (a) 13 (b) 12 (c) 15 (d) 8

129. Flexible cords shall not be used as a substitute for _____ wiring.
 (a) temporary (b) fixed (c) overhead (d) none of these

130. Flexible cords shall be connected to devices and to fittings so that tension will not be transmitted to joints or terminal screws. This shall be accomplished by _____.
    (a) knotting the cord
    (b) winding with tape
    (c) fittings designed for the purpose
    (d) all of these

131. • Flexible cords and cables shall be protected by _____ where passing through holes in covers, outlet boxes, or similar enclosures.
    (a) bushings
    (b) fittings
    (c) a and b
    (d) a or b

Part B. Construction Specifications

132. • A flexible cord conductor intended to be used as a(n) _____ conductor shall have a continuous identifying marker readily distinguishing it from the other conductor or conductors.
    (a) ungrounded
    (b) equipment grounding
    (c) service
    (d) high-leg

Article 402 – Fixture Wires

133. Fixture wires as listed in Table 402–3 are all suitable for a maximum of _____ volts, nominal.
    (a) 150
    (b) 300
    (c) 600
    (d) 1000

134. The smallest size fixture wire permitted in the NEC is No. _____.
    (a) 22
    (b) 20
    (c) 18
    (d) 16

135. Fixture wires shall be permitted for connecting lighting fixtures to the _____ conductors supplying the fixtures.
    (a) service
    (b) branch-circuit
    (c) supply
    (d) none of these

Article 410 – Lighting Fixtures, Lampholders, Lamps, And Receptacles

Article deals with lighting and receptacles. Part L is the receptacle section. When looking up information in this Article, the part headings are very helpful.

Part A. General

136. Fixtures, lamp holders, and receptacles shall have no live parts normally exposed to contact. Cleat-type lampholders and receptacles located at least _____ feet above the floor shall be permitted to have exposed contacts.
    (a) 3
    (b) 6
    (c) 8
    (d) none of these

Part B. Fixture Locations

137. Fixtures shall be permitted to be installed in cooking hoods of _____ occupancies where all of four conditions are met.
    (a) multi-family
    (b) residential
    (c) single family
    (d) nonresidential

138. Pendant or hanging fixtures where located directly within 3 feet horizontally of a bathtub rim, shall be so installed that the fixture is not less than _____ feet vertically from the top of the bathtub rim.
    (a) 4
    (b) 6
    (c) 8
    (d) none of these

139. • A flush recessed fixture equipped with a solid lens when installed in a closet must be on the _____ from a storage area where combustible materials may be stored.
    (a) wall above the closet door at least 18 inches upward and
    (b) ceiling over an area unobstructed to the floor, at least 18 inches upward and horizontally
    (c) outside
    (d) none of these

140. • Incandescent fixtures that have open lamps and pendant type fixtures can be installed in clothes closets, where proper clearance is maintained from combustible products.
    (a) True (b) False

141. Surface-mounted fluorescent fixtures in clothes closets shall be permitted to be installed on the wall above the door or on the ceiling, provided there is a minimum clearance of _____ inches between the fixture and the nearest point of a storage space.
    (a) 3 (b) 6 (c) 9 (d) 12

142. Coves for lighting fixtures shall have adequate space and shall be so located that the lamps and equipment can be properly installed and _____.
    (a) maintained (b) protected from physical damage
    (b) tested (d) inspected

Part D. Fixture Supports

143. The maximum weight of a light fixture that may be mounted by the screw shell of a brass socket is _____ pounds.
    (a) 2 (b) 6 (c) 3 (d) 50

144. The handhole and grounding terminal can be omitted on metal poles _____ feet or less above finish grade if the supply conductors are accessible when the fixture is removed.
    (a) 5 (b) 10 (c) 12 (d) 14

145. Where the outlet box provides adequate support, a fixture weighing up to _____ pounds may be supported by the outlet box.
    (a) 6 (b) 30 (c) 50 (d) 75

146. Trees are not permitted to support _____.
    (a) lighting fixtures (b) temporary wiring
    (c) electrical equipment (d) overhead conductor spans

Part F. Wiring Of Fixtures

147. Fixtures shall be so wired that the _____ of lampholders will be connected to the same fixture or circuit conductor or terminal.
    (a) conductor (b) neutral (c) base (d) screw shell

148. Fixture wires used for fixtures shall not be smaller than No. _____.
    (a) 22 (b) 18 (c) 16 (d) 14

149. No _____ splices or taps shall be made within or on a fixture.
    (a) unapproved (b) untested (c) uninspected (d) unnecessary

150. _____ conductors shall be used for wiring on fixture chains and other movable parts.
    (a) Solid (b) Covered (c) Insulated (d) Stranded

151. • Fixtures that require adjustment or aiming are permitted to be cord-connected without an attachment plug.
    (a) True (b) False

152. Branch circuit conductors within 3 inches of a ballast within the ballast compartment shall be recognized for use at temperatures not lower than 90°C such as:
    I. THHN II. THW III. TW IV. FEP
    (a) I only (b) I and IV only
    (c) I, II and IV (d) I, II, III and IV

153. • Type THW insulation has a _____ °C rating for use in wiring through fixtures.
    (a) 60 (b) 75 (c) 85 (d) 90

Part G. Construction Of Fixtures

154. All fixtures requiring ballasts or transformers shall be plainly marked with their electrical _____ and the manufacturer's name, trademark, or other suitable means of identification.
    (a) voltage (b) rating (c) amperage (d) none of these

155. Portable lamps shall be wired with _____, recognized by Section 400–4, and an attachment plug of the polarized or grounding type.
    (a) flexible cable (b) flexible cord
    (c) nonmetallic flexible cable (d) nonmetallic flexible cord

Part H. Installation Of Lampholders

156. Lampholders with the Edison base screwshell are designed for lampholders and screw-in receptacle adapters.
    (a) True (b) False

Part J. Construction Of Lampholders

157. Switched lampholders shall be of such construction that the switching mechanism interrupts the electrical connection to the _____.
    (a) lampholder (b) center contact (c) branch circuit (d) all of these

Part K. Lamps And Auxiliary Equipment

158. A 1,000-watt incandescent lamp requires a(n) _____ base.
    (a) mogul (b) standard (c) admedium (d) copper

Part L. Receptacles, Cord Connectors, And Attachment Plugs (Caps)

159. Receptacle faceplate covers made of insulating material shall be noncombustible and not less than _____ inch in thickness.
    (a) 0.10 (b) 0.04 (c) 0.01 (d) 0.22

160. Receptacles are not permitted to be mounted to raised (industrial) covers by a single screw.
    (a) True (b) False

161. An enclosure that is weatherproof only when the receptacle cover is closed can be used for receptacles in a wet location, if the receptacle is used only for portable equipment and tools when attended.
    (a) True (b) False

162. The enclosure for a receptacle installed in an outlet box flush-mounted on a wall surface, in a damp or wet location, shall be made weatherproof by means of a weatherproof faceplate assembly that provides a _____ connection between the plate and the wall surface.
    (a) sealed (b) weathertight (c) sealed and protected (d) watertight

Part M. Special Provisions For Flush And Recessed Fixtures

163. A recessed incandescent fixture shall be so installed that adjacent combustible material will not be subjected to temperatures in excess of _____ ºC.
    (a) 75 (b) 90 (c) 125 (d) 150

164. The raceway or cable for tap conductors to recessed lighting fixtures shall be a minimum length of _____ feet.
    (a) 1½ (b) 2 (c) 3 (d) 4

Part N. Construction Of Flush And Recessed Fixtures

165. Fixtures shall be so constructed that adjacent combustible material will not be subjected to temperature in excess of _____ ºC.
    (a) 75 (b) 90 (c) 185 (d) 140

Part P. Electric-Discharge Lighting Systems Of 1000 Volts Or Less

166. Surface-mounted fixtures with ballast must have a minimum clearance of _____ inch(es) from combustible low density fiberboard, unless the fixture is marked "Suitable for Surface Mounting on Combustible Low-Density Cellulose Fiberboard."
    (a) ½ (b) 1 (c) 1½ (d) 2

167. Auxiliary equipment where not installed as part of electric-discharge lighting fixtures (not over 1,000 volts) shall be enclosed in accessible, permanently installed _____.
    (a) nonmetallic cabinets (b) enclosures
    (c) metal cabinets (d) all of these

168. • Equipment having an open-circuit voltage exceeding _____ volts shall not be installed in or on dwelling occupancies.
    (a) 120 (b) 250 (c) 600 (d) 1000

Part Q. Lighting Track

169. Track lighting must be installed exposed in dry locations and cannot be subject to physical damage. Track lighting that operates at over 30 volts RMS must be installed at least _____ feet above the finish grade.
    (a) 3 (b) 5 (c) 6' 6" (d) none of these

170. Lighting track shall not be installed _____.
    (a) where subject to physical damage (b) in wet or damp locations
    (c) both a and b (d) neither a nor b

171. The calculated load for track lighting is _____ VA per foot.
    (a) 75 (b) 30 (c) 45 (d) 60

Article 422 – Appliances

Article 422 contains specific requirements for common appliances. Most branch circuit calculations are contained in Article 220, but Article 422 should be consulted for appliances.

Part B. Branch-Circuit Requirements

172. If a protective device rating is marked on an appliance, the branch circuit overcurrent device rating shall not be greater than _____ percent of the protective device rating marked on the appliance.
    (a) 100 (b) 50 (c) 80 (d) 115

Part C. Installation Of Appliances

173. Dishwashers and trash compactors are permitted to be cord and plug connected. The cord must not be less than 36 inches or more than _____ inches and must be protected from physical damage. The receptacle for the appliances must be accessible and located in the space occupied by the appliances or adjacent thereto.
    (a) 30 (b) 36 (c) 42 (d) 48

174. The ground-fault circuit-interrupter on cord- and plug-connected high-pressure spray washing machines shall be an integral part of the attachment plug or shall be located in the supply cord within _____ inches of the attachment plug.
    (a) 6 (b) 8 (c) 10 (d) 12

175. Electrically heated smoothing irons (flatirons) intended for use in a residence shall be equipped with an identified _____ means.
    (a) disconnecting (b) temperature limiting
    (c) current limiting (d) none of these

176. Listed ceiling fans that do not exceed _____ pounds in weight, with or without accessories, shall be permitted to be supported by outlet boxes identified for such use and supported in accordance with Sections 370–23 and 370–27.
    (a) 20 (b) 25 (c) 30 (d) 35

Chapter 1 Electrical Theory And Code Questions

Part D. Control And Protection Of Appliances

177. The maximum allowable rating of a permanently connected appliance where the branch circuit overcurrent device is used as the appliance disconnecting means is _____ horsepower.
    (a) ⅛ (b) ¼ (c) ½ (d) 1

178. Appliances that have a unit switch with an _____ setting and that disconnects all the ungrounded conductors can serve as the disconnect for the appliance.
    (a) on (b) off (c) on/off (d) all of these

179. The rating or setting of an overcurrent device for a 16.3-ampere single non-motor operated appliance should not exceed _____ amperes.
    (a) 15 (b) 35 (c) 25 (d) 45

Part E. Marking of Appliances

180. Each electric appliance shall be provided with a _____ giving the identifying name and the rating in volts and amperes, or in volts and watts.
    (a) plaque (b) nameplate (c) sticker (d) directory

Article 424 – Fixed Electric Space Heating Equipment

Part B. Installation

181. Baseboard heaters with factory-installed receptacle outlets can be used as the required outlets.
    (a) True (b) False

Part B. Installation

182. Fixed electric space heating equipment requiring supply conductors with over _____ ºC insulation shall be clearly marked.
    (a) 75 (b) 60 (c) 90 (d) all of these

Part C. Control and Protection Of Fixed Electric Space Heating Equipment

183. • Electric heating appliances employing resistance-type heating elements rated more than _____ amperes shall have the heating elements subdivided.
    (a) 60 (b) 50 (c) 48 (d) 35

184. • Resistance-type heating elements in electric space heating equipment shall be protected at not more than _____.
    (a) 95 percent of the nameplate value
    (b) 60 amperes
    (c) 48 amperes
    (d) 150 percent of the rated current

Part E. Electric Space Heating Cables

185. Electric space heating cables shall not be installed over cabinets whose clearance from the ceiling is less than the minimum _____ dimension of the cabinet to the nearest cabinet edge that is open to the room or area.
    (a) horizontal (b) vertical (c) overall (d) depth

186. • The minimum clearance between an electric space heating cable and an outlet box used for surface lighting fixtures shall not be less than _____ inches.
    (a) 8 (b) 14 (c) 18 (d) 6

Part F. Duct Heaters

187. • When any duct heater circuit is energized, means shall be provided to ensure that the fan circuit is energized. This can be accomplished by a _____ to energize the fan circuit.
 (a) fan circuit interlock
 (b) time-controlled delay
 (c) temperature-controlled delay
 (d) motion-controlled relay

Part G. Resistance-Type Boilers

188. A boiler employing resistance-type immersion heating elements contained in an ASME rated and stamped vessel and rated more than 120 amperes shall have the heating elements subdivided into loads not exceeding _____ amperes.
 (a) 70
 (b) 100
 (c) 120
 (d) 150

Part H. Electrode-type Boilers

189. The size of the branch-circuit overcurrent protective devices and conductors for an electrode-type boiler shall be calculated on the basis of _____.
 (a) 125 percent of the total load excluding motors
 (b) 125 percent of the total load including motors
 (c) 150 percent of the nameplate value
 (d) 100 percent of the nameplate value

Article 426 – Outdoor Electric Deicing And Snow-Melting Equipment

Part A. General

190. The ampacity of branch circuit conductors for outdoor deicing equipment shall not be less than _____ percent of the total load of the heaters.
 (a) 100
 (b) 115
 (c) 125
 (d) 150

Part C. Resistance Heating Elements

191. Resistance heating elements of deicing _____ shall not be installed where they bridge expansion joints unless adequately protected from expansion and contraction.
 (a) heating cables
 (b) units
 (c) panels
 (d) all of these

Part E. Skin Effect Heating

192. • For skin effect heating, the provisions of Section 300–20 shall apply to the installation of a single conductor in a ferromagnetic envelope (metal enclosure).
 (a) True
 (b) False

Article 427 – Electric Heating Equipment For Pipelines And Vessels

Part A. General

193. The ampacity of branch-circuit conductors and the rating or setting of overcurrent protective devices supplying fixed electric heating equipment for pipelines and vessels shall be not less than _____ percent of the total load of the heaters.
 (a) 75
 (b) 100
 (c) 125
 (d) 150

Part B. Installation

194. External surfaces of pipeline and vessel heating equipment that operate at temperatures exceeding _____ °F shall be physically guarded, isolated, or thermally insulated to protect against contact by personnel in the area.
 (a) 110
 (b) 120
 (c) 130
 (d) 140

Article 430 – Motors, Motor Circuits, And Controllers

Article 430 is a very important exam preparation chapter. Diagram 430–1 (FPN) is very helpful for determining which Part of Article 430 you need to use. Many people using the NEC have a difficult time with sizing the conductors and overcurrent protection to motors because they try to apply general rules. Article 430 has many exceptions to the general rules, and sizing conductors and overcurrent protection should be done using Article 430 requirements.

Part A. General

195. • For general motor application, the motor branch circuit short-circuit ground-fault protection device must be sized based on the _____ amperes.
    (a) motor nameplate
    (b) NEMA standards
    (c) NEC Table
    (d) Factory Mutual

196. • The motor _____ current as listed in Tables 430–147 through 430–150 must be used for sizing motor circuit conductors and short-circuit-ground-fault protection devices.
    (a) nameplate   (b) full load
    (c) load   (d) none of these

197. Conductors for motors, controllers, and terminals of control circuit devices are required to be of copper wire.
    (a) True   (b) False

198. Torque requirements for a motor control circuit device shall be a minimum of ____ pounds-inches (unless otherwise indicated) for No. 14 and smaller screw-type pressure terminals.
    (a) 7   (b) 9
    (c) 10   (d) 15

199. A motor terminal housing with rigidly mounted motor terminals shall have a minimum of _____ inch(es) between line terminals for a 230-volt motor.
    (a) ¼   (b) ½
    (c) ¾   (d) 1

Part B. Motor Circuit Conductors

200. • Conductors that supply two or more motors shall have an ampacity _____.
    (a) not less than the total sum of the full-load current rating plus 125 percent of the highest motor in the group
    (b) equal to the sum of the full-load current rating of all the motors plus 125 percent of the highest motor in the group
    (c) equal to the sum of the full-load current rating of all the motors plus 25 percent of the highest motor in the group
    (d) not less than 125 percent of the sum of all the motors in the group

CHAPTER 2
NEC Calculations And Code Questions

Scope of Chapter 2

UNIT 5   RACEWAY, OUTLET BOX, JUNCTION BOXES, AND CONDUIT BODY CALCULATIONS

UNIT 6   CONDUCTOR SIZING AND PROTECTION CALCULATIONS

UNIT 7   MOTOR CALCULATIONS

UNIT 8   VOLTAGE DROP CALCULATIONS

UNIT 9   ONE-FAMILY DWELLING-UNIT LOAD CALCULATIONS

Unit 5

Raceway, Outlet Box, And Junction Boxes Calculations

OBJECTIVES

After reading this unit, the student should be able to briefly explain the following concepts:

Part A – Raceway Calculations
Existing raceway calculation
Raceway sizing
Raceway properties
Understanding The National
Electrical Code Chapter 9

Part B – Outlet Box Calculations
Conductor equivalents
Sizing box – conductors all the same size
Volume of box

Part C – Pull And Junction Box Calculations
Depth of box and conduit body sizing
Pull and junction box size calculations

After reading this unit, the student should be able to briefly explain the following terms:

Part A – Raceway Calculations
Alternating current conductor
Resistance
Bare conductors
Bending radius
Compact aluminum building Wires
Conductor properties
Conductor fill
Conduit bodies
Cross-sectional area of insulated conductors
Expansion characteristics of PVC
Fixture wires
Grounding conductors

Lead-covered conductor
NEC errors
Nipple size
Raceway size
Spare space area
Part B – Outlet Box Calculations
Cable clamps
Conductor terminating in the box
Conductor running through the box
Conduit bodies
Equipment bonding jumpers
Extension rings
Fixture hickey
Fixture stud

Outlet box
Pigtails
Plaster rings
Short radius conduit bodies
Size outlet box
Strap
Volume
Yoke
Part C – Pull And Junction Box Calculations
Angle pull calculation
Distance between raceways
Horizontal dimension
Junction boxes

PART A – *RACEWAY FILL*

5–1 UNDERSTANDING THE NATIONAL ELECTRICAL CODE CHAPTER 9

Chapter 9 – Tables And Examples, Part A – Tables

Table 1 – Conductor Percent Fill

The maximum percentage of conductor fill is listed in Table 1 of Chapter 9 and is based on common conditions where the length of the conductor and number of raceway bends are within reasonable limits [Fine Print Note after Note 9], Fig. 5–1.

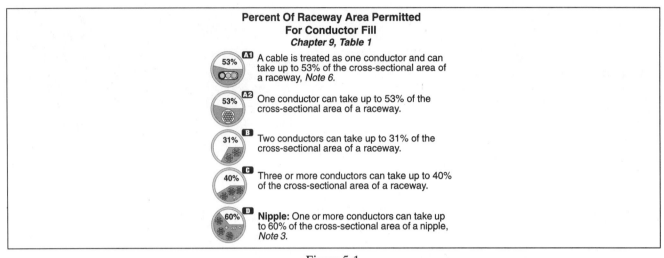

Figure 5-1

Percent Of Raceway Area Permitted For Conductor Fill

Table 1 of Chapter 9, Maximum Percent Conductor Fill	
Number of Conductors	**Percent Fill Permitted**
1 conductor	53% fill
2 conductors	31% fill
3 or more conductors	40% fill
Raceway 24 inches or less	60% fill Chapter 9, Note 3

Table 1, Note 1 – Conductors All The Same Size And Insulation
When all of the conductors are the same size and insulation, the number of conductors permitted in a raceway can be determined simply by looking at the tables located in Appendix C – Conduit and Tubing Fill Tables for Conductors and Fixture Wires of the Same Size.

Tables C1 through C13 are based on maximum percent fill as listed in Table 1 of Chapter 9).
Table C1 – Conductors and fixture wires in electrical metallic tubing
Table C1A – Compact conductors and fixture wires in electrical metallic tubing
Table C2 – Conductors and fixture wires in electrical nonmetallic tubing
Table C2A – Compact conductors and fixture wires in electrical metallic tubing
Table C3 – Conductors and fixture wires in flexible metal conduit
Table C3A – Compact conductors and fixture wires in flexible metal conduit
Table C4 – Conductors and fixture wires in intermediate metal conduit
Table C4A – Compact conductors and fixture wires in intermediate metal conduit
Table C5 – Conductors and fixture wires in liquidtight flexible nonmetallic conduit (gray type)
Table C5A – Compact conductors and fixture wires in liquidtight flexible nonmetallic conduit (gray type)
Table C6 – Conductors and fixture wires in liquidtight flexible nonmetallic conduit (orange type)
Table C6A – Compact conductors and fixture wires in liquidtight flexible nonmetallic conduit (orange type)

Note. The appendix does not have a table for liquidtight flexible nonmetallic conduit of the black type
Table C7 – Conductors and fixture wires in liquidtight flexible metallic conduit
Table C7A – Compact conductors and fixture wires in liquidtight flexible metal conduit
Table C8 – Conductors and fixture wires in rigid metal conduit
Table C8A – Compact conductors and fixture wires in rigid metal conduit
Table C9 – Conductors and fixture wires in rigid PVC Conduit Schedule 80
Table C9A – Compact conductors and fixture wires in rigid PVC Conduit Schedule 80
Table C10 – Conductors and fixture wires in rigid PVC Conduit Schedule 40

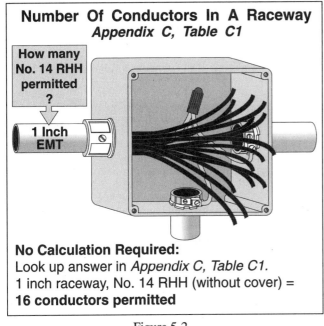

Figure 5-2
Number of Conductors In A Raceway

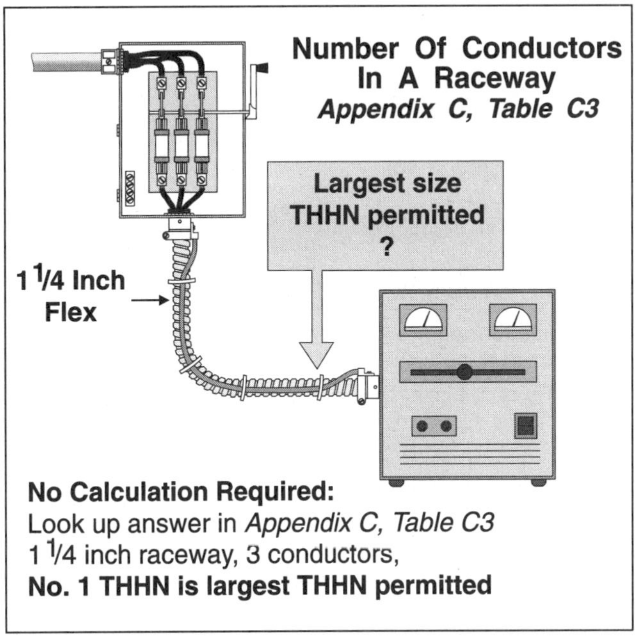

Figure 5-3
Number Of Conductors In A Raceway

Table C10A – Compact conductors and fixture wires in rigid PVC Conduit Schedule 40
Table C11 – Conductors and fixture wires in Type A, PVC Conduit
Table C11A – Compact conductors and fixture wires in Type A, PVC Conduit
Table C12 – Conductors and fixture wires in Type EB, PVC Conduit
Table C12A – Compact conductors and fixture wires in Type EB, PVC Conduit

❏ **Appendix C – Table C1 Example**
How many No. 14 RHH conductors (without cover) can be installed in a one inch electrical metallic tubing, Fig. 5–2?

(a) 25 conductors (b) 16 conductors; (c) 13 conductors (d) 19 conductors

• Answer: (b) 16 conductors, Appendix C, Table C1

❏ **Appendix C – Table C2A – Compact Conductor Example**
How many compact No. 6 XHHW conductors can be installed in a 1¼ inch nonmetallic tubing?

(a) 10 conductors (b) 6 conductors (c) 16 conductors (d) 13 conductors

• Answer: (a) 10 conductors, Appendix C, Table C2A

❏ **Appendix C – Table C3 Example**
If we have a 1¼ inch flexible metal conduit and we want to install three THHN conductors (not compact), what is the largest conductor permitted to be installed, Fig. 5–3?

(a) No. 1 (b) No. 1/0 (c) No. 2/0 (d) No. 3/0

• Answer: (a) No. 1, Appendix C, Table C3

❏ **Appendix C – Table C4 Example**
How many No. 4/0 RHH conductors (with outer cover) can be installed in two inch intermediate metal conduit?

(a) 2 conductors (b) 1 conductors (c) 3 conductors (d) 4 conductors

• Answer: (c) 3 conductors, Appendix C, Table C4

❏ **Appendix C – Table C5 – Fixture Wire Example**
How many No. 18 TFFN conductors can be installed in a ¾ inch liquitight flexible nonmetallic conduit gray type, Fig. 5– 4?

(a) 40 (b) 26 (c) 30 (d) 39

• Answer: (d) 39, Appendix C, Table 5

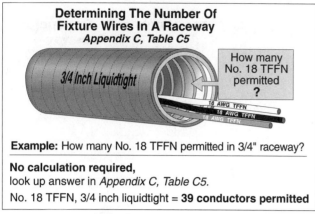

Figure 5-4

Number Of Fixture Wires In A Raceway

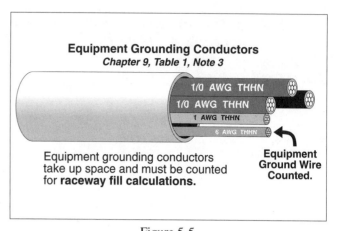

Figure 5-5

Equipment Grouding Conductors

Table 1, Note 3 – Equipment Grounding Conductors

When equipment grounding conductors are installed in a raceway, the actual area of the conductor must be used when calculating raceway fill. Chapter 9, Table 5 can be used to determine the cross-sectional area of insulated conductors, and Chapter 9, Table 8 can be used to determine the cross-sectional area of bare conductors [Note 8 or Table 1, Chapter 9], Fig. 5–5.

Table 1, Note 4 – Nipples, Raceways Not Exceeding 24 Inches

The cross-sectional areas of conduit and tubing can be found in Table 4 of Chapter 9. When a conduit or tubing raceway does not exceed 24 inches in length, it is called a nipple. Nipples are permitted to be filled to 60% of their total cross-sectional area, Fig. 5–6.

Table 1, Note 7

When the calculated number of conductors (all of the same size and insulation) results in 0.8 or larger, the next whole number can be used. But, be careful, this only applies when the conductors are all the same size and insulation. In effect, this note only applies for the development of Appendix C – Tables 1 through 13. and should not be used for exam purposes.

Table 1, Note 8

The dimensions for bare conductor are listed in Table 8 of Chapter 9.

Chapter 9, Table 4 – Conduit And Tubing Cross-Sectional Area

Table 4 of Chapter 9 lists the dimensions and cross-sectional area for conduit and tubing. The cross-sectional area of a conduit or tubing is dependent on the raceway type (cross-sectional area of the area) and the maximum percentage fill as listed in Table 1 of Chapter 9.

❏ **Conduit Cross-sectional Area Example**

What is the total cross-sectional area in square inches of a 1¼" rigid metal conduit that contains three conductors, Fig. 5–7?

(a) 1.063 square inches

(b) 1.526 square inches

(c) 1.098 square inches

(d) any of these

• Answer: (b) 1.526 square inches
  Chapter 9, Table 4

Chapter 9, Table 5 – Dimensions Of Insulated Conductors And Fixture Wires

Table 5 of Chapter 9 lists the cross-sectional area of insulated conductors and fixture wires.

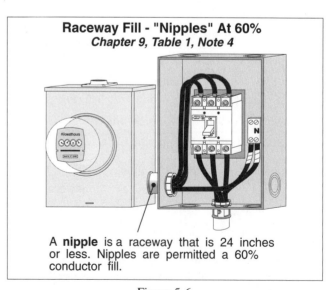

Figure 5-6

Raceway Fill – Nipples At 60%

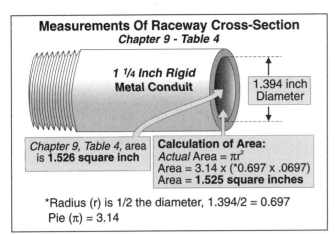

Figure 5-7
Measurements Of Raceway Cross-section

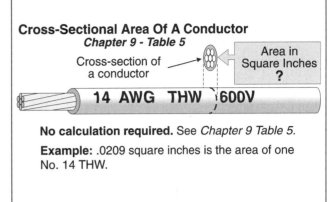

Figure 5-8
Conductor Cross-Sectional Area

Table 5 of Chapter 9 – Conductor Cross-Sectional Area							
	RH	RHH/RHW With Cover	RHH/RHW With *out* Cover or THW	TW	THHN THWN TFN	XHHW	BARE *Stranded Conductors*
Size AWG/kcmil	Approximate Cross-Sectional Area – Square Inches							
Column 1	Column 2	Column 3	Column 4	Column 5	Column 6	Column 7	Chapter 9, Table 8
14	.0209	.0293	.0209	.0139	.0097	.0139	**.004**
12	.0260	.0353	.0260	.0181	.0133	.0181	**.006**
10	.0437	.0437	.0333	.0243	.0211	.0243	**.011**
8	.0835	.0835	.0556	.0437	.0366	.0437	**.017**
6	.1041	.1041	.0726	.0726	.0507	.0590	**.027**
4	.1333	.1333	.0973	.0973	.0824	.0814	**.042**
3	.1521	.1521	.1134	.1134	.0973	.0962	**.053**
2	.1750	.1750	.1333	.1333	.1158	.1146	**.067**
1	.2660	.2660	.1901	.1901	.1562	.1534	**.087**
0	.3039	.3039	.2223	.2223	.1855	.1825	**.109**
00	.3505	.3505	.2624	.2624	.2223	.2190	**.137**
000	.4072	.4072	.3117	.3117	.2679	.2642	**.173**
0000	.4757	.4757	.3718	.3718	.3237	.3197	**.219**

❏ **Table 5 – THW Example**

What is the cross-sectional areas (square inch) for one No. 14 THW conductor, Fig. 5–8?

(a) 0.0206 square inch   (b) 0.0172 square inch
(c) 0.0209 square inch   (d) 0.0278 square inch

• Answer: (c) 0.0209 square inch

❏ **Table 5 – RHW (With Outer Cover) Example**
What is the cross-sectional areas (square inch) for one No. 12 RHW conductor (with outer cover)?
  (a) 0.0206 square inch
  (b) 0.0172 square inch
  (c) 0.0353 square inch
  (d) 0.0278 square inch
  • Answer: (c) 0.0353 square inch

❏ **Table 5 – RHH (Without Outer Cover) Example**
What is the cross-sectional areas (square inch) for one No. 10 RHH (without an outer cover)?
  (a) 0.0117 square inch
  (b) 0.0333 square inch
  (c) 0.0252 square inch
  (d) 0.0278 square inch
  • Answer: (b) 0.0333 square inch

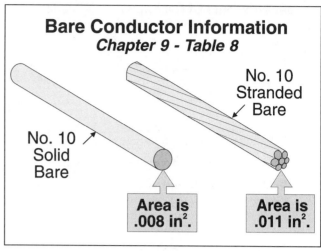

Figure 5-9
Bare Conductor Information

Chapter 9, Table 5 – Compact Aluminum Building Wire Nominal Dimensions And Areas
Tables 5A Chapter 9 list the cross-sectional area for compact aluminum building wires. We will not use these tables for this unit.

Chapter 9, Table 6 – None

Chapter 9, Table 7 – None

Chapter 9, Table 8 – Conductor Properties
Table 8 contains conductor properties such as: cross-sectional area in circular mils, number of strands per conductor, cross-sectional area in square inch for bare conductors, and the conductors resistance at 75°C for direct current for both copper and aluminum wire.

❏ **Bare Conductors Cross-Sectional Area – Example**
What is the cross-sectional area in square inch for one No. 10 bare conductor, Fig. 5–9?
  (a) 0.008 Solid   (b) 0.011 Stranded   (c) 0.038   (d) a or b
  • Answer: (d) 0.008 square inch for solid and 0.011 square inch for stranded

Chapter 9, Table 9 – AC Resistance For Conductors In Conduit Or Tubing
Table 9 contains the alternating current resistance for copper and aluminum conductors.

❏ **Alternating Current Resistance For Conductors Example**
What is the alternating current resistance of No. 1/0 copper installed in a metal conduit? Conductor length 1,000 feet.
  (a) 0.12 ohm   (b) 0.13 ohm   (c) 0.14 ohm   (d) 0.15 ohm
  • Answer: (a) 0.12 ohm

Chapter 9, Table 10 – Expansion Characteristics Of PVC Rigid Nonmetallic Conduit
This table contains the expansion characteristics of PVC.

❏ **Expansion Characteristics Of PVC Rigid Nonmetallic Conduit Example**
What is the expansion characteristics (in inches) for 50 feet of PVC, when the ambient temperature change is 100°F?
  (a) 1 inch   (b) 2 inches   (c) 3 inches   (d) 4 inches
  • Answer: (b) 2 inches

5–2 RACEWAY AND NIPPLE CALCULATION

Appendix C – Tables 1 through 13 cannot be used to determine raceway sizing when conductors of different sizes (or types of insulation) are installed in the same raceway. The following Steps can be used to determine the raceway size and nipple size.

  Step 1: → Determine the cross-sectional area (square inch) for each conductor from Table 5 of Chapter 9 for insulated conductors, and Table 8 of Chapter 9 for bare conductors.

  Step 2: → Determine the total cross-sectional area for all conductors.

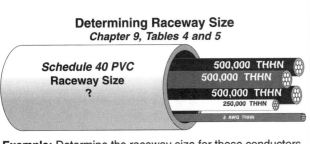
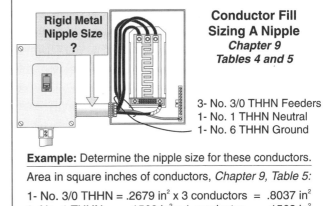

Figure 5-10
Determining Raceway Size

Figure 5-11
Conductor Fill – Sizinf A Nipple

Step 3: → Size the raceway according to Table 4–4 using the percent fill as listed in Table 1 of Chapter 9:
40% for three or more conductors
60% for raceways 24 inches or less in length (nipples).

❑ **Raceway Size Example**

A 400-ampere feeder is installed in schedule 40 rigid nonmetallic conduit. This raceway contains three 500-kcmil THHN conductors, one 250-kcmil THHN conductor, and one No. 3 THHN conductor. What size raceway is required for these conductors, Fig. 5–10?

(a) 2 inch (b) 2½ inch (c) 3 inch (d) 3½ inch

• Answer: (c) 3

Step 1: → Cross- sectional area of the conductors, Table 5 of Chapter 9.
500 kcmil THHN = 0.7073 square inch × 3 wires = 2.1219 square inches
250 kcmil THHN = 0.3970 square inch × 1 wire = 0.3970 square inch
No. 3 THHN = 0.0973 square inch × 1 wire = 0.0973 square inch

Step 2: → Total cross-sectional area of all conductors = 2.6162 square inches

Step 3: → Size the conduit at 40% fill [Chapter 9, Table 1] using Table 4,
3 inch schedule 40 PVC has an cross-sectional area of 2.907 square inches permitted for conductor fill.

❑ **Nipple Size Example**

What size rigid metal nipple is required for three No. 3/0 THHN conductors, one No. 1 THHN conductor and one No. 6 THHN conductor, Fig. 5–11?

(a) 2 inch (b) 1½ inch (c) 1¼ inch (d) None of these

• Answer: (b) 1½ inch

Step 1: → Cross-sectional area of the conductors, Table 5 of Chapter 9.
No. 3/0 THHN = 0.2679 square inch × 3 wires = 0.8037 square inch
No. 1 THHN = 0.1562 square inch × wire = 0.1562 square inch
No. 6 THHN = 0.0507 square inch × 1 = 0.0507 square inch

Step 2: → Total cross-sectional area of the conductors = 1.0106 square inches

Step 3: → Size the conduit at 60% fill [Table 1, Note 4 of Chapter 9] using Table 4.
1¼ inch nipple = 1.526 × 0.6 = 0.912 square inch, too small
1½ inch nipple = 2.07 × 0.6 = 1.242 square inches, *just right*
2 inch nipple = 3.4 08 × 0.6 = 2.0448 square inches, too big

5–3 EXISTING RACEWAY CALCULATION

There are times that you need to add conductors to an existing raceway. This can be accomplished by using the following steps:

Part 1 – Determine the raceways cross-sectional spare space area.

Step 1: → Determine the raceways cross-sectional area for conductor fill [Table 1 and Table 4 of Chapter 9].

Step 2: → Determine the area of the existing conductors [Table 5 of Chapter 9].

Step 3: → Subtract the cross-sectional area of the existing conductors (Step 2) from the area of permitted conductor fill (Step 1).

Part 2 – To determine the number of conductors permitted in spare space area:

Step 4: → Determine the cross-sectional area of the conductors to be added [Table 5 of Chapter 9 for insulated conductors and Table 8 of Chapter 9 for bare conductors].

Step 5: → Divide the spare space area (Step 3 – Part 1) by the conductors cross-sectional area (Step 4 – Part 2).

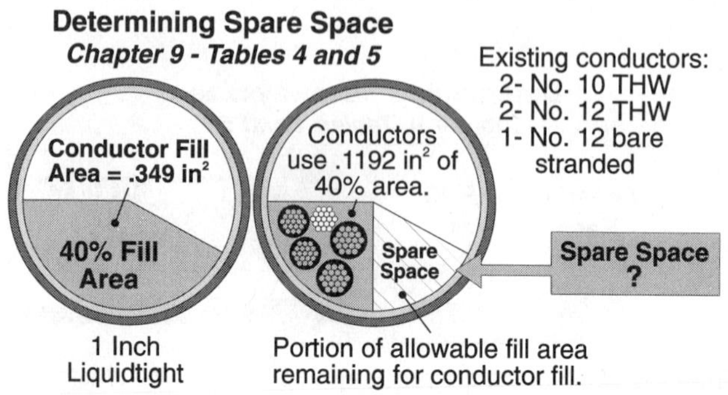

Determining Spare Space
Chapter 9 - Tables 4 and 5

Existing conductors:
2- No. 10 THW
2- No. 12 THW
1- No. 12 bare stranded

Conductor Fill Area = .349 in^2
Conductors use .1192 in^2 of 40% area.
40% Fill Area
Spare Space
Spare Space ?

1 Inch Liquidtight — Portion of allowable fill area remaining for conductor fill.

Example: Determine the area of fill (spare space) remaining.

Spare Space = Allowable fill area - Existing conductor space used
Allowable Fill Area: Chapter 9 Table 4
1 inch liquidtight, 3 or more conductors = 40% = .349 square inch
Space Used: Chapter 9, Table 5
One No. 10 THW = .0333 in^2 x 2 conductors = .0666 square inch
One No. 12 THW = .0260 in^2 x 2 conductors = .0520 square inch
One No. 12 bare stranded, Chapter 9, Tbl 8 = .0060 square inch
Allowable fill area used = **.1192 square inch**

Spare Space = .349 in^2 - .1192 in^2 = **.2264 square inch remaining fill area**

Figure 5-12
Determining Raceway Spare Space Area

❑ **Spare Space Area Example**

An existing one inch liquidtight flexible metallic conduit contains two No. 12 THW conductors, two No. 10 THW conductors, and one No. 12 bare (stranded). What is the area remaining for additional conductors, Fig. 5–12?

(a) 0.2264 square inch (if the raceway is more than 24 inches long)

(b) 0.3992 square inch (if the raceway is less than 24 inches long)

(c) There is no spare space.

(d) a and b are correct

• Answer: (d) a and b

Step 1: → Cross-sectional area permitted for conductor fill [Table 1, Note 4 and Table 4 of Chapter 9]
At 40%) = 0.864 × 0.4 = 0.3456 square inch
Nipple at 60% = 0.864 × 0.6 = 0.5184 square inch

Step 2: → Cross-sectional area of existing conductors.
No. 12 THW 0.0260 square inch × 2 wires = 0.0520 square inch
No. 10 THW 0.0333 square inch × 2 wires = 0.0666 square inch
No. 12 bare 0.006 square inch × 1 wire = 0.006 square inch
(Ground wires must be counted for raceway fill – Table 1, Note 3 of Chapter 9.)
Total cross-sectional area of existing conductors = 0.1192 square inch

Step 3: → Subtract the area of the existing conductors from the permitted area of conductor fill.
Raceway more than 24 inches long = 0.3456 – 0.1192 = 0.2264 square inch
Nipple (less than 24 inches long) = 0.5184 – 0.1192 = 0.3992 square inch

❑ **Number Of Conductors Permitted In Spare Space Area Example**

An existing one-inch liquidtight flexible metallic conduit contains two No. 12 THW conductors, two No. 10 THW conductors, and one No. 12 bare (stranded). How many additional No. 10 THHN conductors can be added to this raceway, Fig. 5–13?

(a) 10 conductors if the raceway is more than 24 inches long

(b) 18 conductors if the raceway is less than 24 inches long

(c) 15 conductors regardless of the raceway length

(d) a and b are correct

• Answer: (d) both a and b are correct

Step 1: → Cross-sectional area permitted for conductor fill [Table 1, Note 4 and Table 4 of Chapter 9]
At 40%)= .864 × 0.4 = 0.3456 square inch
Nipple at 60% = 0.864 × 0.6 = .5184 square inch

Step 2: → Cross-sectional area of existing conductors.
No. 12 THW 0.0260 square inch × 2 wires= 0.0520 square inch
No. 10 THW 0.0333 square inch × 2 wires = 0.0666 square inch
No. 12 bare 0.006 square inch × 1 wire = 0.006 square inch
(Ground wires must be counted for raceway fill – Table 1, Note 3 of Chapter 9.)
Total cross-sectional area of existing conductors = 0.1192 square inch

Step 3: → Subtract the area of the existing conductors from the permitted area of conductor fill.
Raceway more than 24 inches long = 0.3456 – 0.1192 = 0.2264 square inch
Nipple (less than 24 inches long) = 0.5184 – 0.1192 = 0.3992 square inch

Step 3: → Cross-sectional area of the conductors to be installed [Table 5 of Chapter 9]. No. 10 THHN = 0.0211 square inch.

Step 3: → Divide the spare space area (Step 3) by the conductor area.
Raceway at 40% = 0.2264 square inches/0.0211 = 10.7 or 10 conductors
Nipple at 60% = 0.3992 square inch/0.0211 = 18.92 or 18 conductors. We must round down to 18 conductors because Note 8, Table 1 of Chapter 9 only applies if all of the conductors are the same size and same insulation.

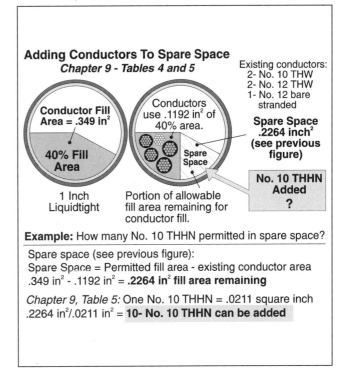

Figure 5-13
Adding Conductors to Spare Space

5–4 TIPS FOR RACEWAY CALCULATIONS

Tip 1: → Take your time
Tip 2: → Use a ruler or straight-edge when using tables
Tip 3: → Watch out for the different types of raceways and conductor insulations, particularly RHH/RHW with or without outer cover.

PART B – OUTLET BOX FILL CALCULATIONS

INTRODUCTION [370-16]

Boxes shall be of sufficient size to provide free space for all conductors. An outlet box is generally used for the attachment of devices and fixtures and has a specific amount of space (volume) for conductors, devices, and fittings. The volume taken up by conductors, devices, and fittings in a box must not exceed the box fill capacity. The volume of a box is the total volume of its assembled parts, including plaster rings, industrial raised covers, and extension rings. The total volume includes only those fitting that are marked with their volume in cubic inches [370-16(a)].

5–5 SIZING BOX – CONDUCTORS ALL THE SAME SIZE [Table 370–16(a)]

When all of the conductors in an *outlet box* are the same size (insulation doesn't matter), Table 370–16(a) of the National Electrical Code can be used to:
(1) Determine the number of conductors permitted in the outlet box, or
(2) Determine the size outlet box required for the given number of conductors.

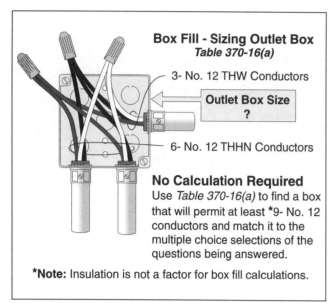

Figure 5-14
Sizing Outlet Box

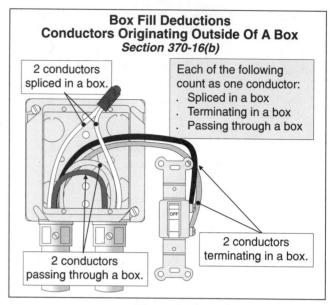

Figure 5-15
Box Fill Deductions

Note. Table 370–16(a) applies only if the outlet box contains no switches, receptacles, fixture studs, fixture hickeys, manufactured cable clamps, or grounding conductors (not likely).

❑ **Outlet Box Size Example**
What size outlet box is required for six No. 12 THHN conductors, and three No. 12 THW conductors? Fig. 5–14.

(a) 4 inches × 1¼ inches square (b) 4 inches × 1½ inches square

(c) 4 inches × 1¼ inches round   (d) 4 inches × 1½ inches round

• Answer: (b) 4 inches × 1½ inches square,
Table 370–16(a) permits nine No. 12 conductors, insulation is not a factor.

❑ **Number of Conductors in Outlet Box Example**
Using Table 370-16(a), how many No. 14 THHN conductors are permitted in a 4" × 1½ inch round box?

(a) 7 conductors   (b) 9 conductors

(c) 10 conductors (d) 11 conductors

• Answer: (c) 7 conductors

5–6 CONDUCTOR EQUIVALENTS [370–16(b)]

Table 370–16(a) does not take into consideration the fill requirements of clamps, support fittings, devices or equipment grounding conductors within the outlet box. In no case can the volume of the box and its assembled sections be less than the fill calculation as listed below:

(1) Conductor Fill – Conductors Running Through. Each conductor that originates outside the box and terminates or is spliced within the box is considered as one conductor, Fig. 5–15.

Conductor Running Through The Box. Each conductor that runs through the box is considered as one conductor, Fig. 5–15. Conductors, no part of which leaves the box, shall not be counted, this includes equipment bonding jumpers and pigtails, Fig. 5–16.

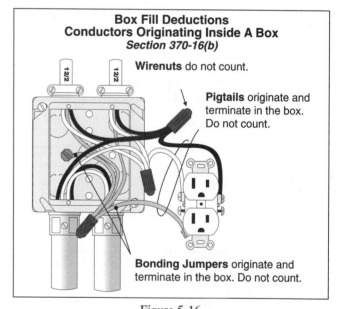

Figure 5-16
Box Fill Deductions

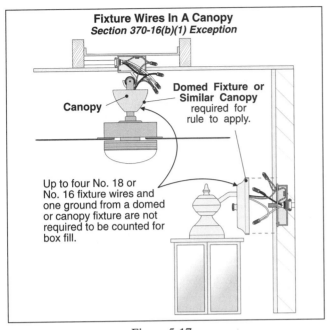

Figure 5-17
Fixture Wires In A Canopy

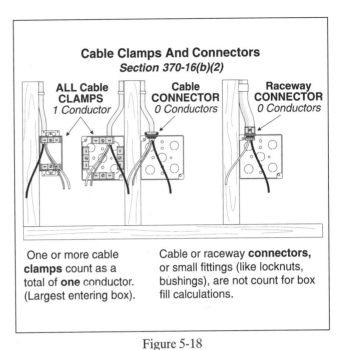

Figure 5-18
Cable Clamps And Connectors

Exception. Fixture wires smaller than No. 14 from a domed fixture or similar canopy are not counted, Fig. 5–17.

(2) Cable Fill. One or more internal cable clamps in the box are considered as one conductor volume in accordance with the volume listed in Table 370-16(b), based on the largest conductor that enters the outlet box, Fig. 5–18.

Note. Small fittings such as locknut and bushings are not counted [370-16(b).

(3) Support Fittings Fill. One or more fixture studs or hickeys within the box are considered as one conductor volume in accordance with the volume listed in Table 370–16(b), based on the largest conductor that enters the outlet box, Fig. 5–19.

(4) Device or Equipment Fill. For each yoke of strap containing one or more devices or equipment is considered as two conductors volume in accordance with the volume listed in Table 370-16(b), based on the largest conductor that terminates on the yoke, Fig. 5–20.

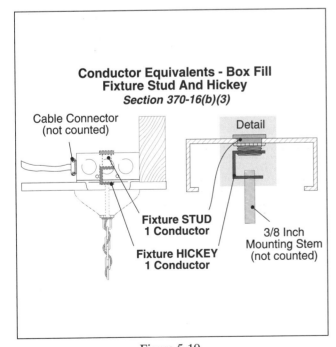

Figure 5-19
Box Fill – Fixture Stud And Hickey

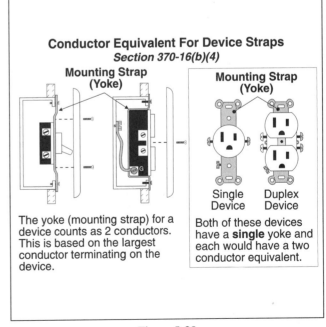

Figure 5-20
Conductor Equivalent For Device Straps

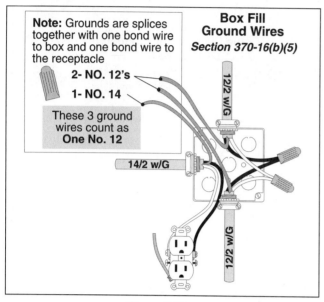

Figure 5-21
Box Fill – Ground Wires

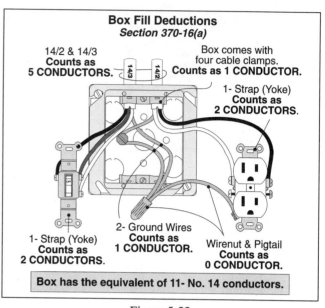

Figure 5-22
Box Fil Deductions

(5) Grounding Conductors. One or more grounding conductors are considered as one conductor volume in accordance with the volume listed in Table 370–16(b), based on the largest grounding conductor that enters the outlet box, Fig. 5–21.

Note. Fixture ground wires smaller than No. 14 from a domed fixture or similar canopy are not counted [370-16(b)(1) Exception].

What's Not Counted

Wirenuts, cable connectors, raceway fittings, and conductors that originate and terminate within the outlet box (such as equipment bonding jumpers and pigtails) are not counted for box fill calculations [370–16(a)].

❑ **Number of Conductors Example**

What is the total number of conductors used for box fill calculations in Figure 5–22?

(a) 5 conductors   (b) 7 conductors

(c) 9 conductors   (d) 11 conductors

• Answer: (d) 11 conductors

Switch – Five No. 14 conductors, two conductors for device and three conductors terminating.
Receptacle – Four No. 12 conductors, two conductors for the device and two conductors terminating.
Ground Wire – One conductor.
Cable Clamps – One conductor.

5–7 SIZING BOX – DIFFERENT SIZE CONDUCTORS [370–16(b)]

To determine the size of the outlet box when the conductors are of different sizes (insulation is not a factor), the following steps can be used:

Step 1: → Determine the number and size of conductors equivalents in the box.
Step 2: → Determine the volume of the conductors equivalents from Table 370–16(b).
Step 3: → Size the box by using Table 370–16(a).

❑ **Outlet Box Sizing Example**

What size outlet box is required for 14/3 romex (with ground) that terminates on a switch, with 14/2 romex that terminates on a receptacle, if the box has internal cable clamps factory installed, Fig. 5–22?

(a) 4 inches × 1¼ inch square   (b) 4 inches × 1½ inch square

(c) 4 inches × 2⅛ inch square   (d) Any of these

• Answer: (c) 4 inches × 2⅛ inch square

Step 1: → Determine the number and size of conductors.
14/3 romex – three No. 14's
14/2 romex – two No. 14's
Cable Clamps – one No. 14
Switch – two No. 14's
Receptacles – two No. 14's
Ground wires – one No. 14
Total – Eleven No. 14's

Step 2: → Determine the volume of the conductors [Table 370–16(b)].
No. 14 = 2 cubic inches
Eleven No. 14 conductors = 11 wires × 2 cubic inches = 22 cubic inches

Step 3: → Select the outlet box from Table 370–16(a).
4 inches × 1¼ inch square, 18 cubic inches, too small
4 inches × 1½ inch square, 21 cubic inches, *just right*
4 inches × 2⅛ inch square, 30.3 cubic inches, larger than necessary

❑ **Domed Fixture Canopy Example [370–16(a), Exception].**

A round 3 inch × ½ inch box has a total volume of 6 cubic inches and has factory internal cable clamps. Can this pancake box be used with a lighting fixture that has a domed canopy? The branch circuit wiring is 14/2 nonmetallic sheath cable and the paddle fan has three fixture wires and one ground wire all smaller than No. 14, Fig. 5–23.

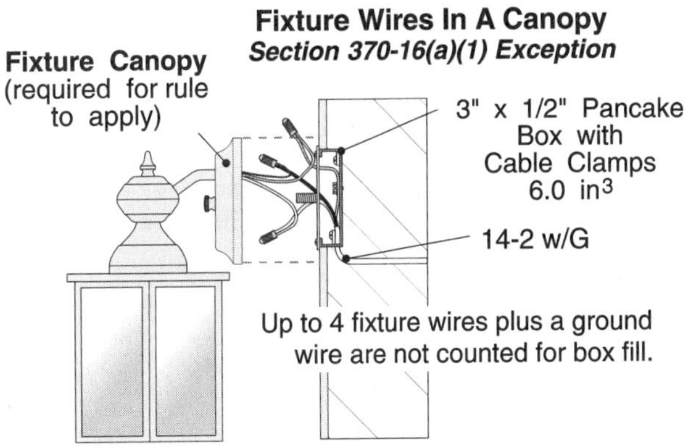

(a) Yes (b) No

• Answer: (a) No,
The box is limited to 6 cubic inches, and the conductors total 8 cubic inches [370–16(a)].

Step 1: → Determine the number and size of conductors within the box.
14/2 romex – two No. 14's
Cable clamps – one No. 14
Ground wire – one No. 14
Total Number four No. 14's

Step 2: → Determine the volume of the conductors [Table 370–16(b)].
No. 14 = 2 cubic inches
Four No. 14 conductors = 4 wires × 2 cubic inches = 8 cubic inches.

Fixture Wires In A Canopy
Section 370-16(a)(1) Exception

Fixture Canopy (required for rule to apply)

3" × 1/2" Pancake Box with Cable Clamps 6.0 in³

14-2 w/G

Up to 4 fixture wires plus a ground wire are not counted for box fill.

Example: Box fill calculation for pancake box.

Number of Conductors in Box: *370-16(a)*
14-2 NM Cable = 2 conductors
14-2 Ground = 1 conductor
Cable Clamps = 1 conductor
Fixture Wires = 0 conductors, *(370-16(a)(1) Exception)*
Total = **4- No. 14 conductors**

Volume of Conductors: *Table 370-16(b)*
1- No. 14 = 2 cubic inches × 4 conductors = **8 cubic inches**

Volume of box is **6 cubic** inches (too small).
Installation is a **VIOLATION.**

Figure 5-23

Fixture Wires In A Canopy

❑ **Conductors Added To Existing Box Example**

How many No. 14 THHN conductors can be pulled through a 4 inch square × 2⅛-inch deep box that has a plaster ring of 3.6 cubic inches? The box already contains two receptacles, five No. 12 THHN conductors, and one No. 12 bare grounding conductor, Fig. 5–24.

(a) 4 conductors (b) 5 conductors (c) 6 conductors (d) 7 conductors

• Answer: (b) 5 conductors

Step 1: → Determine the number and size of the existing conductors.
Two Receptacles – 2 yokes × 2 conductors each = four No. 12 conductors
Five No. 12's – five No. 12 conductors
One ground – one No. 12 conductor
Total conductors Ten No. 12 conductors

Step 2: → Determine the volume of the existing conductors [Table 370–16(b)].
No. 12 conductor = 2.25 cubic inches, 10 wires × 2.25 cubic inches = 22.5 cubic inches

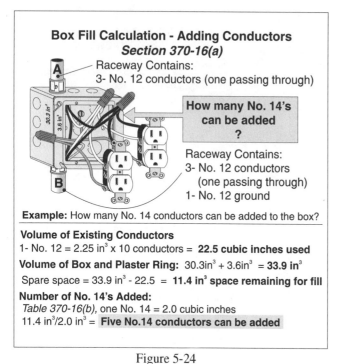

Figure 5-24

Box Fill Calculation – Adding Conductors

Step 3: → Determine the space remaining for the additional No. 14 conductors.
Remaining space = Total Space less existing conductors
Total space = 30.3 cubic inches (box) [Table 370–16(a)] + 3.6 cubic inches (ring) = 33.9 cubic inches
Remaining space = 33.9 cubic inches – 22.5 cubic inches (ten No. 12 conductors)
Remaining space = 11.4 cubic inches

Step 4: → Determine the number of No. 14 conductors permitted in the spare space.
Conductors added = Remaining space/added conductors volume
Conductors added = 11.4 cubic inches/2 cubic inches [Table 370–16(b)]
Conductors added = five No. 14 conductors.

PART C – PULL, JUNCTION BOXES, AND CONDUIT BODIES

INTRODUCTION

Pull boxes, *junction boxes*, and *conduit bodies* must be sized to permit conductors to be installed so that the conductor insulation will not be damaged. For conductors No. 4 and larger, we must size pull boxes, junction boxes, and conduit bodies according to the requirements of Section 370–28 of the National Electrical Code, Fig. 5–25.

5–8 PULL AND JUNCTION BOX SIZE CALCULATIONS

Straight Pull Calculation [370–28(a)(1)]

A straight pull calculation applies when conductors enter one side of a box and leave through the opposite wall of the box. The minimum distance from where the raceway enters to the opposite wall must not be less than eight times the trade size of the largest raceway, Fig. 5–26.

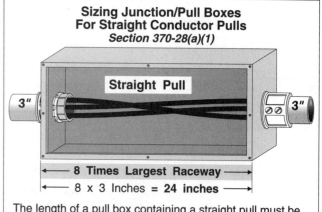

Figure 5-26

Straight Pull Sizing – 370–28(a)(1)

Sizing Junction/Pull Boxes For Angle Conductor Pulls
Section 370-28(a)(2)

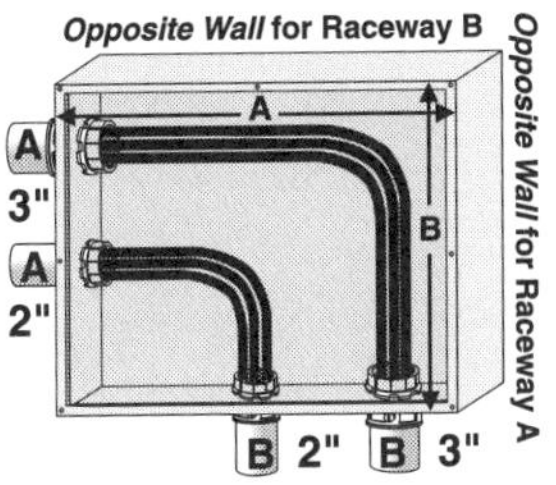

Distance A (Horizontal):
- Measured from the wall where the raceway enters to the opposite wall.
- 6 times the largest raceway plus the sum of any other raceways entering that wall.

Distance B (Vertical):
- Measured from the wall where the raceway enters to the opposite wall.
- 6 times the largest raceway plus the sum of any other raceways entering the same wall.

Example: Determine the minimum size angle pull box.

Dimension A: *370-28(a)(2)*
6 times largest plus sum of other raceways entering the same wall.
"A" = (6 x 3") + 2" = **20 Inches**

Dimension B: *370-28(a)(2)*
6 times largest plus sum of other raceways entering the same wall.
"B" = (6 x 3") + 2" = **20 Inches**

Figure 5-27
Angle Pull Sizing – 270–28(a)(2)

Sizing Junction/Pull Boxes For U-Pulls
Section 370-28(a)(2)

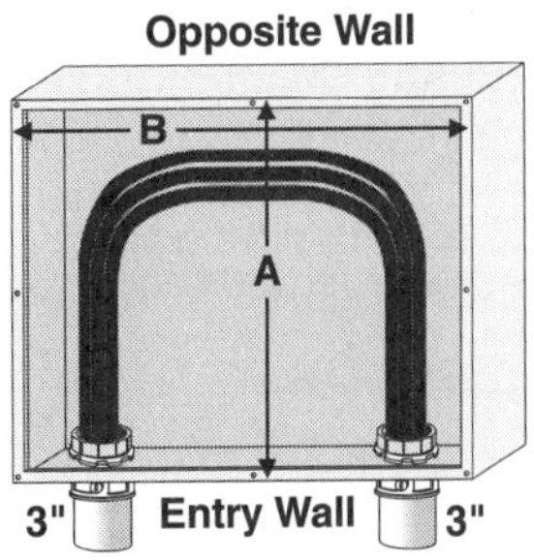

U Pull Sizing:
- The distance for a U Pull is from where a raceway enters a box to the opposite wall.
- A = 6 x the largest raceway plus the sum of the other raceways on the same wall.

Example: Determine the minimum size U-pull box.

Dimension A: *370-28(a)(2)*
6 times largest the largest raceway plus the sum of the other raceways entering the same wall.
"A" = (6 x 3") + 3" = **21 Inches**

Dimension B must be large enough to accomadate the distance between raceways containing the same conductor (6 x 3" = 18") and the size of the connectors (about 6") with room to tighten the locknuts (about 1").

Figure 5-28
Sizing Junction/Pull Boxes For U-Pulls

Angle Pull Calculation [370–28(a)(2)]

An angle pull calculation applies when conductors enter one wall and leave the enclosure not opposite the wall of the conductor entry. The distance for angle pull calculations from where the raceway enters to the opposite wall must not be less than six times the trade diameter of the largest raceway, plus the sum of the diameters of the remaining raceways on the same wall *and row*, Fig. 5–27. When there is more than one *row*, each row shall be calculated separately and the row with the largest calculation shall be considered the minimum angle pull dimension.

U–Pull Calculations [370–28(a)(2)]

A U–pull calculation applies when the conductors enter and leave from the same wall. The distance from where the raceways enter to the opposite wall must not be less than six times the trade diameter of the largest raceway, plus the sum of the diameters of the remaining raceways on the same wall, Fig. 5–28.

Distance Between Raceways Containing The Same Conductor Calculation [370–28(a)(2)]

After sizing the pull box, the raceways must be installed so that the distance between raceways enclosing the same conductors shall not be less than six times the trade diameter of the largest raceway. This distance is measured from the nearest edge of one raceway to the nearest edge of the other raceway. Figure 5–29.

5–9 DEPTH OF BOX AND CONDUIT BODY SIZING [370–28(a)(2), Exception]

When conductors enter an enclosure opposite a *removable cover*, such as the back of a pull box or conduit body, the distance from where the conductors enter to the removable cover shall not be less than the distances listed on Table 373–6(a); one wire per terminal, Fig. 5–30.

Distance Between Raceways Containing The Same Conductor
Section 370-28(a)(2)

The distance between raceways containing the same conductor shall not be less than 6 times the diameter of the larger raceway.

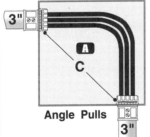

Angle Pulls

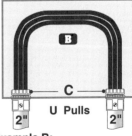

U Pulls

Example A:
C = 6 x 3 inches = **18 inches**

Example B:
C = 6 x 2 inches = **12 inches**

Figure 5-29
Distance Between Raceways Containing The Same Conductor

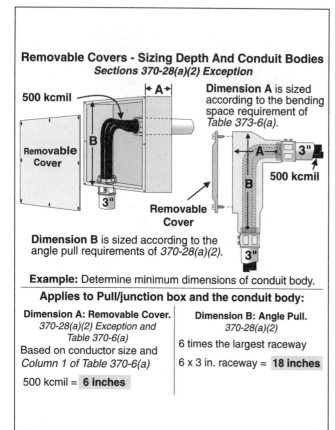

TABLE 373-6(a) - Depth of Box or Conduit Body Distance measured from where conductors enter to removable cover	
Largest Conductor Required To Be Bent	Minimum Depth
No. 4 through No. 2/0	Less than 4 inches
No. 3/0 and No. 4/0	4 inches
250 kcmil	4½ inches
300 - 350 kcmil	5 inches
400 - 500 kcmil	6 inches
600 - 900 kcmil	8 inches
1,000 - 1,250 kcmil	10 inches
1,500 - 2,000 kcmil	12 inches

Figure. 5-30

Removable Covers – Sizing Depth And Conduit Bodies

□ **Depth Of Pull Or Junction Box Example**

A 24 inch × 24 inch pull box has two 2-inch conduits that enter the back of the box with 4/0 conductors. What is the minimum depth of the box?

(a) 4 inches (b) 6 inches (c) 8 inches (d) 10 inches

• Answer: (a) 4 inches, Table 373-6(a)

5–10 JUNCTION AND PULL BOX SIZING TIPS

When sizing pull and junction boxes, the following suggestions should be helpful.

Step 1: → Always draw out the problem.

Step 2: → Calculate the HORIZONTAL distance(s):
- Left to right straight calculation
- Left to right angle or U-pull calculation
- Right to left straight calculation
- Right to left angle or U-pull calculation

Step 3: → Calculate the VERTICAL distance(s):
- Top to bottom straight calculation
- Top to bottom angle or U-pull calculation
- Bottom to top straight calculation
- Bottom to top angle or U-pull calculation

5-11 PULL BOX EXAMPLES

☐ Pull Box Sizing Example No. 1

A junction box contains two 3-inch raceways on the left side and one 3-inch raceway on the right side. The conductors from one of the 3-inch raceways on the left wall are pulled through the 3-inch raceway on the right wall. The other 3-inch raceway's conductors are pulled through a raceway at the bottom of the pull box, Fig. 5–31.

→ **Horizontal Dimension Question A**

What is the horizontal dimension of this box?

(a) 18 inches (b) 21 inches (c) 24 inches (d) none of these

- Answer: (c) 24 inches, Section 370-28, Figure 31, Part A

  Left wall to the right wall angle pull = (6 × 3 inch) + 3 inch = 21 inches

  Left wall to the right wall straight pull = 8 × 3 inch = 24 inches

  Right wall to left wall angle pull = No calculation

  Right wall to the left wall straight pull = 8 × 3 inch = 24 inches

→ **Vertical Dimension Question B**

What is the vertical dimension of this box?

(a) 18 inches (b) 21 inches (c) 24 inches (d) none of these

- Answer: (a) 18 inches, Section 370-28, Figure 31, Part B

  Top to bottom wall angle pull = No calculation

  Top to bottom wall straight pull = No calculation

  Bottom to top wall angle pull = (6 × 3) = 18 inches

  Bottom to top wall straight pull = No calculation

→ **Distance Between Raceways Question C**

What is the minimum distance between the two 3-inch raceways that contain the same conductors?

(a) 18 inches (b) 21 inches

(c) 24 inches (d) None of these

- Answer: (a) 18 inches, (6 × 3 inches) = 18 inches, Section 370-28, Fig. 5–31, Part C

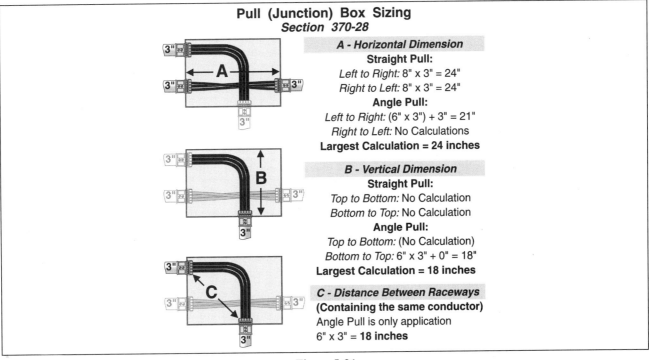

Figure 5-31

Pull (Junction) Box Sizing Example

❑ Pull Box Sizing Example No. 2

A pull box contains two 4-inch raceways on the left side and two 2-inch raceways on the top.

➤ Horizontal Dimension Question A

What is the horizontal dimension of the box?

(a) 28 inches   (b) 21 inches   (c) 24 inches   (d) none of these

- Answer: (a) 28 inches, Section 370-28(a)(2)

  Left wall to the right wall angle pull = (6 × 4 inches) + 4 inches = 28 inches
  Left wall to the right wall straight pull = No calculation
  Right wall to left wall angle pull = No calculation
  Right wall to the left wall straight pull = No calculation

➤ Vertical Dimension Question B

What is the vertical dimension of the box?

(a) 18 inches   (b) 21 inches   (c) 24 inches   (d) 14 inches

- Answer: (d) 14 inches, Section 370-28(a)(2)

  Top to bottom wall angle pull = (6 × 2 inches) + 2 inches = 14 inches
  Top to bottom wall - straight pull = No calculation
  Bottom to top wall angle pull = No calculation
  Bottom to top wall straight pull = No calculation

➤ Distance Between Raceways Question C

What is the minimum distance between the two 4 inch raceways that contain the same conductors?

(a) 18 inches

(b) 21 inches

(c) 24 inches

(d) None of these

- Answer: (c) 24 inches, (6 × 4 inches) = 24 inches, Section 370-28(a)(2)

Unit 5 – Raceway Fill, Box Fill, Junction Boxes, And Conduit Bodies Summary Questions

Part A – Raceway Fill

5–1 Understanding Chapter 9 Tables

1. When all the conductors are the same size and insulation, the number of conductors permitted in a raceway can be determined simply by looking at the Tables listed in _____.
   (a) Chapter 9 (b) Appendix B (c) Appendix C (d) none of these

2. When equipment grounding conductors are installed in a raceway, the actual area of the conductor must be used when calculating raceway fill.
   (a) True (b) False

3. When a raceway does not exceed 24 inches, the raceway is permitted to be filled to _____ percent of its cross–section area.
   (a) 53 (b) 31 (c) 40 (d) 60

4. How many No. 16 TFFN conductors can be installed in a ¾-inch electrical metallic tubing?
   (a) 40 (b) 26 (c) 30 (d) 29

5. How many No. 6 RHH (without outer cover) can be installed in a 1¼-inch nonmetallic tubing?
   (a) 25 (b) 16 (c) 13 (d) 7

6. How many No. 1/0 XHHW can be installed in a 2-inch flexible metal conduit?
   (a) 7 (b) 6 (c) 16 (d) 13

7. How many No. 12 RHH (with outer cover) can be installed in a 1-inch raceway?
   (a) 7 (b) 11 (c) 5 (d) 4

8. If we have a 2½-inch rigid metal conduit and we want to install three THHN compact conductors, what is the largest conductor permitted to be installed?
   (a) No. 4/0 (b) 250 kcmil (c) 300 kcmil (d) 350 kcmil

9. The actual area of conductor fill is dependent on the raceway size and the number of conductors installed. If there are three or more conductor installed in a raceway the total area of conductor fill is limited to _____ percent.
   (a) 53 (b) 31 (c) 40 (d) 60

10. • What is the area in square inches for a No. 10 THW?
    (a) .0333 (b) .0172 (c) .0252 (d) .0278

11. What is the area in square inches for a No. 14 RHW (without cover)?
    (a) .0209 (b) .0172 (c) .0252 (d) .0278

12. What is the area in square inches for a No. 10 THHN?
    (a) .0117 (b) .0172 (c) .0252 (d) .0211

13. What is the area in square inches for a No. 12 RHH (with outer cover)?
    (a) .0117 (b) .0353 (c) .0252 (d) .0327

14. • What is the area in square inches of a No. 8 bare solid?
    (a) .013 (b) .027 (c) .038 (d) .045

162 Unit 5 Raceway, Outlet Boxes, And Junction Boxes   Chapter 2 NEC Calculations And Code Questions

5–2 Raceway And Nipple Calculations

15. The number of conductors permitted in a raceway is dependent on _____.
    (a) the area (square-inch) of the raceway
    (b) the percent area fill as listed in Chapter 9 Table 1
    (c) the area (square-inch) of the conductors as listed in Chapter 9 Table 5 and 8
    (d) all of these

16. A 200-ampere feeder installed in rigid nonmetallic conduit schedule 80, has three – No. 3/0 THHN, one – No. 2 THHN, and one – No. 6 THHN. What size raceway is required?
    (a) 2-inch   (b) 2½-inch   (c) 3-inch   (d) 3½-inch

17. What size rigid metal nipple is required for three– No. 4/0 THHN, one– No. 1/0 THHN and No. 4 THHN?
    (a) 1½-inch   (b) 2-inch   (c) 2½-inch   (d) none of these

5–3 Existing Raceway

18. • An existing rigid metal nipple contains 4 – No. 10 THHN and 1 – No. 10 (bare stranded) ground wire in a ¾-inch raceway. How many additional No. 10 THHN can be installed?
    (a) 5   (b) 7   (c) 9   (d) 11

Part B – Outlet Box Fill Calculations

5–4 Conductors All The Same Size [Table 370–16]

19. What size box is required for six No. 14 THHN and three No. 14 THW?
    (a) 4 × 1¼ inch square   (b) 4 × 1½ inch round
    (c) 4 × 1⅛ inch round   (d) None of these

20. How many No. 10 THHN are permitted in a 4 × 1 1/2 square box?
    (a) 8 conductors   (b) 9 conductors
    (c) 10 conductors   (d) 11 conductors

5–5 Conductor Equivalents [370–16]

21. Table 370–16 does not take into consideration the volume of _____.
    (a) switches and receptacles
    (b) fixture studs and hickeys
    (c) manufactured cable clamps
    (d) all of these

22. When determining the number of conductors for box fill calculations, which of the following statements are true?
    (a) A fixture stud or hickey is considered as one conductor for each type, based on the largest conductor that enters the outlet box.
    (b) Internal factory cable clamps are considered as one conductor for one or more cable clamps, based on the largest conductor that enters outlet box.
    (c) The device yoke is considered as two conductors, based on the largest conductor that terminates on the strap (device mounting fitting).
    (d) All of these.

23. • When determining the number of conductors for box fill calculations, which of the following statements are true?
    (a) Each conductor that runs through the box without loop (without splice) is considered as one conductor.
    (b) Each conductor that originates outside the box and terminates in the box is considered as one conductor.
    (c) Wirenuts, cable connectors, raceway fittings, and conductors that originate and terminate within the outlet box (equipment bonding jumpers and pigtails) are not counted for box fill calculations.
    (d) All of these.

24. It is permitted to omit one equipment grounding conductor and not more than _____ that enter a box from a fixture canopy.
    (a) four fixture wires (b) four No. 16 fixture wires
    (c) four No. 18 fixture wires (d) b and c

25. Can a round 4 inch × ½ inch (pancake) box marked as 8 cubic inches with manufactured cable clamps supplied with 14/2 romex be used with a fixture that has two No. 18 TFN and a canopy cover?
    (a) Yes (b) No

5–7 Different Size Conductors

26. What size outlet box is required for, 12/2 romex that terminates on a switch, 12/3 romex that terminates on a receptacle, and the box has manufactured cable clamps.
    (a) 4 × 1¼ inch square (b) 4 × 1½ inch square
    (c) 4 × 2⅛ inch square (d) any of these

27. • How many No. 14 THHN conductors can be pulled through a 4 inch square × 1½ inch deep box with a plaster ring of 3.6 cubic inches. The box contains a two duplex receptacles, five No. 14 THHN and two grounding conductor.
    (a) One (b) Two (c) Three (d) Four

Part C – Pull, Junction Boxes, And Conduit Bodies

5–8 Pull And Junction Box Size Calculations

28. When conductors No. 4 and larger are installed in boxes and conduit bodies, we must size the enclosure according to which of the following requirements?
    (a) The minimum distance for straight pull calculations from where the conductors enter to the opposite wall must not be less than eight times the trade size of the largest raceway.
    (b) The distance for angle pull calculations from the raceway entry to the opposite wall must not be less than six times the trade diameter of the largest raceway, plus the sum of the diameters of the remaining raceways on the same wall and row.
    (c) The distance for U–pull calculations from where the raceways enter to the opposite wall must not be less than six times the trade diameter of the largest raceway, plus the sum of the diameters of the remaining raceways on the same wall and row.
    (d) The distance between raceways enclosing the same conductors shall not be less than six times the trade diameter of the largest raceway.
    (e) All of the above are correct.

29. When conductors enter an enclosure opposite a removable cover, the distance from where the conductors enter to the removable cover shall not be less than _____.
    (a) six times the largest raceway
    (b) eight times the largest raceway
    (c) a or b
    (d) none of these

The following information applies to Questions 30 and 31.
A junction box contains two 2½-inch raceways on the left size and one 2½-inch raceway on the right side. The conductors from one 2 1/2-inch raceway (left wall) are pulled through the raceway on the right wall. The other 2½-inch raceways conductors are pulled through a raceway at the bottom of the pull box.

30. What is the distance from the left wall to the right wall?
    (a) 18 inches (b) 21 inches (c) 24 inches (d) 20 inches

31. • What is the distance from the bottom wall to the top wall?
    (a) 18 inches (b) 21 inches (c) 24 inches (d) 15 inches

32. What is the distance from between the raceways that contain the same conductors?
    (a) 18 inches (b) 21 inches (c) 24 inches (d) 15 inches

The following information applies to Questions 33, 34, and 35.
A junction box contains two 2-inch raceways on the left side and two 2-inch raceway on the top.

33. What is the distance from the left wall to the right wall?
    (a) 28 inches (b) 21 inches (c) 24 inches (d) 14 inches

34. What is the distance from the bottom wall to the top wall?
    (a) 18 inches (b) 21 inches (c) 24 inches (d) 14 inches

35. What is the distance from between the 2-inch raceways that contain the same conductors?
    (a) 18 inches (b) 21 inches (c) 24 inches (d) 12 inches

Challenge Questions

Part A – Raceway Calculations

36. • A 3-inch schedule 40 PVC raceway contains seven No. 1 RHW without cover. How many number No. 2 THW conductors may be installed in this raceway with the existing conductors?
    (a) 11 (b) 15 (c) 20 (d) 25

Part B – Box Fill Calculations

37. • Determine the minimum cubic inches required for two No. 10 TW passing through a box, four No. 14 THHN terminating, two No. 12 TW terminating to a receptacle, and one No 12 equipment bonding jumper from the receptacle to the box.
    (a) 18.5 cubic inches (b) 22 cubic inches (c) 20 cubic inches (d) 21.75 cubic inches

38. • When determining the number of conductors in a box fill two No. 18 fixture wires, one 14/3 nonmetallic sheathed cable with ground, one duplex switch, and two cable clamps would count as _____.
    (a) 9 conductors (b) 8 conductors (c) 10 conductors (d) 6 conductors

Part C – Pull And Junction Box Calculations

The Figure 5–32 applies to Questions 39, 40, 41, and 42.

39. The minimum horizontal dimension for the junction box shown in the diagram is _____ inches, Fig. 5–32?
    (a) 21 (b) 18
    (c) 24 (d) 20

40. • The minimum vertical dimension for the junction box is _____ inches, Fig. 5–32?
    (a) 16 (b) 18
    (c) 20 (d) 24½

41. The minimum distance between the two 2-inch raceways that contains the same conductor (C) would be _____ inches, Fig. 5–32?
    (a) 12 (b) 18
    (c) 24 (d) 30

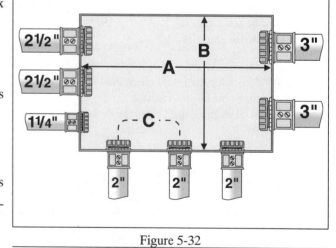

Figure 5-32

42. • If a 3-inch raceway entry (250-kcmils) is in the wall opposite to a removable cover, the distance from that wall to the cover must not be less than _____ inches, Fig. 5–32?
    (a) 4 (b) 4½
    (c) 5 (d) 5½

NEC Questions From Section 430–31 Through Section 800–53

Article 430 – Motors, Motor Circuits, And Controllers

Part C. Motor and Branch-Circuit Overload Protection

43. Motor overload protection is not required where _____.
    (a) conductors are oversized by 125 percent
    (b) conductors are part of a limited-energy circuit
    (c) it might introduce additional hazards
    (d) short-circuit protection is provided

44. Overload devices are intended to protect motors, motor control apparatus, and motor branch circuit conductors against _____.
    (a) excessive heating due to motor overloads
    (b) excessive heating due to failure to start
    (c) short circuits and ground faults
    (d) a and b only

45. Each continuous duty motor _____ horsepower or less, which is not permanently installed, not automatically started, and is within sight of the controller, shall be permitted to be protected against overload by the branch circuit protective device.
    (a) ⅛   (b) ½   (c) ¾   (d) 1

46. Where an overload relay selected by Section 430–32(a)(1) and (c)(1) is not sufficient to start the motor or carry the load, the next higher size overload shall be permitted providing the trip current does not exceed _____ percent of motor full-load current rating if the motor is marked with a service factor of 1.12.
    (a) 100   (b) 110   (c) 120   (d) 130

47. • The minimum number of overload unit(s) required for a three-phase alternating current motor shall be _____.
    (a) one   (b) two   (c) three   (d) any of these

48. • If _____ shutdown is necessary to reduce hazards to persons, the overload sensing devices shall be permitted to be connected to a supervised alarm instead of causing the motors to shut down.
    (a) emergency   (b) normal
    (c) orderly   (d) none of these

Part E. Motor Feeder Short-Circuit And Ground-Fault Protection

49. • Feeder conductors that supply motor and other loads must have overcurrent protection against short-circuit, ground-fault, and overload.
    (a) True   (b) False

Part F. Motor Control Circuits

50. Motor control circuit conductors that extend beyond the motor control equipment enclosure shall be required to have short-circuit and ground-fault protection sized not greater than _____ percent of the conductor ampacity as listed in Table 310–16 for 60ºC conductors.
    (a) 100   (b) 150   (c) 300   (d) 500

51. Overcurrent protection for motor control circuits shall not exceed 400 percent if the conductor does not extend beyond the motor control equipment enclosure.
    (a) True   (b) False

52. If the control circuit transformer is located in the controller enclosure, the transformer must be connected to the _____ side of the control circuit disconnect.
    (a) line   (b) load
    (c) adjacent   (d) none of these

Part G. Motor Controllers

53. The branch-circuit protective device shall be permitted to serve as the controller for a stationary motor rated at _____ horsepower or less that is normally left running and cannot be damaged by overload or failure to start.
    (a) 1 (b) ½ (c) ⅓ (d) ⅛

Part I. Disconnecting Means

54. • A disconnecting means is required to disconnect the _____ from all ungrounded supply conductors.
    (a) motor
    (b) motor or controller
    (c) controller
    (d) motor and controller

55. The disconnect for the controller and motor must open all conductors for a three-phase motor.
    (a) True (b) False

56. The motor disconnecting means shall be a _____.
    (a) circuit breaker
    (b) motor-circuit switch rated in horsepower
    (c) molded case switch
    (d) any of these

57. The branch circuit overcurrent device, such as a plug fuse may serve as the disconnecting means for a stationary motor of ⅛ horsepower or less.
    (a) True (b) False

58. The disconnecting means for a 50-HP three phase, 460-volt induction motor (rated 65 amperes) shall have an ampere rating of not less than _____ amperes.
    (a) 126 (b) 75 (c) 91 (d) 63

59. A switch or inverse time circuit breaker that is _____ shall be permitted to serve as both controller and disconnecting means if it opens all ungrounded conductors to the motor.
    (a) rated at 150 percent of the maximum load of the motor
    (b) operable directly by applying the hand to a lever or handle
    (c) less than 100 amperes
    (d) 30 amperes or less

60. An oil switch used for both a controller and disconnect is permitted on a circuit whose rating _____ volts or 100 amperes.
    (a) is 600
    (b) exceeds 600
    (c) does not exceed 600
    (d) none of these

Article 440 – Air-Conditioning And Refrigerating Equipment

Part B. Disconnecting Means

61. A disconnecting means serving a hermetic refrigerant motor compressor selected on the basis of the nameplate rated load current or branch circuit selection current, whichever is greater, shall have an ampere rating of _____ percent of the nameplate rated load current or branch circuit selection current.
    (a) 125 (b) 80
    (c) 100 (d) 115

62. Which of the following statement(s) about a disconnect for an air-conditioning unit is true?
    (a) The disconnecting means is required to be within sight.
    (b) The disconnecting means is permitted to be mounted directly on the unit.
    (c) The disconnecting means can be inside the unit.
    (d) All of these.

Part E. Controllers For Motor-Compressors

63. A hermetic motor-compressor controller shall have a _____ current rating not less than the nameplate rated-load current or branch-circuit selection current, whichever is greater.
    (a) continuous-duty full-load   (b) locked-rotor   (c) either a or b   (d) both a and b

Part F. Motor-Compressor And Branch-Circuit Overload Protection

64. The attachment plug and receptacle shall not exceed _____ amperes at 250 volts for a cord- and-plug-connected air-conditioning motor-compressor.
    (a) 15   (b) 20   (c) 30   (d) 40

Part G. Provisions for Room Air Conditioners

65. When supplying a nominal 120 volt rated room air conditioner, the length of flexible supply cord shall not exceed _____ feet.
    (a) 4   (b) 6   (c) 8   (d) 10

Article 450 – Transformers And Transformer Vault

For most electricians, transformers are perhaps the most intimidating of all subjects. As with all equipment, it must have overcurrent protection and it must be grounded.

Part A. General Provisions

66. If the primary overcurrent protection device is sized at 250 percent of the primary current, what size secondary overcurrent protection device is required (secondary current of 42 amperes)?
    (a) 40 amperes   (b) 70 amperes   (c) 60 amperes   (d) 90 amperes

Part B. Specific Provisions Applicable To Different Types Of Transformers

67. Transformers of more than _____ kVA rating shall be installed in a transformer room of fire-resistant construction.
    (a) 35,000   (b) 87½   (c) 112½   (d) 75

68. Dry-type transformers installed indoors rated over _____ shall be installed in a vault.
    (a) 1000 volts   (b) 112½ kVA   (c) 50,000 volts   (d) 35,000 volts

Part C. Transformer Vaults

69. The walls and roofs of transformer vaults shall have a minimum fire resistance of _____ hour(s).
    (a) 1   (b) 2   (c) 3   (d) 4

70. Where practicable, transformer vaults containing more than _____ kVA transformer capacity shall be provided with a drain or other means that will carry off any accumulation of oil or water in the vault unless local conditions make this impracticable.
    (a) 100   (b) 150   (c) 200   (d) 250

Article 455 – Phase Converters

Part A. General

71. Means shall be provided to disconnect simultaneously all ungrounded single-phase _____ conductors to the phase converter.
    (a) supply   (b) load   (c) service   (d) neutral

72. The phase converter disconnecting means provided for in Section 455–8 shall be _____ and located in sight from the phase converter.
    (a) protected from physical damage   (b) readily accessible
    (c) easily visible   (d) clearly identified

Article 460 – Capacitors

Part A. 600 Volts, Nominal, And Under

73. The residual voltage of a capacitor shall be reduced to 50 volts or less within _____ after the capacitor is disconnected from the source of supply.
    (a) 15 seconds   (b) 45 seconds   (c) 1 minute   (d) 2 minutes

74. The ampacity of capacitor circuit conductors shall not be less than _____ percent of the rated current of the capacitor.
    (a) 100   (b) 115   (c) 125   (d) 135

Part B. Over 600 Volts, Nominal

75. A capacitor operating at over 600 volts nominal shall be provided with means to reduce the residual voltage to 50 volts or less within _____ after disconnecting it from the source of supply.
    (a) 15 seconds   (b) 45 seconds   (c) 1 minute   (d) 5 minutes

Article 480 – Storage Batteries

76. The nominal battery voltage of a cell is _____ volts for the lead acid type and _____ volts for the alkali type.
    (a) 6, 12   (b) 12, 24   (c) 2, 1.2   (d) 1.2, 2

77. A vented alkaline-type battery operating at less than 250 volts shall be installed with not more than _____ cells in the series circuit in any one tray.
    (a) 10   (b) 12   (c) 18   (d) 20

78. Provisions shall be made for sufficient diffusion and ventilation of the gases from the storage battery to prevent the accumulation of a(n) _____ mixture.
    (a) flammable   (b) explosive   (c) toxic   (d) all of these

79. Each vented cell of a battery shall be equipped with _____.
    (a) pressure relief   (b) a flame arrester   (c) fluid level indicators   (d) none of these

Chapter 5 Special Occupancies

NEC Chapter 5 is a specific locations chapter. Where Chapters 1, 2, and 3 apply throughout the Code and to residential and commercial occupancies in general, Chapter 5 pertains to locations that have special conditions to consider.

Articles 500 through 517 have conditions that can be considered as *hazardous* in nature. Articles 500 through 504 cover the requirements for electrical equipment and wiring for all voltages in locations where fire or explosion hazards may exist due to flammable gasses or vapors, flammable liquids, combustible dust, or ignitable fibers or flyings. Articles 511 through 517 are specific types of hazardous occupancies, such as gas stations and anesthesia areas in health care facilities.

Article 500 – Hazardous (Classified) Locations breaks down and categorizes hazardous area requirements and basically describes and defines those requirements.

Article 500 – Hazardous (Classified) Locations

80. All conduit referred to in hazardous locations shall be threaded with a _____ -inch taper per foot.
    (a) ½   (b) ¾   (c) 1   (d) all of these

81. Through the exercise of ingenuity in the layout of the electrical installation for hazardous (classified) locations, it is frequently possible to locate much of the equipment in less hazardous or in nonhazardous locations and thus to reduce the amount of special equipment required.
    (a) True   (b) False

82. Electrical equipment installed in hazardous (classified) locations must be constructed for the class, division, and group. A group "C" atmosphere contains _____.
    (a) hydrogen   (b) ethylene   (c) propylene oxide   (d) all of these

83. For Class ____ locations, Groups A, B, C, and D, the classification involves determinations of maximum explosion pressure, maximum safe clearance between parts of a clamped joint in an enclosure, and the minimum ignition temperature of the atmospheric mixture.
    (a) I (b) II (c) III (d) all of these

84. A Group "G" atmosphere contains combustible dusts such as flour, grain, wood, plastic, and other chemicals.
    (a) True (b) False

85. Where specifically permitted in Articles 501 through 503, general-purpose equipment or equipment in general-purpose enclosures shall be permitted to be installed in a _____ location if the equipment does not constitute a source of ignition under normal operating conditions.
    (a) Class I (b) Class II (c) Division 1 (d) Division 2

86. Class I equipment installed in hazardous (classified) locations shall not have any exposed surface that operates at a temperature in excess of _____.
    (a) that which the equipment is specifically designed for
    (b) the ignition temperature of the specific gas or vapor to be encountered
    (c) 300°C or 572°F
    (d) all of these

87. Class I, Division 1 locations are those that are hazardous because of the presence of _____.
    (a) combustible dust
    (b) easily ignitable fibers or flyings
    (c) flammable gases or vapors
    (d) flammable liquids or gases

88. When determining a Class I, Division 2 location, _____ are all factors that merit consideration in determining the classification and extent of each location.
    (a) the quantity of flammable material that might escape in case of accident,
    (b) the adequacy of ventilating equipment,
    (c) the record of the industry or business with respect to explosions or fires
    (d) all of these

89. Locations where combustible dust is normally in heavy concentrations are designated as _____ locations.
    (a) Class I, Division 2
    (b) Class II, Division 1
    (c) Class II, Division 2
    (d) Class III, Division 1

90. Class III, Division _____ locations include areas where easily ignitable fibers or combustible flying material is handled, manufactured or used.
    (a) 1 (b) 2 (c) 3 (d) all of these

91. Locations in which combustible fibers are stored are designated as _____.
    (a) Class II, Division 2 (b) Class III, Division 1
    (c) Class III, Division 2 (d) non-hazardous

Article 501 – Class I Locations

92. Wiring methods permitted in Class I, Division 1 locations include _____.
    (a) threaded rigid metal or threaded steel intermediate metal conduit
    (b) explosion proof threaded boxes, fittings, and joints
    (c) liquidtight flexible metal conduit with approved fittings
    (d) a and b

93. Boxes, enclosures, fittings, and joints are not required to be explosion-proof in a Class I, Division 2 location. However, if arcs or sparks (such as from make-or-break contacts) can be made from equipment being utilized, it must be an explosion-proof enclosure meeting the requirements for Class I, Division 1 locations.
    (a) True (b) False

94. • Sealing compound is employed with mineral-insulated cable in a Class I location for the purpose of _____.
    (a) preventing passage of gas or vapor
    (b) excluding moisture
    (c) limiting a possible explosion
    (d) preventing escape of powder

95. For Class I Division 1 locations, a sealoff fitting is required in each conduit of _____-inch(es) in size or larger entering the enclosure or fitting housing terminals, splices, or taps.
    (a) 1   (b) 1.25   (c) 1.5   (d) 2

96. In Class I, Division 1 locations, the Code requires conduit seals adjacent to boxes containing splices if the conduit is equal to or larger than _____-inch(es).
    (a) ¾   (b) 1½
    (c) 1   (d) 2

97. Where seals are required for Class I, Divisions 1 and 2 locations, they shall comply with the following:
    (a) They shall be approved for Class I locations and shall be accessible.
    (b) The sealing compound shall be approved and must not be less than the trade of the conduit, but in no case less than ⅝ inch.
    (c) Splices and taps shall not be made in seal-off fittings.
    (d) All of these.

98. Enclosures for isolating switches in Class I, Division 2 locations _____.
    (a) may be in a general purpose enclosure
    (b) must be in a explosion-proof type enclosure
    (c) must have interlocking devices
    (d) may not have doors in a hazardous area

99. Where flexible metal conduit or liquidtight flexible metal conduit is used as permitted in Class I, Division 2 locations, it shall be installed with an _____ bonding jumper in parallel with each conduit and complying with Section 250–79(f).
    (a) internal   (b) external   (c) either a or b   (d) both a and b

Article 502 – Class II Hazardous Locations

100. In a Class II, Division 1 location where magnesium, aluminum, aluminum bronze, or other powders of hazardous characteristics may be present, fuses, switches, motor controllers, and circuit breakers shall have enclosures specifically approved for such locations.
    (a) True   (b) False

Article 503 – Class III Hazardous Locations

101. A fixture in a Class III location that may be exposed to physical damage shall be protected by a _____ guard.
    (a) substantial   (b) metal   (c) suitable   (d) bronze

Article 511 – Commercial Garages And Storage

102. Parking garages used for parking or storage and where no repair work is done except for exchange of parts and routine maintenance requiring no use of electrical equipment, open flame, welding, or the use of volatile flammable liquids are not classified as a hazardous location.
    (a) True   (b) False

103. Any pit or depression below a garage floor level shall be considered to be a Class I, Division _____ location which shall extend up to said floor level.
    (a) 1   (b) 2   (c) 3   (d) not classified

104. • In a commercial garage, the pit shall be classified as a _____ location, unless provisions are made for six air changes per hour.
    (a) Class I, Division 2   (b) Class II, Division 2
    (c) Class II, Division 1   (d) Class I, Division 1

105. Raceways embedded in a masonry wall or buried beneath a floor shall be considered _____.
    (a) outside the hazardous area if any extensions pass through such areas
    (b) within the hazardous area if any connections lead into such areas
    (c) hazardous if beneath a hazardous area
    (d) outside any hazardous location

106. Equipment that is less than _____ feet above the floor level and that may produce arcs, sparks, or particles of hot metal, such as cutouts, switches, charging panels, generators, motors, or other equipment (excluding receptacles, lamps, and lampholders) having make-and-break contacts, shall be of the totally enclosed-type or so constructed as to prevent escape of sparks or hot metal particles.
    (a) 6　　　　(b) 10　　　　(c) 12　　　　(d) 18

Article 513 – Aircraft Hangars

107. Which of the following areas of an aircraft hanger are not classified as Class I, Division 1 or 2?
    (a) any pit or depression below the level of the hanger floor
    (b) arcas adjacent to and not suitably cut off from the hangar
    (c) areas within the vicinity of aircraft
    (d) adjacent areas where adequately ventilated and where effectively cut off from the hangar

108. Which of the following areas of an aircraft hanger is *not* classified as a Class I location?
    (a) below floor level
    (b) areas not cut off or ventilated
    (c) the vicinity of aircraft
    (d) areas suitably cut off and ventilated

Article 514 – Gasoline Dispensing Stations

109. Each circuit leading to or through a dispensing pump shall be provided with a switch or other acceptable means to disconnect simultaneously from the source of supply all conductors of the circuit, including the _____ conductor, if any.
    (a) grounding　　　　(b) grounded　　　　(c) bonding　　　　(d) all of these

110. In a gas station, an approved seal shall be _____.
    (a) provided in each conduit run entering a dispenser
    (b) provided in each conduit run leaving a dispenser
    (c) the first fitting after the conduit emerges from the ground or concrete
    (d) all of these

111. • Underground gasoline dispenser wiring shall be installed in _____.
    (a) threaded rigid metal conduit
    (b) threaded intermediate metal conduit
    (c) rigid nonmetallic conduit when buried not less than two feet of earth
    (d) all of these

Article 516 – Spray Application, Dipping, And Coating Processes

112. • Locations where flammable paints are dried but in which the ventilating equipment is interlocked with the electrical equipment may be designated as a _____ location.
    (a) Class I, Division 2　　　　(b) Unclassified　　　　(c) Class II, Division 2　　(d) Class II, Division 1

Article 517 – Health Care Facilities

Article 517 applies to all types of health care facilities. This article targets areas that have special requirements. For example, hospitals, doctors' offices, business and administrative offices, kitchens, and cafeterias would be wired according to the requirements of Chapters 1 through 4. Specific areas such as examining rooms, operating rooms, X-ray areas, patient care areas, etc. have special requirements listed in Article 517. As with many articles in the NEC, there are several definitions listed (Section 517–2) which apply to this article only.

Part A. General

113. The patient care area is any portion of a health care facility such as business offices, corridors, lounges, day rooms, dining rooms, or similar areas.
    (a) True   (b) False

Part B. Wiring And Protection

114. • Each general care area patient bed location shall be provided with a minimum number of receptacles such as _____.
    (a) one single or one duplex
    (b) six single or three duplex
    (c) two single or one duplex
    (d) four single or two duplex

115. In Health Care Centers, in the Critical Care Area, each patient bed location shall be provided with a minimum of _____ duplex receptacles.
    (a) 10   (b) 3   (c) 1   (d) 4

Part C. Essential Electrical System

116. • The essential electrical systems in a health care facility shall have sources of power from:
    (a) a normal source generally supplying the entire electrical system.
    (b) one or more alternate sources for use when the normal source is interrupted.
    (c) either a or b
    (d) both a and b

Part D. Inhalation Anesthetizing Locations

117. In a health care facility, receptacles and attachment plugs in a hazardous (classified) location within an anesthetizing area shall be listed for use in Class I, Group _____ locations.
    (a) A   (b) B   (c) C   (d) D

Part G. Isolated Power Systems

118. The maximum internal current permitted to flow through the line isolation monitor, when any point of the isolated system is grounded, shall be _____ when used in a health care facility.
    (a) 15 amperes or less
    (b) no more than 1 ampere
    (c) 1 milliampere
    (d) 10 milliamperes

Article 518 – Places of Assembly

Article 518 applies to buildings or portions of buildings designed or intended for the assembly of 100 or more persons. Section 518–2 lists (but does not limit) several types of occupancies where this article applies. Several of the Articles following 518 are for specific places of assembly.

119. • Electrical nonmetallic tubing can be used in a place of assembly.
    (a) True   (b) False

Article 520 – Theaters, Audience Areas Of Motion Picture And Television Studios, And Similar Locations

Part B. Fixed Stage Switchboard

120. A night club lighting dimmer installed in an ungrounded conductor shall have overcurrent protection rated at no more than _____ percent of its own rating.
    (a) 50   (b) 70   (c) 80   (d) 125

Part C. Stage Equipment – Fixed

121. On fixed stage equipment, portable or strip light fixtures and connector strips shall be wired with conductors having insulation of _____ ºC or suitable for the conditions.
    (a) 75　　(b) 90　　(c) 125　　(d) 200

Part E. Stage Equipment – Portable

122. Flexible conductors used to supply portable stage equipment shall be _____.
    (a) listed　　(b) extra-hard usage　　(c) hard-usage　　(d) a and b

Article 540 – Motion Picture Projectors

Part B. Definitions

123. A professional-type projector employs _____-millimeter film.
    (a) 35　　(b) 70　　(c) a or b　　(d) none of these

Article 545 – Manufactured Building

Part A. General

124. The plans, specifications, and other building details for construction of manufactured buildings are included in the _____ details.
    (a) manufactured buildings　　(b) building components
    (c) building structure　　(d) building system

Article 550 – Mobile Homes And Mobile Home Parks

Article 550 is broken down into three Parts. Part A – General is mostly definitions that apply to this article. Part B – Mobile Homes applies to the power cord, as well as all wiring methods and materials within and on the mobile home, plus calculations. Part C – Services and Feeders applies to the park in general and the individual services on the lots. Miscellaneous buildings such as recreation and meeting rooms, or common laundry facilities fall under the regular requirements of Chapters 1 through 4.

Part A. General

125. With consideration to mobile homes, which of the following (other than built-in dishwashers) are not considered portable appliances if they are cord connected?
    (a) refrigerators　　(b) gas ranges　　(c) clothes washers　　(d) electric ranges

Part B. Mobile Homes

126. Which of the following may be used as a feeder from the service equipment to a mobile home?
    I. a permanently installed circuit
    II. two-50 ampere power supply cords approved for mobile homes
    (a) I only　　(b) II only　　(c) both I and II　　(d) neither I nor II

127. • A mobile home that is factory-equipped with gas or oil-fired heating and cooking appliances shall be permitted to be supplied with a listed mobile home power-supply cord rated _____ amperes.
    (a) 30　　(b) 35　　(c) 40　　(d) 50

128. • The coupler of a mobile home is included when determining the number of branch circuits for a mobile home.
    (a) True　　(b) False

129. The maximum spacing of receptacle outlets on counter tops in a mobile home is _____ feet.
    (a) 6　　(b) 12　　(c) 3　　(d) none of these

Part C. Services and Feeders

130. • A small mobile home park has six mobile homes. What is the total park electrical wiring system load after applying the demand factors permitted in Article 550?
    (a) 4,640 VA (b) 27,840 VA (c) 96,000 VA (d) none of these

Article 551 – Recreational Vehicles And Parks

Part E. Nominal 120- or 120/240-Volt Systems

131. • The working clearance for a distribution panelboard located in a recreational vehicle shall be no less than _____.
    I. 24 inches wide   II. 30 inches deep   III. 30 inches wide
    (a) I only (b) II only (c) I and II (d) II and III

Part G. Recreational Vehicle Parks

132. • Electrical service and feeders of a recreational vehicle park shall be calculated at a minimum of _____ VA per site.
    (a) 1,200 (b) 2,400 (c) 3,600 (d) 9,600

Article 555 – Marinas And Boatyards

133. Receptacles that provide shore power for boats shall be rated not less than _____.
    (a) 20-ampere duplex receptacle
    (b) 20-ampere single receptacle of the locking and grounding type
    (c) 30-ampere single receptacle ground fault
    (d) none of these

Chapter 6 – Special Equipment

NEC Chapter 6 covers the requirements for common special or specific equipment. Most of the equipment in this chapter is a group or system. The equipment can be single objects, such as signs; but, most of the equipment is part of another structure, such as an elevator or overhead crane.

Chapter 6 modifies the general rules of Chapters 1 through 4 as necessary for the individual equipment. Most of the articles in this chapter are short and very specialized. For exam purposes, these are the easiest articles to find answers for.

Many of the articles in this chapter have other documents listed that are related to the subject of that article. For example, Article 610 – Cranes and Hoists, Section 610–1 Scope, (FPN) lists ANSI B–30, Safety Code for Cranes, Derricks, Hoists, Jacks, and Slings. In Article 620 – Elevators, etc., Section 620–1, (FPN) lists ANSI/ASME A17.1–1987, Safety Code for Elevators and Escalators. It is your responsibility to know if these other documents are necessary for you to do the job properly.

Remember, the NEC is only one volume of the NFPA documents. Appendix A of the NEC has some NFPA documents listed that have extracts in the NEC. Section 90–3, Paragraph 4 states that these extracts are identified by the superscript letter x and identified in Appendix A.

Several of the articles in Chapter 6 contain calculations for some of the special equipment concerning wire size, feeder loads, etc. Always check these articles for special requirements and calculations.

Article 600 – Electric Signs And Outline Lighting

Part A. General

134. • Sign circuits containing electric-discharge lighting transformers exclusively shall not be rated in excess of _____ amperes.
    (a) 20 (b) 30 (c) 40 (d) 50

135. Signs operated by electronic or electromechanical controllers located externally to the sign shall have a disconnecting means located _____.
    (a) within sight of the sign
    (b) within sight of the controller location
    (c) only in the controller
    (d) only externally to the controller

136. • Terminal enclosures in signs made of sheet metal shall have a minimum thickness of No. _____ MSG.
    (a) 20　　(b) 24　　(c) 28　　(d) 34

137. • In the exposed type of show-window signs over 600 volts, electrode connections shall be _____.
    (a) supported to maintain separation of not less than 1½ inches between conductors
    (b) enclosed by receptacles
    (c) suitably guarded by grounded metallic enclosure
    (d) inaccessible to qualified personnel

Article 604 – Manufactured Wiring Systems

138. • Manufactured wiring systems shall be constructed with _____.
    (a) listed AC or MC cable
    (b) No. 10 or 12 AWG copper-insulated conductors
    (c) conductors suitable for nominal 600 volts
    (d) all of these

Article 610 – Cranes And Hoists

Part B. Wiring

139. • A crane has two motors: one rated 106 amperes for 30 minutes and the other rated 72 amperes for 60 minutes. The minimum calculated motor load for the two motors is _____ amperes.
    (a) 150　　(b) 204.5　　(c) 142　　(d) 110

Article 620 – Elevators, Escalator

Part B. Conductors

140. The conductors to the hoistway door interlocks from the hoistway riser shall be suitable for a temperature of not less than _____ ºC and shall be type SF or equivalent.
    (a) 200　　(b) 60
    (c) 90　　(d) 110

Part F. Disconnecting Means And Control

141. Where multiple driving machines are connected to a single elevator, escalator, moving walk, or pumping unit, there shall be one disconnecting means to disconnect the _____.
    I. motor(s)　　　　　　II. control valve operating magnets
    (a) I only　　　　　　(b) II only
    (c) I or II　　　　　　(d) I and II

Part K. Overspeed

142. • An elevator shall have a single means for disconnecting all ungrounded main power supply conductors for each unit.
    (a) This does not include the emergency power service.
    (b) This includes the emergency power service.
    (c) This does not include the emergency power service if the system is automatic.
    (d) No elevators are to operate on emergency power systems.

Article 630 – Electric Weldersns

Part D. Resistance Welders

143. The ampacity for the supply conductors for a resistance welder with a duty cycle of 15 percent and a primary current of 21 amperes is _____ amperes.
    (a) 9.45   (b) 8.19   (c) 6.72   (d) 5.67

Article 640 – Sound-Recording Equipment

144. The maximum fill in wireways or gutters used for sound-recording installations is _____ percent.
    (a) 20   (b) 58   (c) 50   (d) 75

145. In sound recording equipment, terminals for conductors other than power-supply conductors shall be separated from the terminals of the power-supply conductors by a _____.
    (a) minimum of 2 inches
    (b) spacing at least as great as the spacing between power-supply terminals of opposite polarity
    (c) solid nonmetallic barrier
    (d) none of these

Article 645 – Data Processing Equipment

146. • Liquidtight flexible conduit may be permitted to enclose branch circuit conductors for computers or data communication equipment.
    (a) True   (b) False

147. • In data processing rooms, a disconnecting means is required _____.
    (a) to disconnect the air conditioning to the room
    (b) to disconnect the power to electronic computer/data processing equipment
    (c) to disconnect all power and lighting
    (d) for each of A and B and a single disconnecting means is permitted to control both A and B

Article 660 – X-Ray Equipment

Part A. General

148. The ampacity requirements for a disconnecting means of x-ray equipment shall be based on _____ percent of the input required for the momentary rating of the equipment if greater than the long-time rating.
    (a) 125   (b) 100
    (c) 50   (d) none of these

149. • The ampacity of supply branch circuit conductors and the overcurrent protection devices for x-ray equipment shall not be less than _____.
    (a) 50 percent of the momentary rating
    (b) 100 percent of the long-time rating
    (c) the larger of a or b
    (d) the smaller of a or b

Article 665 – Induction/Dielectric Heating

Part B. Guarding, Grounding, and Labeling

150. Interlocks on access doors or panels shall not be required if the applicator is an induction heating coil at dc ground potential or operating at less than _____ volts ac.
    (a) 50   (b) 120
    (c) 150   (d) 240

Article 668 – Electrolytic Cells

151. An assembly of electrically interconnected electrolytic cells supplied by a source of dc power is called a _____.
    (a) cell line attachment   (b) cell line
    (c) electrolytic cell bank   (d) battery storage bank

Article 675 – Electrical Irrigation Machines

Part A. General

152. A center pivot irrigation machine may have hand portable motors.
    (a) True   (b) False

Article 680 – Pools, Spas, And Fountains

Part A. General

153. The scope of Article 680 includes.
    (a) wading and decorative pools   (b) fountains
    (c) hydromassage bathtubs   (d) all of these

154. A pool capable of holding water to a maximum depth of _____ inches is a storable pool.
    (a) 18   (b) 36   (c) 42   (d) none of these

155. Ground-fault circuit-interrupters for the pool or spa 120-volt wet-niche light shall only be of the circuit breaker type.
    (a) True   (b) False

156. The conductors on the load side of a transformer or any GFCI protection device that supplies power to a pool light, shall not be in the same raceway or enclosure with other conductors.
    (a) True   (b) False

157. • At dwelling units, a 125-volt receptacle is required to be installed within 20 feet of the inside wall of the pool.
    (a) True   (b) False

158. Paddle fans must not be installed within _____ feet from the edge of outdoor pools, spas, and hot tubs.
    (a) 3   (b) 5   (c) 10   (d) 12

159. Overhead branch circuit, feeder, or utility service conductors that operate at 480 volts line to line can be installed over pools and other structures. The clearance from the water must be at least _____ feet.
    (a) 14   (b) 16   (c) 18   (d) 20

160. • Underground outdoor pool or spa equipment rooms or pits must have adequate drainage to prevent water accumulation during abnormal operation.
    (a) True   (b) False

Part B. Permanently Installed Pools

161. The maximum voltage between conductors to an outdoor underwater pool lighting fixture supplied by a transformer is _____ volts.
    (a) 6   (b) 14   (c) 15   (d) 18

162. When rigid nonmetallic conduit extends from the pool light forming shell to a suitable junction box, a No. 8 _____ conductor shall be installed in the raceway.
    (a) solid bare   (b) solid insulated   (c) stranded insulated   (d) b or c

163. • When bonding pool and spa equipment, a solid No. 8 copper conductor must be run back to the service equipment. This conductor must be unbroken.
    (a) True   (b) False

164. • When bonding together pool reinforcing steel and welded wire fabric (wire-mesh) with tie-wire, the tie-wires must be _____.
    (a) stainless steel (b) accessible
    (c) made up tight (d) none of these

165. The items required to be connected to the common bonding grid of a pool shall be done with a minimum No. 8 _____ conductor.
    (a) solid insulated (b) solid bare (c) solid covered (d) any of these

166. Wet niche lighting fixtures shall be connected to an equipment grounding conductor not smaller than No. _____.
    (a) 10 (b) 6 (c) 8 (d) 12

167. • Pool-associated motors must be grounded by a minimum No. 12 insulated copper conductor. This grounding conductor must be installed in _____.
    (a) rigid nonmetallic conduit
    (b) Electrical metallic tubing shall be permitted to be used to protect conductors where installed on or within buildings.
    (c) flexible metal conduit
    (d) a or b

168. A pool deck area radiant heat cable shall _____.
    (a) not be installed within 5 feet horizontally from the inside wall of the pool
    (b) be mounted at least 12 feet vertically above the pool deck
    (c) not be permitted
    (d) none of these

Part D. Spas And Hot Tubs

169. The wiring for spas and hot tubs installed outdoors, such as receptacle, switches, lighting locations, grounding, and bonding must comply with the same requirements as permanently installed pools.
    (a) True (b) False

Part E. Fountains

170. The maximum length of exposed cord in a fountain shall be _____ feet.
    (a) 3 (b) 4 (c) 6 (d) 10

Part G. Hydromassage Bathtubs

171. For hydromassage bathtubs, the following applies.
    (a) All associated electric components must be protected by GFCI protection devices.
    (b) Switches must be located a minimum of 5 feet from the tub.
    (c) Fixtures located within 5 feet, must be a minimum of 7 feet 6 inches over the maximum water level and GFCI protected.
    (d) Wiring methods are limited to rigid metal, intermediate metal, or rigid nonmetallic conduit.

Chapter 7 – Special Conditions

NEC Chapter 7 can be thought of as a special power chapter. It contains the requirements for emergency and standby systems, signaling systems, plus a few other different power systems.

Article 700 – Emergency Systems

Article 700 pertain to systems that are essential for safety to human life. These systems are often comprised of backup power for emergency power and lighting. This article covers the requirements for the installation of emergency systems where required by other codes or laws. It does not specify where emergency systems are to be installed. See NFPA 101, Life Safety Codes for where emergency systems are required and the locations of emergency or exit lights. The FPNs after 700–1 list other documents related to this subject.

Part A. General

172. Audible and visual signal devices for an emergency system shall be provided, where practicable, for the purpose(s) of indication _____.
    (a) that the battery is carrying load
    (b) of derangement of the emergency source
    (c) that the battery charger is not functioning
    (d) all of these

Part C. Sources of Power

173. Where storage batteries are used as source of power for emergency systems, they shall be _____.
    (a) provided with automatic battery charging means
    (b) of the alkali or acid type
    (c) able to maintain the total load for $1\frac{1}{2}$ hours
    (d) all of these

174. • Battery pack units may supply emergency power if connected _____.
    (a) on the lighting circuit of the area
    (b) on any receptacle circuit
    (c) on any branch circuit
    (d) ahead of the main

Article 701 – Legally Required Standby Systems

Article 701 Legally Required Standby Systems are systems intended to provide electrical power to aid in fire fighting, rescue operations, control of health hazards, and similar operations. See Section 701–2 FPN.

The requirements for legally required standby systems are similar to those for emergency systems. When normal power is lost, *legally required systems* are required to come on in 60 seconds or less, and the wiring can be mixed with general wiring. *Emergency systems* are required to come on in 10 seconds or less, and the wiring is completely separated from the general wiring.

Part A. General

175. A legally required standby system is intended to automatically supply power to _____.
    (a) those systems classed as emergency systems
    (b) selected loads
    (c) both a and b
    (d) none of these

Article 702 – Optional Standby Systems

Optional Standby Systems are those whose failure could cause physical discomfort, or serious interruption of an industrial process, damage to process equipment, or disruption of business. See Section 702–2 FPN.

Part A. General

176. Optional standby systems are typically installed to provide an alternate source of power for _____.
    (a) data processing and communication systems
    (b) emergency systems for health care facilities
    (c) emergency systems for hospitals
    (d) none of these

177. Optional standby systems shall have adequate capacity and rating for the supply of _____.
    (a) all emergency lighting and power loads
    (b) all equipment to be operated at one time
    (c) 100 percent of the appliance loads and 50 percent of the lighting loads
    (d) 100 percent of the lighting loads and 75 percent of the appliance loads

Article 710 – Over 600 Volts, Nominal General

Part F. Tunnel Installations

178. The equipment grounding conductor for the over 600-volt circuit conductors in a tunnel shall be _____.
  I. insulated
  II. bare
  III. inside the raceway
  IV. outside the raceway
  (a) I and III
  (b) II and III
  (c) I, III, and IV
  (d) I, II, and III

Article 720 – Circuits and Equipment Operating At Less Than 50 Volts

179. Circuits and equipment operating at less than 50 volts shall use receptacles that are not rated less than _____ amperes.
  (a) 10
  (b) 15
  (c) 20
  (d) 30

Article 725 – Remote-Control, Signaling, And Power-Limited Circuits

Article 725 covers remote-control, signaling, and power-limited circuits that are not an integral part of a device or appliance. See Section 725–1 (FPN). This Article covers signaling type systems such as burglar alarms and coaxial wiring associated with the interconnection of electronic data processing and computer equipment not within a data processing room as covered in Article 645.

The 1990 NEC rearranged Articles 725, 760, 770, 800 and 820 to have similar formats. You will see the similarities as you study these articles.

A **signaling circuit** is any electrical circuit that supplies energy to an appliance or device that gives a visual and/or audible signal. Examples are doorbells, fire or smoke detectors, and fire or burglar alarms.

A **remote control circuit** is any circuit which has as its load device the operating coil of a magnetic motor starter, a contactor, or relay. These circuits control one or more other circuits. Low voltage relay switching of lighting and power loads can be a remote-control circuit.

Power-limited circuits are circuits used for functions other than signaling or remote-control, but in which the source of the energy supply is limited in its power (volts x amps) to specified maximum levels. Low voltage lighting using 12-volt lamps in fixtures fed from 120/12-volt transformers, is a typical power-limited circuit application.

Class 1 systems include all signaling and remote-control systems which do not have the special current limitations of Class II and III systems.

Class 2 and 3 systems are those in which the current is limited to certain specified low values by fuses or circuit breakers, and by supply transformers which will deliver only small currents, or by other approved means. All Class 2 and 3 circuits must have a power source with power-limiting characteristics as described in Tables 725–31(a) and (b), in addition to overcurrent protection.

Part A. General

180. Remote-control circuits to safety-control equipment shall be _____ if the failure of the equipment to operate introduces a direct fire or life hazard.
  (a) Class 1
  (b) Class 2
  (c) Class 3
  (d) Class 1, Division 1

Part B. Class 1 Circuits

181. Class 1 power limited circuits shall be supplied from a source having a rated output of not more than 30 volts. If, in fact, the voltage rating is 25 volts, the maximum volt-ampere of the circuit would be _____ VA.
  (a) 750
  (b) 700
  (c) 1,000
  (d) 1200

182. A Class 1 signaling circuit shall not exceed _____ volts.
  (a) 130
  (b) 150
  (c) 220
  (d) 600

183. Power supply and Class 1 circuit conductors shall be permitted in the same cable, enclosure, or raceway only where the _____.
  (a) equipment powered is functionally associated
  (b) circuits involved are not a mixture of alternating and direct current
  (c) Class I circuits are never permitted with the power supply
  (d) none of these answers are correct

Chapter 2   NEC Calculations And Code Questions   Unit 5   Raceway, Outlet Boxes, And Junction Boxes   181

Part C. Class 2 And Class 3 Circuits

184. The power for Class 2 or Class 3 circuits shall be _____.
 (a) inherently limited requiring no overcurrent protection
 (b) limited by a combination of a power source and overcurrent protection
 (c) inherently limited and requires overcurrent protection
 (d) a and b only

185. • In Class 2 alternating current circuit installations, the maximum voltage for wet contact that is most likely to occur is _____ volts peaks.
 (a) 10.5   (b) 21.2   (c) 100   (d) none of these

186. The maximum overcurrent protection in amperes, when required, for a Class 2 remote-control circuit is _____ amperes.
 (a) 1.0   (b) 2.0   (c) 5.0   (d) 7.5

187. Conductors of Class 2 and Class 3 circuits shall not be placed in any enclosure, raceway, cable, or similar fittings with conductors of Class 1 or electric light or power conductors. Except when they are _____.
 (a) insulated for the maximum voltage
 (b) totally comprised of aluminum conductors
 (c) separated by a barrier
 (d) all of these

188. • Doorbell wiring, rated as a Class 2 control circuit in a dwelling unit _____ run in the same raceway or enclosure with light and power conductors.
 (a) is permitted with 600-volt insulation to be
 (b) shall not be
 (c) shall be
 (d) is permitted, if the insulation is equal to the highest installed, to be

189. Class 2 control circuits and power circuits _____.
 I. may occupy the same raceway   II. shall be put in different raceways
 (a) I   (b) II
 (c) I and II   (d) none of these

190. • Raceways shall not be used as a means of support for Class 2 or Class 3 circuit conductors.
 (a) True   (b) False

Article 760 – Fire Alarm Signaling Systems

Article 760 – Fire Alarm Signaling Systems covers the installation of wiring and equipment of fire alarm signaling systems operating at 600 volts, nominal, or less. Examples of this type of system could include fire alarm, guard tour, sprinkler water flow, and sprinkler supervisory systems. A document used in close association with this type of system would be NFPA 72.

Part A. General

191. Wiring methods permitted in ducts or plenums used for environmental air include _____. This is permitted only if the devices or equipment in the duct are used to sense or act upon the contained air.
 I. IMC
 II. EMT
 III. Rigid nonmetallic conduit
 IV. Rigid metal conduit
 V. Flexible metal conduit up 4 feet
 (a) I, II, and IV
 (b) I, IV, and V
 (c) I, II, IV, and V
 (d) none of these

192. Fire alarm wiring is permitted to be installed in ducts that contain loose stock or vapors, if installed in rigid metal or intermediate metal conduit. Be sure to read Section 300–22.
(a) True (b) False

Part B. Nonpower-Limited Fire Alarm Signaling Circuits

193. For nonpower limited fire alarm signaling circuits, a No. 18 conductor is considered protected if the overcurrent device protecting the system is not over _____ amperes.
(a) 15 (b) 10
(c) 20 (d) 7

Part C. Power-Limited Fire Alarm Signaling Circuits

194. Equipment shall be durably marked where plainly visible to indicate each circuit that it is _____.
(a) on a power-limited fire alarm signaling circuit
(b) a power-limited fire alarm signaling circuit
(c) a fire alarm circuit
(d) none of these

195. Conductors of light or power may occupy the same enclosure or raceway with conductors of power-limited fire alarm signaling circuits.
(a) True (b) False

196. Power-limited fire alarm signaling circuit cables in a building shall be _____.
(a) labeled
(b) Type FPL
(d) identified
(e) none of these

197. Coaxial cables used in power-limited fire alarm systems shall have an overall insulation rated at _____ volts.
(a) 100 (b) 300
(c) 600 (d) 1000

Article 770 – Optical Fiber Cables And Raceways

Article 770 – Optical Fiber Cables covers the use of optical fiber cables where used in conjunction with electrical conductors for communication, signaling, and control circuits. This is a relatively new technology and you will see this article change and grow. It does not cover optical cables under the control of utility companies. One advantage for using optical fiber cable in lieu of metallic conductors is that they are not affected by electrical noise.

Part C. Cables Within Buildings

198. A conductive general use type of optical fiber cable installed as wiring within buildings shall be Type _____.
(a) FPLP (b) CL2P
(c) OFPN (d) OFC

Chapter 8 – Communications Systems

NEC Chapter 8 covers communications systems and is independent of the other chapters except where they are specifically referenced therein.

Article 800 – Communication Circuits covers telephone, telegraph (except radio), outside wiring for fire and burglar alarms and similar central station systems, and telephone systems not connected to a central station system, by using similar types of equipment, methods of installation, and maintenance.

Article 810 – Radio and Television Equipment covers radio and television receiving equipment and amateur radio transmitting and receiving equipment, but not equipment and antennas used for coupling carrier current to power line conductors.

Article 820 – Community Antenna Television and Radio Distribution Systems covers coaxial cable distribution of radio frequency signals typically employed in community antenna television (CATV) systems.

Article 800 – Communications Circuits

Part B. Conductors Outside and Entering Buildings

199. Where practical, a separation of at least _____ feet shall be maintained between open conductors of communication systems on buildings and lightning conductors.
    (a) 6
    (b) 8
    (c) 10
    (d) 12

Part C. Conductors Within Buildings

200. Floor penetrations requiring Types CMR communication cable shall contain only cables suitable for riser or plenum use except where the listed cables are _____.
    (a) encased in metal raceways
    (b) located in a fireproof shaft with firestops at each floor
    (c) a or b
    (d) none of these

Unit 6

Conductor Sizing And Protection Calculations

OBJECTIVES

After reading this unit, the student should be able to briefly explain the following concepts:

Conductor allowable ampacity
Conductor ampacity
Conductor bunching derating factor, Note 8(a)
Conductor sizing summary
Conductor insulation property

Conductor size – voltage drop
Conductors in parallel
Current-carrying conductors
Equipment conductors size and protection examples
Minimum conductor size

Overcurrent protection of equipment conductors
Overcurrent protection
Overcurrent protection of conductors
Terminal ratings

After reading this unit, the student should be able to briefly explain the following terms:

Ambient temperature
American wire gauge
Ampacity
Ampacity derating factors
Conductor bundled
Conductor properties

Continuous load
Current-carrying conductors
Fault current
Interrupting rating
Overcurrent protection device
Overcurrent protection

Parallel conductors
Temperature correction factors
Terminal ratings
Voltage drop

6–1 CONDUCTOR INSULATION PROPERTY [Table 310–13]

Table 310–13 of the NEC provides information on conductor properties such as, permitted use, maximum operating temperature, and other insulation details, Fig. 6–1. See table 310–13 on the next page.

The following aberrations and explanations should be helpful in understanding Table 310–13 as well as Table 310–16.

-2 - Conductor is permitted to be used at a continuous 90°C operation temperature
F - Fixture wire (solid or 7 strand) [Table 402–3]
FF - Flexible Fixture wire (19 strands) [Table 402–3] 60°C Insulation rating
H - 75°C Insulation rating
HH - 90°C Insulation
N - Nylon outer cover
T - Thermoplastic insulation
W - Wet or Damp

For details on fixture wires, see Article 402, Table 402–3 and Table 402–5.

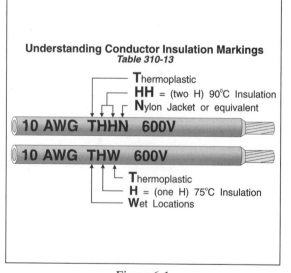

Figure 6-1

Conductor Insulation Markings

TABLE 310–13 CONDUCTOR INFORMATION

Type Letter	Column 2 Trade Name	Column 3 Maximum Operating Temperature	Column 4 Applications Provisions	Column 5 Sizes Available	Column 6 Outer Covering
THHN	Heat resistant thermoplastic	90ºC	Dry and damp locations	14 – 1000	Nylon jacket or equivalent
THHW	Moisture- & heat-resistant thermoplastic	75ºC 90ºC	Wet locations Dry and damp locations	14 – 1000	None
THW	Moisture- & heat-resistant thermoplastic	75ºC 90ºC	Dry, damp, and wet locations Within electrical discharge lighting equipment See Section 410–31	14 – 2000	None
THWN	Moisture- & heat-resistant thermoplastic	75ºC	Dry and wet locations	14 – 1000	Nylon jacket or equivalent
TW	Moisture- resistant thermoplastic	60ºC	Dry and wet locations	14 – 2000	None
XHHW	Moisture- & heat-resistant cross-linked synthetic polymer	90ºC 75ºC	Dry and damp locations Wet locations	14 – 2000	None

❏ **Table 310–13 Example**

TW can be described as _____ Fig. 6–2?

(a) thermoplastic insulation
(b) suitable for wet locations
(c) suitable for dry locations
(d) maximum operating temperature of 60ºC
(e) all of the above

• Answer: (e) all of the above

❏ **Table 402–3 Example**

TFFN can be described as _____?

(a) 19-strand fixture wire
(b) thermoplastic insulation with a nylon outer cover
(c) suitable for dry and wet locations
(d) maximum operating temperature of 90ºC
(e) all of the above except (c)

• Answer: (e) all of the above except (c)

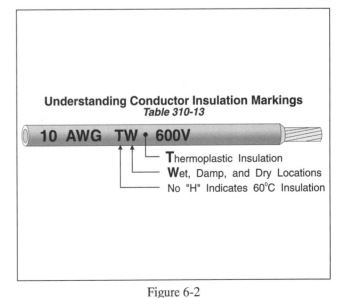

Figure 6-2

Conductor Insulation Marking Example

6–2 CONDUCTOR ALLOWABLE AMPACITY [310–15]

The ampacity of a conductor is the current the conductors can carry continuously under the specific condition of use [Article 100 definition]. The ampacity of a conductor is listed in Table 310-16 under the condition of no more than three current carrying conductors bundled together in an ambient temperature of 86ºC. The ampacity of a conductor changes if the ambient temperature is not 86ºF or if more than three current-carrying conductors are bundled together for more than two feet.

Section 310–15 lists 2 ways of determining conductor ampacity.

Ampacity Method 1: Simply use the values listed in Tables 310–16 through 310–19 of the NEC according to the specific conditions of use.

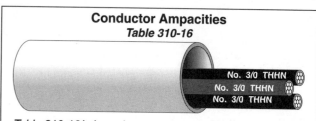

Table 310-16 is based on an ambient temperature of 86ºF and 3 current-carrying conductors in a raceway or cable.

Figure 6-3

Conductor Ampacity Table 310–16

Ampacity Method 2: Engineer supervised formula: $I = \sqrt{\dfrac{TC - (TA + DELTATD)}{RDC(1 + YC) RCA}}$

For all practical purposes, the electrical trade uses the ampacities as listed in Table 310–16 of the National Electrical Code.

TABLE 310–16. ALLOWABLE AMPACITIES OF INSULATED CONDUCTORS
Based On Not More Than Three Current-Carrying Conductors
And Ambient Temperature of 30ºC (86ºF)

Size AWG kcmil	Temperature Rating of Conductor, See Table 310–13						Size AWG kcmil
	60ºC (40ºF) †TW	75ºC (167ºF) †THHN †THW †THWN †XHHW Wet Location	90ºC (194ºF) †THHN †THHW †XHHW Dry Location	60ºC (40ºF) †TW	75ºC (167ºF) †THHN †THW †THWN †XHHW Wet Location	90ºC (194ºF) †THHN †THHW †XHHW Dry Location	
	COPPER			ALUMINUM/COPPER-CLAD ALUMINUM			
14	20†	20†	25†				12
12	25†	25†	30†	20†	20†	25†	10
10	30	35†	40†	25	30†	35†	8
8	40	50	55	30	40	45	8
6	55	65	75	40	50	60	6
4	70	85	95	55	65	75	4
3	85	100	110	65	75	85	3
2	95	115	130	75	90	100	2
1	110	130	150	85	100	115	1
1/0	125	150	170	100	120	135	1/0
2/0	145	175	195	115	135	150	2/0
3/0	165	200	225	130	155	175	3/0
4/0	195	230	260	150	180	205	4/0
250	215	255	290	170	205	230	250
300	240	285	320	190	230	255	300
350	260	310	350	210	250	280	350
400	280	335	380	225	270	305	400
500	320	380	430	260	310	350	500

† The maximum *overcurrent protection device* permitted for copper No. 14 is 15 amperes, No. 12 is 20 amperes, and No. 10 is 30 amperes. The maximum overcurrent protection device permitted for aluminum No. 12 is 15 amperes and No. 10 is 25 amperes. This † note does not establish the conductor ampacity but limits the maximum size protection device that may be placed on the conductor, Fig. 6–4.

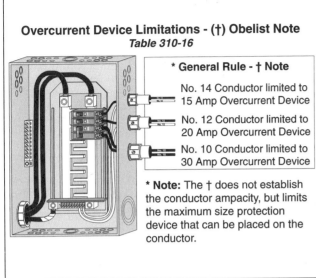

Figure 6-4
† Note of Table 310–16

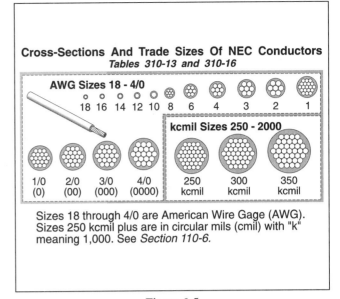

Figure 6-5
Conductor Size

CAUTION: The overcurrent protection device limitations of the † Note to Table 310–16 does not apply to the following: See Section 240–3(d) through (m).

Air conditioning conductor	Section 440–22
Capacitor conductors	Article 460
Class 1 remote control conductors	Article 725
Cooking equipment taps	210–19(b) Exception No. 1
Feeder taps	Section 240–21
Fixture wires and taps	210–19(c) Exception
Motor branch circuit conductors	Section 430–52
Motor feeder conductor	Section 430–62
Motor control conductors	Section 430–72
Motor taps branch circuits	Section 430–53(d)
Motor taps feeder conductors	Section 430–28
Power loss hazard conductors	Section 240–3(a)
Two-wire transformers	Section 240–3(i)
Transformer Tap Conductor	Section 240–21(b)(d) and (m)
Welders	Article 630

6–3 CONDUCTOR SIZING [110–6]

Conductors are sized according to the American Wire Gage (AWG) from Number 40 through Number 0000. The smaller the AWG size, the larger the conductor. Conductors larger than 0000 (4/0) are identified according to their circular mil area, such as 250,000, 300,000, 500,000. The circular mil size is often expressed in kcmil, such as 250 kcmil, 300 kcmil, 500 kcmil etc, Fig. 6–5.

Smallest Conductor Size

The smallest size conductor permitted by the National Electrical Code for branch circuits, feeders, or services is No. 14 copper or No. 12 aluminum [Table 310–5].Some local codes require a minimum No. 12 for commercial and industrial installations. Conductors smaller than No. 14 are permitted for:

    Class 1 circuits [402–11, Exception, and 725–16(a)]
    Class 2 circuits [725–51(e) and 725–53]
    Class 3 circuits [725–51(e) and 725–51(g)]
    Cords [Table 400–5(a)]
    Fixture wire [402–5 and 410–24]
    Flexible cords [400–12]
    Motor control circuits [430–72]

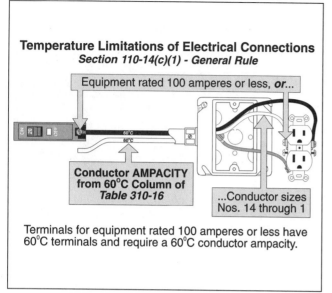

Figure 6-6

Conductor Size, Terminals rated 60ºC

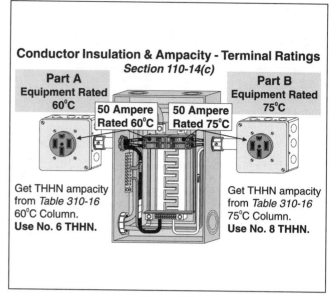

Figure 6-7

Part A Conductor Sized, Terminals Rated 60ºC
Conductor Sized, Terminals Rated 75ºC

Nonpower limited fire alarm circuits [760–16 and 760–17(c)(1), Exception 1]
Power limited fire alarm circuits [760–28(a) Exception 2, Exception 3, 760–51(b), and 760–53].

6–4 TERMINAL RATINGS [110–14(c)]

When selecting a conductor for a circuit, we must always select the conductors not smaller than the terminal ratings of the equipment.

Circuits Rated 100 Amperes And Less

Equipment terminals rated 100 amperes or less (and pressure connector terminals for No. 14 through No. 1 conductors), shall have the conductor sized no smaller than the 60ºC temperature rating listed in Table 310–16, unless the terminals are marked otherwise, Fig. 6–6.

❑ **Terminal Rated 60ºC Example**

What size THHN conductor is required for a 50-ampere circuit if the equipment is listed for use at 60ºC, Fig. 6–7 Part A?

(a) No. 10 (b) No. 8
(c) No. 6 (d) Any of these

• Answer: (c) No. 6

Conductors must be sized to the lowest temperature rating of either the equipment (60ºC) or the conductor (90ºC). THHN insulation can be used, but the conductor size must be selected based on the 60ºC terminal rating of the equipment, not the 90ºC rating of the insulation. Using the 60ºC column of Table 310–16, this 50 ampere circuit requires a No. 6 THHN conductor (rated 55 amperes at 60ºC).

❑ **Terminal Rated 75ºC Example**

What size THHN conductor is required for a 50-ampere circuit if the equipment is listed for use at 75ºC, Fig. 6–7 Part B?

(a) No. 10

(b) No. 8

(c) No. 6

(d) Any of these

• Answer: (b) No. 8

Conductors must be sized according to the lowest temperature rating of either the equipment (75ºC) or the conductor (90ºC). THHN conductors can be used, but the conductor size must be selected according to the 75ºC terminal rating of the equipment, not the 90ºC rating of the insulation. Using the 75ºC column of Table 310–16, this installation would permit No. 8 THHN (rated 50 amperes 75ºC) to supply the 50-ampere load.

Circuits Over 100 Amperes

Terminals for equipment rated over 100 amperes and pressure connector terminals for conductors larger than No. 1, shall have the conductor sized according to the 75ºC temperature rating listed in Table 310–16, Fig. 6–8.

☐ Over 100 Ampere Example

What size THHN conductor is required to supply a 225-ampere feeder?

(a) No. 1/0 (b) No. 2/0
(c) No. 3/0 (d) No. 4/0

- Answer: (d) No. 4/0

The conductors in this example must be sized to the lowest temperature rating of either the equipment (75ºC) or the conductor (90ºC). THHN conductors can be used, but the conductor size must be selected according to the 75ºC terminal rating of the equipment. Using the 75ºC column of Table 310–16, this would require a No. 4/0 THHN (rated 230 amperes at 75ºC) to supply the 225 ampere load. No. 3/0 THHN is rated 225 amperes at 90ºC, but we must size the conductor to the terminal rating at 75ºC.

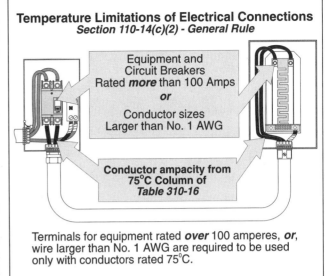

Terminals for equipment rated *over* 100 amperes, *or*, wire larger than No. 1 AWG are required to be used only with conductors rated 75°C.

Figure 6-8
Temperature Limitations Of Electrical Connections

Minimum Conductor Size Table

When sizing conductors the following table must always be used to determine the minimum size conductor.

CAUTION: When sizing conductors, we must consider conductor voltage drop, ambient temperature correction and conductor bunching derating factors. These subjects are covered later in this book.

Terminal Size And Matching Copper Conductor		
Terminal ampacity	60ºC Terminals Wire Size	75ºC Terminals Wire Size
15	14	14
20	12	12
30	10	10
40	8	8
50	6	8
60	4	6
70	4	4
100	1	3
125	1/0	1
150	–	1/0
200	–	3/0
225	–	4/0
250	–	250 kcmil
300	–	350 kcmil
400	–	2 – 3/0
500	–	2 – 250 kcmil

What Is The Purpose Of THHN [110-14(c)]?

In general, 90ºC rated conductor ampacities cannot be used for sizing a circuit conductor. However, THHN offers the opportunity of having a greater conductor ampacity for conductor ampacity derating. The higher ampacity of THHN can permit a conductor to be used without having to increase it's size because of conductor ampacity derating. Remember, the advantage of THHN is not to permit a smaller conductor, but to prevent you from having to install a larger conductor when ampacity derating factors are applied, Fig. 6–9.

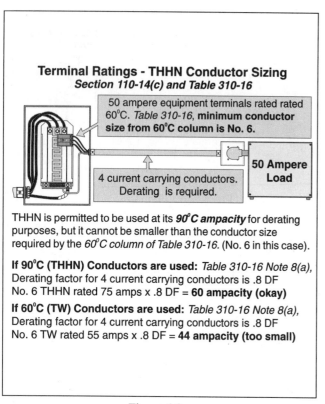

Figure 6-9
Purpose of THHN

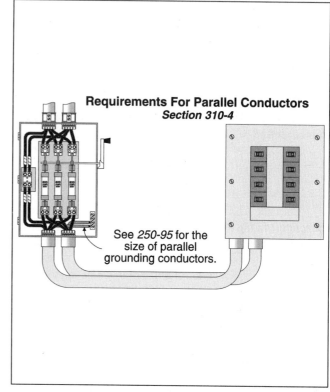

Figure 6-10
Requirements for Conductors in Parallel

6–5 CONDUCTORS IN PARALLEL [310–4]

As the conductor size increases, the cross-sectional area required to carry one ampere increases. Parallel conductors (electrically joined at both ends) permit a reduced cross-sectional area per ampere which can result in significant cost saving, Fig. 6–10.

Conductor Size	Circular Mils Chapter 9, Table 8	Ampacity 75ºC	Circular Mils Per Amperes
No. 1/0	105,600 cm	150 amperes	704 cm/per amp
No. 3/0	167,600 cm	200 amperes	838 cm/per amp
250 kcmil	250,000 cm	255 amperes	980 cm/per amp
500 kcmil	500,000 cm	380 amperes	1,316 cm/per amp
750 kcmil	750,000 cm	475 amperes	1,579 cm/per amp

❑ **Conductors In Parallel Example**

What size 75ºC conductor is required for a 600-ampere service that has a calculated demand load of 550 amperes, Fig. 6–11?

(a) 500 kcmil   (b) 750 kcmil   (c) 1,000 kcmil   (d) 1,250 kcmil

• Answer: (d) 1,250 kcmil rated 590 amperes [Table 310–16]

❑ **Parallel Conductor Example**

What size conductor would be required in each of two raceways for a 600 ampere service?. The calculated demand load is 550 amperes, Fig. 6–12?

(a) two – 300 kcmil   (b) two – 250 kcmil   (c) two – 500 kcmil   (d) two– 750 kcmil

• Answer: (a) 2– 300 kcmil each rated 285 amperes.

285 amperes × 2 conductors = 570 amperes [Table 310–16]

When we can install two 300-kcmil conductors (600,000 circular mils total) instead of one 1,250-kcmil conductor, see previous example.

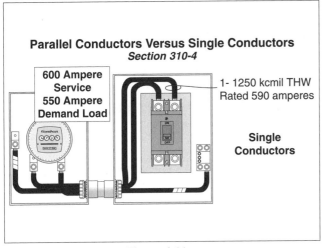

Figure 6-11
Non-parallel Example

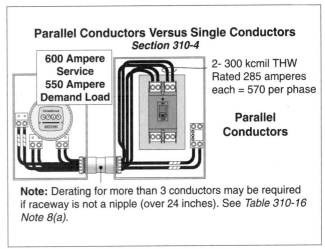

Figure 6-12
Parallel Example

Grounding Conductors In Parallel

When equipment grounding conductors are installed with circuit conductors that are run in parallel, each raceway must have an equipment grounding conductor. The equipment grounding conductor installed in each raceway must be sized according to the overcurrent protection device rating that protects the circuit [250–95]. Parallel equipment grounding conductors are not required to be a minimum No. 1/0 [310–4], Fig. 6–13.

❑ **Grounding Conductors In Parallel Example**

What size equipment grounding conductor is required in each of two raceways for a 600-ampere feeder, Fig. 6–14?

(a) No. 3 (b) No. 2 (c) No. 1 (d) No. 1/0

• Answer: (c) No. 1, Section and Table 250–95

There must be a total of two No. 1 equipment grounding conductors. One No. 1 in each raceway.

6–6 CONDUCTOR SIZE – VOLTAGE DROP [210–19(a), FPN No. 4, And 215–2(b), FPN No. 2]

Conductor voltage drop is the result of the conductors resistance opposing the current flow, $E_{vd} = R \times I$. The NEC does not required that we limit the voltage drop on conductors, but the Code does suggest that we consider its effects. The subject of conductor voltage drop is covered in detail in Unit 8.

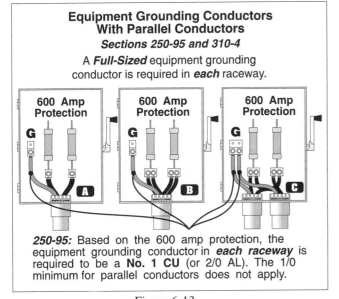

Figure 6-13
Parallel Grounding Conductor

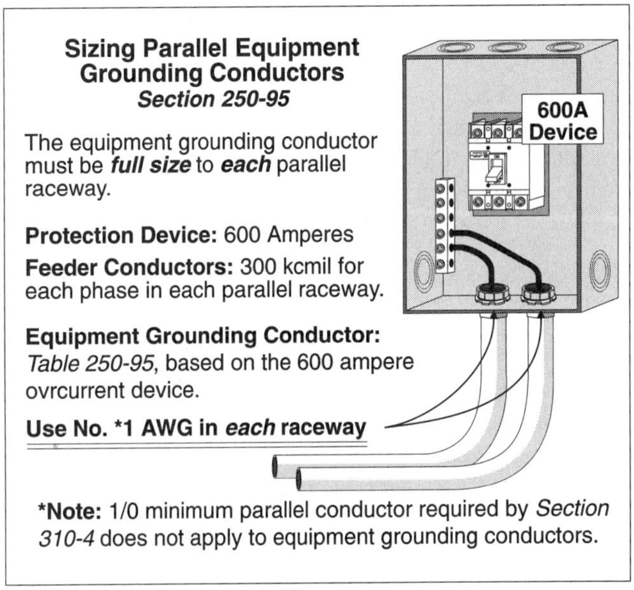

Figure 6-14
Parallel Grounding Conductor Example

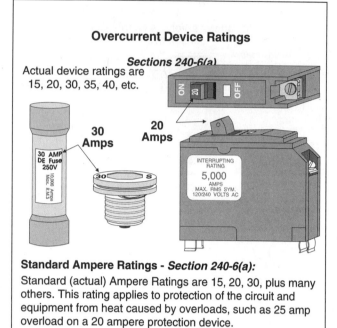

Figure 6-15
Overcurrent Protection Device Rating

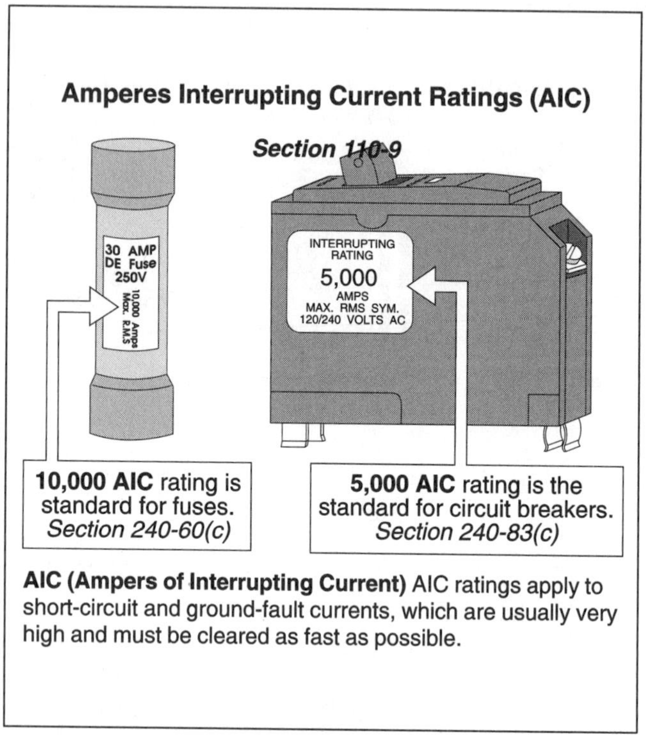

Figure 6-16
Interrupting Rating

6–7 OVERCURRENT PROTECTION [Article 240]

One purpose of an overcurrent protection device is to open a circuit to prevent damage to persons or property due to excessive or dangerous heat. Overcurrent protection devices have two ratings, overcurrent and interrupting (AIC).

Overcurrent Rating

This is the actual ampere rating of the protection device, such as 15, 20, or 30 amperes. If the current flowing through the protection device exceeds the device setting for a specific period of time, the protection device will open to remove the danger of excess heat, Fig. 6–15.

Interrupting Rating (Short-circuit) [110–9]

The overcurrent protection device is designed to clear fault-current and it must have a short-circuit interrupting rating sufficient for the available fault current. The minimum interruption rating for circuit breakers is 5,000 amperes [240–83(c)] and 10,000 amperes for fuses [240–60(c)]. The overcurrent protection device must clear fault-current and prevent the increase of damage to other equipment or other parts of the circuit, Fig. 6–16.

If the overcurrent protection device does not have an interrupting rating above the available fault current, it could explode while attempting to clear the fault. In addition, downstream equipment must have a withstand rating so that it will not suffer serious damage during a ground-fault [110–10].

Standard Size Protection Devices [240–6(a)]

The National Electrical Code list standard sized overcurrent protection devices: 15, 20, 25, 30, 35, 40, 45, 50, 60, 70, 80, 90, 100, 110, 125, 150, 175, 200, 225, 250, 300, 350, 400, 450, 500, 600, 700, 800, 1000, 1200, 1600, 2000, 2500, 3000, 4000, 5000, and 6000 amperes. *Exception:* Additional standard ratings for fuses include : 1, 3, 6, 10, and 601 amperes.

Continuous Load [384–16(c)]

Overcurrent protection devices are sized no less than 125 percent of the continuous load, plus 100 percent of the noncontinuous load. Conductors for continuous loads are sized based on the ampacities as listed on Table 310-16 "before any ampacity derating factors are applied" [110-14(c), 210–22(c), 220–3(a), 220–10(b), and 384–16(c)], Fig. 6–17.

❑ **Continuous Load Example**

What size protection device is required for a 100-ampere continuous load, Fig. 6–18?

(a) 150 ampere   (b) 100 ampere   (c) 125 ampere   (d) any of these

• Answer: (c) 125 ampere, 100 amperes × 1.25 = 125 amperes [240–6(a)]

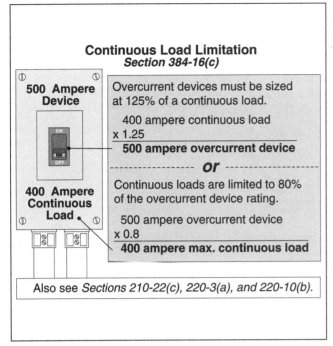

Figure 6-17

Continuous Load on Protection Device

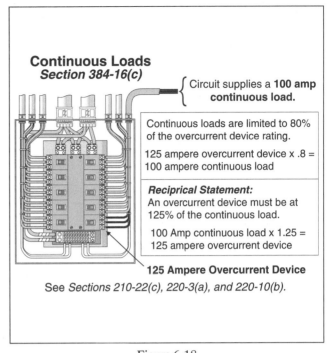

Figure 6-18

Continuous Load Example

6–8 OVERCURRENT PROTECTION OF CONDUCTORS – GENERAL REQUIREMENTS [240–3]

There are many different rules for sizing and protecting conductors. It is not simply a 20-ampere breaker on No. 12 wire. The general rule for conductor overcurrent protection is that a conductor must be protected against overcurrent in accordance with its ampacity [240–3].

However, if the ampacity of a conductor does not correspond with the standard ampere rating of a fuse or circuit breaker as listed in 240–6(a), the next larger overcurrent protection device (not over 800 amperes) is permitted [240–3(b)]. The next size up rule only applies if the conductor does not supply multioutlet receptacles for cord-and-plug connected loads and does not exceed 800 amperes, Fig. 6–19.

❏ **Overcurrent Protection Of Conductors Example**

What size conductor is required for a 130-ampere load that is protected with a 150-ampere breaker, Fig. 6–20?

(a) No. 1/0 (b) No. 1
(c) No. 2   (d) any of these

• Answer: (b) No. 1

The conductor must be sized according to the 75°C column of Table 310–16, which is No. 1 [110-14(c)(2)]. This No. 1 conductor is rated 130 amperes at 75°C column of Table 310–16 and is permitted to be protected by a 150-ampere protection device. [240–3(b)]

6–9 OVERCURRENT PROTECTION OF CONDUCTORS – SPECIFIC REQUIREMENTS

When sizing and protecting conductors for equipment, be sure to apply the specific NEC requirement.

Equipment
Air Conditioning [440–22, and 440–32]
Appliances [422–4, 422–5 and 422–28]

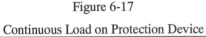

Figure 6-19

Next Size Larger Protection

Cooking Appliances [210–19(b), 210–20, Exception No. 1, Note 4 of Table 220–19]
Electric Heating Equipment [424–3(b)]
Fire Protective Signaling Circuits [760–23]
Motors:
 Branch-Circuits [430–22(a), and 430–52]
 Feeders [430–24, and 430–62]
 Remote Control [430–72]
Panelboard [384–16(a)]
Transformers [240–21 and 450–3]

Feeders And Services
Dwelling Unit Feeders and Neutral [215–2 and Note 3 of Table 310–16]
Feeder Conductor [215–2 and 215–3]
Not Over 800 Amperes [240–3(b)]
Feeders Over 800 Amperes [240–3(c)]
Service Conductors [230–90(a)]
Temporary Conductors [305–4]

Grounded Conductor
Neural Calculations [220–22]
Grounded Service Size [230–23(b)]

Tap Conductors
Tap – Ten foot [240–21(b)]
Tap – Twenty-five foot [240–21]
Tap – One hundred foot [240–21(e)]
Tap – Outside Feeder [240–21(m)]

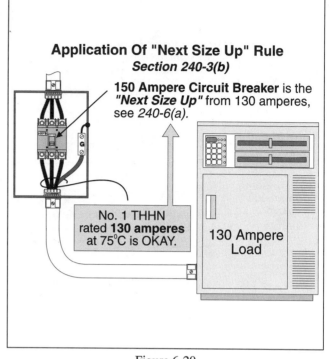

Figure 6-20

Next Size Larger Protection Example

6–10 EQUIPMENT CONDUCTORS SIZE AND PROTECTION EXAMPLES

☐ **Air Conditioning Example**

An air conditioner nameplate indicates the minimum circuit ampacity of 28 amperes and maximum fuse size of 45 amperes. What is the minimum size branch circuit conductor and the maximum size overcurrent protection device, Fig. 6–21?

(a) No. 10 with a 45-ampere fuse (b) No. 8 with a 60-ampere fuse
(c) No. 8 with a 50-ampere fuse (d) No. 10 with a 30-ampere fuse

• Answer:(a) No. 6 with a 60 ampere Fuse

Conductor: The conductors must be sized based on the 60°C column of Table 310–16, which is No. 10.

Overcurrent Protection: The protection device must not be greater than a 45-ampere fuse, preferably a dual-element fuse because of high inrush starting current.

Note. The obelisk (†) note on the bottom of Table 310-16 limiting the overcurrent protection device to 30 amperes for No. 10 wire does not apply to air conditioning equipment [240-3(h)].

☐ **Water Heater Example**

What size conductor and protection device is required for a 4,500 VA water heater, Fig. 6–22?

(a) No. 10 wire with 20-ampere protection (b) No. 12 wire with 20-ampere protection
(c) No. 10 wire with 25-ampere protection (d) No. 10 wire with 30-ampere protection
(e) c or d

• Answer: (e) C or D
 I = VA/E, I = 4,500 VA/240 volts, I = 18.75 amperes

Conductor: The conductor is sized at 125 percent of the water heater rating [422–14(b)], Minimum conductor = 18.75 amperes × 1.25 = 23.4 amperes

Conductor is sized according to the 60°C column of Table 310–16, = No. 10 rated 30 amperes

Overcurrent Protection: The overcurrent protection device is sized no more than 150 percent of the appliance rating, [422–28(e)], 18.75 amperes × 1.50 = 28.1 amperes, next size up = 30 amperes. Please note that the exception to 422–28(e) permits the next size up.

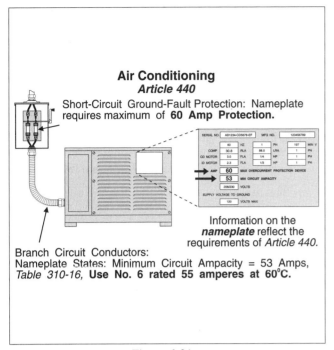

Figure 6-21
Air Condition Example

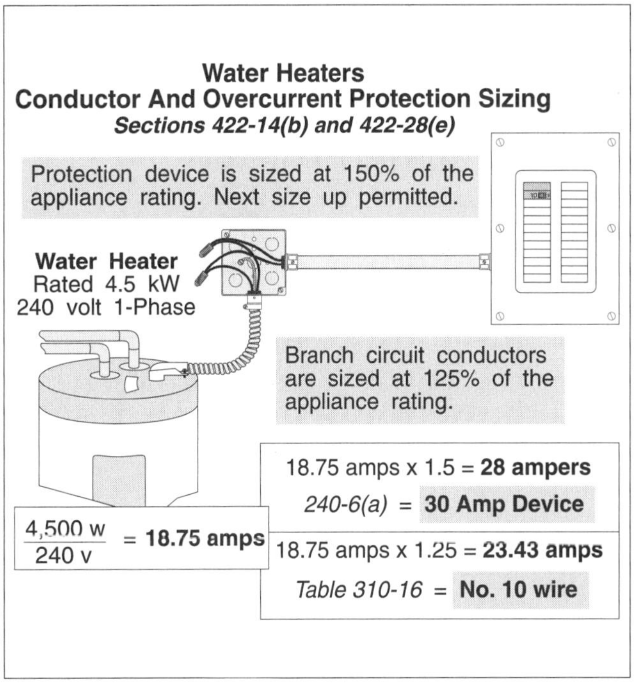

Figure 6-22
Water Heater Example

❑ **Motor Example**

What size branch circuit conductor and short-circuit protection (circuit breaker) is required for a 2-horsepower motor rated 230 volts, Fig. 6–23?

(a) No. 14 with a 15-ampere breaker

(b) No. 12 with a 20-ampere breaker

(c) No. 12 with a 30-ampere breaker

(d) No. 14 with a 30-ampere breaker

• Answer (d) No. 14 with a 30-ampere protection device

Conductors: Conductors are sized no less than 125 percent of the motor full load current [430–22(a)].
12 amperes × 1.25 = 15 amperes, Table 310–16, No. 14 is rated 20 amperes 60°C.

Protection: The short-circuit protection (circuit breaker) is sized at 250 percent of the motor full load current. 12 amperes × 2.5 = 30 amperes [240-6(a) and 430-52(c)(1) Exception No. 1].

Note. The obelisk (†) note on the bottom of Table 310-16 limiting the overcurrent protection device to 15 amperes for No. 14 wire does not apply to motors [240-3(f)]. This concept is explained in detail in the unit on motors.

6–11 CONDUCTOR AMPACITY [310–10]

The insulation temperature rating of a conductor is limited to a operating temperature that prevents serious heat damage to the conductors insulation. If the conductor carries excessive current, the I²R heating within the conductor can destroy the conductor insulation. To limit elevated conductor operation temperatures, we must limit the current flow (ampacity) on the conductors.

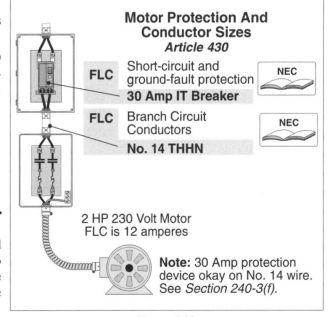

Figure 6-23
Motor Example

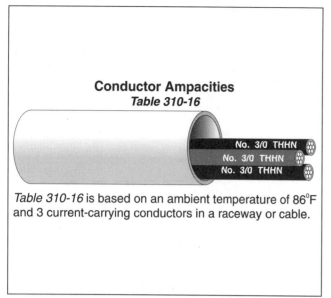

Figure 6-24

Conductor Ampacity, Table 310–16

Ambient Temperature 150°F	Ambient Temperature 86°F (30°C)	Ambient Temperature 75°F
90°C Table Ampacity	90°C Table Ampacity	90°C Table Ampacity
Correction Factor = 58% (.58)	Correction Factor = 100% (1.00)	Correction Factor = 104% (1.04)
30 Table amps x .58 Ampacity = 17.4 amperes	30 Table amps x 1.00 Ampacity = 30 amperes	30 Table amps x 1.04 Ampacity = 31.2 amperes
Ambient temperature over 86°C, ***ampacity is lower***	Ambient temperature 86°C, ***ampacity remains the same***	Ambient temperature below 86°C, ***ampacity is higher***

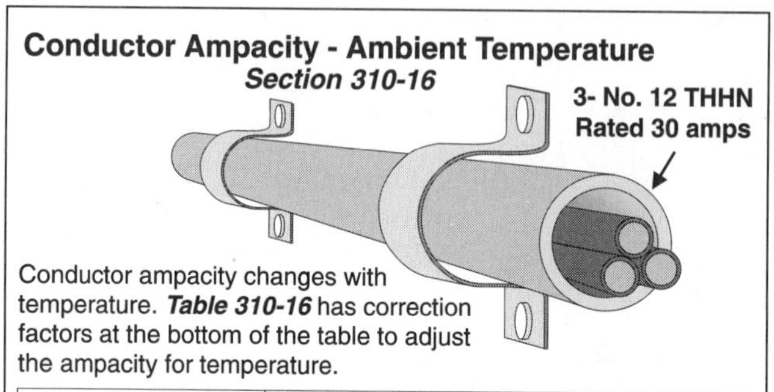

Figure 6-25

Conductor Ampacity Affected by Temperature

Allowable Ampacities

The ampacity of a conductor is the current the conductors can carry *continuously* under the specific condition of use [Article 100 definition]. The ampacity of a conductor is listed in Table 310-16 under the condition of no more than three current-carrying conductors bundled together in an ambient temperature of 86ºC. The ampacity of a conductor changes if the ambient temperature is not 86ºF or if more than three current carrying conductors are bundled together in anyway, Fig. 6-24.

6–12 *AMBIENT TEMPERATURE DERATING FACTOR [Table 310–16]*

The ampacity of a conductor as listed on Table 310–16 is based on the conductor operating at an ambient temperature of 86ºF, (30ºC). When the ambient temperature is different than 86ºF (30ºC), for a prolonged period of time, the conductor ampacity as listed in Table 310–16 must be adjusted to a new ampacity, Fig. 6–25.

In general, 90ºC rated conductor ampacities cannot be used for sizing a circuit conductor. However higher insulation temperature rating offers the opportunity of having a greater conductor ampacity for conductor ampacity derating and a reduced temperature correction factor. The temperature correction factors used to determine the new conductor ampacity are listed at the bottom of Table 310–16. The following formula can be used to determine the conductors new ampacity when the ambient temperature is not 86ºF (30ºC).

Ampacity = Allowable Ampacity × Temperature Correction Factor

Note. If the length of the conductor in the different ambient is 10 feet or less and does not exceed 10 percent of the total length of the conductor, then the ambient temperature correction factors of Table 310–16 do not apply [310–15(c) Exception].

❑ **Ambient Temperature Below 86ºF (30ºC) Example**

What is the ampacity of No. 12 THHN when installed in a walk-in cooler that has an ambient temperature of 50ºF, Fig. 6–26?

(a) 31 amperes (b) 35 amperes (c) 30 amperes (d) 20 amperes

- Answer: (a) 31 amperes

  New Ampacity = Table 310–16 Ampacity × Temperature Correction Factor

  Table 310–16 ampacity for No. 12 THHN is 30 amperes at 90ºC.

  Temperature Correction Factor, 90ºC conductor rating installed at 50ºC is 1.04

  New Ampacity = 30 amperes × 1.04 = 31.2 amperes

  Ampacity increases when the ambient temperature is less than 86ºF (30ºC).

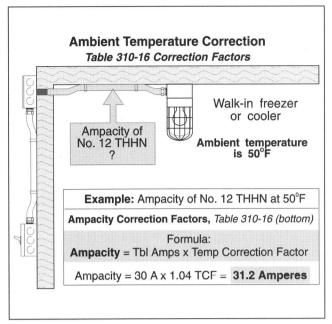

Figure 6-26
Ambient Temperature Correction – Table 310–16

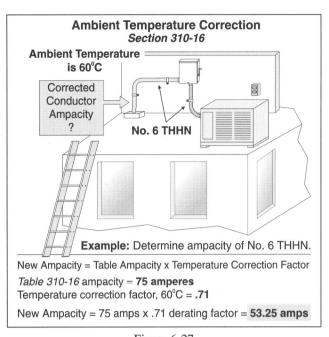

Figure 6-27
Conductor Ampacity – Temperature Below 86°F

AMBIENT TEMPERATURE CORRECTION FACTORS				
When the ambient temperature is other than 86°F (30°C), multiply the ampacities of 310–16 by the appropriate factor shown below				
Ambient Temp °F Fahrenheit	60°C Insulation	75°C Insulation	90°C Insulation	Ambient Temp°C Centigrade
below 78	1.08	1.05	1.04	below 26
78–86	1.00	1.00	1.00	26–30
87–95	.91	.94	.96	31–35
96–104	.82	.88	.91	36–40
105–113	.71	.82	.87	41–45
114–122	.58	.75	.82	46–50
123–131	.41	.67	.76	51–55
132–140	*	.58	.71	56–60
141–158	*	.33	.58	61–70
159–176	*	*	.41	71–80
* Conductor with this insulation can not be installed in this ambient temperature location.				

❑ **Ambient Temperature Above 86°F (30°C) Example**

What is the ampacity of No. 6 THHN when installed on a roof that has an ambient temperature of 60°C, Fig. 6–27?

(a) 53 amperes (b) 35 amperes (c) 75 amperes (d) 60 amperes

• Answer: (a) 53 amperes

New Ampacity = Table 310–16 Ampacity × Ambient Temperature Correction Factor

Table 310–16 ampacity for No. 6 THHN is 75 amperes at 90°C.

Temperature Correction Factor, 90°C conductor rating installed at 60°C is .71

New Ampacity = 75 amperes × 0.71 = 53.25 amperes

Ampacity decreases when the ambient temperature is more than 86°F (30°C).

Conductor Ampacity - Ambient Temperature
Table 310-16

Example: Determine THHN conductor size.

New Ampacity = Table Ampacity x Temperature Correction Factor

Table 310-16:

No. 10 THHN, 40 amperes x .91 = 36.4 amperes, too small
No. 8 THHN, 55 amperes x .91 = 50.0 amperes, Okay
No. 6 THHN, 75 amperes x .91 = 68.8 amperes, too large

Figure 6-28
Conductor Ampacity, Ambient Temp Above 86°F

Ampacity Correction For Conductor Bunching
Table 310-16 Note 8(a)

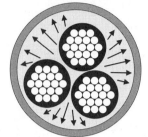
Ampacity at 100%
No Derating for 3 Conductors

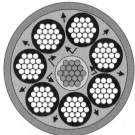
Derated Ampacity at 70%
for 8 Conductors

Conductors have surface space for heat dissipation.

Conductors in the *bunch* have heat held in by other conductors.

Figure 6-29
Conductor Size Example

❏ **Conductor Size Example**

What size conductor is required to supply a 40 ampere load? The conductors pass through a room where the ambient temperature is 100°F, Fig. 6–28?

(a) No. 10 THHN (b) No. 8 THHN (c) No. 6 THHN (d) any of these

• Answer: (b) No. 8 THHN

The conductor to the load must have an ampacity of 40 amperes after applying the ambient temperature derating factor.

New Ampacity = Table 310–16 Amperes × Ambient Temperature Correction.

No. 10 THHN, 40 amperes × 0.91 = 36.4 amperes

No. 8 THHN, 55 amperes × .91 = 50.0 amperes

No. 6 THHN, 75 amperes × .91 = 68.0 amperes

6–13 CONDUCTOR BUNCHING DERATING FACTOR, NOTE 8(a) OF TABLE 310–16

When many conductors are bundled together, the ability of the conductors to dissipate heat is reduced. The National Electrical Code requires that the ampacity of a conductor be reduced whenever four or more current-carrying conductors are bundled together, Fig. 6–29. In general, 90°C rated conductor ampacities cannot be used for sizing a circuit conductor. However higher insulation temperature rating offers the opportunity of having a greater conductor ampacity for conductor ampacity derating.

The ampacity derating factors used to determine the new ampacity is listed in Note 8(a) of Table 310–16. The following formula can be used to determine the new conductor ampacity when more than three current-carrying conductors are bundled together.

New Ampacity = Table 310–16 Ampacity × Note 8(a) Correction Factor

Note. Note 8(a) derating factors do not apply when bundling conductor in a nipple (not exceeding 24 inches) [See Exception No. 3 to Note 8(a)], Fig. 6–30.

❏ **Conductor Ampacity Example**

What is the ampacity of four current carrying No. 10 THHN conductors installed in a raceway or cable, Fig. 6–31?

(a) 20 amperes (b) 24 amperes (c) 32 amperes (d) None of these

• Answer: (c) 32 amperes.

New Ampacity = Table 310–16 Ampacity × Note 8(a) Derating Factor

Table 310–16 ampacity for No. 10 THHN is 40 amperes at 90°C

Note 8(a) derating factor for four current-carrying conductors is 0.8

New Ampacity = 40 amperes × 0.8 = 32 amperes

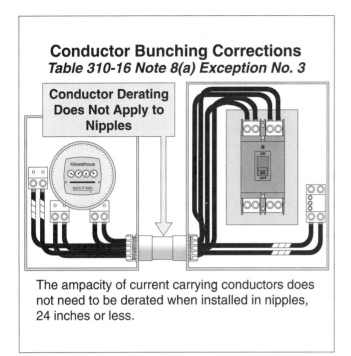

Figure 6-30
Conductor Bunching

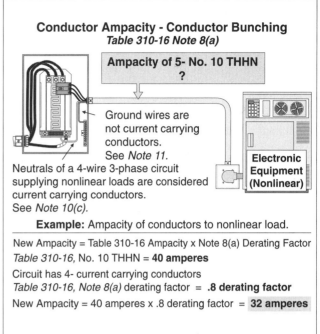

Figure 6-31
Note 8(a) Derating Factors Do Not Apply to Nipples

Conductor Bundling Derating Factors Note 8(a) of Table 310–16	
Number of Current Carrying Conductors	Conductor Ampacity Derating Factor
4 –6	80% or .80
7 – 9	70% or .70
10 – 20	050% or .5
21 – 30	45% or .45
31 – 40	40% or .40
41 – 60	35% or .35

❏ **Conductor Size Example**

A raceway contains four current-carrying conductors. What size conductor is required to supply a 40-ampere load, Fig. 6–32?

(a) No. 10 THHN (b) No. 8 THHN
(c) No. 6 THHN (d) None of these

• Answer: (b) No. 8 THHW

The conductor must have an ampacity of 40 amperes after applying Note 8 derating factor.

Ampacity = 310–16 Ampacity × Note 8 Derating Factor.

No. 10 THHN, 40 amperes × 0.8 = 32 amperes

No. 8 THHN, 55 amperes × 0.8 = 44 amperes

No. 6 THHN, 75 amperes 0.8 = 60 amperes

6–14 *AMBIENT TEMPERATURE AND CONDUCTOR BUNDLING DERATING FACTORS*

If the ambient temperature is different than 86°F (30°C) and there are more than three current-carrying conductors bundled together, then the ampacity listed in Table 310–16 must be adjusted for both conditions, Fig. 6–33.

The following formula can be used to determine the new conductor ampacity when both ambient temperature and Note 8(a) derating factors apply:

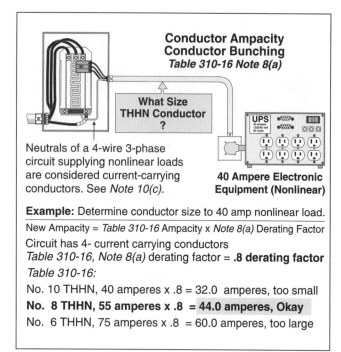

Figure 6-32
Conductor Bundle Ampacity Example

Ampacity = Table 310–16 Ampacity × Temperature Correction Factor × Note 8(a) Correction Factor

Note. If the length of conductor bunching or ambient temperature is 10 feet or less and does not exceed 10 percent of the conductor length, then neither ampacity correction factor applies [310–15(c), Exception].

❑ **Conductor Ampacity Example**
What is the ampacity of four current-carrying No. 8 THHN conductors installed in ambient temperature of 100°F, Fig. 6–34?

(a) 25 amperes (b) 40 amperes (c) 55 amperes (d) 60 amperes

• Answer: (b) 40 amperes

New Ampacity = Table 310–16 Ampacity × Temperature Factor × Note 8(a) Factor

Table 310–16 ampacity of No. 8 THHN is 55 amperes at 90°C.

Temperature Correction factor for 90°C conductor insulation at 100°F is 0.91.

Note 8(a) derating factor for four conductors is 0.8.

New Ampacity = 55 amperes × 0.91 × 0.8

New Ampacity = 40 amperes

6–15 CURRENT-CARRYING CONDUCTORS

Note 8(a) of Table 310–16 derating factors only apply when there is more than three current-carrying conductors bundled. Naturally all phase conductors are considered current-carrying and the following should be helpful in determine which other conductors are considered current carrying.

Grounded Conductor – Balanced Circuits, Note 10(a) of Table 310–16

The grounded (neutral) conductor of a balanced 3-wire circuit, or a balanced 4-wire wye circuit is not considered a current-carrying conductor, Fig. 6–35.

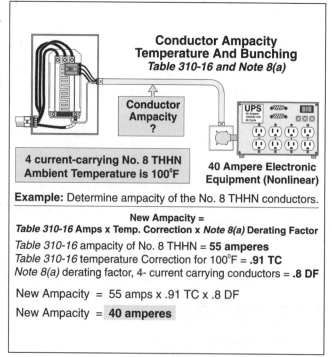

Figure 6-34
Ampacity Correction, Both Temperature and Bundling

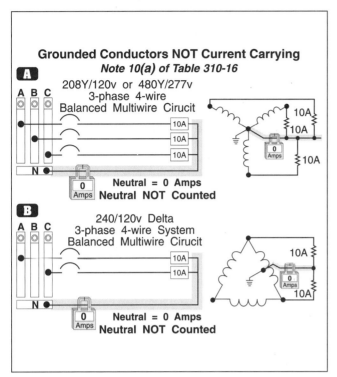

Figure 6-35

Conductor Ampacity Correction Example

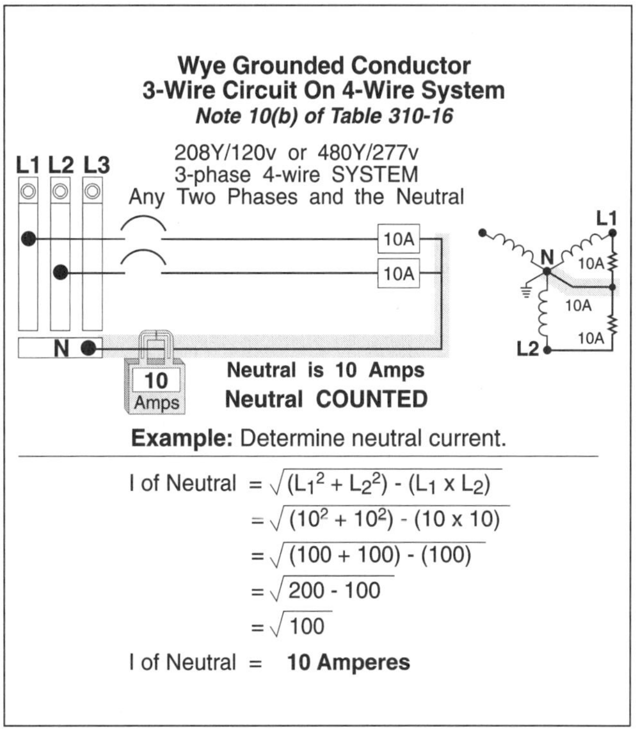

Figure 6-36

Balanced Circuits

Grounded Conductor – Unbalanced 3-wire Wye Circuit, Note 10(b) of Table 310–16

The grounded (neutral) conductor of balanced 3-wire wye circuit is considered a current-carrying conductor, Fig. 6–36. This can be proven with the following formula:

$I_n = \sqrt{(L_1 + L_2) - (L_1 \times L_2)}$ L_1 = Current of one phase L_2 = Current of the other phase

❑ **Grounded Conductor Example**

What is the neutral current for a balanced 16-ampere, 3-wire, 208Y/120-volt branch circuit that supplies fluorescent lighting, Fig. 6–37?

(a) 8 amperes   (b) 16 amperes

(c) 32 amperes   (d) none of these

• Answer: (b) 16 amperes

$I_n = \sqrt{(L_1 + L_2) - (L_1 \times I_2)}$

$I_n = \sqrt{(16^2 + 16^2) - (16 \times 16)}$

$I_n = \sqrt{512 - 256}$ $I_n = \sqrt{256}$ $I_n = 16$ amperes

Grounded Conductor – Nonlinear Loads, Note 10(c) of Table 310–16

The grounded (neutral) conductor of a balanced 4-wire wye circuit that is at least 50 percent loaded with nonlinear loads (computers, electric discharge lighting, etc.) is considered a current-carrying conductor, Fig. 6–38.

CAUTION: Because of nonlinear loads, triplen harmonic currents add on the neutral conductor. The actual current on the neutral can be almost 200 percent of the ungrounded conductor's current, Fig. 6–39.

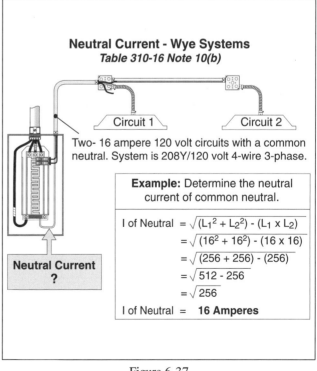

Figure 6-37

Wye 3-wire Unbalanced Circuit

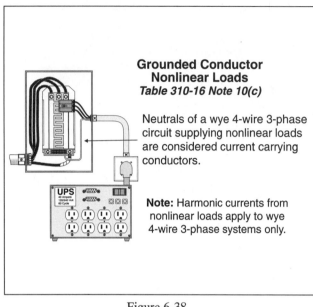

Figure 6-38

Unbalanced Wye Example

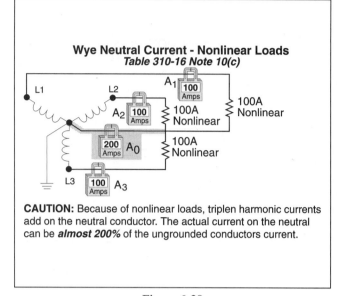

Figure 6-39

Nonlinear Loads

Two-wire Circuits

Both the grounded and ungrounded conductor of a two-wire circuit carries current and both are considered current-carrying, Fig. 6–40.

Grounding and Bonding Conductors – Note 11 of Table 310–16

Grounding and bonding conductors do not normally carry current and are not considered current-carrying, Fig. 6–41.

6–16 CONDUCTOR SIZING SUMMARY

Conductor Ampacity, Section 310–10

The ampacity of a conductor is dynamic and changes with changing conditions. The factors that affect conductor ampacity are, Fig. 6–42:

1) The allowable ampacity as listed in Table 310–16.
2) The ambient temperature correction factors, if the ambient temperature is not 86ºF.
3) Conductor bunching derating factors, if more than three current-carrying conductors are bundled together.

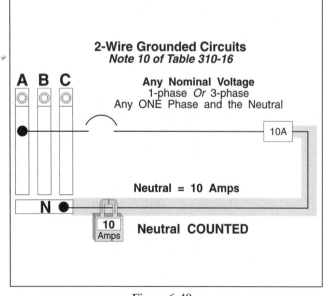

Figure 6-40

Neutral Overload

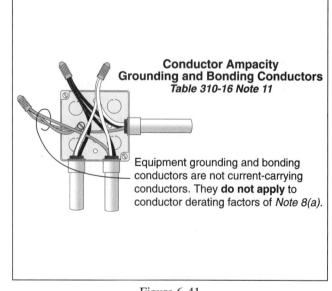

Figure 6-41

Two Wire Circuits

Terminal Ratings, Section 110–14(c)

Equipment rated 100 amperes or less must have the conductor sized no smaller than the 60°C column of Table 310–16. Equipment rated over 100 amperes must have the conductors sized no smaller than the for 75°C column of Table 310–16. However higher insulation temperature rating offers the opportunity of having a greater conductor ampacity for conductor ampacity derating.

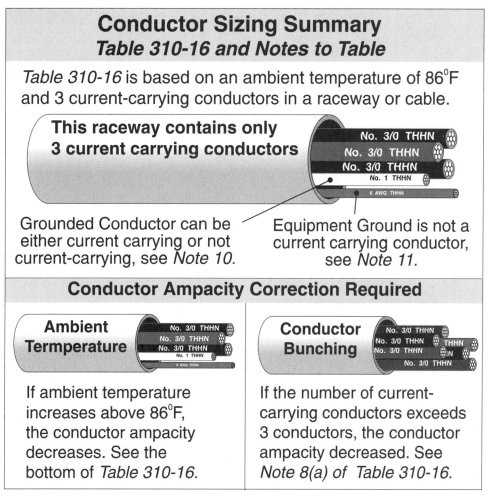

Figure 6-42

Conductor Sizing Summary

Unit 6 – Conductor Sizing And Protection Summary Questions

6–1 Conductor Insulation Property [Table 310-13]

1. THHN can be described as _____?
   (a) thermoplastic insulation with an nylon outer cover
   (b) suitable for dry and wet locations
   (c) maximum operating temperature of 90ºC
   (d) a and c

6–2 Conductor Allowable Ampacity [310–15]

2. • The maximum overcurrent protection device size for No. 14 is 15 amperes, No. 12 is 20 amperes, and No. 10 is 30 amperes. This is a general rule, but it does not apply to motors or air conditioners according to Section 240-3.
   (a) True   (b) False

6–3 Conductor Sizing

3. • Conductor sizes are expressed in American Wire Gage (AWG) from No. 40 through 0000. Conductors larger than _____ are express in circular mils.
   (a) No. 1/0   (b) No. 1
   (c) No. 3/0   (d) No. 4/0

4. The smallest size conductor permitted for branch circuits, feeders, and services for residential, commercial, and industrial locations is _____.
   (a) No. 14 copper   (b) No. 12 Aluminun
   (c) No. 12 copper   (d) a and b

6–4 Terminal Rating [110-14(c)]

5. Equipment terminals rated 100 amperes or less (circuit breakers, fuses, etc.) and pressure connector terminals for No. 14 through No. 1 conductors, shall have the conductor sized according to 60ºC temperature rating as listed in Table 310–16.
   (a) True   (b) False

6. • What is the minimum size THHN conductor that is permitted to terminate on a 70-ampere circuit breaker of fuse? Be sure to comply whith the requirements of Section 110-14(c)(1)?
   (a) No. 8   (b) No. 6
   (c) No. 4   (d) none of these

7. • What size THHN conductor is required for a 70-ampere branch-circuit if the circuit breaker and equipment is listed for 75ºC terminals and the load does not exceed 65 amperes?
   (a) No. 10   (b) No. 8   (c) No. 6   (d) No. 4

8. Terminals for equipment rated over 100 amperes and pressure connector terminals for conductors larger than No. 1, shall have the conductor sized according to 75ºC temperature rating as listed in Table 310–16.
   (a) True   (b) False

9. • What size THHN conductor is required for an air-conditioning unit, if the nameplate requires a conductor ampacity of 34 amperes? Terminals of all the equipment and circuit breakers are rated 75ºC.
   (a) No. 12   (b) No. 10
   (c) No. 8   (d) No. 14

10. What is the minumum size THHN conductor is required for a 150-ampere circuit breaker or fuse? Be sure to comply with the requirements of Section 110-14(c)(2).
    (a) No. 1/0   (b) No. 2/0
    (c) No. 3/0   (d) No. 4/0

11. In general, THHN (90ºC) conductor ampacities cannot be used when sizing conductor; but when more than three current-conductors conductors are bundled together, or if the ambient temperature is greater than 86ºF, the allowable conductor ampacity must be decreased. THHN offers the opportunity of having a greater ampacity for conductor derating purposes, thereby permitting the same conductor to be used without having to increase the conductor size.
    (a) True   (b) False

12. • What size conductor is required to supply a 190-ampere load in a dry location? Terminals are rated 75ºC.
    (a) 300 kcmils THHN   (b) No. 4/0 XHHW
    (c) No. 3/0 THWN   (d) none of these

6–5 Conductors In Parallel [310–4]

13. • Phase and grounded (neutral) conductors sized No. 1 AWG and larger are permitted to be connected in parallel.
    (a) True   (b) False

14. To insure that the currents are evenly distributed between the parallel conductors, each conductor within a parallel set must be installed in the same type of raceway (metal or nonmetallic) and must be the same length, material, mils area, insulation type, and must terminate in the same method.
    (a) True   (b) False

15. • Paralleling of conductors is done by sets. One phase or neutral is not required to be paralleled the same as those of another phase or neutral.
    (a) True   (b) False

16. When an electric relay (coil) is energized, the initial current can be very high causing significant voltage drop. The reduce voltage at the coil (because of voltage drop) can cause the coil contacts to chatter (open and close like a buzzer) or not close at all. Paralleling of control wiring conductors is often necessary to reduce the effects of voltage drop for long control runs.
    (a) True   (b) False

17. When equipment grounding conductors are installed in parallel, each raceway must have a full-size equipment grounding conductor sized according to the overcurrent protection device rating of that circuit.
    (a) True   (b) False

18. All parallel equipment grounding conductors must be the same length, material, circular mils, insulation, terminate in the same manner, and are required to be a minimum No. 1/0.
    (a) True   (b) False

19. What size equipment grounding conductor is required in each raceway for an 800-ampere, 500-kcmil feeder parallel in two raceways?
    (a) No. 3   (b) No. 2   (c) No. 1   (d) No. 1/0

20. If we have an 800-ampere service, with a calculated demand load of 750 amperes, what size 75ºC conductors would be required if parallel in two raceways?
    (a) two – No. 4/0   (b) two – 250 kcmil   (c) two – 500 kcmil   (d) two – 750 kcmil

21. • What are the circular mils required for a 250-ampere feeder paralleled in two raceways?
    (a) two – No. 3   (b) two – No. 2   (c) two – No. 2/0   (d) two – No. 1/0

6–6 Conductor Size – Voltage Drop

22. There is no mandatory rule in the NEC to limiting the voltage drop on conductors, but the NEC recommends that you consider it effects.
    (a) True   (b) False

6–7 Overcurrent Protection [Article 240]

23. One of the purposes of conductor overcurrent protection is to protect the conductors against excessive or dangerous heat.
    (a) True   (b) False

24. The overcurrent device must be designed and rated to clear fault current and must have a short-circuit interrupting rating sufficient for the available fault levels. The minimum interruption rating for circuit breakers is _____ amperes and _____ amperes for fuses.
    (a) 10,000; 10,000 (b) 5,000; 5,000 (c) 5,000; 10,000 (d) 10,000; 5,000

25. The following are the standard sized circuit breakers and fuses _____ ampere.
    (a) 25
    (b) 90
    (c) 350
    (d) any of these

26. The maximum continuous load permitted on a overcurrent protection device is limited to _____ percent of the device rating.
    (a) 80 (b) 100 (c) 125 (d) 150

6–8 Conductor Sizing And Protection [240-3]

27. If the ampacity of a conductor does not correspond with the standard ampere rating of a fuse or circuit breaker, the next size up protection device is permitted. This applies only if the conductors supply multioutlet receptacles for portable cord- and plug-connected loads.
    (a) True (b) False

28. What size conductor (75ºC) is required for a 70-ampere breaker that supplies a 70-ampere load?
    (a) No. 8 (b) No. 6 (c) No. 4 (d) Any of these

6–11 Conductor Ampacity [310-10 And 310-15]

29. The temperature rating of a conductor is the maximum operating temperature the conductor insulation can withstand (without serious damage) over a prolonged period of time. The _____ provide guidance for adjusting the conductors ampacities for the different conditions.
    (a) conductor allowable ampacities
    (b) ambient temperature correction factors
    (c) Notes correction factors
    (d) continuous load factor
    (e) a, b and c

6–12 Ambient Temperature Correction Factor [Table 310–16]

30. The ampacities listed in Table 310–16 apply only when the ambient temperature is 40ºC and there are no more than two current-carrying conductors bundled together. If the ambient temperature is not 40ºC, or there are more than two current-carrying conductors in a raceway, the allowable ampacities must be adjusted to reflect the ampacity under the condition of use.
    (a) True (b) False

31. • What is the ampacity of No. 8 THHN conductors when installed in a walk-in cooler if the ambient temperature is 50ºF?

    (a) 40 amperes (b) 50 amperes (c) 55 amperes (d) 57 amperes

32. • What size conductor is required to feed a 16-ampere load, when the conductors (in a raceway) pass over a roof, ambient temperature of 100ºF? The circuit is protected with a 20-ampere protection device.
    (a) No. 14 THHN (b) No. 12 THHN (c) No.10 THHN (d) No. 8 THHN

6–13 Conductor Bunching Derating Factor, Note 8(a) Of Table 310–16

33. When four or more current-carrying conductors are bundled together for more than _____ inches, the conductor allowable ampacity must be reduced according to the factors listed in Note 8(a) of Table 310–16.
    (a) 12 (b) 24 (c) 36 (d) 48

34. What is the ampacity of four No. 1/0 THHN conductors?
    (a) 110 amperes (b) 135 amperes (c) 155 amperes (d) 175 amperes

35. • A raceway contains eight current-carrying conductors. What size conductor is required to feed a 21 ampere noncontinuous lighting load? Overcurrent protection device rated 30 amperes.
    (a) No. 14 THHN (b) No. 12 THHN (c) No. 10 THHN (d) Any of these

Chapter 2 NEC Calculations And Code Questions   Unit 6 Conductor Sizing And Protection 207

6–14 Ambient Temperature And Conductor Bunching Derating Factors

36. What is the ampacity of eight current-carrying No. 10 THHN conductors installed in ambient temperature of 100°F?
    (a) 21 amperes   (b) 26 amperes   (c) 32 amperes   (d) 40 amperes

6–15 Current-Carrying Conductors

37. • The neutral conductor of a balanced three-wire, delta circuit, or four-wire, wye circuit is considered a current-carrying conductor for the purpose of applying Note 8 derating factors.
    (a) True   (b) False

38. The neutral conductor of a balanced four-wire wye circuit that is at least 50 percent loaded with nonlinear loads (electric-discharge lighting, electronic ballast, dimmers, controls, computers, laboratory test equipment, medical test equipment, recording studio equipment, etc.) is not considered a current-carrying conductor for the purpose of applying Note 8 derating factors.
    (a) True   (b) False

39. • The neutral conductor of balanced three-wire wye circuit is not considered a current-carrying conductor for the purpose of applying Note 8(a) derating factors.
    (a) True   (b) False

6–16 Conductor Sizing Summary

40. • The ampacity of a conductor can be different along the length of the conductor. The higher ampacity is permitted to be used for the lower ampacity, if the length of the lower ampacity is no more than 10 feet or no more than 10 percent of the length of the circuit conductors.
    (a) True   (b) False

41. Most terminals are rated 60°C for equipment 100 amperes and less and 75°C for equipment terminals rated over 100 amperes. Despite the conductor ampacity, conductors must be sized no smaller than the terminal temperature rating.
    (a) True   (b) False

Challenge Questions

6–7 Overcurrent Protection - Continuous Load [220–3(a), 220–10(b), 384–16(a)]

42. • A continuous load of 27 amperes requires the circuit overcurrent device to be sized at _____ amperes.
    (a) 20   (b) 30   (c) 40   (d) 35

43. • What size overcurrent protection device is required for a 45 ampere continuous load? The circuit is in a raceway with 14 current-carrying conductors.
    (a) 45 amperes   (b) 50 amperes   (c) 60 amperes   (d) 70 amperes

44. • A 65 ampere continuous load requires a _____ ampere overcurrent protection device.
    (a) 60   (b) 70   (c) 75   (d) 90

45. • A department store (continuous load) feeder supplies a lighting load of 103 amperes. The minimum size overcurrent protection device permitted for this feeder is _____ amperes.
    (a) 110   (b) 125   (c) 150   (d) 175

6–12 Ambient Temperature Derating Factor [Table 310–16]

46. • A No. 2 TW conductor is installed in a location where the ambient temperature is expected to be 102°F. The temperature correction factor for conductor ampacity in this location is _____.
    (a) .96   (b) .88   (c) .82   (d) .71

47. • If the ambient temperature is 71°C, the minimum insulation that a conductor must have and still have the capacity to carry current is _____.
    (a) 60°C   (b) 105°C   (c) 90°C   (d) any of these

6–13 Conductor Bunching Derating Factor, Note 8(a) Of Table 310–16

48. • The ampacity of six current-carrying No. 4/0 XHHW aluminum conductors installed in a ground floor slab (wet location) is _____ amperes.
 (a) 135 (b) 185 (c) 144 (d) 210

6–14 Ambient Temperature And Conductor Bundling Derating Factors

49. • The ampacity of fifteen current-carrying No. 10 RHW aluminum conductors, in an ambient temperature of 75ºF, would be _____ amperes.
 (a) 30 (b) 22 (c) 16 (d) 12

50. • A No. _____ THHN conductor is required for a 19.7-ampere load if the ambient temperature is 75ºF, and there are nine current-carrying conductors in the raceway.
 (a) 14 (b) 12 (c) 10 (d) 8

51. • The ampacity of the nine current-carrying No. 10 THW conductors installed in a 20-inch-long raceway is _____ amperes.
 (a) 25 (b) 30 (c) 35 (d) none of these

52. • The ampacity of ten current-carrying No. 6 THHW conductors installed in an 18-inch conduit, with an ambient temperature of 39ºC is _____ amperes.
 (a) 47 (b) 68 (c) 66 (d) 75

6–15 Current-Carrying Conductors

53. A raceway contains the following: One four-wire multiwire circuit that supplies a balanced incandescent 120 volt lighting load; one four-wire multiwire circuit that supplies a balanced 120 volt fluorescent lighting load; two conductors that supply a receptacle; and there is one equipment grounding conductor. The system is three phase 208Y/120 volts. How many of these conductors are considered current-carrying?
 (a) 7 conductors (b) 8 conductors (c) 9 conductors (d) 11 conductors

54. • There are a total of nine No. 10 THW conductors in a raceway. The system voltage is three-phase, 208Y/120 volts. One of the conductors is an equipment grounding conductor; four conductors supply a four-wire multiwire 120-volt branch circuit for balanced fluorescent lighting, and the remaining conductors supply a four-wire multiwire 120-volt branch circuit for balanced incandescent lighting. Taking all of these factors into consideration, how many of these conductors are considered current-carrying?
 (a) 6 conductors (b) 7 conductors (c) 9 conductors (d) 10 conductors

NEC Questions

Article 810 – Radio and Television Equipment

Part B. Receiving Equipment – Antenna Systems

55. Soft-drawn or medium-drawn copper lead in conductors for television equipment antenna systems shall be permitted where the maximum span between points of support is less than _____ feet.
 (a) 35 (b) 30 (c) 20 (d) 10

56. An outdoor antenna of a receiving station with a 75 foot span using a copper-clad steel conductor shall not be less than No. _____.
 (a) 10 (b) 12 (c) 14 (d) 17

Part C. Amateur Transmitting And Receiving Stations – Antenna Systems

57. • For antenna conductors for an amateur transmitting and receiving station where of hard-drawn copper; the maximum open span length is 200 feet and requires a minimum of No. _____.
 (a) 14 (b) 12 (c) 10 (d) 8

Article 820 – Community Antenna Television And Radio Distribution Systems

Part B. Cables Outside and Entering Buildings

58. Where practical, coaxial cables for a CATV system shall be separated by _____ from lightning conductors.
    (a) 3 inches   (b) 6 inches   (c) 6 feet   (d) 2 feet

Part E. Cables Within Buildings

59. Coaxial cable is permitted to be placed in a raceway, compartment, outlet box, or junction box, with the conductors of light or power circuits, or Class 1 circuits when _____.
    (a) installed in rigid conduit   (b) separated by a permanent barrier
    (c) insulated   (d) none of these

Chapter 9 – Tables And Examples

NEC Chapter 9 has two distinct sections. Part A. Tables and Part B. Examples.

Part A. Tables are related to conductors and conduits in some form.

Table 1 list the maximum percentage of conductor fill based on common conditions where the length of the conductor and number of raceway bends are within reasonable limits [Fine Print Note after Note 9].

Table 1, Note 1 – When all of the conductors are the same size and insulation, the number of conductors permitted in a raceway can be determined simply by looking at the Tables located in Appendix C – Conduit and Tubing Fill Tables for Conductors and Fixture Wires of the Same Size. Tables C1 through C13 are based on maximum percent fill as listed in Table 1 of Chapter 9.

Table 1, Note 3 – When equipment grounding conductors are installed in a raceway, the actual area of the conductor must be used when calculating raceway fill. Chapter 9, Table 5 can be used to determine the cross-sectional area of insulated conductors, and Chapter 9, Table 8 can be used to determine the cross-sectional area of bare conductors [Note 8 or Table 1, Chapter 9].

Table 1, Note 4 – The cross-sectional areas of conduit and tubing can be found in Table 4 of Chapter 9. When a conduit or tubing raceway does not exceed 24 inches in length, it is permitted to be filled to 60 percent of its total cross-sectional area.

Table 1, Note 7 – When the calculated number of conductors (all of the same size and insulation) results in .8 or larger, the next whole number can be used. This note only applies for the development of Appendix C and should not be used for exam purposes.

Table 1, Note 8 – The dimensions for bare conductor are listed in Table 8 of Chapter 9.

Table 4 lists the dimensions and cross-sectional area for conduit and tubing dependent on the raceway type and the maximum percentage fill as listed in Table 1 of Chapter 9.

Table 5 lists the cross-sectional area of insulated conductors and fixture wires and Table 5A list the cross-sectional area for compact aluminum building wires.

Table 8 contains conductor properties such as: cross-sectional area in circular mils, number of strands per conductor, cross-sectional area in square inch for bare conductors, and the conductors resistance at 75ºC for direct current for both copper and aluminum wire.

Table 9 the conductors resistance at 75ºC for alternating current for both copper and aluminum wire.

Table 10 contains the expansin characteristics of PVC.

Part A. Tables, Notes To Tables

60. A 1½ inch nipple with three conductors can be filled to an area of _____ square inches.
    (a) .88   (b) 1.07   (c) 1.224   (d) 1.34

61. The number of No. 12 THW conductors permitted in ¾-inch conduit will be _____ the number of No. 12 TW permitted in a ¾-inch conduit. *Note.* THW insulation is thicker than TW insulation
    (a) equal to   (b) greater than   (c) less than   (d) none of these

62. When conduit nipples having a maximum length not to exceed 24 inches are installed between boxes:
    I. The nipple can be filled 75 percent.   II. Note 8 (a) derating does apply.
    III. Note 8 (a) derating does not apply.   IV. The nipple can be filled 60 percent.
    (a) I and II   (b) II and IV   (c) III and IV   (d) I and II

210 Unit 6 Conductor Sizing And Protection   Chapter 2 NEC Calculations And Code Questions

63. A bare No. 10 AWG solid copper wire has a cross section area of _____ square inches.
    (a) .008   (b) .101   (c) .012   (d) .106

64. What length of nipple may utilize the 60 percent conductor fill?
    (a) 12 inch   (b) 18 inch   (c) 24 inch   (d) all of these

65. You may install _____ No. 8 TW conductors in a 1½-inch conduit.
    (a) 11   (b) 22   (c) 17   (d) 8

66. The area of square inches for a Aluminum No. 1/0 bare conductor is _____.
    (a) .087   (b) .109   (c) .137   (d) .173

67. Five lead-covered conductors are fill-limited to _____ percent of the conduit or tubing area.
    (a) 35   (b) 60   (c) 40   (d) 55

Random Sequence, Section 90 Through Section 200–10

68. • The Code covers all of the following electrical installations except _____.
    (a) industrial substations
    (b) floating dwellings
    (c) in or on private and public buildings
    (d) public utilities generation and transmission

69. The National Electrical Code covers installation of _____.
    (a) conductors and equipment within or on public and private buildings or other structures
    (b) conductors and equipment that connect to the supply of electricity
    (c) outside conductors and equipment on the premises, and fiber optic cable
    (d) all of the above

70. • An overload may be caused by _____.
    I. a short circuit   II. a ground fault
    III. operation of equipment in excess of normal full load amps
    (a) I only   (b) II only
    (c) III only   (d) I, II or III

71. When the term exposed is used by the Code it refers to _____.
    (a) capable of being inadvertently touched or approached nearer than a safe distance by a person
    (b) parts are not suitably guarded, isolated, or insulated
    (c) wiring on, or attached to, the surface or behind panels being designed to allow access
    (d) all of these

72. A _____ rated box may also be used as a weatherproof box.
    (a) raintight   (b) rainproof   (c) watertight   (d) all of these

73. • The purpose of this Code is the practical safeguarding of persons and property from hazards arising from the use of electricity. This Code contains provisions considered necessary for safety regardless of _____.
    (a) efficient use
    (b) convenience
    (c) good service or future expansion of electrical use
    (c) all of these

74. • On circuits over 600 volts, nominal, where energized live parts are exposed, the minimum clear work space shall not be less than _____ feet high for over 600 volts.
    (a) 3   (b) 5   (c) 6¼   (d) 6½

75. Plans and specifications that provide ample space in raceways, spare raceways, and additional spaces will allow for future increases in the use of electricity. Distribution centers located in readily accessible locations will provide convenience and safety of operation. See Sections 110-16 and 240-24 for clearances and accessibility.
    (a) True   (b) False

76. At least one entrance not less than 24 inches wide and 6½ feet high shall be provided to give access to the working space about electric equipment. On switchboard and control panels exceeding 6 feet in width, there shall be one entrance at each end of such board, except where the work space required in Section 110-34(a) is doubled.
    (a) True   (b) False

77. _____ shall be provided to give safe access to the working space around electric equipment over 600 volts, installed on platforms, balconies, mezzanine floors, or in attic or roof rooms or spaces.
    (a) Ladders
    (b) Platforms or ladders
    (c) Permanent ladders or stairways
    (d) Openings

78. The dimension of working clearance for access to live parts operating at 300 volts, nominal, to ground where there are exposed live parts on both sides of the work space (not guarded) with the operator between is _____ feet according to Table 110–16(a).
    (a) 3   (b) 3½   (c) 4   (d) 4 ½

79. Only wiring methods recognized as _____ are included in the Code.
    (a) identified   (b) efficient   (c) suitable   (d) cost effective

80. An enclosure or device so constructed that a beating rain will not enter is rainproof.
    (a) True   (b) False

81. Soldered splices must be joined mechanically so as to be electrically secure before soldered.
    (a) True   (b) False

82. Nonmetallic sheath cable, when run in the space between studs are defined as _____.
    (a) inaccessible   (b) concealed   (c) hidden   (d) enclosed

83. • When one electrical circuit controls another circuit through a relay, that first circuit is called a _____.
    (a) control circuit   (b) remote control circuit
    (c) signal circuit   (d) controller

84. An insulated conductor larger than No. 6 AWG that is not white or gray may be reidentified and used as a grounded conductor.
    (a) True   (b) False

85. A contact device installed at the outlet for the connection of a single attachment plug is known as a _____.
    (a) receptacle outlet   (b) duplex receptacle   (c) receptacle   (d) plug

86. A type of surface or flush raceway designed to hold conductors and receptacles, assembled in the field or at the factory is called _____.
    (a) wiremold   (b) multioutlet assembly
    (c) surface raceways   (d) plugmold

87. Equipment is required to be installed and used according to its _____ instructions.
    (a) listed or published   (b) labeled or design
    (c) listed or labeled   (d) any of the above

88. • In applications where a cable containing an insulated conductor with a white or natural gray outer finish is used for single-pole, three-way, or four-way switch loops, the white or natural gray conductor can be used for the supply to the switch and reidentification of the white or natural gray conductor is not required.
    (a) True   (b) False

89. A solderless pressure connector is a device that _____ between two or more conductors, or between one or more conductors and a terminal by means of mechanical pressure and without the use of solder.
    (a) provides access
    (b) protects the wiring
    (c) is never needed
    (d) establishes a connection

90. A qualifying term (as applied to circuit breakers) indicating that there is a purposely introduced delay in the tripping action of the circuit breaker, which delay decreases as the magnitude of the current increases, is defined by the Code as _____.
    (a) adverse time　　(b) inverse time　　(c) time delay　　(d) a timed unit

91. A junction box above a suspended ceiling that has removable panels is considered to be _____.
    (a) concealed　　(b) accessible　　(c) readily accessible　　(d) recessed

92. • The necessary equipment, usually a circuit breaker or switch and fuse, and their accessories, is referred to as _____.
    (a) service equipment　　(b) service
    (c) service disconnect　　(d) service overcurrent device

93. Electrical equipment that depends on the _____ principles for cooling of exposed surfaces shall be installed so that room airflow over such surfaces is not prevented by walls or by adjacent installed equipment.
    (a) Peter
    (b) natural circulation of air and convection
    (c) artificial cooling and circulation
    (d) air-conditioning

94. The Code defines a(n) _____ as a structure that stands alone or that is cut off from adjoining structures by fire walls; with all openings therein protected by approved fire doors.
    (a) unit　　(b) apartment　　(c) building　　(d) utility

95. • It is the intent of this Code that factory-installed internal wiring or the construction of equipment need not be inspected at the time of installation of the equipment, except to detect alterations or damage, if the equipment has been listed by a qualified electrical testing laboratory.
    (a) True　　(b) False

96. For the purpose of Article 200 connected so as to be capable of carrying current as distinguished from connection through electromagnetic induction defines _____.
    (a) effectively grounded　　(b) electrically connected
    (c) a grounded system　　(d) none of these

97. • Circuits for lighting and power shall not be connected to any system containing _____.
    (a) hazardous material
    (b) trolley wires with ground returns
    (c) poor wiring methods
    (d) dangerous chemicals or gasses

98. A(n) _____ is a point of a wiring system where current or power is taken for equipment that uses electricity.
    (a) box　　(b) receptacle　　(c) outlet　　(d) device

99. A 7 foot high fence (or wall) is considered acceptable for preventing access to an over 600 volts, nominal installation.
    (a) True　　(b) False

100. Parts of electrical equipment which, under ordinary operation produce _____, shall be enclosed or separated and isolated from all combustible material.
    (a) arcs or sparks　　(b) flames　　(c) molten metal　　(d) all of these

101. A _____ is a system where the premises wiring system whose power is derived from transformer and that has no direct electrical connection, including a solidly connected grounded circuit conductor, to supply conductors originating in another system.
    (a) separately derived　　(b) classified
    (c) direct　　(d) emergency

102. The Code prohibits damage to the internal parts of electrical equipment by foreign material such as paint, plaster, cleaners, etc. Precautions must be taken to provide protection from the detrimental effects of paint, plaster, cleaners, etc. on the internal parts of _____, unless specified by the manufacture or listed for the purpose.
    (a) panelboards　　(b) receptacles
    (c) lampholders　　(d) all of the above

103. An isolating switch is one that is _____.
  (a) not readily accessible to persons unless special means for access are used
  (b) capable of interrupting the maximum operating overload current of a motor
  (c) intended for use in general distribution and branch circuits
  (d) intended for isolating an electrical circuit from the source of power

104. Material identified by the superscript letter "x" includes text extracted from _____.
  (a) other Code articles and sections (b) other NFPA documents
  (c) other IOWA publications (d) none of these

105. A value assigned to a circuit or system for the purpose of conveniently designating its voltage class is named _____.
  (a) root-mean-square (b) circuit voltage (c) nominal voltage (d) source voltage

106. A _____ is a device, or group of devices, or other means by which the conductors of a circuit can be disconnected from their source of supply.
  (a) circuit breaker (b) fuse (c) disconnecting means (d) switch

107. • The conductor used to connect the grounded circuit of a wiring system to a grounding electrode is the _____.
  (a) grounded conductor (b) grounding conductor
  (c) bonding jumper (d) bonding jumper, main

108. • The case or housing of apparatus, or the fence surrounding an installation to prevent accidental contact of persons to energized parts is called _____.
  (a) guarded (b) covered (c) protection (d) an enclosure

109. Connected to earth or to some conducting body that serves in place of the earth is called _____.
  (a) grounding (b) bonded (c) grounded (d) all of the above

110. The scope of Article 200 provides requirements for _____.
  (a) identification of terminals
  (b) grounded conductors in premises wiring systems
  (c) identification of grounded conductors
  (d) all of these

111. The premises wiring shall not be electrically connected to a supply system unless the latter (supply system) contains, for any grounded conductor of the interior system, a corresponding conductor that is grounded.
  (a) True (b) False

112. The National Electrical Code is _____.
  (a) intended to be a design manual
  (b) meant to be used as an instruction guide for untrained persons
  (c) for the practical safeguarding of persons and property
  (d) published by the Bureau of Standards

113. A system or circuit conductor that is intentionally grounded is a _____.
  (a) grounding conductor (b) unidentified conductor
  (c) grounded conductor (d) none of the above

114. A hoistway is any _____ in which an elevator or dumbwaiter is designed to operate.
  (a) hatchway or well hole (b) vertical opening or space
  (c) shaftway (d) all of these

115. _____ means so constructed or protected that exposure to the weather will not interfere with successful operation.
  (a) Weatherproof (b) Weathertight (c) Weather-resistant (d) All Weather

116. Varying duty is defined as _____.
  (a) intermittent operation in which the load conditions are regularly recurrent
  (b) operation at a substantially constant load for an indefinite length of time
  (c) operation for alternate intervals of load and rest; or load, no load, and rest
  (d) operation at loads, and for intervals of time, both of which may be subject to wide variations

117. According to the Code, automatic is self-acting, operating by its own mechanism when actuated by some impersonal influence such as _____.
    (a) change in current strength (b) temperature
    (c) mechanical configuration (d) all of these

118. Electrical wiring and equipment that is capable of being reached quickly for operation, renewal, or inspections, without resorting to portable ladders and such is known as _____.
    (a) accessible (equipment) (b) accessible (wiring methods)
    (c) accessible, readily (d) all of these

Random Sequence, Section 200–11 Through Section 230–23

119. The rating of a branch circuit serving continuous loads shall be not less than _____.
    (a) 125 percent of the noncontinuous loads plus 100 percent of the continuous loads
    (b) 100 percent of the noncontinuous loads plus 125 percent of the continuous loads
    (c) 80 percent of the branch circuit rating
    (d) 125 percent of all loads involved

120. • What is the feeder demand load for three household ranges rated 9-kW and three household ranges rated 14-kW?
    (a) 78 kW (b) 21 kW (c) 56 kW (d) 22.05 kW

121. • The minimum feeder demand load for fifteen 8 kW cooking units according to Column A of Table 220–19 would be _____ kW.
    (a) 30 (b) 15 (c) 21 (d) 38.4

122. • GFCI-protected grounding-type receptacles installed in an existing outlet box without an equipment grounding conductor must be _____ to identify that they are GFCI protected.
    (a) listed (b) labeled (c) identified (d) marked

123. Autotransformers may supply any type of branch circuit.
    (a) True (b) False

124. The ampacity of feeder conductors shall not be less than _____ amperes where the load supplied consists of any of the following number and types of circuits: (1) two or more 2-wire branch circuits supplied by a 2-wire feeder; (2) more than two 2-wire branch circuits supplied by a 3-wire feeder; (3) two or more 3-wire branch circuits supplied by a 3-wire feeder; or (4) two or more 4-wire branch circuits supplied by a 3-phase 4-wire feeder.
    (a) 30 (b) 100 (c) 60 (d) 20

125. • Table 220–19 applies to household cooking appliances of ____ kW and larger.
    (a) 3½ (b) 1½ (c) 2 (d) 8

126. • When applying a demand factor to commercial kitchen equipment, a demand factor of _____ percent applies to the total connected (nameplate) load of two ranges, two water heaters, and two cooking equipment exhaust fans.
    (a) 100 (b) 90 (c) 80 (d) 70

127. Service conductors run above the top level of a window _____.
    (a) must be kept three feet away from the window
    (b) are permitted to be less than three feet from the window
    (c) are considered accessible
    (d) none of these

128. • What is the minimum 240-volt branch circuit conductor ampacity for one wall-mounted oven-rated 8 kW?
    (a) 66.7 amperes (b) 33.3 amperes (c) 27 amperes (d) none of these

129. Overhead service conductors shall _____.
    (a) normally withstand exposure to atmospheric conditions
    (b) be insulated with an extruded thermoplastic or thermosetting insulating material
    (c) be covered with an extruded thermoplastic or thermosetting insulating material
    (d) all of these

130. In dwelling unit garages, GFCI protection is not required for receptacles that are not readily accessible.
(a) True (b) False

131. Inaccessible receptacles for appliances fastened in place or appliances in dedicated spaces (dishwasher, microwave, etc.) are considered the receptacle outlets required by 210–52(c).
(a) True (b) False

132. The equipment grounding conductor of a branch circuit shall be identified by a _____ color.
I. natural gray II. continuous green
III. continuous white IV. green with yellow stripe
(a) I and III (b) II and IV (c) III only (d) II only

133. One- and two-family dwelling units at grade level shall have _____ outlets installed outdoors for each unit.
(a) an optional number of (b) one (c) a required number of (d) two

134. • Where two dwelling units are supplied by a single feeder and the computed load under Part B of Article 220 exceeds that for _____ identical units computed under Section 220–32, the lesser of the two loads shall be permitted to be used.
(a) two (b) three (c) four (d) five

135. The maximum unbalanced load for household electric ranges, wall-mounted ovens, or counter-mounted cooking units, shall be calculated at _____ percent of the demand load as determined by Table 220–19.
(a) 50 (b) 70 (c) 100 (d) 125

136. Conductors for feeders, sized to prevent a voltage drop exceeding 3 percent at the farthest outlet of power, or combinations of such loads and where the maximum total voltage drop on both feeders and branch circuits to the farthest outlet does not exceed _____ percent, will provide reasonable efficiency of operation.
(a) 1 (b) 3 (c) 5 (d) 7

137. • The load for 100 receptacles in a commercial building after applying the demand factors of Table 220–13 is _____ VA.
(a) 15,000 (b) 10,000 (c) 8,000 (d) 14,000

138. • A feeder neutral load is permitted to be calculated at 70 percent of _____.
(a) that portion of the load on the ungrounded conductors in excess of 200 amperes
(b) that portion of the load on the grounded conductors in excess of 200 amperes
(c) the demand load of household ranges and dryers as calculated in Article 220
(d) b and c

139. In dwelling-unit bathrooms, at least one receptacle outlet is required adjacent to each _____ sink.
(a) wet bar (b) kitchen (c) bath (d) all of these

140. • Tap conductors for household cooking equipment supplied from a 50 ampere branch circuit shall have an ampacity of not less than _____.
(a) 50 (b) 70 (c) 20 (d) 80

141. On a four-wire delta connected secondary with the midpoint of one phase grounded, the leg with the highest voltage to ground shall be identified by a(n) _____ color.
(a) orange (b) red (c) brown (d) yellow

142. The receptacle outlets required by Article 210 for dwelling units shall be in addition to any receptacle that is part of any lighting fixture or appliance, located within _____, or located over 5½ feet above the floor.
(a) cabinets or cupboards (b) wall units or wet bars
(c) both a and b (d) neither a nor b

143. • Receptacles installed under exceptions to Section 210–8(a)(2) shall be considered as meeting the requirements of Section 210–52(g).
(a) True (b) False

144. When the number of receptacles for an office building is unknown, an additional load of _____ volt-amperes per square foot is required.
(a) 1 (b) 2 (c) ½ (d) 3

145. Trees cannot be used for the support of overhead conductor spans.
    (a) True (b) False

146. Receptacles rated 15 amperes may be connected to 20-ampere multioutlet branch circuits.
    (a) True (b) False

147. A small appliance circuit in a dwelling unit shall not supply an outlet for the _____.
    (a) dining room (b) pantry (c) hallway (d) refrigerator

148. • The wall space afforded by _____, such as freestanding bar-type counters or railings, shall be included in the 6-foot measurement of receptacles for dwelling units.
    (a) room dividers (b) partial walls (c) fixed room dividers (d) wet bars

149. For purposes of section 210–8(a)(4), unfinished basements are defined as portions or areas of the basement not intended as _____ rooms and limited to storage areas, work areas, and the like.
    (a) finished (b) play (c) inhabitable (d) habitable

150. When determining the minimum number of branch circuits for a building, the _____ shall be used.
    (a) total computed load (b) size or rating of circuits used
    (c) expected load (d) a and b

151. Twenty ampere, 120 volt grounding type receptacles may be installed only on circuits of the voltage class and current for which they have been rated.
    (a) True (b) False

152. For circuits supplying lighting units having ballasts, transformers, or autotransformers, the computed load shall be based on the _____.
    (a) total ampere ratings of such units and not on the total watts of the lamps
    (b) wattage of the lamps
    (c) volt-amperes (VA) of the lamps
    (d) none of these

153. All 125-volt, single-phase, 15- or 20-ampere receptacles installed in boathouses of dwelling units shall have ground-fault circuit-interrupter protection for personnel.
    (a) True (b) False

154. • The overload protection of the _____ conductors for a new restaurant shall be in accordance with Sections 230–90 and 240–3.
    (a) grounded (b) feeder (c) service-entrance (d) grounding

155. Permanently installed electric baseboard heaters equipped with factory-installed receptacle outlets shall not be connected to the heater circuits.
    (a) True (b) False

156. The neutral feeder conductor must be capable of carrying the maximum _____ load.
    (a) connected (b) unbalanced (c) demand (d) grounded

157. Which of the following wiring methods are permitted for outside wiring?
    (a) MC Cable (b) flexible metal conduit
    (c) EMT (d) all of these

158. Dwelling unit appliance outlets are required to be located within _____ feet of the intended location of the appliance to be served.
    (a) 6 (b) 3 (c) 4 (d) 2

159. If a building has more than one voltage system, each ungrounded multiwire branch circuit conductor must _____.
    (a) be identified by phase and voltage
    (b) have identification permanently posted at each panelboard
    (c) a and b
    (d) none of these

Chapter 2 NEC Calculations And Code Questions   Unit 6 Conductor Sizing And Protection   217

160. All general-use receptacle outlets of 20-ampere or less rating in one-family, two-family, and multifamily dwellings and in guest rooms of hotels and motels [except those connected to the receptacle circuits specified in Sections 220–4(b) and (c)] shall be considered as outlets for _____, and no additional load calculations shall be required for such outlets.
 (a) habitable rooms
 (b) unfinished basements
 (c) receptacles
 (d) general illumination

161. The minimum size branch circuit conductor permitted by the Code is _____ .
 (a) No. 14   (b) No. 12   (c) No. 10   (d) none of these

162. General lighting load feeder demand factors apply to apartment buildings without provisions for cooking.
 (a) True   (b) False

163. Table 220–19 demand factors apply only to household cooking appliances over _____ kW.
 (a) 3   (b) 1¾   (c) 2   (d) 8

164. A 230 volt, single-phase, 21.85 kW commercial cooking appliance is to be supplied by TW copper branch circuit conductors. The minimum size conductor permitted for this load is a No. _____.
 (a) 3   (b) 2   (c) 1   (d) 1/0

165. When non-grounding type receptacles are replaced in locations where ground-fault circuit-interrupter protection is required, the replacement receptacles must be _____.
 (a) of the grounding type in all cases
 (b) of the nongrounding type in all cases
 (c) GFCI protected
 (d) none of these

166. Overhead spans of open conductors and open multiconductor cables of not over 600 volts, nominal, shall have a minimum of _____ feet over commercial areas not subject to truck traffic where the voltage is limited to 300 volts to ground.
 (a) 11   (b) 12   (c) 15   (d) 18

Random Sequence, Section 230–23 Through Section 250–79

167. • A service is supplied by three metal raceways. Each raceway contains 600 kcmil phase conductors. Determine the size of a single-service bonding jumper for all three raceways.
 (a) 3/0   (b) 4/0   (c) 225 kcmil   (d) 500 kcmil

168. • Frames of electric ranges and clothes dryers may be grounded to the grounded (neutral) conductor of a service-entrance cable having an uninsulated grounded conductor, where the branch circuit _____.
 (a) originates at the service equipment
 (b) originates at a subpanel
 (c) is a three-wire with ground cable
 (d) uninsulated neutrals are not permitted

169. • Where an alternating current system is grounded at any point, a grounded (neutral) conductor is required and must be bonded to each service disconnect. If the service conductors are 600 kcmil, the grounded service conductor must be sized not smaller than _____ .
 (a) No. 1   (b) No. 1/0   (c) No. 2   (d) neutral not required

170. Direct buried service lateral conductors emerging from underground must be protected by enclosures or raceways to a point _____ feet above ground.
 (a) 6   (b) 8   (c) 10   (d) 12

171. • For grounded systems, a main bonding jumper shall be used to connect the _____ to the grounded conductor of the system at each service disconnect.
 (a) equipment grounding conductor
 (b) service disconnect enclosure
 (c) a and b
 (d) none of these

172. • Under certain conditions, tap conductors in lengths up to _____ feet are permitted in high bay manufacturing buildings that have walls over 35 feet high.
 (a) 25   (b) 50   (c) 75   (d) 100

173. • Ungrounded service conductors shall not be smaller than _____.
   I. 100 ampere for a three-wire service to a one-family dwelling with six or more two-wire branch circuits
   II. 100 ampere for a three-wire service to a one-family dwelling with an initial net computed load of 10 kVA or more
   III. 60 ampere for nondwelling loads
   (a) I only   (b) II only   (c) III only   (d) I, II and III

174. • _____ is defined as properly localizing a fault condition to restrict outages to the equipment affected, accomplished by choice of selective fault protective devices.
   (a) Monitoring   (b) Coordination   (c) Choice selection   (d) Fault device

175. For circuits over _____ volts to ground, bonding jumpers are required around ringed knockouts for metal raceways.
   (a) 240   (b) 277   (c) 250   (d) 480

176. Temporary ground-fault currents that occur on the grounding conductor, only while the grounding conductors are performing their intended function are considered to be _____ currents.
   (a) objectionable   (b) not objectionable
   (c) abnormal   (d) none of these

177. The important function of a type S fuse is that it _____.
   (a) is noninterchangeable
   (b) is slow burning
   (c) provides motor protection
   (d) is fast acting

178. Which of the following appliance installed in residential occupancies need to be grounded?
   (a) toaster   (b) can opener   (c) blender   (d) aquarium

179. • Conductive materials such as raceways that contain electrical conductors are grounded to.
   I. prevent lightning surges   II. limit the voltage to ground
   III. to aid the overcurrent device to clear the short or ground fault
   (a) I only   (b) II only   (c) III only   (d) II and III

180. • Where extensive metal in or on buildings may become energized and is subject to personal contact, the Code suggests that _____.
   (a) adequate bonding and grounding will provide additional safety
   (b) grounding is required
   (c) bonding is required
   (d) none of these

181. • The service shall have an accessible and external means of connecting _____ conductors of different systems, such as CATV and telephone.
   (a) bonding   (b) grounding   (c) a and b   (d) none of these

182. Outside conductors tapped to a feeder or connected to a transformer secondary shall be permitted to be protected by complying with four conditions. One of those conditions is that the _____ for the conductors are installed at a readily accessible location either outside of a building or structure, or inside nearest the point of entrance of the conductors.
   (a) overcurrent device   (b) disconnecting means
   (c) directory   (d) all of these

183. Circuit breakers shall be _____.
   (a) capable of being opened by manual operation
   (b) capable of being closed by manual operation
   (c) trip free
   (d) all of these

184. • Under certain conditions, service conductors are permitted above _____.
   (a) pools   (b) diving structures   (c) roofs   (d) all of these

185. • Type S fuseholders and adapters shall be so designed that either the fuseholder itself or the fuseholder with a Type S adapter inserted can be used for any fuse other than a Type S fuse.
    (a) True   (b) False

186. Where an ac system operating at less than 1,000 volts is grounded at any point, the grounded conductor shall be run to each service disconnecting means and shall be bonded to each disconnecting means enclosure. This conductor shall be routed with the phase conductors and shall not be smaller than the required grounding electrode conductor specified in _____.
    (a) Table 310-16   (b) Table 250-94
    (c) Table 250-95   (d) none of these

187. Providing all bonding requirements are met, the frame of the vehicle that has a vehicle mounted generator may serve as the _____ for a system supplied by the vehicle-mounted generator.
    (a) neutral   (b) grounded conductor
    (c) grounding electrode   (d) none of these

188. Service equipment shall be suitable for fault currents that are available at the _____.
    (a) load terminals   (b) circuit breakers   (c) supply terminals   (d) overcurrent device

189. • The Code prohibits the use of nonstandard ampere ratings for fuses and inverse time circuit breakers.
    (a) True   (b) False

190. • Supplementary overcurrent protection is not permitted for the protection of flexible cords and fixture wires.
    (a) True   (b) False

191. • When a separate building is supplied from a common service, a bonding jumper from the grounded conductor to the grounding electrode (at the separate building disconnect) is not required if a(n) _____ is run with the circuit conductors for the required grounding.
    (a) equipment grounding conductor
    (b) grounded conductor
    (c) neutral
    (d) grounding electrode conductor

192. No. 18 TFF fixture wire used for a 120-volt control circuit (less than 50 feet) can be protected by a _____-ampere overcurrent protection device.
    (a) 15   (b) 20   (c) 30   (d) none of these

193. Insulated equipment grounding conductors larger than No. _____ are not required to have a green outer finish, if permanently identified at each end and at every point where the conductor is accessible.
    (a) 6   (b) 8   (c) 4   (d) 1

194. Individual open service conductors not exceeding 600 volts must maintain a _____-inch spacing between conductors when installed not exposed to the weather.
    (a) 4½   (b) 4   (c) 3½   (d) 2½

195. The _____ must be installed as close as practical to (and preferably in the same area as) the grounding conductor connection at a separately derived system.
    (a) grounding electrode   (b) grounding conductor
    (c) grounded conductor   (d) none of these

196. Main and equipment bonding jumpers shall be attached by which of the following methods?
    (a) exothermic welding
    (b) inserting the proper screw
    (c) listed clamps
    (d) any of these

197. A 55-ampere feeder (nonmotor) conductor is permitted to be protected by a(n) _____ ampere overcurrent protection device.
    (a) 60   (b) 70   (c) 80   (d) 90

198. Because of _____, fuses and circuit breakers shall be so located or shielded that persons will not be burned or otherwise injured by their operation.
   I. arcing   II. suddenly moving parts   III. explosions
   (a) I only   (b) I or II
   (c) I, II, or III   (d) II only

199. Where a change occurs in the size of the ungrounded conductor, a similar change shall _____ to be made in the size of the grounded conductor.
   (a) be required   (b) be permitted
   (c) not be permitted   (d) none of these

200. Overcurrent protection for _____ services shall be selected or set to carry locked-rotor current of the motor(s) indefinitely.
   (a) hospital emergency (b) large motor
   (c) fire pump   (d) air conditioning

Unit 7

Motor Calculations

OBJECTIVES

After reading this unit, the student should be able to briefly explain the following concepts:

- Branch circuit short-circuit
- Ground-fault protection
- Feeder-protection
- Feeder conductor size
- Highest rated motor
- Motor overcurrent protection
- Motor branch circuit conductors
- Motor VA calculations
- Motor calculation steps
- Overload protection

After reading this unit, the student should be able to briefl explain the following terms:

- Code letter
- Dual-element fuse
- Ground-fault
- Inverse time breaker
- Heaters
- Motor full-load current
- Motor nameplate current rating
- Next smaller device size
- One-time fuse
- Overcurrent protection
- Overcurrent
- Overload
- Service factor (SF)
- Short-circuit
- Temperature rise

INTRODUCTION

When sizing conductors and *overcurrent protection* for motors, we must comply with the requirements of the National Electrical Code, specifically Article 430 [240–3(f)].

7–1 MOTOR BRANCH CIRCUIT CONDUCTORS [430–22(a)]

Branch circuit conductors to a single motor must have an ampacity of not less than 125 percent of the motor's full-load current as listed in Tables 430–147 through 430–150 [430–6(a)]. The actual conductor size must be selected from Table 310–16 according to the terminal temperature rating (60º or 75ºC) of the equipment [110–14(c)], Fig. 7–1.

☐ **Motor Branch Circuit Conductors Example**

What size THHN conductor is required for a 2-horespower, 240-volt single-phase motor, Fig. 7–2?

(a) No. 14   (b) No. 12
(c) No. 10   (d) No. 8

• Answer: (a) No. 14

Motor Full-Load Current – Table 430–148:
2-horsepower, 240-volts single-phase = 12 ampere.

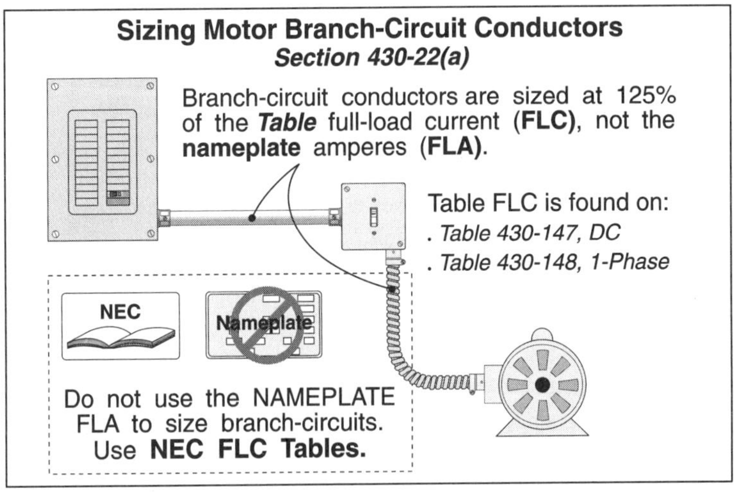

Sizing Motor Branch-Circuit Conductors
Section 430-22(a)

Branch-circuit conductors are sized at 125% of the *Table* full-load current (**FLC**), not the **nameplate** amperes (**FLA**).

Table FLC is found on:
. Table 430-147, DC
. Table 430-148, 1-Phase

Do not use the NAMEPLATE FLA to size branch-circuits.
Use **NEC FLC Tables.**

Figure 7-1

Branch Circuit Conductor Size

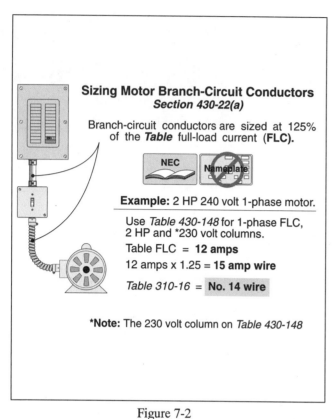

Figure 7-2
Conductor Size Example

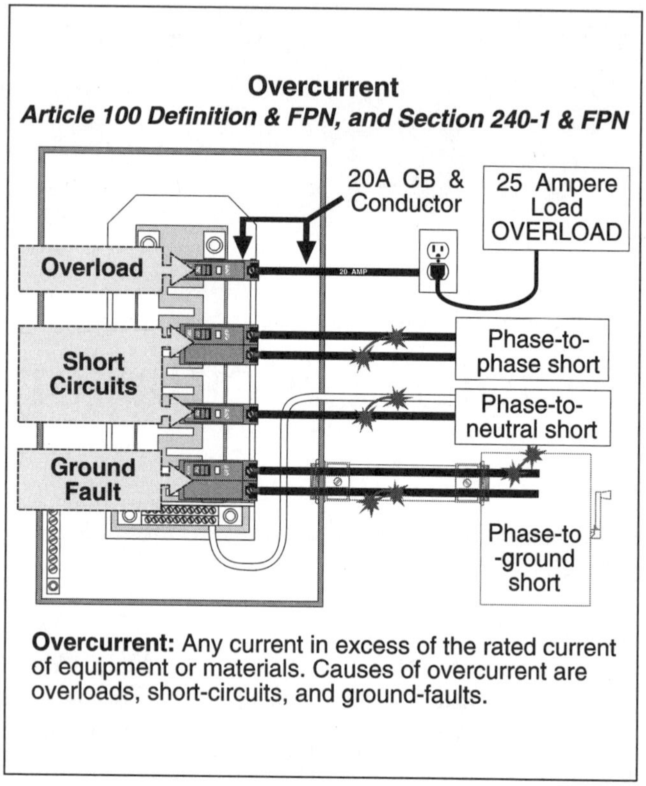

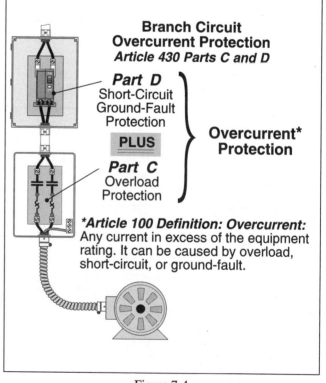

Figure 7-3
Overcurrent–Definition

Conductor Sized at 125 percent Of Motor Full-Load Current:
12 amperes × 1.25 = 15 amperes, Table 310–16, No. 14 THHN rated 20 amperes at 60ºC.

Note. The minimum size conductor permitted for building wiring is No. 14 [310–5], however local codes and many industrial facilities contains requirements that No. 12 be used as the smallest building wire.

7–2 MOTOR OVERCURRENT PROTECTION

Motors and their associated equipment must be protected against overcurrent (overload, short-circuit or ground-fault) [Article 100], Fig. 7–3. Due to the special characteristics of induction motors, overcurrent protection is generally accomplished by having the overload protection separated from the *short-circuit* and *ground-fault protection* device, Fig. 7–4. Article 430, Part C contains the requirements for motor overload protection and Part D of Article 430 contains the requirements for motor short-circuit and ground-fault protection.

Overload Protection

Overload is the condition in which current exceeds the equipment ampere rating, which can result in equipment damage due to dangerous overheating [Article 100]. Overload protection devices, sometimes called heaters, are intended to protect the motor, the motor-control equipment, and the branch-circuit conductors from excessive heating due to motor overload [430–31]. Overload protection is not intended to protect against short-circuits or ground-fault currents, Fig. 7–5.

Figure 7-4
Motor Overcurrent Protection

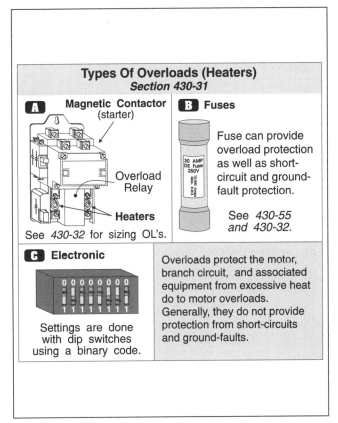

Figure 7-5
Types of Overloads

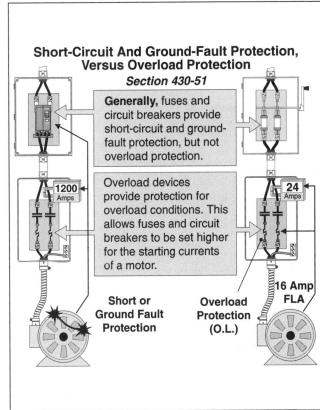

Figure 7-6
Short Circuit and Ground-Fault Protection

If overload protection is to be accomplished by the use of fuses, a fuse must be installed to protect each ungrounded conductor [430–36 and 430–55].

Short-Circuit And Ground-Fault Protection

Branch-circuit short-circuit and ground-fault protection devices are intended to protect the motor, the motor control apparatus, and the conductors against short-circuit or ground-faults, but they are not intended to protect against an overload [430–51], Fig. 7–6.

Note. The ground-fault protection device required for motor circuits is not the type required for personnel [210–8], feeders [215–9 and 240–13], services [230–95], or temporary wiring for receptacles [305–6].

7–3 OVERLOAD PROTECTION [430–32(a)]

In addition to short-circuit and ground-fault protection, motors must be protected against overload. Generally, the motor overload device is part of the motor starter; however, a separate overload device like a dual-element fuse can be used. Motors rated more than 1-horsepower without integral thermal protection, and motors 1-horsepower or less (automatically started) [430–32(c)], must have an overload device sized in response to the motor nameplate current rating [430–6(a)]. The overload device must be sized no larger than the requirements of Section 430–32. However, if the overload is sized according to Section 430–32 and it is not capable of carrying the motor starting or running current, the next larger size overload can be used [430–34]. The sizing of the overload device is dependent on the following factors.

Service Factor

Motors with a nameplate *service factor* (SF) rating of 1.15 or more, shall have the overload protection device sized no more than 125 percent of the motor nameplate current rating. If the overload is sized at 125 percent and it is not capable of carrying the motor starting or running current, the next size larger protection device can be used. But when we use the next size protection, it cannot exceed 140 percent of the motor nameplate current rating [430–34].

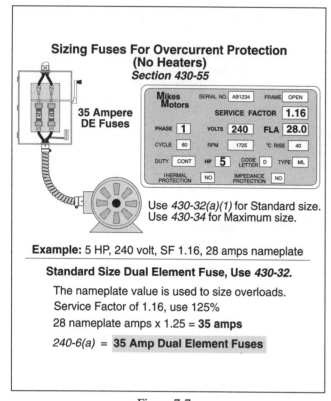

Figure 7-7
Overload Size with Service Factor

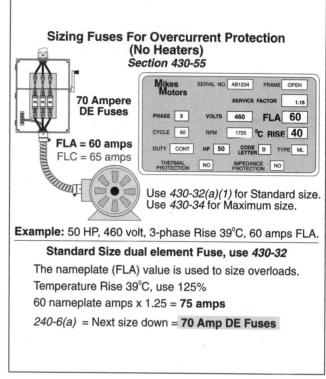

Figure 7-8
Overload Size–Temperature Factor

❑ **Service Factor Example**

If a dual-element fuse is used for overload protection, what size fuse is required for a 5-horsepower, 230-volt, single-phase motor, service factor 1.16 if the motor nameplate current rating is 28 amperes, Fig. 7–7?

    (a) 25 amperes   (b) 30 amperes   (c) 35 amperes   (d) 40 amperes

• Answer: (c) 35-ampere dual-element fuse

Overload protection is sized to motor nameplate current rating [430–6(a) and 430–32(a)(1)]. Standard Size, 28 amperes × 1.25 = 35 amperes [240–6(a)].

Temperature Rise

Motors with a nameplate temperature rise rating not over 40ºC, shall have the overload protection device sized no more than 125 percent of motor nameplate current rating. If the overload is sized at 125 percent and it is not capable of carrying the motor starting or running current, then the next size larger protection device can be used. When we use the next size protection, it cannot exceed 140 percent of the motor nameplate current rating [430–34].

❑ **Temperature Rise Example**

If a dual-element fuse is used for the overload protection, what size fuse is required for a 50-horsepower, 460-volt, three-phase motor, Temperature Rise 39ºC, motor nameplate current rating of 60 amperes (FLA), Fig. 7–8?

    (a) 40 amperes   (b) 50 amperes   (c) 60 amperes   (d) 70 amperes

• Answer: (d) 70 ampere

Overloads are sized according to the motor nameplate current rating of 60 amperes, not the 65 amperes Full Load Current rating listed in Table 430-150. 60 amperes × 1.25 = 75 amperes, 70 ampere [240–6(a) and 430–32(a)(1)]. If the 70 ampere fuse blows, the maximum size overload is limited to 140 percent of the motor nameplate current rating. 60 amperes × 1.4 = 84 ampere, next size down protection is 80 amperes [430–34].

All Other Motors

Motors that do not have a service factor rating of 1.15 and up, or a temperature rise rating of 40ºC and less, must have the overload protection device sized at not more than 115 percent of the motor nameplate ampere rating. If the overload is sized at 115 percent and it is not capable of carrying the motor starting or running current, the next larger protection device can be used. But when we use the next size protection, it cannot exceed 130 percent of the motor nameplate current rating [430–34].

Number of Overloads [430–37]

An overload protection device must be installed in each *ungrounded conductor* according to the requirements of Table 430-37. If a fuse is used for overload protection, a fuse must be installed in each ungrounded conductor [430-36].

7-4 BRANCH CIRCUIT SHORT-CIRCUIT GROUND-FAULT PROTECTION [430–52(c)(1)]

In addition to overload protection each motor and its accessories requires short-circuit and ground-fault protection. NEC Section 430–52(c) requires the motor branch circuit short-circuit and ground-fault protection (except torque motors) to be sized no greater than the percentages listed in Table 430–152. When the short-circuit ground-fault protection device value determined from Table 430–152 does not correspond with the standard rating or setting of overcurrent protection devices as listed in Section 240–6(a), the next higher protection device size may be used [430-52(c)(1) Exception No. 1], Fig. 7–9.

To determine the percentage from Table 430–152 to be used to size the motor branch circuit short-circuit ground-fault protection device, the following steps should be helpful.

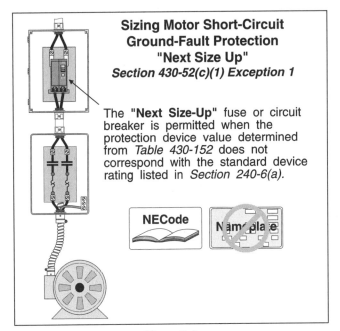

Figure 7-9

Short-Circuit and Ground-Fault Protection

Step 1: → Locate the motor type on Table 430–152: Wound rotor, direct-current or all other motors.

Step 3: → Select the percentage from Table 430–152 according to the type of protection device, such as nontime delay (one-time) fuse, dual element fuse, or inverse time circuit breaker.

Note. Where the protection device rating determined by Table 430-152 does not correspond with the standard size or rating of fuses or circuit breakers listed in Section 240-6(a), the next higher standard size rating shall be permitted [430-52(c)(1) Exception No. 1].

NEC Table 430–152			
	Percent of Full-Load Current (FLC) Tables 430–147, 148 and 150		
Type of Motor	One-Time Fuse	Dual-Element Fuse	Circuit Breaker
Direct Current and Wound-Rotor Motors	150	150	150
All Other Motors	300	175	250

❏ **Branch Circuit Example**

Which of the following statements are true?

(a) The branch circuit short-circuit protection (nontime delay fuse) for a 3-horespower, 115-volt motor, single-phase shall not exceed 110 amperes.

(b) The branch circuit short-circuit protection (dual element fuse) for a 5-horsepower, Design E, 230-volt motor, single-phase shall not exceed 50 amperes.

(c) The branch circuit short-circuit protection (inverse time breaker) for a 25-horsepower, synchronous, 460-volt motor, three-phase shall not exceed 70 amperes.

(d) All of these are correct.

• Answer: (d) All of these are correct.

Short-circuit and ground-fault protection, 430-53(c)(1) Exception No. 1 and Table 430-152:

Table 430-148 – 34 amperes × 3.00 = 102 amperes, next size up permitted, 110 amperes.

Table 430-148 – 28 amperes × 1.75 = 49 amperes, next size up permitted, 50 amperes.

Table 430-150 – 26 amperes × 2.50 = 65 amperes, next size up permitted, 70 amperes.

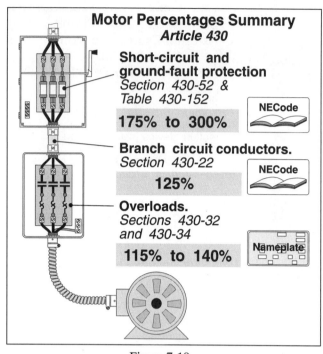

Figure 7-10
Motor Percentages Summary

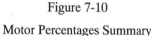

Figure 7-11
Motor Protection And Conductor Sizes

WARNING: Conductors are sized at 125 percent of the motor full-load current [430-22(a)], overloads are sized from 115 percent to 140 percent of the motor *nameplate* current rating [430-32(a)(1)], and the short-circuit ground-fault protection device is sized from 175 percent to 300 percent of the *motor full-load* current [Table 430-152]. As you can see, there is no relationship between the branch circuit conductor ampacity (125 percent) and the short-circuit ground-fault protection device (175 percent up to 300 percent), Fig. 7–10.

❑ **Branch Circuit Example**

Which of the following statements are true for a 1-horsepower, 120-volt motor, nameplate current rating of 14 amperes, Fig. 7–11?

(a) The branch circuit conductors can be No. 14 THHN.

(b) Overload protection is from 16.1 amperes to 18.2 amperes.

(c) Short-circuit and ground-fault protection is permitted to be a 40 ampere circuit breaker.

(d) All of these are correct.

• Answer: (d) All of these are correct.

Conductor Size [430–22(a)]: 16 amperes × 1.25 = 20 amperes = No. 14 at 60ºC, Table 310–16.

Overload Protection Size: Standard – 14 amperes (nameplate) × 1.15 = 16.1 amperes [430–32(a)(1)].

Maximum – 14 amperes × 1.3 = 18.2 amperes [430–34].

Short-Circuit And Ground-Fault Protection: 16 amperes (FLC) × 2.50 = 40 ampere circuit breaker, [430–52(c)(1), Tables 430–152 and 240–6].

This bothers many electrical people, but you need to realize that the No. 14 THHN conductors and motor is protected against overcurrent by the 16-ampere overload device and the 40-ampere short-circuit protection device.

7–5 FEEDER CONDUCTOR SIZE [430–24]

Motors Only

Conductors that supply several motors must have an ampacity of not less than:

(1) 125 percent of the highest rated motor full-load current (FLC) [430–17], plus

(2) the sum of the full-load currents of the other motors (on the same phase) [430–6(a)].

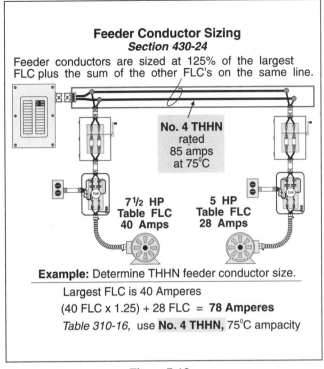

Figure 7-12
Feeder Conductor Sizing

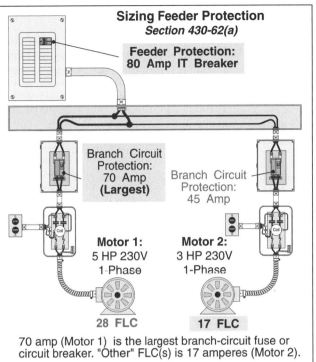

Figure 7-13
Sizing Feeder Protection

❑ **Feeder Conductor Size – Motors Only Example**

What size feeder conductor (in amperes) is required for two motors. Motor 1 – 7½-horespower, single-phase 230-volts (40 amperes) and motor 2 – 5-horespower, single-phase 230-volts (28 amperes)? Terminals rated for 75ºC, Fig. 7–12.

(a) 50 amperes (b) 60 amperes (c) 70 amperes (d) 80 amperes

• Answer: (d) 80 amperes, (40 amperes × 1.25) + 28 amperes = 78 amperes

No. 4 conductor at 75ºC, Table 310-16 is rated for 85 amperes.

7–6 FEEDER PROTECTION [430–62(a)]

Motors Only

Motor *feeder* conductors must have protection against short-circuits and ground-faults but not overload. The protection device must be sized not greater than: The largest branch-circuit short-circuit ground-fault protection device [430–52(c)] of any motor of the group, plus the sum of the full-load currents of the other motors (on the same time and phase).

❑ **Feeder Protection – Motors Only Example**

What size feeder protection (inverse time breaker) is required for a 5-horsepower, 230-volts single-phase motor and a 3-horespower, 230-volt single-phase motor, Fig. 7–13?

(a) 30 ampere breaker (b) 40 ampere breaker
(c) 50 ampere breaker (d) 60 ampere breaker

• Answer: (d) 60-ampere breaker

Motor Full Load Current [Table 430–148]
5-horespower, motor FLC = 28 amperes [Table 430–148]
3-horespower, motor FLC = 17 amperes [Table 430–148]
Branch Circuit Protection [430–52(c)(1), Table 430–152 and 240–6(a)]
5-horsepower: 28 amperes × 2.5 = 70 amperes
3 horsepower: 17 amperes × 2.5 = 42.5 amperes; Next size up, 45 amperes

Feeder Conductor [430–24(a)]
28 amperes × 1.25 + 17 amperes = 52 amperes, No. 6 rated 55 amperes at 60ºC, Table 310-16

Feeder Protection [430–62]
Not greater than 70-amperes protection, plus 17 amperes = 87 amperes; Next size down, 80 amperes

7-7 HIGHEST RATED MOTOR [430-17]

When selecting the feeder conductors or feeder short-circuit ground-fault protection device, the highest rated motor shall be the highest rated motor full-load current, not the highest-rated horsepower [430-17].

❑ **Highest Rated Motor Example**

Which is the highest-rated motor of the following, Fig. 7-14.

(a) 10-horsepower, three-phase, 208-volt

(b) 5-horsepower, single-phase, 208-volt

(c) 3-horsepower, single-phase, 120-volt

(d) any of these

• Answer: (c) 3-horsepower, single-phase, 120-volt, Full Load Current of 34 amperes

   10-horsepower = 30.8 amperes [Table 430–150]
   5-horsepower = 30.8 amperes [Table 430–148]
   3-horsepower = 34.0 amperes [Table 430–148]

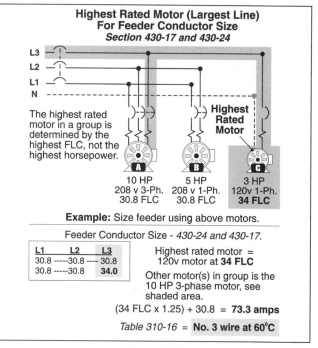

Figure 7-14

Highest Rated Motor For Feeder Conductor Size

7-8 MOTOR CALCULATION STEPS

Steps & NEC Rules	M1	M2	M3
Step 1 Motor FLC Tables 430–147, 148 & 150	____FLC	____FLC	____FLC
Step 2 Overloads (Heaters) Based On Motor Nameplate Current Rating			
Standard 430–32(a)(1)	___ × 1.___ = ___	___ × 1.___ = ___	___ × 1.___ = ___
Maximum 430–34	___ × 1.___ = ___	___ × 1.___ = ___	___ × 1.___ = ___
Step 3 Branch Circuit Conductor. 430–22(a), Table 310–16	___ × 1.25 = ___	___ × 1.25 = ___	___ × 1.25 = ___
Step 4 Branch Circuit Protection. Table 430–152, 430–52(c), 240–6(a)	___ × ___ = ___ Next size UP	___ × ___ = ___ Next size Up	___ × ___ = ___ Next size UP
Step 5 Feeder Conductor 430–24 and Table 310–16	\multicolumn{3}{c}{___ × 1.25 + ___ + ___ + ___ = ___ Table 310–16, Use No. ___ (60 or 75ºC?)}		
Step 6 Feeder Protection 430–62, Table 430–152, and 240–6(a)	\multicolumn{3}{c}{___ + ___ + ___ + ___ = ___ Next size down}		

❑ **Motor Calculation Steps Example**

Given: One 10-horsepower, 208-volt, three-phase motor and three – 1-horespower, 120-volt motors. Determine the standard and maximum overload sizes, branch circuit conductor (THHN) branch circuit short-circuit ground-fault protection device (Iinverse Time Breaker), for all motors; and then determine the feeder conductor and protection size, Fig. 7–15.

SOLUTION	M1	M2	M3
Step 1 Motor FLC Tables 430–147, 148 or 150	Table 430–150 30.8 Full Load Current	Table 430–147 16 Full Load Current	
Step 2 Overloads (Heaters) Based On Motor Nameplate Current Rating			
Step 2A Standard Overload Nameplate 430–32(a)(1)	No Name Plate, Use FLC 30.8 amperes × 1.15 = 35.4 amperes	No Name Plate, Use FLC 16 amperes × 1.15 = 18.4 amperes	
Step 2B Maximum Overload Nameplate 430–34	30.8 amperes × 1.3 = 40 amperes	16 amperes × 1.3 = 20.8 amperes	
Step 3 Branch Circuit Conductors 125 percent of FLC 110–14(c), 430–22(a), Table 310–16	30.8 amperes × 1.25 = 38.5 amperes Table 310–16 No. 8 THHN 60°C	16 amperes × 1.25 = 20 amperes Table 310–16 No. 14 THHN 60°C	
Step 4 Branch Circuit Protection FLC × Table 430–152 percent 430–52(c), 240–6(a)	30.8 amperes × 2.5 = 77 amperes 240–6(a) = 80 amps	16 amperes × 2.5 = 40 amperes 240–6(a) = 40 amps	
Step 5 Feeder Conductor FLC × 125 percent + FLC others 430–24, Table 310–16	(30.8 amperes × 1.25) + 16 amperes = 54.5 amperes Table 310–16, Use No. 6 THHN, rated 55 amperes 60°C		
Step 6 Feeder Protection Largest Branch protection + FLCs of other motors 430–62, Table 430–152, 240–6(a)	Inverse Time Breaker 80 amperes + 16 amperes + = 96 amperes, next size down, 90 amps		

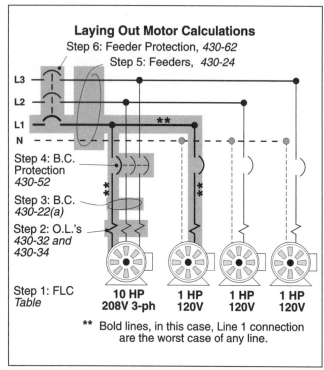

Figure 7-15
Laying Out Motor Calculations

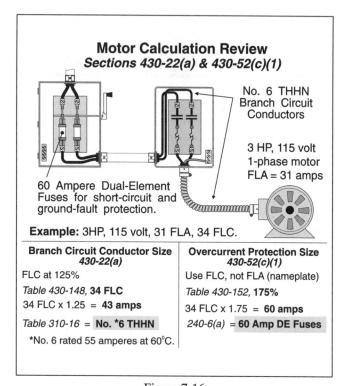

Figure 7-16
Motor Calculation Review

7–9 MOTOR CALCULATION REVIEW

❑ Branch Circuit Review Example

Size the branch circuit conductors (THHN) and short-circuit ground-fault protection device for a 3-horsepower, 115-volt, single-phase motor. The motor nameplate full-load amperes is 31 amperes and dual-element fuses to be used for short-circuit and ground-fault protection, Fig. 7–16.

Branch Circuit Conductors [430–22(a)]: Branch circuit conductors to a single motor must have an ampacity of not less than 125 percent of the motor full-load *current* as listed in Tables 430–147 through 430–150 [430–6(a)]. Three-horespower, full-load current is 34 amperes, 34 amperes × 125 percent = 43 amperes. Table 310–16, at 60ºC terminals, the conductor must be a No. 6 THHN rated 55 amperes [110–14(c)(1)].

Branch Circuit Short-Circuit Protection [430–52(c)(1)]: The branch-circuit short-circuit and ground-fault protection device protects the motor, the motor control apparatus, and the conductors against overcurrent due to short-circuits or ground-faults, but not overload [430–51]. The branch-circuit short-circuit and ground-fault protection device is sized by considering the type of motor and the type of protection device according to the motor full-load current listed in Table 430–152. When the protection device values determined from Table 430–152 do not correspond with the standard rating of overcurrent protection devices as listed in Section 240–6(a), the next higher overcurrent protection device must be installed. The branch-circuit short-circuit protection is sized at; 34 amperes × 175 percent = 60 amperes dual-element fuse [240–6(a)]. See Example No. 8 in the back of the NEC Code Book.

❑ Feeder and Branch Circuit Review Example

Size the feeder conductor (THHN) and protection device (inverse time breakers 75ºC terminal rating) for the following motors: Three 1-horsepower, 115-volt single-phase motors, and three 5-horsepower, 208-volt, single-phase motors, and one 15-horespower, wound-rotor, 208-volt, three-phase motor, Fig. 7–17.

Branch Circuit Conductors [110–14(c) 75ºC terminals, Table 310–16 and 430–22(a)]:

15 horsepower: 46.2 amperes [Table 430–150] × 125 percent = 58 amperes, No. 6 THHN, rated 65 amperes

5 horsepower: 30.8 amperes [Table 430–148] × 125 percent = 39 amperes, No. 8 THHN, rated 50 amperes

1 horsepower: 16 amperes [Table 430–148] × 125 percent = 20 amperes, No. 14 THHN, rated 20 amperes

Branch circuit short-circuit protection [240–6(a), 430–52(c)(1), and Table 430–152]:

15 horsepower: 46.2 amperes × 150 percent (wound-rotor) = 69 amperes; Next size up = 70 amperes.

5 horsepower: 30.8 amperes × 250 percent = 77 amperes; Next size up = 80 amperes

1 horsepower: 16 amperes × 250 percent = 40 amperes

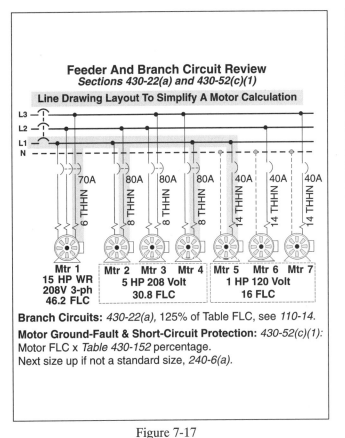

Figure 7-17
Branch Circuit Review

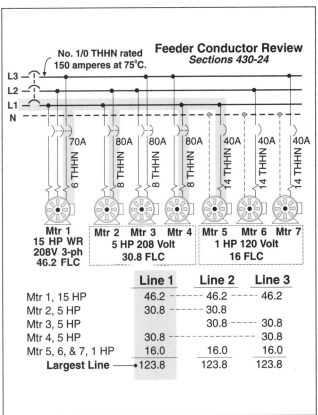

Figure 7-18
Feeder Conductor Review

Feeder Conductor [430–24]: Conductors that supply several motors must have an ampacity of not less than 125 percent of the highest rated motor full-load current [430–17], plus the sum of the other motor full-load currents [430–6(a)]. These conductors must be protected against short-circuits and ground-faults according to Section 430–62. The feeder conductor is sized at (46.2 amperes × 1.25) + 30.8 amperes + 30.8 amperes + 16 amperes = 136 amperes, [Table 310–16, No. 1/0 THHN], rated 150 amperes, Fig. 7–18

> **Note.** When sizing the feeder conductor, be sure to only include the motors that are on the same phase. For that reason, only four motors are used for the feeder conductor size.

Feeder Protection [430–62]: Feeder conductors must be protected against short-circuits and ground-faults sized not be greater than the maximum branch-circuit short-circuit ground-fault protection device [430–52(c)(1)] plus the sum of the full-load currents of the other motors (on the same phase). The protection device must not be greater than: (80 amperes + 30.8 amperes + 46.2 amperes + 16 amperes) = 173 amperes; Next size down, 150 ampere circuit breaker [240–6(a)]. See example No. 8 in the back of the NEC Code Book, Fig. 7–19.

> **Note.** When sizing the feeder protection, be sure to only include the motors that are on the same phase. For that reason, only four motors are used for the feeder conductor size.

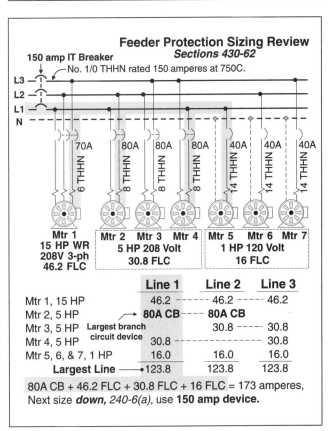

Figure 7-19
Feeder Protection Sizing Review

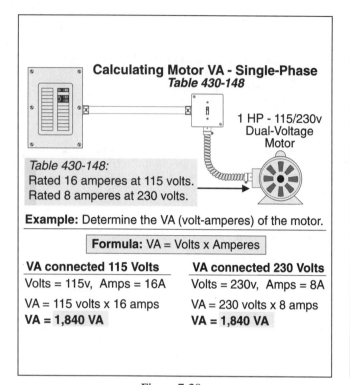

Figure 7-20
Calculating Motor VA – Single-Phase

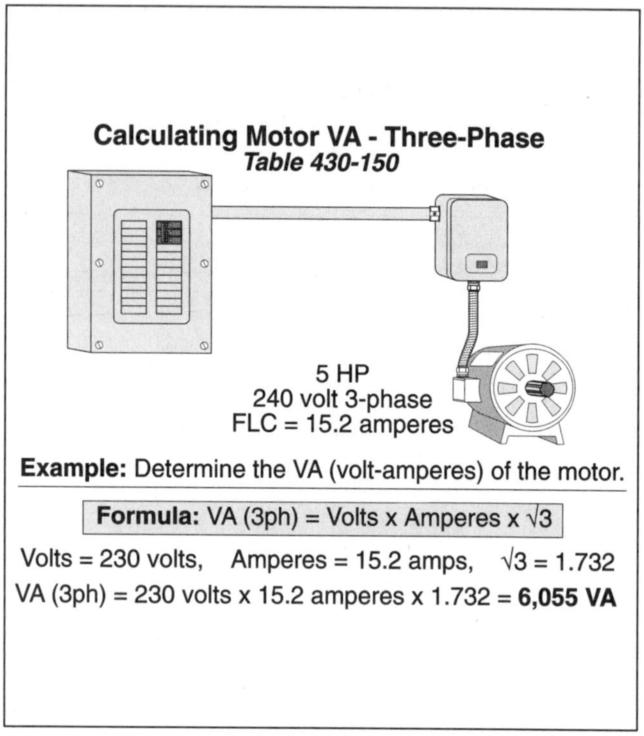

Figure 7-21
Calculating Motor VA – Three-Phase

7–10 MOTOR VA CALCULATIONS

The input VA of a motor is determined by multiplying the motor volts time the motor amperes. To determine the *motor VA rating*, the following formulas can be used:

Single-Phase Motor VA = Volts × Motor Amperes

Three-Phase motor VA = Volts × Motor Amperes × $\sqrt{3}$

Note. Many people believe that a 230-volt motor will consume less power than a 115-volt motor. A motor will consume approximately the same amount of power whether its wired at the 115 volts or 230 volts.

❑ **Single-Phase Motor VA Example**

What is the motor input VA of a 1-horsepower, motor rated 115/230-volts? The system voltage is 115/230-volts, Fig. 7–20.

(a) 1,840 VA at 115 volts   (b) 1,840 VA at 230 volts   (c) a and b   (d) none of these

• Answer: (c) a and b

Motor VA = Volts × Full Load Current

VA of 1-horesepower 230 volts = 230 volts × 8 amperes = 1,840 VA

VA of 1-horesepower 115 volts = 115 volts × 16 amperes = 1,840 VA

CAUTION: Some exam testing agencies use the table amperes times the given voltage. Under this condition, the answer would be 120 volts × 16 amperes or 240 volts × 8 amperes = 1,920 VA

❑ **Three-Phase Motor VA Example**

What is the input VA of a 5--horesepower, 240-volt, three-phase motor, Fig. 7–21?

(a) 6,055 VA   (b) 3,730 VA   (c) 6,440 VA   (d) 8,050 VA

• Answer: (a) 6,055 VA Table 430–150

Motor VA = Volts × Table Full-Load Current × $\sqrt{3}$

Table 430–150 Full-Load Current = 15.2 amperes

Motor VA = 230 volts × 15.2 amperes × 1.732 = 6,055 VA or

240 volts × 15.2 amperes = 6,318 VA

Unit 7 – Motor Calculations Summary Questions

Introduction

1. When sizing conductors and overcurrent protection for motors, we must comply with the requirements of NEC Article 430, not Article 240.
   (a) True
   (b) False

7–1 Motor Branch Circuit Conductors [430–22(a)]

2. What size THHN conductor is required for a 5-horsepower, 230-volt, single-phase motor? Terminals are rated 75°C.
   (a) No. 14
   (b) No. 12
   (c) No. 10
   (d) No. 8

7–2 Motor Overcurrent Protection

3. Motors and their associated equipment must be protected against overcurrent (overload, short-circuit or ground-fault), but because of the special characteristics of induction motors, overcurrent protection is generally accomplished by having the overload protection separate from the short circuit and ground fault protection.
   (a) True
   (b) False

4. • Which Parts of Article 430 contain the requirements for motor overcurrent protection?
   (a) Overload protection – Part C
   (b) Short-circuit ground-fault protection Part D
   (c) a and b
   (d) None of these

5. • Overload is the condition where current is greater than the equipment ampacity rating resulting in equipment damage due to dangerous overheating [Article 100]. Overload protection devices, sometimes called heaters, are intended to protect the _____ from dangerous overheating.
   (a) motor
   (b) motor-control equipment
   (c) branch-circuit conductors
   (d) all of these

6. The branch-circuit short-circuit and ground-fault protection device is intended to protect the motor, the motor control apparatus and the conductors against overcurrent due to _____.
   (a) short-circuits
   (b) ground-faults
   (c) overloads
   (d) a and b

7–3 Overload Protection

7. • The NEC requires motor overloads to be sized according to the motor full-load current rating as listed in Tables 430–147, 148, or 150.
   (a) True
   (b) False

8. The standard size overload must be sized according to the requirements of Section 430–32 of the NEC. If the overload sized according to Section 430–32 is not capable of carrying the motor starting and running current, the next size up, overload can be used if sized according to the requirements of Section 430–34.
   (a) True
   (b) False

9. Motors with a nameplate service factor (SF) rating of 1.15 or more must have the overload protection device sized at no more than _____ percent of the motor nameplate current rating. If the overload is not capable of carrying the motor starting or running current, the next size up protection device is permitted, but it cannot exceed _____ percent of the motor nameplate current rating.
   (a) 100, 125
   (b) 115, 130
   (c) 115, 140
   (d) 125, 140

10. Motors with a nameplate temperature rise rating not over 40°C, must have the overload protection device sized at no more than ____ percent of motor nameplate current rating. If the overload is not capable of carrying the motor starting or running current of the motor, the next size up is permitted, but it cannot exceed ____ percent of the motor nameplate current rating.
    (a) 100, 125   (b) 115, 130   (c) 115, 140   (d) 125, 140

11. Motors that do not have a service factor of 1.15 and up or a temperature rise rating not over 40°C, must have the overload protection device sized at not more than ____ percent of the motor nameplate ampere rating. If the overload is not capable of carrying the motor starting or running current, the next size up device is permitted, but it cannot exceed ____ percent of the motor nameplate current rating.
    (a) 100, 125   (b) 115, 130   (c) 115, 140   (d) 125, 140

12. If a dual-element fuse is used for overload protection, what size fuse is required for a 5-horsepower, 208-volt, three-phase motor, service factor 1.16, motor nameplate current rating of 16 amperes (FLA)?
    (a) 20 amperes   (b) 25 amperes   (c) 30 amperes   (d) 35 amperes

13. If a dual-element fuse is used for the overload protection, what size fuse is required for a 30-horsepower, 460-volt, three-phase synchronous motor, temperature rise 39°C?
    (a) 20 amperes
    (b) 25 amperes
    (c) 30 amperes
    (d) 40 amperes

7–4 Branch Circuit Short-Circuit Ground-Fault Protection [430–52(a)]

14. In addition to overload protection, each motor and its accessories requires short-circuit and ground-fault protection according to the requirements of Section 430–52. When sizing the branch circuit protection device, we must consider which of the following factors?
    (a) the motor type, such as induction, synchronous, wound-rotor, etc.
    (b) the motor Code letter starting characteristics
    (c) the type of protection device to be used, fuse or breaker
    (d) all of these

15. The NEC requires motor branch circuit short-circuit and ground-fault protection to be sized not greater than the percentages listed in Table 430–152. When the short-circuit ground-fault protection device value determined from Table 430–152 does not correspond with the standard rating of overcurrent protection devices as listed in Section 240–6(a), the next _____ device size must be used.
    (a) smaller   (b) larger   (c) a or b   (d) none of these

16. To determine the percentage of the motor FLC from Table 430–152 that is to be used to size the motor branch circuit short-circuit ground-fault protection device, the following steps should be used:
    (a) Locate the motor type on Table 430–152, such as direct current, wound rotor, high-reactance, autotransformer start, or all other motors.
    (b) Locate the motor starting conditions, such as Code letter or no Code letter.
    (c) Select the percentage from Table 430–152 according to the type of protection device, such as one-time fuse, dual-element fuse, or circuit breaker.
    (d) all of these

17. If the branch circuit short-circuit ground-fault protection dual-element fuse selected (sized not greater than the percentages listed in Table 430–152,) is not capable of carrying the load, the next larger size dual-element fuse can be used. The next size dual-element fuse cannot exceed ____ percent of the motor full-load current rating.
    (a) 125   (b) 150   (c) 175   (d) 225

18. Conductors are sized at ____ percent of the motor full-load currents [430–6 and 430–22], overloads from ____ percent, and the motor short-circuit ground-fault protection device (invere time circuit breaker) is sized up to ____. As you can see, there is no relationship between the branch circuit conductor ampacity and the short-circuit ground-fault protection device!
    (a) 125, 115, 250
    (b) 100, 125, 150
    (c) 125, 125, 125
    (d) 100, 100, 100

19. Which of the following statements are true for a 10-horsepower, 208-volt, three-phase motor, nameplate current 29 amperes?
    (a) The branch circuit conductors can be No. 8 THHN.
    (b) Overload protection is from 33 amperes to 38 amperes.
    (c) Short-circuit and ground-fault protection is an 80-ampere circuit breaker.
    (d) All of these are correct.

7–5 Feeder Conductor Size [430–24]

20. Feeder conductors that supply several motors must have an ampacity of not less than:
    I. 125 percent of the highest rated motor FLC
    II. The sum of the full-load currents of the other motors on the same phase [430–6(a)].
    (a) I only   (b) II
    (c) I or II   (d) I and II

21. Motor feeder conductors (sized according to 430–24) must have a feeder protection device to protect against short-circuits and ground-faults (not overloads), sized not greater than:
    I. The largest branch-circuit short-circuit ground-fault protection device [430–52] of any motor of the group
    II. The sum of full-load currents of the other motors on the same time phase
    (a) I only   (b) II
    (c) I or II   (d) I and II

7–6 Feeder – Protection [430–62(a)]

22. • Which of the following statements about a 30-horsepower, 460-volt, three-phase synchronous motor and a 10-horsepower, 460-volt, three-phase motor are true?
    (a) The 30-horsepower motor has No. 8 THHN with a 70-ampere breaker.
    (b) The 10-horsepower motor has No. 14 THHN with a 35-ampere breaker.
    (c) The feeder conductors must have a No. 6 THHN with a 80-ampere breaker.
    (d) All of these.

7–7 Highest Rated Motor [430–17]

23. When selecting the feeder conductors and short-circuit ground-fault protection device, the highest rated motor shall be the highest rated _____.
    (a) horsepower
    (b) full-load current
    (c) nameplate current
    (d) any of these

24. • Which is the highest rated motor of the following?
    I. 25-horsepower synchronous, three-phase, 460-volt
    II. 20-horsepower, three-phase, 460-volt
    III. 15-horsepower, three- phase, 460-volt
    IV. 3-horsepower, 120-volt
    (a) I only   (b) II only   (c) III only   (d) IV only

7–10 Motor VA Calculations

25. What is the VA input of a dual-voltage 5-horsepower, three-phase motor rated 460/230 volts?
    (a) 3,027 VA at 460 volts
    (b) 6,055 VA at 230 volts
    (c) 6,055 VA at 460 volts
    (d) b and c

26. What is the input VA of a 3-horsepower, 208-volt, single-phase motor?
    (a) 3,890 VA   (b) 6,440 VA   (c) 6,720 VA   (d) none of these

Challenge Questions

7–1 Motor Branch Circuit Conductors [430–22(a)]

27. • The branch circuit conductors of a 5-horsepower, 230-volt, motor (with a nameplate rating of 25 amperes), shall have an ampacity of not less than _____ amperes. Note: The motor is used for intermittent duty and cannot run for more than 5 minutes at any one time due to the nature of the apparatus it drives.
    (a) 33   (b) 37   (c) 21   (d) 23

7–3 Motor Overload [430–32(a)(1) And 430–34]

28. The standard overload protective device for a 2-horsepower, 115-volt motor that has a full-load current rating of 24 amperes, and a nameplate rating of 21.5 amperes shall not exceed _____ amperes.
    (a) 20.6   (b) 24.7   (c) 29.9   (d) 33.8

29. • The maximum overload protective device for a 2-horsepower, 115-volt motor with a nameplate rating of 22 amperes is _____ amperes. The Service Factor is 1.2.
    (a) 30.8   (b) 33.8   (c) 33.6   (d) 22.6

Ultimate Trip Setting [430–32(a)(2)]

30. The ultimate trip overload device of a thermally protected 1½-horsepower, 120-volt motor would be rated no more than _____ amperes.
    (a) 31.2   (b) 26   (c) 28   (d) 23

7–4 Branch Circuit Protection [430–52 And Table 430–152]

31. A 2-horsepower, 120-volt wound rotor motor would require a _____ branch circuit short circuit protection device.
    (a) 15 ampere   (b) 20 ampere   (c) 25 ampere   (d) 40 ampere

32. • The branch circuit short circuit protection device for a 10-horsepower, 230-volt, motor shall not exceed _____ amperes. *Note:* Use an inverse time breaker for protection.
    (a) 60   (b) 70   (c) 75   (d) 80

33. The branch circuit protection for a 125-horsepower, 240-volt direct current motor is _____ amperes.
    (a) 400   (b) 600   (c) 700   (d) 800

7–5 Motor Feeder Conductor Size [430–24]

34. • The motor controller (below) would require a No. _____ THHN for the feeder, if the motor terminals are rated for 75°C.

15-horsepower, 208v 3-Phase	15-horsepower, 208v 3-Phase	15-horsepower, 208v 3-Phase	Motor Controller 208/120-volt, 3-Phase
1-horsepower, 120v 1-Phase	1-horsepower, 120v 1-Phase	1-horsepower, 120v 1-Phase	
3-horsepower, 208v 1-Phase	3-horsepower, 208v 1-Phase	3-horsepower, 208v 1-Phase	

(a) 2   (b) 3/0   (c) 4/0   (d) 250 kcmils

7–6 Motor Feeder Protection [430–62(a)]

35. • There are three motors: Motor 1 is 5-horsepower, 230-volt, with service factor of 1.2; Motors 2 and 3 are 1½-horsepower, 120-volt. This group is fed with a 3-wire feeder. Using an inverse time breaker, the feeder conductor protection device after balancing all three motors would be _____ amperes.
    (a) 60   (b) 70   (c) 80   (d) 90

Chapter 2 NEC Calculations And Code Questions   Unit 7 Motor Calculations 237

36. • If an inverse time breaker is used for the feeder short circuit protection, what size protection is required for the following three-phase motors?
    Motor 1 = 40-horsepower, 52 FLC
    Motor 2 = 20-horsepower, 27 FLC
    Motor 3 = 10-horsepower, 14 FLC
    Motor 4 = 5-horsepower, 7.6 FLC
    (a) 225 amperes   (b) 200 amperes   (c) 125 amperes   (d) 175 amperes

37. • If dual-element fuses are used to protect a three wire 115/230-volt feeder conductor for twenty-two $\frac{1}{2}$-horsepower, single-phase, 115-volt motors, the fuse size selected should be sized not greater than _____ amperes.
    (a) 125   (b) 90   (c) 100   (d) 110

38. • The feeder protection for a 25-horsepower, 208-volt, 3-phase and three 3-horsepower, 120-volt motors would be _____ amperes, after balancing. *Note:* Use Inverse time breakers (ITB).
    (a) 225   (b) 200   (c) 300   (d) 250

NEC Questions
Random Sequence, Section 230–31 Through Section 328–17

39. The service disconnecting means shall be _____.
    (a) accessible   (b) readily accessible (c) outdoors   (d) indoors

40. Single pole breakers can be used for three- and four-wire line to line circuits, but handle tie bars are required.
    (a) True   (b) False

41. Service-entrance conductors are required to be insulated except when they are _____.
    (a) bare copper installed in a raceway
    (b) bare copper and part of the cable assembly is identified for underground use
    (c) copper-clad aluminum with individual insulation
    (d) a and b

42. Underground copper service conductors shall not be smaller than No. _____ AWG.
    (a) 3   (b) 4   (c) 6   (d) 8

43. Conductors are permitted to be tapped from feeders or the transformer secondary if the total length of the tap conductors are no longer than _____ feet. The tap conductors must have an ampacity not less than $\frac{1}{3}$ the ampacity of the overcurrent protection and must terminate in a single circuit breaker, or set of fuses having a rating not greater than the secondary conductor ampacity.
    (a) 10   (b) 25   (c) 100   (d) no limit

44. In the case of transformer feeder taps with primary plus secondary not over 25 feet long, conductors supplying a transformer shall be permitted to be tapped, without overcurrent protection at the tap, from a feeder where five conditions are met. One of those conditions is that the conductors supplying the _____ of a transformer have an ampacity at least $\frac{1}{3}$ of the rating of the overcurrent device protecting the feeder conductors.
    (a) primary   (b) secondary   (c) tertiary   (d) none of these

45. DC systems to be grounded shall have the grounding connection made at _____.
    (a) one supply station only   (b) one or more supply stations
    (c) the individual services   (d) any point on the premise wiring

46. • Metal enclosures for conductors added to existing installations of _____, which do not provide an equipment ground, are not required to be grounded if they are less than 25 feet long and free from probable contact with grounded conductive material.
    (a) nonmetallic-sheathed cable
    (b) open wire
    (c) knob-and-tube wiring
    (d) all of these

47. A grounded circuit conductor shall not be used for grounding noncurrent-carrying metal parts of equipment on the load side of _____.
    (a) the service disconnecting means
    (b) separately derived system disconnecting means
    (c) overcurrent devices for separately derived system not having a main disconnecting means
    (d) all of these

48. Main and equipment bonding jumpers shall be a _____.
    (a) screw   (b) wire   (c) bus   (d) any of these

49. • An _____ shall be used to connect the grounding terminal of a grounding-type receptacle to a grounded box.
    (a) equipment bonding jumper   (b) equipment grounding jumper
    (c) either a or b   (d) both a and b

50. • Metal parts that serve as the grounding conductor must be _____ together to ensure electrical continuity and have the capacity to conduct safely any fault current likely to be imposed.
    (a) grounded   (b) effectively bonded
    (c) attached   (d) any of these

51. • For field installations where the tap conductors leave the enclosure or vault in which the tap is made, the rating of the overcurrent device on the line side of the tap conductors shall not exceed 1000 percent of the tap conductor's ampacity. This requirement applies to tap conductors that are have a maximum lenght of _____ feet.
    (a) 10   (b) 15
    (c) 100   (d) no limit on length of tap

52. A device which, when interrupting currents in its range, will reduce the current flowing in the faulted circuit to a magnitude substantially less than that obtainable in the same circuit if the device were replaced with a solid conductor having comparable impedance is defined as a _____ protection device.
    (a) short-circuit   (b) short-circuit overcurrent
    (c) ground-fault   (d) current-limiting overcurrent

53. • Where ground-fault protection is provided for the _____ disconnecting means and interconnection is made with another supply system by a transfer device, means or devices may be needed to ensure proper ground-fault sensing by the ground-fault protection equipment.
    (a) circuit   (b) service
    (c) switch and fuse combination (d) both b and c

54. The ampacity of a No. 1 bare copper conductor in free air is _____ amperes.
    (a) 150   (b) 200   (c) 225   (d) 250

55. • A metal conduit used to protect the grounding electrode conductor need not be physically (mechanically) continuous to the grounding electrode if the metal raceway is electrically continuous.
    (a) True   (b) False

56. • Aluminum cable trays shall not be used as equipment grounding conductors for circuits with ground-fault protection above _____ amperes.
    (a) 1,600   (b) 60   (c) 600   (d) 1,200

57. Thermoplastic and fibrous outer braid Type TBS insulation would be used for wiring _____.
    (a) switchboards only   (b) in a dry location
    (c) in a wet location   (d) fixtures

58. • Electrodes of stainless steel or other nonferrous rod shall have a diameter of not less than _____ inch.
    (a) ½   (b) ¾   (c) 1   (d) ⅝

59. • The temperature limitation of MI cable is based _____. (See notes to Table 310–16).
    (a) on the insulation of the cable
    (b) on the insulation of individual conductors
    (c) on the insulating materials used in the end seal
    (d) according to the ambient temperature of the area

60. • Wiring methods permitted in the drop ceiling area used for environmental air include _____.
    (a) electrical metallic tubing
    (b) flexible metal conduit of any length
    (c) armored cable (BX) of any length
    (d) all of these

61. In designing circuits, the current-carrying capacity of conductors should be corrected for heat at room temperatures above _____ °F.
    (a) 30
    (b) 86
    (c) 94
    (d) 75

62. • The largest size grounding electrode conductor required for any service is a _____ copper.
    (a) No. 6
    (b) No. 1/0
    (c) No. 3/0
    (d) 250,000

63. On a four-wire, three-phase wye circuit, where the major portion of the load consists of electrical discharge lighting, data processing equipment, or other harmonic current inducting loads, the neutral conductor shall be counted when applying Note 8 derating factors.
    (a) True
    (b) False

64. • All metal raceways, cable armor, boxes, fittings, cabinets, and other metal enclosures for conductors must be _____ joined together to form a continuous electric conductor.
    (a) electrically
    (b) permanently
    (c) metallically
    (d) none of the above

65. • No. 12 service entrance conductors are permitted for a single circuit. This would require a No. _____ grounding electrode conductor.
    (a) 6
    (b) 12
    (c) 8
    (d) 10

66. The ampacities as listed in the tables of Article 310 are based on temperature alone and do not take _____ into consideration.
    (a) continuous loads
    (b) voltage drop
    (c) insulation
    (d) wet locations

67. No more than _____ layers of flat conductor cable can cross at any one point.
    (a) 2
    (b) 3
    (c) 4
    (d) no limit

68. • Because of the increasing use of nonmetallic repairs to the interior metal water pipes, interior metal water pipe located more than _____ feet from the point of entrance to the building shall not be used to serve as a conductor for the purpose of the bonding of the different building electrodes.
    (a) 2
    (b) 4
    (c) 5
    (d) 6

69. A No. 1/0 single copper conductor isolated in air with an operating temperature of 90 degrees C shall have an ampacity of _____ amperes at 15 kV.
    (a) 260
    (b) 300
    (c) 185
    (d) 215

70. • Insulated conductors used in wet locations shall be _____.
    I. lead covered
    II RHW, TW, THW, THHW, THWN, XHHW
    III. a type listed for wet locations
    (a) I only
    (b) II only
    (c) III only
    (d) I, II or III

71. The maximum voltage permitted between ungrounded conductors of flat conductor cable systems is _____ volts.
    (a) 600
    (b) 300
    (c) 250
    (d) 150

72. • All receptacles for temporary branch circuits are required to be electrically connected to the _____ conductor.
    (a) grounded
    (b) grounding
    (c) equipment grounding
    (d) grounding electrode

73. When a metal underground water pipe is to be used as the grounding electrode, it must be _____.
    (a) supplemented by an additional electrode
    (b) protected from physical damage
    (c) in direct contact with the earth for a minimum of 10 feet
    (d) a and c

74. • Raceways shall be provided with _____ where necessary to compensate for thermal expansion and contraction.
    (a) expansion couplings (b) expansion joints
    (c) bonding jumpers (d) none of these

75. Cable tray systems shall not be used _____.
    (a) in hoistways (b) where subject to severe physical damage
    (c) in hazardous locations (d) a and b

76. The _____ is defined as the area between the top of direct burial cable and the finished grade.
    (a) notch (b) cover (c) gap (d) none of these

77. Type FCC cable consists of _____ conductors.
    (a) three or more square (b) two or more round
    (c) three or more flat (d) two or more flat

78. Conductors in metal raceways and enclosures shall be so arranged as to avoid heating the surrounding metal by alternating current induction. To accomplish this, the _____ conductor(s) shall be grouped together.
    I. phase II. neutral III. equipment grounding
    (a) I only (b) I and II (c) I, II, and III (d) I and III

79. Cable trays shall _____.
    (a) include fittings for changes in direction and elevation
    (b) have side rails or equivalent structural members
    (c) be made of corrosion-resistant material
    (d) all these

80. • Type MV is defined as a single or multiconductor solid dielectric insulated cable rated _____ volts or higher.
    (a) 601 (b) 1001 (c) 2001 (d) 6001

81. • When ungrounded conductors are increased in size to compensate for voltage drop, the equipment grounding conductor is not required to be increased, because it is not a current-carrying conductor.
    (a) True (b) False

82. It is not the intent of Section 310–4 to require that conductors of one phase, neutral or grounded circuit conductor be the same as those of another phase, neutral, or grounded circuit conductor to achieve _____.
    (a) polarity (b) balance (c) grounding (d) none of these

83. • Ceiling support wire in fire-rated ceiling assemblies can be used for the support of branch circuit raceways that provide power to equipment and fixtures within or attached to the suspended ceiling.
    (a) True (b) False

84. There is a relationship between lettering on the conductors and what the lettering means, with this in mind, THWN is rated _____.
    (a) 75º (b) for wet locations
    (c) both a and b (d) not enough information

85. Lightning rods are not permitted to be used as a grounding electrode.
    (a) True (b) False

86. Which of the following applies to the temporary wiring of branch circuits?
    (a) No open wiring conductors shall be laid on the floor.
    (b) All circuits shall originate in an approved panelboard.
    (c) All conductors shall be protected by overcurrent devices in accordance with Article 240.
    (d) all of these

87. • Where moisture could enter a raceway and contact energized live parts, _____ are required at one or both ends of the raceway.
    (a) seals (b) plugs (c) a or b (d) none of these

88. Temporary wiring must be _____ immediately upon the completion of the purpose for which it was installed.
    (a) disconnected (b) removed (c) deenergized (d) any of the above

89. Conductors smaller than No. 1/0 AWG are permitted to be connected in parallel to supply control power, provided _____.
    (a) they are all contained within the same raceway or cable
    (b) each parallel conductor has an ampacity sufficient to carry the entire load
    (c) the circuit overcurrent protection device rating does not exceed the ampacity of any individual parallel conductor
    (d) all of these

90. Insulated grounded conductors of No. _____ or smaller, shall have an outer identification of a white or natural gray color.
    (a) 2 (b) 3 (c) 4 (d) 6

91. An exposed wiring support system using a messenger wire to support insulated conductors is known as _____.
    (a) open wiring
    (b) messenger supported wiring
    (c) field wiring
    (d) none of these

92. • The provisions required for mounting conduits on indoor walls or rooms that must be hosed down frequently is _____ between the mounting surface and the electrical equipment.
    (a) placed so a permanent ¼ air space separates them from supporting surface
    (b) separated by insulated bushings
    (c) separated by noncombustible tubing
    (d) none of these

93. • Article 300 on wiring methods covers all wiring installations except _____.
    (a) Class 1 and Class 2 control circuits
    (b) fire alarm circuits
    (c) CATV systems
    (d) all of these

94. A vertical run of 4/0 copper must be supported at intervals not exceeding _____ feet.
    (a) 80 (b) 100 (c) 120 (d) 40

95. For applications where burial depths of underground circuits must be deeper than shown in the underground ampacity Tables 310–69 through 310–84, the following ampacity derating factor shall be permitted to be used: _____ percent per increased foot of depth for all values of Rho.
    (a) 3 (b) 6 (c) 9 (d) 12

96. When a service has two or more enclosures, a grounding electrode tap is permitted from the grounded conductor of each service disconnect.
    (a) True (b) False

97. A _____ shall be permitted in lieu of a box or terminal fitting at the end of a conduit where the raceway terminates behind an unenclosed switchboard or similar equipment.
    (a) bushing (b) bonding bushing (c) coupling (d) connector

98. Where screws are used to mount knobs for the support of open wiring on insulators, or where nails or screws are used to mount cleats, they shall be of a length sufficient to penetrate the wood to a depth equal to at least _____ the height of the knob and the full thickness of the cleat.
    (a) one-eighth (b) one-quarter (c) one-third (d) one-half

99. The grounding conductor connection to the grounding electrode shall be made by _____.
    (a) listed lugs
    (b) exothermic welding
    (c) listed pressure connectors
    (d) any of these

100. The points of attachment of the interior metal water pipe bonding conductor shall be _____.
    (a) readily accessible (b) sealed (c) buried (d) accessible

101. A box shall not be required where cables or conductors from cable trays are installed in bushed conduit and tubing used for support or for protection against _____.
    (a) abuse
    (b) unauthorized access
    (c) physical damage
    (d) tampering

102. Conductors in raceways must be _____ between outlet devices and there shall be no splice or tap within a raceway itself.
(a) continuous (b) installed (c) copper (d) in conduit

103. The grounding conductor for secondary circuits of instrument transformers and for instrument cases shall not be smaller than No. _____ copper.
(a) 18 (b) 16 (c) 14 (d) 12

Random Sequence, Section 250–125 Through Section 365–3

104. Bus runs of over 600 volts having sections located both inside and outside of buildings shall have a _____ at the building wall to prevent interchange of air between indoor and outdoor sections.
(a) waterproof rating (b) vapor seal (c) fire barrier (d) b and c

105. • It shall be permissible to extend busways vertically through dry floors if totally enclosed (unventilated) where passing through and for a minimum distance of _____ feet above the floor to provide adequate protection from physical damage.
(a) 6 (b) 6 ½ (c) 8 (d) 10

106. • Type NM and Type NMC cables shall be permitted to be used in _____ .
(a) one family dwellings (b) multifamily dwellings
(c) other structures (d) all of these

107. • The use of AC cable is permitted for _____ .
(a) wet locations (b) embedding in concrete
(c) exposed work (d) all of these

108. Conductors larger than that for which the wireway is designed shall be permitted to be installed in any wireway.
(a) True (b) False

109. • Where devices or plug-in connections for tapping off feeder or branch circuits from busways consist of an externally operable fusible switch, _____ shall be provided for operation of the disconnecting means from the floor.
(a) ropes (b) chains (c) hook sticks (d) any of these

110. • MC cable shall be permitted for systems _____ .
(a) of 600 volts, nominal or less only
(b) over 600 volts, only
(c) in excess of 600 volts, nominal
(d) in excess of 5000 volts

111. Where NMC or NM cable is run across the top of floor joists, or within 7 feet (2.13 m) of floor or floor joists across the face of rafters or studding, in attics and roof spaces that are accessible, the cable shall be protected by substantial guard strips that are at least as high as the cable. Where this space is not accessible by permanent stairs or ladders, protection shall only be required within _____ feet of the nearest edge of the scuttle hole or attic entrance.
(a) 7 (b) 6 (c) 5 (d) 3

112. • Flexible metallic tubing shall be permitted to be used _____ .
(a) in dry and damp locations (b) for direct burial
(c) in lengths over six feet (d) for a maximum of 1,000 volts

113. • The use of surface nonmetallic raceways shall be permitted _____ .
(a) in dry locations (b) where concealed (c) in hoistways (d) all of these

114. The radius of a 3-inch conduit containing conductors with lead sheath shall not be less than _____ inches.
(a) 18 (b) 31 (c) 21 (d) 36

115. • Where liquidtight flexible conduit is used to connect to equipment and flexibility is required, such as to prevent the transmission of vibrations, a separate _____ conductor must be installed.
(a) bond jumper (b) bonding
(c) equipment grounding conductor (d) nothing is required

116. The maximum number of No. 14 THHN permitted in ⅜ liquidtight flexible metal conduit with inside fittings is _____.
 (a) 3 (b) 7 (c) 5 (d) 6

117. In addition to the marking requirements of Section 310–11, AC cable shall _____ by distinctive external markers on the cable sheath throughout its entire length.
 (a) have ready identification of the manufacturer
 (b) be marked with not suitable for exposed work
 (c) be marked with not for use in corrosive locations
 (d) all of these

118. Threadless couplings approved for use with intermediate metal conduit in wet locations shall be of the _____ type.
 (a) rainproof (b) raintight (c) moistureproof (d) concrete-tight

119. A run of electrical metallic tubing between outlet boxes shall not exceed _____ offsets close to the box.
 (a) 360 degrees plus (b) 360 degrees total including
 (c) four quarter bends plus (d) 180 degrees total including

120. • The minimum radius of a field bend on 1¼ rigid metallic conduit is _____ inches.
 (a) 7¾ (b) 8 (c) 14 (d) 10

121. The minimum size conductor permitted for MC cable is No. _____.
 I. 18 copper II. 14 copper III. 12 copper-clad aluminum
 (a) I only (b) II only (c) III only (d) I and III

122. The metallic sheath of metal-clad cable shall be continuous and _____.
 (a) flame-retardant (b) weatherproof (c) close fitting (d) all of these

123. • USE or SE cable must have a minimum of _____ conductors (including the uninsulated one) in order for one of the conductors to be uninsulated.
 (a) one (b) two (c) three (d) four

124. • The cross sectional area of 1 inch IMC is approximately _____ square inch(es).
 (a) 1.22 (b) 0.62 (c) 0.96 (d) 2.13

125. Switch, outlet, and tap devices of insulating material shall be permitted to be used without boxes in exposed NMC cable.
 (a) True (b) False

126. Inserts set in fiber underfloor raceways after the floor is laid shall be _____ into the raceway.
 (a) taped (b) glued (c) screwed (d) mechanically secured

127. The outer sheath of MI cable is made up of _____.
 (a) aluminum (b) steel alloy (c) copper (d) b or c

128. The maximum size flexible metallic tubing permitted is _____ inch(es).
 (a) ½ (b) ¾ (c) 2 (d) 4

129. SE cable with an uninsulated grounded conductor can be used for interior wiring for _____; according to Section 250–60.
 I. feeders II. branch circuits III. ranges and dryers
 (a) I only (b) II only (c) III only (d) none of these

130. Electrical nonmetallic tubing is composed of a material that is resistant to moisture, chemical atmospheres, and is _____.
 (a) flame-resistant (b) flame-retardant (c) fireproof (d) nonflammable

131. Electrical metallic tubing shall not be used _____.
 (a) where, during installation or afterward, it will be subject to severe physical damage
 (b) where protected from corrosion solely by enamel
 (c) both a and b
 (d) none of these

132. Aerial cable used for nonmetallic extensions and its tap connectors shall be provided with an approved means for _____.
    (a) operation (b) polarization (c) grounding (d) disconnect

133. The header on a cellular concrete floor raceway shall be installed _____ to the cells.
    (a) in a straight line
    (b) at right angles to the cells
    (c) a and b
    (d) none of these

134. Although exposed AC cable is required to closely follow the surface of the building, it can be used _____.
    (a) in lengths of 24 inches or less at terminals where flexibility is necessary
    (b) on the underside of floor joists in basements where supported at each joist where not subject to damage
    (c) both a and b are permitted
    (d) none of these

135. • Electrical nonmetallic tubing is not permitted in theaters and places of assembly.
    (a) True (b) False

136. The definition of a first floor, as it applies to nonmetallic sheath cable, is the level where _____ percent or more of the outside wall area is at or above finish grade.
    (a) 25 (b) 50 (c) 75 (d) 100

137. Type MI cable conductors shall be of _____ with a cross-sectional area corresponding to standard AWG sizes.
    (a) solid copper
    (b) solid or stranded copper
    (c) stranded copper
    (d) solid copper or aluminum

138. The outer sheath of copper sheath MI cable is to provide _____.
    (a) an adequate path for grounding purposes
    (b) a moisture seal
    (c) mechanical protection
    (d) all of these

139. Cablebus is ordinarily assembled at the _____.
    (a) point of installation
    (b) manufacturer's location
    (c) distributor's location
    (d) none of these

140. MC cable shall be supported and secured at intervals not exceeding _____ feet.
    (a) 3 (b) 6
    (c) 4 (d) 2

141. The radius of the curve of the inner edge of any bend shall not be less than _____ of AC cable.
    (a) five times the largest conductor within the cable
    (b) three times the diameter of the cable
    (c) five times the diameter of the cable
    (d) six times the outside diameter of the conductors

142. Splices and taps shall be permitted within a wireway provided they are accessible. The conductor, including splices and taps, shall not fill the wireway to more than _____ percent of its area at that point.
    (a) 25 (b) 80 (c) 125 (d) 75

143. • Nonmetallic sheath cable conductors must be rated _____ °C.
    (a) 60 (b) 75
    (c) 90 (d) any of these

144. The classification of nonmetallic extensions include _____.
    (a) surface extensions intended for mounting directly on the surface of walls
    (b) surface extensions intended for mounting directly on the surface of ceilings
    (c) aerial cable containing a supporting messenger cable as part of the assembly
    (d) all of these

145. Liquidtight where used as a fixed raceway must be secured within _____ inch(es) on each side of the box and at intervals not exceeding _____ feet.
    (a) 12, 4½   (b) 18, 3   (c) 12, 3   (d) 18, 4

146. Loop wiring in an underfloor raceway _____ to be a splice or tap.
    (a) is considered
    (b) shall not be considered
    (c) is permitted
    (d) shall not be connected

147. • Rigid nonmetallic conduit shall not be used _____.
    (a) where subject to severe corrosive influences
    (b) in cinder fill
    (c) for the support of fixtures and equipment
    (d) all of these

148. Straight runs of 1 inch rigid metal conduit using threaded couplings may be secured at intervals not exceeding _____ feet.
    (a) five   (b) ten   (c) twelve   (d) fourteen

149. • The maximum number of conductors permitted in any surface raceway shall be _____.
    (a) no more than 30 percent of the inside diameter
    (b) no greater than the number for which it was designed
    (c) no more than 75 percent of the cross-sectional area
    (d) that which is permitted on the Table

150. When installing conductors in electrical metallic tubing, splices and taps shall be made only in _____.
    I. junction boxes only   II. conduit bodies only   III. outlet or device boxes
    (a) I only   (b) III only   (c) I or III   (d) I, II, or III

151. Each run of nonmetallic extensions shall terminate in a fitting that covers the _____.
    (a) device
    (b) box
    (c) end of the extension
    (d) end of the assembly

152. The cablebus assembly is designed to carry _____ current and to withstand the magnetic forces of such current.
    (a) service   (b) load   (c) fault   (d) grounded

153. Inserts for cellular metal floor raceways shall be leveled to the floor grade and sealed against the entrance of _____.
    (a) concrete   (b) water   (c) moisture   (d) all of these

Random Sequence, Section 370–15 Through Section 422–18

154. A fixture marked, "Suitable for Damp Locations" _____ be used in a wet location.
    (a) can   (b) cannot

155. • In no case shall conductors within flexible cords and cables be associated in such a way with respect to the kind of circuit, the wiring method used, or the number of conductors such that the _____ temperature of the conductors is exceeded.
    (a) operating   (b) governing   (c) ambient   (d) limiting

156. When determining the number of conductors in a box, a reduction for cable clamps applies to boxes that have _____ clamps inside the box.
    (a) user installed   (b) single   (c) manufactured   (d) none of these

157. When determining the number of conductors in a box, which of the following items that originate and terminate within the box are not counted?
    (a) wirenuts   (b) pigtails
    (c) cable connectors   (d) all of these

158. • The insulating fitting or insulating material shall be permitted to have a temperature rating less than the insulation temperature rating of the installed conductors.
(a) True (b) False

159. Bare current-carrying metal parts shall be securely and rigidly supported so that the minimum clearance to any metal surface of the auxiliary gutter is not less than _____ inch(es).
(a) ½ (b) ¾ (c) 1 (d) 2½

160. The NEC requires a lighting outlet in clothes closets.
(a) True (b) False

161. The area of a 3 inches × 2 inches × 2 inches device box is _____ cubic inches.
(a) 12 (b) 14 (c) 10 (d) 8

162. Pilot lights, instruments, potential, current transformers and other switchboard devices with potential coils shall be supplied by a circuit that is protected by overcurrent devices rated _____.
(a) 15 amperes or more (b) 15 amperes or less
(c) 20 amperes or less (d) 10 amperes or less

163. • Conduit bodies with conductors larger than No. 6 shall have a cross-sectional area at least twice that of the largest conduit to which they are connected.
(a) True (b) False

164. Surface extensions from recessed boxes can be made by mechanically attaching an extension ring over the recessed box and attaching the extension wiring method to the extension ring.
(a) True (b) False

165. An outlet box contains two cable clamps, six No. 12 wires, and one single pole switch. The minimum size box for this combination would be _____ cubic inches.
(a) 12 (b) 13½ (c) 14½ (d) 20.25

166. Grounding conductors installed in panelboards (not part of service equipment) shall not terminate on the _____ terminal.
(a) grounding (b) grounded (c) supply (d) secondary

167. Boxes, conduit bodies, and fittings installed in wet locations need not be listed for use in wet locations.
(a) True (b) False

168. • Where nonmetallic-sheathed cable is used with nonmetallic boxes no larger than 2¼ x 4 inches, the cable is not required to be secured to the box if the cable is fastened within _____ inches.
(a) 6 (b) 8 (c) 10 (d) 12

169. When sizing a junction box, a straight pull shall not be less than _____ for a junction box under 600 volts.
(a) 8 times the diameter of the largest raceway
(b) 6 times the diameter of the largest raceway
(c) 48 times the outside diameter of the largest shielded conductor
(d) 36 times the largest conductor

170. • Metal plugs or plates used with nonmetallic boxes shall be recessed _____ inch.
(a) ⅜ (b) ½ (c) ¼ (d) ⅛

171. • Individual appliances that are continuously loaded shall have the branch circuit rating sized no less than _____ percent of the appliance marked ampere rating.
(a) 150 (b) 100 (c) 125 (d) 80

172. Where a wood brace is used for mounting a box, it shall have a cross section not less than nominal _____ inches.
(a) 1 × 2 (b) 2 × 2 (c) 2 × 3 (d) 2 × 4

173. The minimum distance an outlet box (that contains the fixture tap supply conductors) can be placed from a recessed fixture is _____ feet.
(a) 1 (b) 2 (c) 3 (d) 4

Chapter 2 NEC Calculations And Code Questions   Unit 7 Motor Calculations 247

174. A straight pull shall not be less than _____ for a junction box over 600 volts.
 (a) 18 times the diameter of the largest raceway
 (b) 48 times the diameter of the largest raceway
 (c) 48 times the outside diameter of the largest shielded conductor
 (d) 36 times the largest conductor

175. If constructed of steel, the metal shall not be less than _____ inch uncoated for a cabinet or cutout box.
 (a) 0.53   (b) 0.035   (c) 0.053   (d) 1.35

176. Panelboards equipped with snap switches rated at 30 amperes or less shall have overcurrent protection not in excess of _____ amperes.
 (a) 30   (b) 50   (c) 100   (d) 200

177. An insulated conductor used within a switchboard shall be _____.
 (a) listed
 (b) flame-retardant
 (c) rated for the highest voltage it may contact
 (d) all of these

178. Flexible cords and cables can be used for _____.
 (a) wiring of fixtures
 (b) portable lamps or appliances
 (c) stationary equipment facilitate frequently interchange
 (d) all of these

179. Tubing having cut threads and used as arms or stems on light fixtures shall have a wall thickness not less than _____ inch.
 (a) 0.020   (b) 0.025
 (c) 0.040   (d) 0.015

180. Electric discharge lighting over 1000 volts (neon) is not permitted _____ dwelling-units.
 (a) on   (b) in   (c) both a and b   (d) none of these

181. In completed installations, each outlet box shall have a _____.
 (a) cover   (b) faceplate   (c) canopy   (d) any of these

182. Metal poles that support lighting fixtures must meet the following requirements: _____.
 (a) they must have an accessible handhole (sized 2 x 4 inch) with a raintight cover
 (b) an accessible grounding terminal must be installed accessible from the handhole
 (c) both a and b
 (d) neither a nor b

183. Boxes can be supported from a multiconductor cord or cable, provided the conductors are protected from _____.
 (a) strain   (b) temperature   (c) sunlight   (d) swaying

184. Wiring on fixture chains and other movable parts shall be _____.
 (a) rated for 110°C (b) stranded   (c) hard usage rated (d) none of these

185. • Auxiliary gutters shall not contain more than _____ conductors at any cross section.
 (a) thirty   (b) forty current-carrying
 (c) twenty   (d) thirty current-carrying

186. The flexible cord conductor identification required in the Code shall consist of one of five methods. One of those methods is a tracer in a braid of any color contrasting with that of the braid and _____ in the braid of the other conductor or conductors.
 (a) a solid color   (b) no tracer   (c) two tracers   (d) none of these

187. A surface-mounted or recessed fluorescent fixture shall not be permitted to be installed within closets.
 (a) True   (b) False

188. Knife switches rated for more than 1200 amperes at 250 volts _____
    (a) are used only as isolating switches
    (b) may be opened under load
    (c) should be placed so that gravity tends to close them
    (d) should be connected in parallel

189. Recessed incandescent fixtures with a completely enclosed lamp in clothes closets shall be permitted to be installed in the wall or on the ceiling, provided there is a minimum clearance of _____ inches between the fixture and the nearest point of a storage space.
    (a) 3 (b) 6 (c) 9 (d) 12

190. Receptacles, cord connectors, and attachment plugs shall be constructed so that the receptacle or cord connectors will not accept an attachment plug with a different _____ or current rating than that for which the device is intended.
    (a) voltage (b) amperage (c) heat (d) all of these

191. • Cord-equipped fixtures shall terminate at the _____ end of the cord in a grounding-type attachment plug (cap) or busway plug.
    (a) outer (b) inner (c) supply (d) termination

192. Lampholders installed over highly combustible material shall be of the _____ type.
    (a) industrial (b) switched (c) unswitched (d) residential

193. • A cablebus system shall include approved fittings for _____.
    (a) bends (b) U-turns (c) dead ends (d) all of these

194. The repair of hard-service cord (see Column 1, Table 400–4) No. _____ and larger shall be permitted if conductors are spliced in accordance with Section 110–14(b) and the completed splice retains the insulation, outer sheath properties, and usage characteristics of the cord being spliced.
    (a) 16 (b) 15 (c) 14 (d) 12

195. A lighting and appliance branch circuit panelboard contains six three-pole breakers and eight two-pole breakers. The maximum allowable number of single-pole breakers permitted to be added to this panelboard is _____.
    (a) 8 (b) 16 (c) 28 (d) 42

196. Paddle fans are not permitted to be supported directly by lighting fixture outlet boxes _____.
    (a) True (b) False

197. Fixtures shall be wired with conductors having insulation suitable for the environmental conditions and _____ to which the conductors will be subjected.
    (a) temperature (b) voltage (c) current (d) all of these

198. Branch circuit conductors to individual appliances shall not be sized _____ than required by the appliance markings or instructions.
    (a) larger (b) smaller

199. • A receptacle outlet installed outdoors shall be located so that _____ is not likely to touch the outlet cover or plate.
    (a) a person (b) water accumulation
    (c) metal (d) none of these

200. When counting the number of conductors in a box, a conductor running through the box is counted as _____ conductor(s).
    (a) one (b) two (c) zero (d) none of these

Unit 8

Voltage Drop Calculations

OBJECTIVES

After reading this unit, the student should be able to briefly explain the following concepts:

Alternating current resistance as compared to direct current
Conductor resistance
Conductor resistance – alternating current circuits
Conductor resistance – direct current circuits
Determining circuit voltage drop
Extending circuits
Limiting current to limit voltage drop
Limiting conductor length to limit voltage drop
Resistance – alternating current
Sizing conductors to prevent Excessive voltage drop
Voltage drop considerations
Voltage drop recommendations

After reading this unit, the student should be able to briefly explain the following terms:

American wire gauge
CM = circular mils
Conductor
Cross-sectional area
D = distance
Eddy currents
E_{VD} = Conductor voltage drop expressed in volts
I = load in amperes at 100 percent
K = direct current constant
Ohm's law method
Q = alternating current adjustment factor
R = resistance
Skin effect
Temperature coefficient
VD = volts dropped

8–1 CONDUCTOR RESISTANCE

Metals that carry electric current are called conductors or wires and oppose the flow of electrons. Conductors can be solid, stranded, copper or aluminum. The conductors opposition to the flow of current (resistance) is determined by the material type (copper/aluminum), the cross-sectional area (wire size) and the conductor's length and operating temperature. The *resistance* of any property is measured in ohms.

Material

Silver is the best conductor because it has the lowest resistance, but its high cost limits its use to special applications. *Aluminum* is often used when weight or cost are important considerations, but *copper* is the most common type of metal used for electrical conductors, Fig. 8–1.

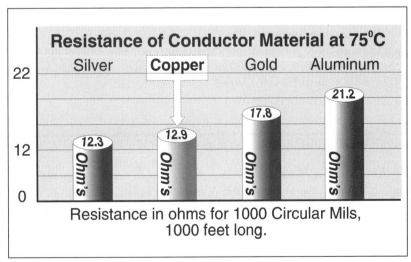

Figure 8-1

Resistance of Conductor Materials

249

Cross-sectional Area

The *cross-sectional area* of a conductor is the conductors surface area expressed in circular mils. The greater the conductor cross-sectional area, the greater the number of available electron paths, and the lower the conductor resistance. Conductors are sized according to the *American Wire Gauge* which ranges from a small of No. 40 to the largest of No. 0000 (4/0). The conductors resistance varies inversely with the conductor's diameter; that is, the smaller the wire size the greater the resistance and the larger the wire size the lower the resistance, Fig. 8–2.

Conductor Length

The resistance of a conductor is directly proportional to the conductor length. The following table provides examples of conductor resistance and *area circular mils* for conductor lengths of 1,000 feet. Naturally, longer or shorter lengths will result in different conductor resistances.

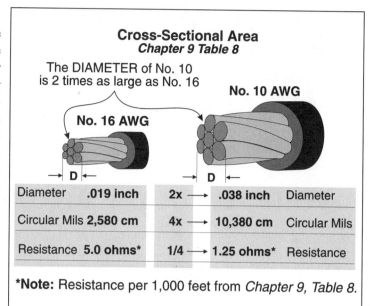

Figure 8-2
Conductor Cross-Sectional Area

Conductor Properties – NEC Chapter 9, Table 8			
Conductor Size American Wire Gauge	Conductor Resistance Per 1,000 Feet At 75ºC	Conductor Diameter In Mils	Conductor Area Circular Mils
No. 14	3.1 ohms	64.084	4,107
No. 12	2.0 ohms	80.808	6,530
No. 10	1.2 ohms	101.89	10,380
No. 8	.78 ohm	128.49	16,510
No. 6	.49 ohm	162.02	26,240

Temperature

The resistance of a conductors changes with changing temperature, this is called *temperature coefficient*. Temperature coefficient describes the affect that temperature has on the resistance of a conductor. Positive temperature coefficient indicates that as the temperature rises, the conductor resistance will also rise. Examples of conductors that have a positive temperature coefficient are silver, copper, gold, and aluminum conductors. Negative temperate coefficient means that as the temperature increases, the conductor resistance decreases.

The conductor resistances listed in the National Electrical Code Table 8 and 9 of Chapter 9, are based on an operating temperature of 75ºC. A three-degree change in temperature will result in a 1 percent change in conductor resistance for both copper and aluminum conductors. The formula to determine the change in conductor resistance with changing temperature is listed at the bottom of Table 8, Chapter 9 in the NEC.

8–2 CONDUCTOR RESISTANCE – DIRECT CURRENT CIRCUITS, [Chapter 9, Table 8 of the NEC]

The National Electric Code lists the resistance and area circular mils for both direct current and alternating current circuit conductors. Direct current circuit conductor resistances are listed in Chapter 9, Table 8, and alternating current circuit conductor resistances are listed in Chapter 9, Table 9 of the NEC.

The *direct current conductor resistances* listed in Chapter 9, Table 8 apply to conductor lengths of 1,000 feet. The following formula can be used to determine the conductor resistance for conductor lengths other than 1,000 feet:

$$\text{Direct current Conductor Resistance} = \frac{\text{Conductor Resistance Ohms}}{1,000 \text{ Feet}} \times \text{Conductors Length}$$

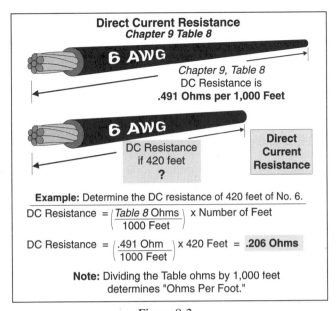

Figure 8-3

Direct Current Conductor Resistance

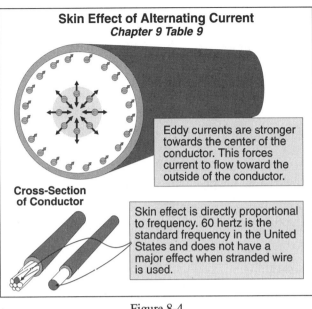

Figure 8-4

Skin Effect

❑ **Direct Current Conductor Resistance Examples**

What is the resistance of 420 feet of No. 6 copper, Fig. 8–3?

- Answer: .206 ohms

The resistance of No. 6 copper 1,000 feet long is .491 ohms, Chapter 9, Table 8. The resistance for 420 feet is:

(.491 ohms /1,000 feet) × 420 feet = .206 ohms

What is the resistance of 1,490 feet of No. 3 aluminum?

- Answer: .600 ohms

The resistance of No. 3 aluminum 1,000 feet long is .403 ohms, Chapter 9, Table 8. The resistance for 1,490 feet is:

(.403 ohms/1,000 feet) × 1,490 feet = .600 ohms

8–3 CONDUCTOR RESISTANCE – ALTERNATING CURRENT CIRCUITS

In direct current circuits, the only property that opposes the flow of electrons is resistance, In alternating current circuits, the expanding and collapsing magnetic field within the conductor induces an electromotive force that opposes the flow of alternating current. This opposition to the flow of alternating current is called *inductive reactance* which is measured in ohms.

In addition, alternating current flowing through a conductor generates small erratic independent currents called *eddy currents*. Eddy currents are greatest in the center of the conductors and repel the flowing electrons toward the conductor surface; this is known as *skin effect*. Because of skin effect, the effective cross-sectional area of an alternating current conductor is reduced, which results in an increase of the conductor resistance. The total opposition to the flow of alternating current (resistance and inductive reactance) is called *impedance* and is measured in ohms, Fig. 8–4.

8–4 ALTERNATING CURRENT RESISTANCE AS COMPARED TO DIRECT CURRENT

The opposition to current flow is greater for alternating current as compared to direct current circuits, because of inductive reactance, eddy currents and skin effect. The following two tables give examples of the difference between alternating current circuits are as compared to direct current circuits, Fig. 8–5.

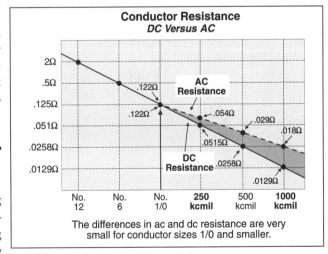

Figure 8-5

Conductor Resistance AC versus DC

COPPER – Alternating Current versus Direct Current Resistance at 75°C			
Conductor Size	Alternating Current Chapter 9, Table 9	Direct Current Chapter 9, Table 8	AC resistance greater than DC resistance by %
250,000	0.054 ohm per 1,000 feet	0.0515 ohm per 1,000 feet	4.85%
500,000	0.029 ohm per 1,000 feet	0.0258 ohm per 1,000 feet	12.40%
1,000,000	0.018 ohm per 1,000 feet	0.0129 ohm per 1,000 feet	39.50%

ALUMINUM – Alternating Current versus Direct Current Resistance at 75°C			
Conductor Size	Alternating Current Chapter 9, Table 9	Direct Current Chapter 9, Table 8	AC resistance greater than DC resistance by %
250,000	0.086 ohm per k feet	0.0847 ohm per k feet	1.5%
500,000	0.045 ohm per k feet	0.0424 ohm per k feet	6.13%
1,000,000	0.025 ohm per k feet	0.0212 ohm per k feet	17.12%

8–5 RESISTANCE ALTERNATING CURRENT [Chapter 9, Table 9 Of The NEC]

Alternating current conductor resistances are listed in Chapter 9, Table 9 of the NEC. The alternating current resistance of a conductor is dependent on the conductors material (copper or aluminum) and on the magnetic property of the raceway.

❑ **Chapter 9, Table 9 Alternating Current Resistance Example**
- Answer: What is the alternating current resistance for a 250,000 circular mils conductor 1,000 feet long?

   Copper conductor in nonmetallic raceway = 0.052 ohm per 1,000 feet
   Copper conductor in aluminum raceway = 0.057 ohm per 1,000 feet
   Copper conductor in steel raceway = 0.054 ohm per 1,000 feet
   Aluminum conductor in nonmetallic raceway = 0.085 ohm per 1,000 feet
   Aluminum conductors in aluminum raceway = 0.090 ohm per 1,000 feet
   Aluminum conductors in steel raceway = 0.086 ohm per 1,000 feet

Alternating Current Conductor Resistance Formula
The following formula can be used to determine the resistance of different lengths of conductors.

Alternating Current Resistance =
$$\frac{\text{Conductor Resistance Ohms}}{1,000 \text{ Feet}} \times \text{Conductor Length}$$

❑ **Alternating Current Conductor Resistance Examples**
What is the alternating current resistance of 420 feet of No. 2/0 copper installed in a metal raceway, Fig. 8–6?

- Answer: 0.042 ohm

   The resistance of No. 2/0 copper is .1 ohms per 1,000 feet, Chapter 9, Table 9. The resistance of 420 feet of No. 2/0 is: (0.1 ohm/1,000 feet) × 420 feet = 0.042 ohm

What is the alternating current resistance of 169 feet of 500 kcmil installed in aluminum conduit?

- Answer: 0.0054 ohms

   The resistance of 500 kcmils installed in aluminum conduit is 0.032 ohm per 1,000 feet. The resistance of 169 feet of 500 kcmils in aluminum conduit is:

   (0.032 ohm/1,000 feet) × 169 feet = 0.0054 ohm

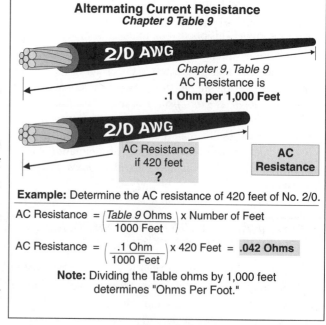

Figure 8-6
Alternating Current Resistance

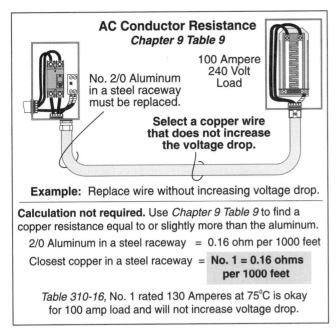

Figure 8-7
Alternating Current Conductor Resistance Example

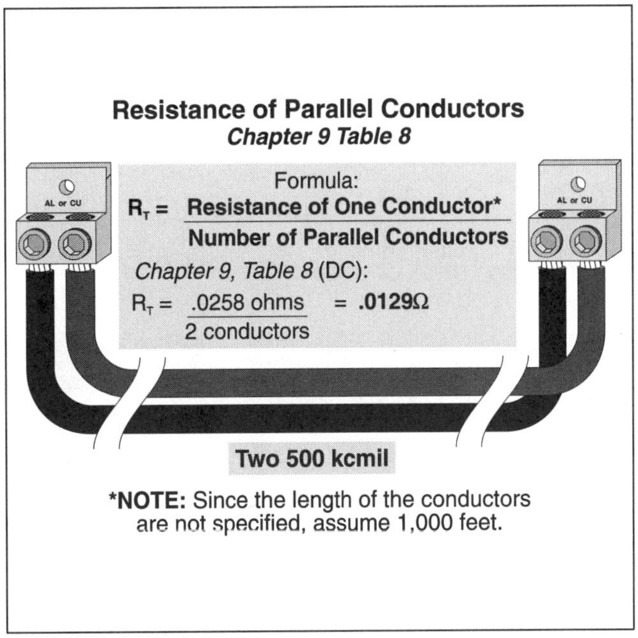

Figure 8-8
Resistance Parallel Conductors Example

Converting Copper To Aluminum Or Aluminum To Copper

When requested to determine the replacement conductor for copper or aluminum the following steps should be helpful.
Step 1: → Determine the resistance of the existing conductor using Table 9, Chapter 9 for 1,00 0 feet.
Step 2: → Using Table 9, Chapter 9 to locate a replacement conductor that has a resistance of not more than the existing conductors.
Step 2: → Verify that the replacement conductor has an ampacity [Table 310–16] sufficient for the load.

❏ **Converting Copper To Aluminum Or Aluminum To Copper Example**

A 100 ampere, 240-volt, single-phase load is wired with No. 2/0 aluminum conductors in a steel raceway. What size copper wire can we use to replace the aluminum wires and not have a greater voltage drop. Note: The wire selected must have an ampacity of at least 100 amperes, Fig. 8–7?

(a) No. 1/0   (b) No. 1   (c) No. 2   (d) No. 3

• Answer: (b) No. 1 copper

The resistance of No. 2/0 aluminum in a steel raceway is 0.16 ohm per 1,000 feet. No. 1 copper in a steel raceway has a resistance of 0.16 ohms and it has an ampacity of 130 amperes at $75°C$ [Table 310–16 and 110–14(c)].

Determining The Resistance Of Parallel Conductors

The resistance total in a parallel circuit is always less than the smallest resistor. The equal resistors formula can be used to determine the resistance total of parallel conductors:

$$\text{Resistance Total} = \frac{\text{Resistance of One Conductor} *}{\text{Number of Parallel Conductors}}$$

*Resistance according to Chapter 9 Table 8 or Table 9 of the NEC, assuming 1,000 feet unless specified otherwise.

❏ **Resistance of Parallel Conductors Example**

What is the direct current resistance for two 500 kcmil conductors in parallel, Fig. 8–8?

(a) 0.0129 ohm   (b) 0.0258 ohm   (c) 0.0518 ohm   (d) 0.0347 ohm

• Answer: (a) 0.0129 ohms

$$\text{Resistance Total} = \frac{\text{Resistance of One Conductor}}{\text{Number of Parallel Conductors}}$$

$$\text{Resistance Total} = \frac{0.0258 \text{ ohm}}{2 \text{ conductors}} = 0.0129 \text{ ohm}$$

Note. You could also look up the resistance of 1,000 kcmils in Chapter 9, Table 8, which is 0.0129 ohm.

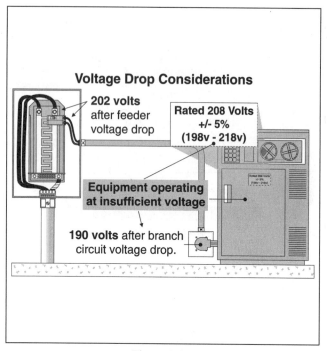

Figure 8-9
Voltage Drop Considerations

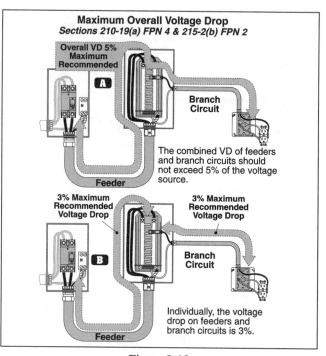

Figure 8-10
Maximum Recommended Voltage Drop

8–6 VOLTAGE DROP CONSIDERATIONS

The resistance of a conductors opposition to current flow results in conductor *voltage drop*; E (voltage drop) = I (amperes) × R (ohms). Because of the conductor's voltage drop, the operating voltage of the equipment will be reduced. Inductive loads, such as motors and ballasts, can overheat at reduced *operating voltage*, which can result in reduced equipment operating life as well as inconvenience for the customer. Electronic equipment such as computers, laser printers, copy machines, etc., can suddenly power down resulting in data loss if the operating voltage is too low. Resistive equipment simply will not provide the power output at reduced operating voltage, Fig. 8–9.

8–7 NEC VOLTAGE DROP RECOMMENDATIONS

Contrary to many beliefs, the NEC does not contain any requirements for sizing ungrounded conductors for voltage drop. It does recommend in many areas of the NEC that we consider the effects of conductor voltage drop when sizing conductors. See some of these recommendations in the Fine Print Notes to sections 210–19(a), 215–2(b), 230–31(c) and 310–15. Please be aware that Fine Print Notes in the NEC are recommendations, they are not requirements [90–5]. The NEC recommends that the maximum combined voltage drop for both the feeder and branch circuit should not exceed 5 percent, and the maximum on the feeder *or* branch circuit should not exceed 3 percent. Figure 8–10.

❏ **NEC Voltage Drop Recommendation Example**

What are the minimum NEC recommended operating volts for a 115-volt rated load that is connected to a 120/240-volt source, Fig. 8–11?

(a) 120 volts (b) 115 volt (c) 114 volts (d) 116 volts

• Answer: (c) 114 volts

The maximum conductor voltage drop recommended for both the feeder and branch circuit is 5 percent of the voltage source (120 volts). The total conductor voltage drop (feeder and branch circuit) should not exceed; 120 volts × .05 = 6 volts. The operating voltage at the load is calculated by subtracting the conductors voltage drop from the voltage source; 120 volts – 6 volts drop = 114 volts.

8–8 DETERMINING CIRCUIT CONDUCTORS VOLTAGE DROP

When the circuit conductors have already been installed, the *voltage drop* of the conductors can be determined by the Ohm's Law Method or by the formula method.

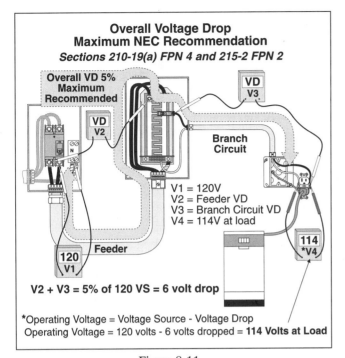

Figure 8-11

Overall Voltage Drop Example

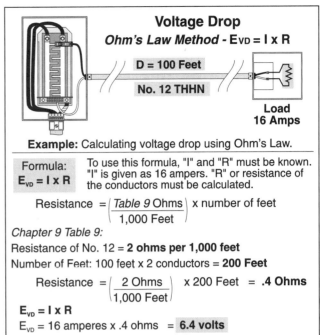

Figure 8-12

Voltage Drop – Ohm's Law, 120 Volt Example

Ohm's Law Method – Single-Phase Only

$E_{vd} = I \times R$

E_{vd} = Conductor voltage drop expressed in volts

I = The load in amperes at 100 percent, not at 125 percent for motors or continuous loads

R* = Conductor Resistance, Chapter 9, Table 8 for direct current circuits, or Chapter 9, Table 9 for alternating current circuits

*For conductors No. 1/0 and smaller, the difference in resistance between direct current and alternating current circuits is so little that it can be ignored. In addition, we can ignore the small difference in resistance between stranded or solid wires.

❏ **Ohm's Law Conductor Voltage Drop 120 Volt Example**

What is the voltage drop of two No. 12 THHN conductors that supply a 16-ampere, 120 volt continuous load located 100 feet from the power supply, Fig. 8–12?

(a) 3.2 volts   (b) 6.4 volts   (c) 9.6 volts   (d) 12.8 volts

• Answer: (b) 6.4 volts

$E_{vd} = I \times R$

I = Amperes = 16 amperes

R = 2 ohms per 1,000 feet, Chapter 9, Table 9. The resistance No. 12 wire, 200 feet long is:

(2 ohms/1,000 feet) × 200 feet = 0.4 ohm

E_{vd} = 16 amperes × 0.4 ohms E_{vd} = 6.4 volts

❏ **Ohm's Law Conductor Voltage Drop 240 Volt Example**

A single-phase, 24-ampere, 240 volt load is located 160 feet from the panelboard and is wired with No. 10 THHN. What is the voltage drop of the circuit conductors, Fig. 8–13?

(a) 4.5 volts   (b) 9.25 volts   (c) 3.6 volts   (d) 5.5 volts

• Answer: (b) 9.25 volts is the closest answer

$E_{vd} = I \times R$

I = 24 amperes

R = 1.2 ohms per 1,000 feet, Chapter 9, Table 9. The resistance for 320 feet of No. 10 is:

(1.2 ohms/1,000 feet) × 320 feet = 0.384 ohm

E_{vd} = 24 amperes × 0.384 ohm E_{vd} = 9.216 volts drop

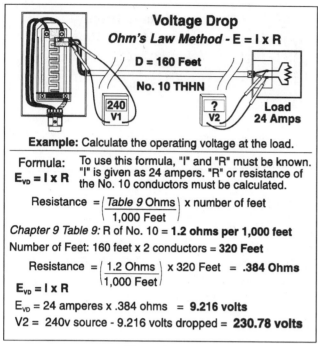

Figure 8-13

Voltage Drop – Ohm's Law, 240 Volt Example

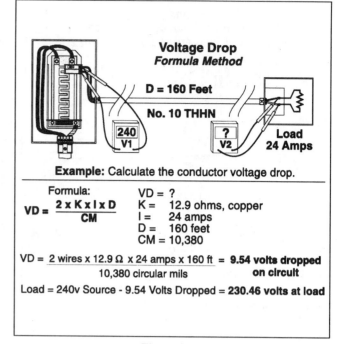

Figure 8-14

Voltage Drop – Formula Method, Single-Phase Example

Formula Method

In addition to the Ohm's Law Method, the following formula can be used to determine the conductor voltage drop.

Single–Phase $E_{VD} = \dfrac{2 \times (K \times Q) \times I \times D}{CM}$ Three–Phase $E_{VD} = \dfrac{\sqrt{3} \times (K \times Q) \times I \times D}{CM}$

VD = Volts Dropped: The voltage drop of the circuit expressed in volts. The NEC recommends a maximum 3 percent voltage drop for either the branch circuit or feeder.

K = Direct Current Constant: This constant K represents the direct current resistance for a one thousand circular mils conductor that is one thousand feet long, at an operating temperature of 75°C. The constant K value is 12.9 ohms for copper, and 21.2 ohms for aluminum.

I = Amperes: The load in amperes at 100 percent, not at 125 percent for motors or continuous loads!

Q = Alternating Current Adjustment Factor: For alternating current circuits with conductors No. 2/0 and larger, the direct current resistance constant K must be adjusted for the effects of self-induction (eddy currents). The Q-Adjustment Factor is calculated by dividing the alternating current resistance listed in Chapter 9, Table 9, by the direct current resistance listed in Chapter 9, Table 8 in the NEC. For all practical exam purposes, this resistance adjustment factor can be ignored because exams rarely give alternating current voltage drop questions with conductors larger than No. 1/0.

D = Distance: The distance the load is from the power supply.

CM = Circular-Mils: The circular mils of the circuit conductor as listed in Chapter 9, Table 8 NEC.

❑ **Voltage Drop Formula Single-Phase Example**

A 24-ampere, 240-volt load is located 160 feet from a panelboard and is wired with No. 10 THHN. What is the approximate voltage drop of the branch circuit conductors, Fig. 8–14?

(a) 4.25 volts   (b) 9.5 volts   (c) 3 percent   (d) 5 percent

• Answer: (b) 9.5 volts is the closest

$VD = \dfrac{2 \times K \times I \times D}{CM}$

K = 12.9 ohms, Copper   I = 24 amperes

D = 160 feet   CM = No. 10, 10,380 circular mils, Chapter 9, Table 8

$VD = \dfrac{2 \text{ wires} \times 12.9 \text{ ohms} \times 24 \text{ amps} \times 160 \text{ feet}}{10,380 \text{ circular mils}}$

VD = 9.54 volts dropped

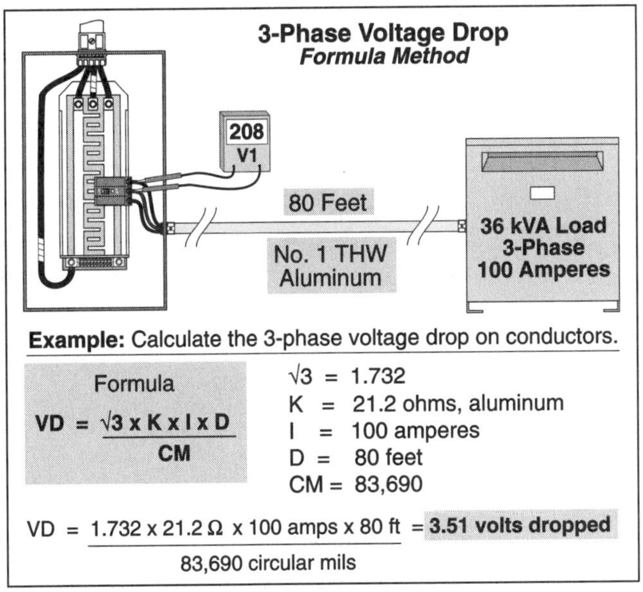

Figure 8-15: 3-Phase Voltage Drop Formula Method

Example: Calculate the 3-phase voltage drop on conductors.

Formula:
$$VD = \frac{\sqrt{3} \times K \times I \times D}{CM}$$

- $\sqrt{3} = 1.732$
- K = 21.2 ohms, aluminum
- I = 100 amperes
- D = 80 feet
- CM = 83,690

$$VD = \frac{1.732 \times 21.2\,\Omega \times 100\,\text{amps} \times 80\,\text{ft}}{83{,}690\,\text{circular mils}} = \textbf{3.51 volts dropped}$$

Figure 8-15
Voltage Drop – Formula Method, Three-Phase Example

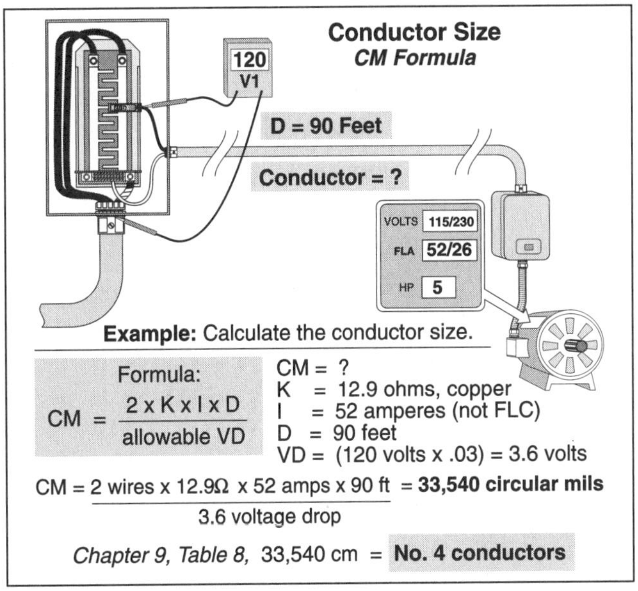

Figure 8-16: Conductor Size CM Formula

- D = 90 Feet
- Conductor = ?
- VOLTS 115/230
- FLA 52/26
- HP 5

Example: Calculate the conductor size.

Formula:
$$CM = \frac{2 \times K \times I \times D}{\text{allowable VD}}$$

- CM = ?
- K = 12.9 ohms, copper
- I = 52 amperes (not FLC)
- D = 90 feet
- VD = (120 volts × .03) = 3.6 volts

$$CM = \frac{2\,\text{wires} \times 12.9\,\Omega \times 52\,\text{amps} \times 90\,\text{ft}}{3.6\,\text{voltage drop}} = \textbf{33,540 circular mils}$$

Chapter 9, Table 8, 33,540 cm = **No. 4 conductors**

Figure 8-16
Conductor Size Example

❏ Voltage Drop Formulas Three-Phase Example

A three-phase, 36-kVA load rated 208-volts is located 80 feet from the panelboard and is wired with No. 1 THHN Aluminum. What is the approximate voltage drop of the feeder circuit conductors, Fig. 8–15?

(a) 3.5 volts (b) 7 volts (c) 3 percent (d) 5 percent

- Answer: (a) 3.5 volts is the closest

$$VD = \frac{\sqrt{3} \times K \times I \times D}{CM}$$

$\sqrt{3} = 1.732$

K = 21.2 ohms, aluminum

I = 100 amperes, $I = \frac{VA}{(E \times \sqrt{3})}, = \frac{36{,}000\,VA}{(208\,\text{volts} \times 1.732)}$

D = 80 feet

CM = No. 1, 83,690 circular mils, Chapter 9, Table 8

$$VD = \frac{1.732 \times 21.2\,\text{ohms} \times 100\,\text{amps} \times 80\,\text{feet}}{83{,}690\,\text{circular mils}}$$

VD = 3.51 volts dropped

8–9 SIZING CONDUCTORS TO PREVENT EXCESSIVE VOLTAGE DROP

The size of a conductor (actually its resistance) affects the conductor voltage drop. If we want to decrease the voltage drop of a circuit, we can increase the size of the conductor (reduce its resistance). When sizing conductors to prevent excessive voltage drop, the following formulas can be used:

Single–Phase $CM = \dfrac{2 \times K \times I \times D}{\textbf{Voltage Dropped}}$ **Three–Phase** $CM = \dfrac{\sqrt{3} \times K \times I \times D}{\textbf{Voltage Dropped}}$

❏ Size Conductor Single-Phase Example

A 5 horsepower motor is located 90 feet from a 120/240-volt panelboard. What size conductor should we use if the motor nameplate indicates 52 amperes at 115 volts? Terminals rated for 75ºC, Fig. 8–16.

(a) No. 10 THHN (b) No. 8 THHN
(c) No. 6 THHN (d) No. 4 THHN

- Answer: (d) No. 4 THHN

$$CM = \frac{2 \times K \times I \times D}{VD}$$

K = 12.9 ohms, copper
I = 52 amperes at 115 volts
D = 90 feet
VD = 3.6 volts, = 120 volts × 0.03

$$CM = \frac{2 \text{ wires} \times 12.9 \text{ ohms} \times 52 \text{ amperes} \times 90 \text{ feet}}{3.6 \text{ volts}}$$

CM = 33,540 circular mils
CM = No. 4, Chapter 9, Table 8

Note

The NEC Section 430–22(a), requires that the motor conductors be sized not less than 125 percent of the motor full load currents as listed in Table 430–148. The motor full-load current is 56 amperes, and the conductor must be sized at; 56 amperes × 1.25 = 70 amperes. The No. 4 THHN required for voltage drop is rated for 75 amperes at 75°C according to Table 310–16 and 110–14(c).

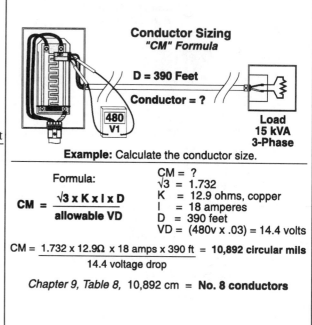

Figure 8-17
Conductor Size, Three-Phase Example

❑ **Size Conductor Three-Phase Example**

A three-phase, 15-kVA load rated 480 volts is located 390 feet from the panelboard. What size conductor is required to prevent the voltage drop from exceeding 3 percent, Fig. 8–17?

(a) No. 10 THHN (b) No. 8 THHN (c) No. 6 THHN (d) No. 4 THHN

- Answer: (b) No. 8 THHN

$$CM = \frac{\sqrt{3} \times K \times I \times D}{VD}$$

K = 12.9 ohms, copper

I = 18 amperes, $\frac{VA}{(E \times 1.732)} = \frac{15,000 \text{ VA}}{(480 \text{ volts} \times 1.732)} = 18$ amperes

D = 390 feet
VD = 14.4 volts, = 480 volts × 0.03

$$CM = \frac{1.732 \times 12.9 \text{ ohms} \times 18 \text{ amperes} \times 390 \text{ feet}}{14.4 \text{ volts}}$$

CM = 10,892 = No. 8, Chapter 9, Table 8

8-10 LIMITING CONDUCTOR LENGTH TO LIMIT VOLTAGE DROP

Voltage drop can also be reduced by limiting the length of the conductors. The following formulas can be used to help determine the maximum conductor length to limit the voltage drop to NEC suggestions.

$$\text{Single-Phase} = D = \frac{CM \times VD}{2 \times K \times I}$$

$$\text{Three-Phase} = D = \frac{CM \times VD}{\sqrt{3} \times K \times I}$$

❑ **Limit Distance Single-Phase Example**

What is the maximum distance a single-phase, 10-kVA, 240-volt load can be located from the panelboard so the voltage drop does not exceed 3 percent? The load is wired with No. 8 THHN, Fig 8–18.

(a) 55 feet (b) 110 feet (c) 165 feet (d) 220 feet

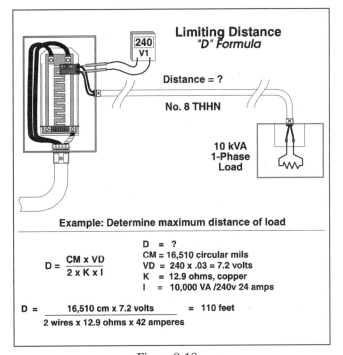

Figure 8-18

Maximum Distance, Single-Phase Example

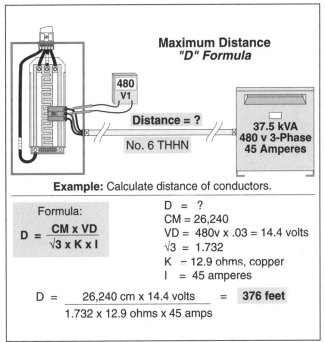

Figure 8-19

Maximum Distance, Three-Phase Example

- Answer: (b) 110 feet

$$D = \frac{CM \times VD}{2 \times K \times I}$$

CM = No. 8, 16,510 circular mils, Chapter 9, Table 8

VD = 7.2 volts, = 240 volts × 0.03

K = 12.9 ohms, copper

I = 42 amperes, I = VA/E, I = 10,000 VA/240 volts

$$D = \frac{16,510 \text{ circular mils} \times 7.2 \text{ volts}}{2 \text{ wires} \times 12.9 \text{ ohms} \times 42 \text{ amperes}}$$

D = 110 feet

❑ **Limit Distance Three-Phase Example**

What is the maximum distance a three-phase, 37.5-kVA, 480-volt transformer, wired with No. 6 THHN, can be located from the panelboard so that the voltage drop does not exceed 3 percent, Fig. 8–19?

(a) 275 feet (b) 325 feet (c) 375 feet (d) 425 feet

- Answer: (c) 375 feet

$$D = \frac{CM \times VD}{\sqrt{3} \times K \times I}$$

CM = No. 6, 26,240 circular mils, Chapter 9, Table 8

K = 12.9 ohms, copper

$\sqrt{3}$ = 1.732

VD = 14.4 volts, = 480 volts × 0.03

I = 45 amperes, = $\frac{VA}{(Volts \times 1.732)}$, = $\frac{37,500 \text{ VA}}{(480 \text{ volts} \times 1.732)}$

$$D = \frac{26,240 \text{ circular mils} \times 14.4 \text{ volts}}{1.732 \times 12.9 \text{ ohms} \times 45 \text{ amperes}}$$

D = 376 feet

8–11 LIMITING CURRENT TO LIMIT VOLTAGE DROP

Sometimes the only method of limiting the circuit voltage drop is to limit the load on the conductors. The following formulas can be used to determine the maximum load.

$$\text{Single-Phase} = I = \frac{CM \times VD}{2 \times K \times D} \qquad \text{Three-Phase} = I = \frac{CM \times VD}{\sqrt{3} \times K \times D}$$

❑ **Maximum Load Single-Phase Example**

An existing installation contains No. 1/0 THHN aluminum conductors in a nonmetallic raceway to a panelboard located 220 feet from a 230-volt power source. What is the maximum load that can be placed on the panelboard so that the NEC recommendations for voltage drop are not exceeded, Fig. 8–20?

(a) 50 amperes (b) 75 amperes (c) 100 amperes (d) 150 amperes

- Answer: (b) 75 amperes is the closest answer

$$I = \frac{CM \times VD}{2 \times K \times D}$$

CM = No. 1/0, 105,600 circular mils, Chapter 9, Table 8
VD = 6.9 volts, = 230 volts × 0.03
K = 21.2 ohms, aluminum
D = 220 feet

$$I = \frac{105{,}600 \text{ circular mils} \times 6.9 \text{ volts drop}}{2 \text{ wires} \times 21.2 \text{ ohms} \times 220 \text{ feet}} =$$

I = 78 amperes

Note. The maximum load permitted on 1/0 THHN Aluminum at 75°C is 120 amperes; Table 310–16.

❑ **Maximum Load Three-Phase Example**

An existing installation contains No. 1 THHN conductors in an aluminum raceway to a panelboard located 300 feet from a three-phase 460/230 volt power source. What is the maximum load the conductors can carry so that the NEC recommendation for voltage drop is not exceeded, Fig. 8–21?

(a) 170 amperes (b) 190 amperes (c) 210 amperes (c) 240 amperes

- Answer: (a) 170 amperes

$$I = \frac{CM \times VD}{\sqrt{3} \times 2 \times K \times D}$$

CM = No. 1, 83,690 circular mils, Chapter 9, Table 8
VD = 13.8 volts, = 460 volts × 0.03
$\sqrt{3}$ = 1.732
K = 12.9 ohms, copper
D - 220 feet

$$I = \frac{83{,}690 \text{ circular mils} \times 13.8 \text{ volts}}{1.732 \times 12.9 \text{ ohms} \times 300 \text{ feet}}$$

I = 172 amperes

Note. The maximum load permitted on No. 1 THHN at 75°C is 130 amperes; Table 310–16 and 110–14(c).

8–12 EXTENDING CIRCUITS

If you want to extend an existing circuit and you want to limit the voltage drop, follow these steps.

Step 1: → Determine the voltage drop of the existing conductors.

$$\text{Single-Phase VD} = \frac{2 \times K \times I \times D}{CM}$$

$$\text{Three-Phase VD} = \frac{\sqrt{3} \times K \times I \times D}{CM}$$

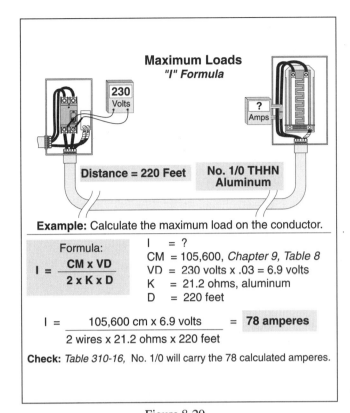

Figure 8-20
Maximum Load, Single-Phase Example.

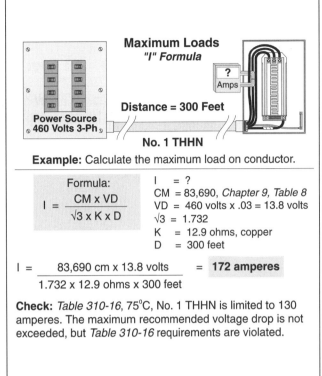

Figure 8-21
Maximum Distance, Three-Phase Example.

Step 2: → Determine the voltage drop permitted for the extension by subtracting the voltage drop of the existing conductors from the permitted voltage drop.

Step 3: → Determine the extended conductor size

$$\text{Single-Phase CM} = \frac{2 \times K \times I \times D}{VD}$$

$$\text{Three-Phase CM} = \frac{\sqrt{3} \times K \times I \times D}{VD}$$

❏ **Single-Phase Example**

An existing junction box is located 55 feet from the panelboard and contains No. 4 THW aluminum. We want to extend this circuit 65 feet and supply a 50 ampere, 240-volt load. What size copper conductors can we use for the extension, Fig. 8–22?

(a) No. 8 THHN   (b) No. 6 THHN   (c) No. 4 THHN   (d) No. 5 THHN

• Answer: (b) No. 6 THHN

Step 1 → Determine the voltage drop of the existing conductors:

$$\text{Single Phase VD} = \frac{2 \times K \times I \times D}{CM}$$

K = 21.2 ohms, aluminum
D = 55 feet
I = 50 amperes
CM = No. 4, 41,740 circular mils, Chapter 9, Table 8

$$VD = \frac{2 \text{ wires} \times 21.2 \text{ ohms} \times 50 \text{ amps} \times 55 \text{ feet}}{41,740 \text{ circular mils}}$$

VD = 2.79 volts

Step 2 → Determine the voltage drop permitted for the extension by subtracting the voltage drop of the existing conductors from the total permitted voltage drop. Total permitted voltage drop = 240 volts × 0.03 = 7.2 volts less existing conductor voltage drop of 2.79 volts = 4.41 volts.

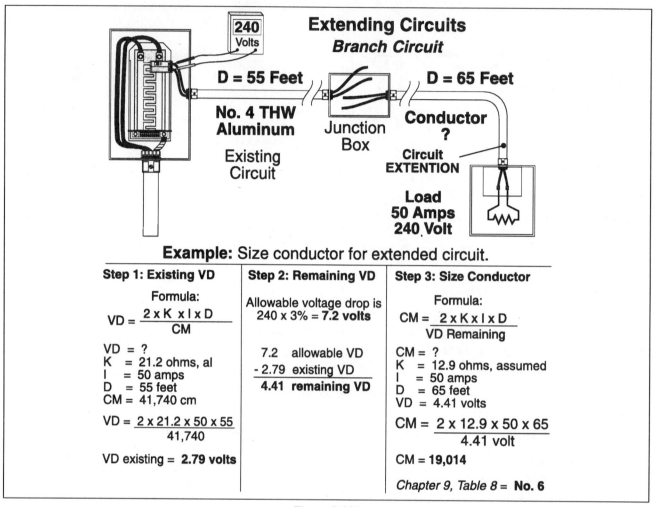

Figure 8-22
Extending Circuit Example.

Step 3 → Determine the extended conductor size.

$$CM = \frac{2 \times K \times I \times D}{VD}$$

K = 12.9 ohms, copper
I = 50 amperes
D = 65 feet
VD = 7.2 volts total less 2.79 volts = 4.41 volts

$$CM = \frac{2 \text{ wires} \times 12.9 \text{ ohms} \times 50 \text{ amps} \times 65 \text{ feet}}{4.41 \text{ volts drop}}$$

CM = 19,014 circular mils
CM = No. 6 conductor, Chapter 9, Table 8

Unit 8 – Voltage Drop Summary Questions

8–1 Conductor Resistance

1. The _____ the number of free electrons, the better the conductivity of the conductor.
   (a) greater (b) fewer

2. Conductor resistance is determined by the _____.
   (a) material type
   (b) cross-sectional area
   (c) conductor length
   (d) conductor operating temperature
   (e) all of these

3. _____ is the best conductor, better than gold, but the high cost limits its use to special applications, such as fuse elements and some switch contacts.
   (a) Silver (b) Copper
   (c) Aluminum (d) None of these

4. _____ conductors are often used when weight or cost are important consideration.
   (a) Silver (b) Copper
   (c) Aluminum (d) None of these

5. • Conductor cross-sectional area is the surface area expressed in _____,
   (a) square inches (b) mils
   (c) circular mils (d) none of these

6. The resistance of a conductor is directly proportional to the conductor length.
   (a) True (b) False

7. The resistance of a conductor changes with temperature. Temperature coefficient describes the effect that temperature has on the resistance of a conductor. Conductors with a _____ temperature coefficient have an increase in resistance with an increase in temperature.
   (a) positive (b) negative (c) neutral (d) none of these

8–2 Conductor Resistance – Direct Current [Chapter 9, Table 8 Of The NEC]

8. The National Electric Code list the resistance and circular mils for both direct current and alternating current conductors. direct current conductor resistances are listed in Chapter 9 Table _____ and alternating current conductor resistance are listed in Chapter 9 Table _____.
   (a) 1, 5 (b) 3, 4
   (c) 9, 8 (d) 8, 9

9. What is the direct current resistance of 400 feet of No. 6?
   (a) .2 ohm (b) .3 ohm (c) .4 ohm (d) .5 ohm

10. What is the direct current resistance of 150 feet of No. 1?
    (a) .023 ohm (b) .031 ohm (c) .042 ohm (d) .056 ohm

11. What is the direct current resistance of 1,400 feet of No. 3 AL?
    (a) .23 ohm (b) .31 ohm (c) .42 ohm (d) .56 ohm

12. What is the direct current resistance of 800 feet of No. 1/0 AL?
    (a) .23 ohm (b) .16 ohm (c) .08 ohm (d) .56 ohm

13. What is the direct current resistance of 120 feet of No. 1?
    (a) .23 ohm (b) .16 ohm (c) .02 ohm (d) .56 ohm

14. What is the direct current resistance of 1,249 feet of No. 3 AL?
    (a) .23 ohm (b) .16 ohm (c) .02 ohm (d) .5 ohm

8–3 Conductor Resistance – Alternating Current Circuits

15. The intensity of the magnetic field is dependent on the intensity of alternating current current. The greater the current flow, the greater the overall magnetic field.
    (a) True (b) False

16. • The expanding and collapsing magnetic field within the conductor, exerts a force on the moving electrons. This force is called counter-electromotive-force.
    (a) True (b) False

17. _____ currents are small independent currents that are produced as a result of the expanding and collapsing magnetic field. They flow erratically within the conductor opposing current flow and consuming power.
    (a) Lentz (b) Ohms (c) Eddy (d) Kerchoffs

18. The expanding and collapsing magnetic field induces a counter voltage within the conductors which repels the flowing electrons towards the conductor surface. This is known as _____ effect.
    (a) inductive (b) skin (c) surface (d) watt

8–4 Alternating Current Resistance As Compared To Direct Current

19. The opposition to current flow is greater for alternating current circuits because of _____ than for direct current circuits.
    (a) eddy currents (b) skin effect (c) cemf (d) all of these

8–5 Conductor Resistance – Alternating Current [Chapter 9, Table 9 Of The NEC]

20. The alternating current conductor resistances listed in Chapter 9 Table 9 of the NEC are different for copper and aluminum and for nonmagnetic and magnetic raceways.
    (a) True (b) False

21. What is the alternating current resistance of 320 feet of No. 2/0?
    (a) .03 ohm (b) .04 ohm (c) .05 ohm (d) .06 ohm

22. What is the alternating current resistance of 1,000 feet of 500 kcmil when installed in an aluminum raceway?
    (a) .032 ohm (b) .027 ohm (c) .029 ohm (d) .030 ohm

23. What is the alternating current resistance of 220 feet of No. 2/0?
    (a) .012 ohm (b) .022 ohm (c) .33 ohm (d) .43 ohm

24. What is the alternating current resistance of 369 feet of 500 kcmil installed in nonmetallic conduit?
    (a) .01 ohm (b) .02 ohm (c) .03 ohm (d) .04 ohm

25. A 36 ampere load is located 80 feet from the panelboard and is wired with No. 1 THHN aluminum. What is the total resistance of the circuit conductors?
    (a) .04 ohm (b) .25 ohm (c) .5 ohm (d) all of these

26. What size copper conductors can be used to replace No. 1/0 aluminum that supplies a 110 ampere load. We do not want to increase the circuit voltage drop.
    (a) No. 3 (b) No. 2 (c) No. 1 (d) No. 1/0

27. What is the direct current resistance in ohms for three 300 kcmil conductors in parallel?
    (a) .014 ohm (b) .026 ohm (c) .052 ohm (d) .047 ohm

28. What is the ac resistance in ohms for three 1/0 THHN aluminum conductors in parallel?
    (a) .1 ohms (b) .2 ohms (c) .4 ohms (d) .067 ohms

Chapter 2 NEC Calculations And Code Questions   Unit 8 Voltage Drop Calculations 265

8–6 Voltage Drop Consideration

29. Because of the great demand for electricity, utilities sometimes are required to reduce their output voltage. In addition, the utility and customers transformers, services, feeders and branch circuit conductors oppose the flow of current. The opposition to current flow results in voltage drop. All circuits have voltage drop, simply because all conductors have resistance.
    (a) True   (b) False

30. When sizing conductors for feeders and branch circuits, the NEC _____ that we take voltage drop into consideration [210–19(a) FPN No. 4, 215–2 FPN No. 2 and 310–15 FPN].
    (a) permits   (b) suggest   (c) requires   (d) demands

31. _____ equipment such as motors and electromagnetic ballast can overheat at reduced voltage. This results in reduced equipment operating life and inconvenience to the customer.
    (a) Inductive   (b) Electronic   (c) Resistive   (d) All of these

32. • _____ equipment such as computers, laser printer, copy machine, etc., can suddenly power down because of reduce voltage, resulting in data losses.
    (a) Inductive   (b) Electronic   (c) Resistive   (d) All of these

33. Resistive loads such as incandescent lighting and electric space heating can have their power output decreased by the square of the voltage.
    (a) True   (b) False

34. What is the power consumed of a 4.5 kW 230 volt water heater operating at 200 volts?
    (a) 2,700 watts   (b) 3,400 watts   (c) 4,500 watts   (d) 5,500 watts

35. How can conductor voltage drop be reduced?
    (a) reduce the conductor resistance
    (b) increase conductor size
    (c) decrease conductor length
    (d) all of these

8–7 NEC Voltage Drop Recommendations

36. If the branch circuit supply voltage is 208 volts, the maximum recommended voltage drop of the circuit should not be more than _____ volts.
    (a) 3.6   (b) 6.24   (c) 6.9   (d) 7.2

37. If the feeder supply voltage is 240 volts, the maximum recommended voltage drop of the feeder should not be more than _____ volts.
    (a) 3.6   (b) 6.24   (c) 6.9   (d) 7.2

8–8 Determining Circuit Conductors Voltage Drop

38. What is the voltage drop of two No. 12 THHN conductors supplying a 12-ampere continuous load. The continuous load is located 100 feet from the power supply.
    (a) 3.2 volts   (b) 4.75 volts   (c) 6.4 volts   (d) 12.8 volts

39. A single-phase, 24-amperes, 240 volt-load is located 160 feet from the panelboard. The load is wired with No. 10 THHN. What is the approximate voltage drop of the branch circuit conductors?
    (a) 4.25 volts   (b) 9.5 volts   (c) 3 percent   (d) 5 percent

40. A three-phase, 36-kVA load, rated 208-volts is located 100 feet from the panelboard and is wired with No. 1 THHN Aluminum. What is the approximate voltage drop of the circuit conductors?
    (a) 3.5 volts   (b) 5 volts   (c) 3 volts   (d) 4.4 volts

8–9 Sizing Conductors To Prevent Excessive Voltage Drop

41. A single-phase, 5-horsepower motor is located 110 feet from the panelboard. The nameplate indicates that the voltage is 115/230 and the FLA is 52/26 amperes. What size conductor is required if the motor winding are connected in parallel and operates at 115 volts?
    (a) No. 10 THHN (b) No. 8 THHN (c) No. 6 THHN (d) No. 3 THHN

42. A single-phase, 5-horsepower motor is located 110 feet from the panelboard. The nameplate indicates that the voltage is 115/230 and the FLA is 52/26 amperes. What size conductor is required if the motor winding are connected in series and operates at 230 volts?
    (a) No. 10 THHN (b) No. 8 THHN (c) No. 6 THHN (d) No. 4 THHN

43. A three-phase, 15-kW, 480-volt load is located 300 feet from the panelboard. What size conductor is required to prevent the voltage drop from exceeding 3 percent?
    (a) No. 10 THHN (b) No. 8 THHN (c) No. 6 THHN (d) No. 4 THHN

8–10 Limiting Conductor Length To Limit Voltage Drop

44. What is the approximate distance a single phase 7.5-kVA, 240-volt load can be located from the panelboard so the voltage drop does not exceed 3 percent? The load is wired with No. 8 THHN.
    (a) 55 feet (b) 110 feet (c) 145 feet (d) 220 feet

45. What is the approximate distance a three phase 37.5-kVA, 460-volt transformer, wired with No. 6 THHN, can be located from the panelboard so the voltage drop does not exceed 3 percent?
    (a) 250 feet (b) 300 feet (c) 345 feet (d) 400 feet

8–11 Limiting Current To Limit Voltage Drop

46. An existing installation contains No. 1/0 THHN aluminum conductors in a nonmetallic raceway to a panelboard located 190 feet from a 240 volt power source. What is the maximum load that can be placed on the panelboard so that the NEC recommendations for voltage drop are not exceeded?
    (a) 90 amperes (b) 110 amperes (c) 70 amperes (d) 175 amperes

47. An existing installation contains No. 2 THHN conductors in an aluminum raceway to a panelboard located 275 feet from a three-phase 460/230 volt power source. What is the maximum load the conductors can carry without exceeding the NEC recommendation for voltage drop.
    (a) 110 amperes (b) 175 amperes (c) 190 amperes (d) 149 amperes

8–12 Extending Circuits

48. An existing junction box is located 65 feet from the panelboard and contains No. 4 THHN Aluminum. We want to extend this circuit 85 feet and supply a 50-ampere, 208-volt load. What size copper conductors can we use for the extension?
    (a) No. 8 THHN (b) No. 6 THHN (c) No. 4 THHN (d) No. 5 THHN

8–13 Miscellaneous

49. What is the circuit voltage if the conductor voltage drop is 3.3 volts? Assume 3 percent voltage drop.
    (a) 110 volts (b) 115 volts (c) 120 volts (d) none of these

Challenge Questions

8–1 Conductor Resistance

50. The resistance of a conductor is affected by temperature change. This is called the _____.
    (a) temperature correction factor (b) temperature coefficient
    (c) ambient temperature factor (d) none of these

8–3 Conductor Resistance – Alternating Current Circuits

51. The total opposition to current flow in an alternating current circuit is expressed in ohms and is called _____.
    (a) impedance   (b) conductance   (c) reluctance   (d) resistance

8–5 Alternating Current Conductor Resistance [Chapter 9, Table 9 Of The NEC]

52. A 40-ampere, 240-volt, single-phase load is located 150 feet from an existing junction box. The junction box is located 50 feet from the panelboard and is wired with No. 4 THHN aluminum wire. The total resistance of the two No. 4 conductors from the panelboard to the junction box is approximately _____ ohm.
    (a) .03   (b) .09   (c) .05   (d) .04

53. A load is located 100 feet from a 230-volt power supply and is wired with No. 4 THHN aluminum conductors. What size copper conductor can be used to replace the aluminum conductors and not increase the conductor voltage drop?
    (a) No. 6   (b) No. 8   (c) No. 1   (d) No. 1/0

8–7 NEC Voltage Drop Recommendations

54. A 40-ampere, 240-volt rated, single-phase load is wired 150 feet from a junction box, and the junction box is located 50 feet from a panelboard (for a total of 200 feet). If the voltage at the panelboard is 240 volts, what is the minimum voltage recommended by the NEC at the 40-ampere load?
    (a) 228.2 volts   (b) 232.8 volts   (c) 236.2 volts   (d) 117.7 volts

8–8 Determining Circuit Conductors Voltage Drop

55. What is the voltage drop of two No. 4 aluminum conductors that supply a 5-horsepower, single-phase, 208-volt motor that has a nameplate rating of 55 amperes? The motor is located 95 feet from the power supply.
    (a) 3.25 volts   (b) 5.31 volts   (c) 6.24 volts   (d) 7.26 volts

8–9 Sizing Conductors To Prevent Excessive Voltage Drop

56. A 40-ampere, 240-volt, single-phase load is located 150 foot from an existing junction box. The junction box is located 50 feet from the panelboard. When the 40-ampere load is on, the voltage at the junction box would be calculated to be 236 volts. The NEC recommends voltage drop for this branch circuit not exceed 3 percent of the 240-voltage source, which is 7.2 volts. What size conductor could be installed from the junction box to the load and still meet the NEC recommendations?
    (a) No. 3   (b) No. 1   (c) No. 1/0   (d) No. 6

8–10 Limiting Conductor Length To Limit Voltage Drop

57. How far can a 50-ampere, three-phase, 230-volt load be located from the panel, if fed with No. 3 THHN, and still meet the NEC recommendations for voltage drop?
    (a) 275 feet   (b) 300 feet   (c) 325 feet   (d) 350 feet

8–11 Limiting Current To Limit Voltage Drop

58. Two No. 8 THHN supply a 120-volt load that is located 225 feet from the panelboard. What is the maximum load, in amperes, that can be applied to these conductors without exceeding the NEC recommendation on conductor voltage drop?
    (a) 0 amperes   (b) 5 amperes   (c) 10 amperes   (d) 15 amperes

8–13 Miscelleous Questions

59. An 8-ohm resistor is connected to a 120-volt power supply. Using a voltmeter, we measure 112 volts across the resistor. What is the current of the 8 ohm resistor in amperes?
    (a) 14.00 amperes   (b) 13.37 amperes   (c) 15.73 amperes   (d) 19.41 amperes

60. A 480-volt, single-phase feeder carries 400 amperes and has a 7.2 voltage drop. What is the total resistance of the conductors in this circuit?
    (a) .1880 ohm   (b) .1108 ohm   (c) .0190 ohm   (d) .0180 ohm

61. An 8 ohm resistor operates at 112 volts and is connected to a 115 volt power supply. The percent of voltage drop for the circuit is _____ percent.
    (a) 2.6 (b) 3.0 (c) 5.0 (d) 7.0

NEC QUESTIONS
Random Sequence, Section 370–29 Through Section 514–5

62. • Electric heaters of the cord- and plug-connected immersion type shall be so constructed and installed that current-carrying parts are effectively _____ from electrical contact with the substance in which they are immersed.
    (a) isolated (b) protected (c) insulated (d) all of these

63. Listed boxes designed for underground installation can be directly buried where covered by _____.
    (a) concrete (b) gravel
    (c) noncohesive granulated soil (d) b and c

64. The distance required in Section 370–71 for angle or U-Pulls between each cable or conductor entry inside the box (pull and junction boxes for use on systems over 600 volts, nominal) and the opposite wall of the box shall _____ for additional entries by the amount of the sum of the outside diameters, over sheath, of all other cables or conductor entries through the same wall of the box.
    (a) be increased (b) be decreased (c) remain the same (d) none of these

65. No. 18 TFFN wire is rated for _____ amperes.
    (a) 14 (b) 10 (c) 8 (d) 6

66. Where provision must be made for limited flexibility, such as at motor terminals, in a Class I, Division 2 location, flexible conduit with approved fittings may be used.
    (a) True (b) False

67. Sealing compound is employed with MI cable terminal fittings in Class I locations for the purpose of _____.
    (a) preventing the passage of gas or vapor (b) excluding moisture and other fluids
    (c) limiting a possible explosion (d) preventing the escape of powder

68. Which of the following wiring methods are permitted in a Class I, Division 1 location?
    (a) threaded rigid metal conduit (b) threaded IMC
    (c) MI cable (d) all of these

69. Motor tap conductors not over _____ feet must be enclosed in a raceway.
    (a) 10 (b) 15 (c) 20 (d) 25

70. A motor for usual use shall be marked with a time rating of _____.
    I. continuous   II. 30 or 60 minutes   III. 5 or 15 minutes
    (a) I (b) I or II (c) II and III (d) I, II, or III

71. Each conduit leaving a Class I, Division 1 location requires one seal on either side of the hazardous location boundary. Unions, couplings, boxes, or fittings are permitted between the seal and the point where the conduit leaves the Division 1 location.
    (a) True (b) False

72. What size overcurrent protectin is required for a 7.5 kVA transformer that has a primary current rating of 21 amperes?
    (a) 15 (b) 25 (c) 35 (d) 45

73. • The rating of the branch circuit short circuit ground fault protection device for an individual motor-compressor shall not exceed _____ percent of the rated-load current or branch circuit selection current, whichever is greater, if the protection device will carry the starting current of the motor.
    (a) 100 (b) 125 (c) 175 (d) None of these

74. Where all the conditions of the Code are met, several motors not over one horsepower are permitted on one branch-circuit providing that the full-load rating of each motor does not exceed _____ amperes.
    (a) 6 (b) 10 (c) 15 (d) 20

Chapter 2 NEC Calculations And Code Questions   Unit 8 Voltage Drop Calculations 269

75. In Class II, Division 1 and 2 locations, an approved method of bonding is the use of _____.
    (a) bonding jumpers with approved fittings
    (b) double locknut types of contacts
    (c) locknut-bushing types of contacts
    (d) any of the above are approved methods of bonding

76. A standard circuit breaker mounted in a Class I, Division 2 location with make-and-break contacts and not hermetically sealed or oil immersed is required to be installed in a Class I, Division 1 rated enclosure.
    (a) True   (b) False

77. Areas adjacent to defined locations in commercial garages which flammable vapors are not likely to be released such as stock rooms, switchboard rooms, and other similar locations shall not be classified when mechanically ventilated at a rate of _____ or more air changes per hour or when effectively cut off by walls or partitions.
    (a) two   (b) four   (c) six   (d) none of these

78. Connectors for electric vehicle charging in a commercial garage shall _____.
    (a) not be located in Class I locations
    (b) be installed so they will disconnect readily
    (c) have live parts guarded from accidental contact
    (d) all of these

79. The motor overload protection shall not be shunted or cut out during the starting period if the motor is _____.
    (a) not automatically started
    (b) automatically started
    (c) manually started
    (d) all of these

80. • Meters, instruments, and relays installed in Class I Division 2 locations can have switches, circuit breakers, and make-and-break contacts of push buttons, relays, alarm bells, and horns installed in enclosures approved for general purpose if current-interrupting contacts are _____.
    (a) immersed in oil   (b) enclosed within a hermetically sealed chamber
    (c) either a or b   (d) both a and b

81. Infrared lamps for industrial heating appliances shall have overcurrent protection not exceeding _____ amperes.
    (a) 30   (b) 40   (c) 50   (d) 60

82. • Overload relays for motors that are not capable of _____ shall be protected by fuses or circuit breakers.
    (a) opening short circuits   (b) clearing overloads
    (c) ground-fault protection   (d) all of these

83. For Class _____ locations, Groups E, F, and G, the classification involves the tightness of the joints of assembly and shaft openings, to prevent entrance of dust in the dust-ignition-proof enclosure. The blanketing effect of layers of dust on the equipment may cause overheating, electrical conductivity of the dust and the ignition temperature of the dust.
    (a) I   (b) II   (c) III   (d) all of these

84. • A motor feeder supplying specific fixed motor loads having conductors sized by Section 430–24 shall have a protective device with a rating or setting of _____ branch circuit short circuit and ground fault protective device for any motor in the group plus the sum of the full-load currents of the other motors of the group.
    (a) not greater than the largest
    (b) 125 percent of the largest rating
    (c) equal to the largest rating
    (d) none of these

85. In a Class II, Division 1 location where magnesium, aluminum, aluminum bronze, or other powders of hazardous characteristics may be present, transformers and capacitors _____.
    (a) must be dust-tight
    (b) may be pipe ventilated
    (c) must be approved for Class II, Division 1
    (d) shall not be used

86. Examples of resistance heaters could be _____.
    (a) tubular heaters and strip heaters
    (b) immersion heaters and heating blankets
    (c) heating cables or heating tape
    (d) all of these

87. • Motor disconnecting means shall be a motor-circuit switch rated in _____, be a molded case switch, or a circuit breaker.
    (a) horsepower
    (b) watts
    (c) amperes
    (d) locked-rotor current

88. When selecting the conductors according to Section 430–24, the motor with the highest _____ shall be the highest-rated motor.
    (a) FLC
    (b) horsepower
    (c) nameplate
    (d) code letter

89. The total rating of a plug-connected room air conditioner, where lighting units or other appliances are also supplied, shall not exceed _____ percent.
    (a) 80
    (b) 70
    (c) 50
    (d) 40

90. In a Class II location, where electrically-conducting dust is present, flexible connections can be made with _____.
    (a) flexible metal conduit
    (b) type AC armored cable
    (c) hard usage cord
    (d) liquidtight flexible metal conduit with approved fittings

91. A permanently mounted lighting fixture in a commercial garage and located over lanes on which vehicles are commonly driven shall be located not less than _____ feet above floor level.
    (a) 10
    (b) 12
    (c) 14
    (d) none of these

92. Class _____ locations usually include places such as sawmills, textile mills, and similar locations where easily ignitable fibers or combustible flyings are present from the manufacturing process.
    (a) I
    (b) II
    (c) III
    (d) all of these

93. It is necessary, that equipment be approved not only for the class, but also for the specific group of the gas, vapor or dust that will be present.
    (a) True
    (b) False

94. Stock rooms and similar areas adjacent to aircraft hangars but effectively isolated and adequately ventilated shall be designated as _____ locations.
    (a) Class I, Division 2
    (b) Class II, Division 1
    (c) Class II, Division 2
    (d) nonhazardous

95. Motor control circuit transformers rated less than 2 amperes can have the primary protection device set no more than _____ percent of the rated primary current rating.
    (a) 150
    (b) 200
    (c) 400
    (d) 500

96. The disconnecting means for a torque motor shall have an ampere rating of at least _____ percent of the motor nameplate current.
    (a) 100
    (b) 115
    (c) 125
    (d) 175

97. Types of resistive heaters are heating _____.
    (a) blankets
    (b) tape
    (c) barrels
    (d) a and b

98. If the disconnect is not within sight of the fixed electric space heater without supplementary overcurrent protection devices, the disconnect must be capable of being _____.
    (a) locked
    (b) locked closed
    (c) locked open
    (d) within sight

99. On space heating cables, the blue leads would represent a cable rated for a branch circuit with a voltage of _____.
    (a) 120
    (b) 240
    (c) 208
    (d) 277

100. Motor control circuits shall be so arranged that they will be disconnected from all sources of supply when the disconnecting means is in the open position. Where separate devices are used for the motor and control circuit, they shall be located immediately adjacent one to each other.
    (a) True
    (b) False

Chapter 2 NEC Calculations And Code Questions   Unit 8 Voltage Drop Calculations 271

101. The frames of stationary motors (over 600 volts) shall be grounded _____.
(a) where supplied by metal-enclosed wiring
(b) where in a wet location and not isolated or guarded
(c) if in a hazardous (classified) location
(d) all of these

102. The disconnecting means for air-conditioning and refrigeration equipment must be _____ from the air-conditioning or refrigerating equipment.
(a) readily accessible   (b) within sight   (c) either a or b   (d) both a and b

103. If the electric appliance is to be used on a specific _____, it shall be so marked.
(a) frequency or frequencies
(b) service or voltage
(c) conductor or conductors
(d) type of cable

104. For electrode-type boilers, each boiler shall be designed so that in normal operation there is no change in state of the heat transfer medium, and shall be equipped with a temperature sensitive _____.
(a) protective device   (b) limiting means   (c) shut-off device   (d) all of these

105. Each generator shall be provided with a _____ giving the maker's name, the rated frequency, power factor, number of phases if of alternating current, the rating in kilowatts or kilovolt amperes, the normal volts and amperes corresponding to the rating, rated revolutions per minute, insulation system class and rated ambient temperature or rated temperature rise, and time rating.
(a) list   (b) faceplate   (c) nameplate   (d) sticker

106. A heating panel is a complete assembly provided with a junction box or a length of flexible conduit for connection to a(n) _____.
(a) wiring system   (b) service   (c) branch circuit   (d) approved conductor

107. Exposed elements of impedance heating systems shall be physically guarded, isolated, or thermally insulated with a _____ jacket to protect against contact by personnel in the area.
(a) corrosion resistant   (b) waterproof   (c) weatherproof   (d) flame retardant

108. When a controller is not within sight of the motor location, the motor disconnecting means shall be capable of _____.
(a) ground-fault operation
(b) opening all ungrounded conductors
(c) being locked in the open position
(d) being locked in the closed position

109. All fixed outdoor deicing and snow-melting equipment shall be provided with a means for disconnection from all _____ conductors.
(a) grounded   (b) grounding   (c) ungrounded   (d) neutral

110. Electric space heating cables shall not extend beyond the room or area in which they _____.
(a) provide heat   (b) originate   (c) terminate   (d) are connected

111. The ultimate trip current of a thermally protected motor (full-load current not exceeding 9 amperes) shall not exceed _____ percent of motor full-load current given in Tables 430–148, 430–149, and 430–150.
(a) 140   (b) 156   (c) 170   (d) none of these

112. Fixed electric space heating equipment shall be installed to provide the _____ spacing between the equipment and adjacent combustible material, unless it has been found to be acceptable where installed in direct contact with combustible material.
(a) required   (b) minimum   (c) maximum   (d) safest

113. The controller shall have a horsepower rating _____ the horsepower rating of the motor.
(a) lower than
(b) not lower than
(c) equal to
(d) none of these

114. For permanently connected appliances rated over _____ volt-amperes or ⅛-horsepower, the branch-circuit switch or circuit breaker shall be permitted to serve as the disconnecting means where the switch or circuit breaker is within sight from the appliance or is capable of being locked in the open position.
    (a) 200 (b) 300 (c) 400 (d) 500

115. An overload device used to protect continuous-duty motors (rated more than 1-horsepower) shall be selected to trip or rated at no more than _____ percent of the motor nameplate full-load current rating for motors with a marked service factor not less than 1.15.
    (a) 110 (b) 115 (c) 120 (d) 125

Random Sequence, Section 515–2 Through Chapter 9

116. Service-entrance conductors for a manufactured building (prefab) shall be installed _____.
    (a) after erection at the building site
    (b) before erection at the building site
    (c) before erection where the point of attachment is known prior to manufacture
    (d) a or c

117. Each general care area patient bed location shall be provided with a minimum number of receptacles such as _____ duplex.
    (a) one single or one (b) six single or three (c) two single or one (d) four single or two

118. The power source maximum nameplate rating would be _____ amperes for a Class II, 25-volt control circuit with inherently limited power source. See the formula in Table 725–31(a)
    (a) 3 (b) 4 (c) 6 (d) 7

119. Receptacles must not be within 10 feet of the water's edge for _____.
    (a) pools (b) outdoor spas
    (c) outdoor hot tubs (d) all of these

120. The maximum number of No. 4 RHH with outer covering permitted in a one inch electrical metallic tubing is _____.
    (a) one (b) two (c) three (d) four

121. A 250 kcmil bare cable has a conductor diameter of _____ inch.
    (a) 0.557 (b) 0.755 (c) 0.575 (d) 0.690

122. Where the calculated number of conductors, all of the same size, includes a decimal fraction, the next higher whole number shall be used where this decimal is _____ or larger.
    (a) .5 (b) .08 (c) .008 (d) .8

123. A pool transformer is required to _____.
    (a) be of the isolated winding type
    (b) have a grounded metal barrier between the primary and secondary windings
    (c) be identified for the purpose
    (d) all of these

124. Conductors of two or more Class II circuits shall be permitted within the same cable, raceway, or enclosure provided all the conductors are _____.
    (a) insulated for the maximum of any conductor
    (b) separated by a permanent partition
    (c) all of the same voltage and amperage class
    (d) all of these

125. Above ground bulk storage tanks shall be classified as _____ for the space between 5 feet and 10 feet from open end of vent, extending in all directions.
    (a) Class I, Division 1
    (b) Class I, Division 2
    (c) Class II, Division 1
    (d) Class II, Division 2

Chapter 2 NEC Calculations And Code Questions Unit 8 Voltage Drop Calculations 273

126. An aboveground tank in a bulk storage plant is a Class I, Group D, Division 1 location within _____ feet of open end of vent, extending in all directions.
(a) 12 (b) 10 (c) 6 (d) 5

127. Feeders to floating buildings can be _____ if listed for wet locations and sunlight resistance.
(a) rigid nonmetallic conduit
(b) liquidtight metallic conduit
(c) liquidtight nonmetallic conduit
(d) portable power cables

128. Exposed power-limited fire alarm cables on the load side of overcurrent protection, transformers, and current-limiting devices for power-limited circuits, are permitted if _____.
(a) adequately supported
(b) terminated in approved fittings
(c) the building construction is utilized for physical protection
(d) all of these

129. Exposed power-limited fire alarm circuit cables are required to be installed in _____ when passing through a floor or wall to a height of 7 feet above the floor unless adequate protection can be afforded by the building construction or unless an equivalent solid guard is provided.
(a) rigid metal conduit (b) rigid nonmetallic conduit
(c) EMT (d) any of these

130. The alternating current resistance of 1,000 feet of 4/0 aluminum would have the same resistance as 1,000 feet of _____ copper when installed in a steel raceway.
(a) No. 1 (b) No. 1/0 (c) No. 2/0 (d) 250 kcmil

131. Electrical fire alarm installations in hollow spaces, vertical shafts, and ventilation or air-handling ducts shall be made so that the spread of fire or products of combustion (such as smoke, or gases) will not be _____.
(a) possible (b) substantially increased
(c) permitted (d) none of these

132. All electric equipment, including power supply cords, used with storable pools shall be protected by _____.
(a) fuses (b) circuit breakers (c) double-insulation (d) GFCI

133. Transfer equipment, including automatic transfer switches, shall be _____.
(a) automatic
(b) identified for emergency use
(c) approved by the authority having jurisdiction
(d) all of these

134. • Conductors for an appliance circuit supplying more than one appliance or appliance receptacle in an installation operating at less than 50 volts shall not be smaller than No. _____ copper or equivalent.
(a) 18 (b) 14 (c) 12 (d) 10

135. Where nipples 24 inches or less are used, the ampacity adjustment factors for more than 3 current-carrying conductors _____ to this condition.
(a) do not apply (b) shall be applied

136. Where exposed to contact with electric light or power conductors, the noncurrent-carrying metallic members of optical fiber cables entering buildings shall be _____.
(a) grounded at the point of emergence through an exterior wall
(b) grounded at the point of emergence through a concrete floor slab
(c) interrupted as close to the point of entrance as practicable by an insulating joint (d) any of these

137. In Class 2 or Class 3 control circuit systems where overcurrent protection is used, the overcurrent protection shall not be _____.
(a) a fuse or breaker (b) interchangeable with devices of higher ratings
(c) over 15 amperes (d) rated for Class 1 installations

138. Fire alarm systems, multiconductor control, or power cable which is specifically listed for the use can be installed in the space over a hung ceiling used for environmental air.
    (a) True   (b) False

139. Cables and conductors of two or more power-limited fire alarm circuits are permitted to be installed in the same cable, enclosure, or raceway.
    (a) True   (b) False

140. The metallic sheath of communications cables entering buildings shall be _____.
    (a) grounded at the point of emergence through an exterior wall
    (b) grounded at the point of emergence through a concrete floor slab
    (c) interrupted as close to the point of entrance as practicable by an insulating joint
    (d) any of these

141. Type TFN and other approved insulations used on a nonpower-limited fire alarm circuit is required for _____.
    (a) conductor sizes Nos. 16 and 18
    (b) all conductors up to 600 volts
    (c) conductors sizes Nos. 18, 16, and 14
    (d) conductor sizes No. 14 and larger

142. Metal conduit and metal piping within _____ feet of the inside walls of the pool, and that are not separated from the pool by a permanent barrier, are required to be bonded.
    (a) 4   (b) 5   (c) 8   (d) 10

143. The maximum load permitted on a heating element in a pool heater shall not exceed _____ amperes.
    (a) 20   (b) 35   (c) 48   (d) 60

144. Wet-niche lighting fixtures must be grounded by a minimum No. 12 insulated copper conductor. This grounding conductor must be installed in _____.
    (a) rigid nonmetallic conduit
    (b) electrical metallic conduit when on or in the building
    (c) flexible metal conduit
    (d) a or b

145. Class 1 control circuit overcurrent protection for tapped conductors larger than No. 14 shall be protected by overcurrent devices rated at not more than _____.
    (a) the ampacity of the conductors as per Tables 310–16 through 310–19
    (b) the ampacity of the remote control conductors
    (c) 125 percent of the ampacity of the conductors
    (d) 300 percent of the ampacity of the remote control conductors

146. A 16-kW nameplate rating of a free-standing range in a mobile home would use _____ VA as the computed load.
    (a) 8,000   (b) 8,800   (c) 9,600   (d) 10,000

147. Access to equipment shall not be denied by an accumulation of wires and cables that prevent removal of panels, including suspending ceiling tiles. The intention of this Section is to limit the quantity of wires and cables that can be installed (laid) directly upon lay-in type suspended ceilings. This applies to _____.
    (a) Class 2 and 3 Cables
    (b) communication cables
    (c) CATV coaxial cables
    (d) all of these

148. • A motel conference room is designed for the assembly of more than 100 persons. The fixed wiring methods require _____.
    (a) rigid nonmetallic conduit   (b) MC cable
    (c) nonmetallic sheathed cable   (d) type AC cable

149. A forming shell shall be provided with a number of grounding terminals that shall be _____ the number of conduit entries.
    (a) one more than   (b) two more than   (c) the same as   (d) none of these

150. An electrically driven or controlled machine with one or more motors, not hand portable, and used primarily to transport and distribute water for agricultural purposes is called a(n) _____.
(a) irrigation machine
(b) electric water distribution system
(c) center pivot irrigation machine
(d) automatic water distribution system

151. An outdoor portable electric sign shall have a ground-fault interrupter _____.
(a) located on the sign
(b) located in the power supply cord within 12 inches of the attachment plug
(c) as an integral part of the attachment plug of supply cord
(d) b or c

152. All exposed metal parts of cranes, hoists, and accessories shall _____ into a continuous electrical conductor.
(a) be bonded with No. 6 or larger conductors to be made
(b) be metallically joined together
(c) be grounded
(d) not be grounded or made

153. Unshielded lead-in antenna conductors of amateur transmitting stations attached to building surfaces, shall clear the building surface which is wired over by a distance not less than _____ inches.
(a) 1   (b) 2   (c) 3   (d) 4

154. The inside diameter of a one inch intermediat metal conduit is _____ inch.
(a) 1.0   (b) 1.105   (c) .86   (d) .34

155. Two conductors are permitted to fill the percent area of a conduit _____ percent.
(a) 40   (b) 31   (c) 53   (d) 30

156. The conductor used to ground the outer cover of a coaxial cable shall be _____.
I. insulated   II. No. 14 AWG minimum
III. guarded from physical damage when necessary
(a) I only   (b) II only   (c) III only   (d) I, II, and III

157. Soft-drawn or medium-drawn copper lead in conductors for receiving antenna systems shall be permitted where the maximum span between points of support is less than _____ feet.
(a) 35   (b) 30   (c) 20   (d) 10

158. A pool light junction box that has a raceway that extends directly to underwater pool light forming shells shall be located not less than _____ feet from the outdoor pool or spa.
(a) 2   (b) 3   (c) 4   (d) 6

159. A lighting fixture located within 5 feet from the inside wall of an indoor spa or hot tub must be _____.
(a) a minimum of 7 feet 6 inches over maximum water level
(b) protected by a ground-fault circuit-interrupter
(c) either a or b
(d) both a and b

160. Underground rigid nonmetallic wiring when located less than 5 feet from the inside wall of the pool or spa must be buried not less than _____ inches.
(a) 6   (b) 10   (c) 12   (d) 18

161. Interconnecting cables under raised computer room floors provide a significant fire and smoke load and must be listed as Type _____ Cable when used under raised floors of a computer room.
(a) RF   (b) UF   (c) LS   (d) DP

162. • A minimum of 70 percent of all recreational vehicle sites with electrical supply shall each be equipped with a _____ ampere receptacle.
(a) 15   (b) 20   (c) 30   (d) 50

163. Dielectric heating equipment auxiliary rectifiers used with filter capacitors in the output for bias supplies, tube keyers, etc., bleeder resistors shall be used even though the DC voltage may not exceed _____ volts.
(a) 24   (b) 50   (c) 115   (d) 240

164. 15 and 20 ampere receptacles located in pediatric areas shall be _____.
    (a) tamper resistant (b) isolated (c) GFI (d) specification grade

165. Class 1 control circuits using No. 18 AWG shall use insulation types _____.
    (a) RFH–2, RFHH–2 or RFHH–3
    (b) TF, TFF, TFN, or TFFN
    (c) RHH, RHW, THWN, or THHN
    (d) a and b

Random Sequence, Section 90–2 Through Section 225–8

166. Loads computed for dwelling unit small appliance circuits shall be permitted to be included with the _____ load and subjected to the demand factors permitted in Table 220–11 for the general lighting load.
    (a) general lighting (b) feeder (c) appliance (d) receptacle

167. The demand factors of Table 220–20 apply to space heating, ventilating, and air-conditioning equipment.
    (a) True (b) False

168. Under the optional method for calculating a single-family dwelling, all other loads beyond the initial 10 kW is to be assessed at a _____ percent demand factor.
    (a) 40 (b) 50 (c) 60 (d) 75

169. _____ in dwelling units shall supply only loads within that dwelling unit or loads associated only with that dwelling unit.
    (a) Service entrance conductors (b) Ground fault protection
    (c) Branch circuits (d) none of these

170. The small appliance circuit applies to the _____ as well as the kitchen.
    (a) dining room (b) refrigerator (c) breakfast room (d) all of these

171. Where more than one building or other structure is on the same property and under single management, each building or other structure served shall be provided with means for disconnecting all _____ conductors.
    (a) grounded (b) grounding (c) ungrounded (d) underground

172. • Heating, air-conditioning, and refrigeration equipment on rooftops or in attics and crawl spaces, shall be a receptacle outlets for use in servicing these units. The receptacle shall be _____.
    (a) GFCI protected (b) at an accessible location
    (c) within 25 feet of the equipment (d) All of the above

173. • The demand factors listed in Table 220–11 shall not be applied in determining the number of _____ for general illumination.
    (a) receptacles (b) light outlets (c) branch circuits (d) switches

174. The neutral conductor of a 3-wire branch circuit supplying a household electric range, a wall-mounted oven, or a counter-mounted cooking unit shall be permitted to be smaller than the ungrounded conductors where the maximum demand of a range of 8 ¾ kW or more rating has been computed according to Column A of Table 220–19, but shall have an ampacity of not less than _____ percent of the branch-circuit rating and shall not be smaller than No. _____.
    (a) 50, 6 (b) 70, 6 (c) 50, 10 (d) 70, 10

175. Conductors larger than No. 4/0 are measured in _____.
    (a) inches (b) circular mils (c) square inches (d) AWG

176. One receptacle outlet shall be installed at each island or peninsular counter top with a long dimension of _____ inches or greater and a short dimension of _____ inches or greater.
    (a) 12, 24 (b) 24, 12 (c) 24, 48 (d) 48, 24

177. The 24-inch dimension required in a dwelling unit counter top is measured along the wall line or center line, and the intent is that there be a receptacle outlet for every _____ linear feet or fraction thereof of counter length.
    (a) 2 (b) 3 (c) 4 (d) 5

178. At least one receptacle outlet shall be installed directly above a show window for each _____.
    (a) 12 square feet (b) 12 linear feet (c) 10 linear feet (d) 15 horizontal feet

179. All services, feeders, and branch circuits must be legibly marked to indicate their intended purpose unless located and arranged in such a manner so that the purpose is clearly evident.
(a) Yes (b) No

180. Where switches, cutouts, or similar equipment operating at 600 volts, nominal, or less, are installed in a room or enclosure where there are exposed energized parts or exposed wiring operating at over 600 volts, nominal, the high-voltage equipment shall be effectively separated from the space occupied by the low-voltage equipment by a suitable _____.
(a) partition (b) fence (c) screen (d) any of these

181. _____ means that an object is not readily accessible to persons unless special means for access are used.
(a) Isolated (b) Secluded (c) Protected (d) Locked

182. To guard live parts over 50 volts but less than 600 volts, the equipment can be _____.
(a) isolated in a room accessible to qualified persons only
(b) located on a balcony
(c) elevated 8 feet or more above the floor
(d) any of these

183. The Code defines a(n) _____ as one familiar with the construction and operation of the equipment and the hazards involved.
(a) inspector (b) master electrician
(c) journeyman electrician (d) qualified person

184. At least one receptacle outlet is required in each _____.
(a) basement (b) attached garage
(c) detached garage with electric power (d) all of these

185. A system intended to provide protection of equipment from damaging line-to-ground fault currents by operating to cause a disconnecting means to open all ungrounded conductors of the faulted circuit, is defined as _____.
(a) ground-fault protection of equipment
(b) protection of equipment
(c) ground-fault protection of wiring systems
(d) protection of wiring systems

186. The dimension of working clearance for access to live parts operating at 300 volts, nominal, to ground where there are exposed live parts on one side and no live or grounded parts on the other side of the working space, or exposed live parts on both sides effectively guarded by suitable wood or other insulating materials is _____ feet according to Table 110–16(a).
(a) 3 (b) 3½ (c) 4 (d) 4½

187. Circuit protective devices are used to clear a fault without the occurrence of extensive damage to the electrical components of the circuit. This fault shall be assumed to be either between two or more of the _____, or between any circuit conductor and the grounding conductor or enclosing metal raceway.
(a) bonding jumpers (b) grounding jumpers
(c) wiring harness (d) circuit conductors

188. The Code covers underground installations in mines and self-propelled mobile surface mining machinery, and its attendant electrical trailing cable.
(a) True (b) False

189. The minimum working clearance on a circuit 120 volts to ground, with exposed live parts on one side and no live or grounded parts on the other side of the working space is _____ feet.
(a) 3 (b) 3 ½ (c) 4 (d) 6

190. A unit of an electrical system which is intended to carry but not utilize electrical energy is a(n) _____.
(a) raceway (b) fitting (c) device (d) enclosure

191. Acceptable to the authority having jurisdiction means that the equipment has been _____.
(a) identified (b) listed (c) approved (d) labeled

192. Receptacles and cord connectors having grounding terminals must have those terminals _____.
    (a) grounded (b) bonded (c) labeled (d) none of these

193. Factory-installed internal wiring of equipment need not be inspected at the time of installation of the equipment, except to _____.
    (a) modify the equipment (b) reconstruct the equipment
    (c) detect alterations or damage (d) none of these

194. A branch circuit that supplies only one utilization equipment is a(n) _____ branch circuit.
    (a) individual (b) general purpose (c) isolated (d) special purpose

195. The authority having jurisdiction of enforcement of the Code will have the responsibility _____.
    (a) for making interpretations of the rules of the Code
    (b) for deciding upon the approval of equipment and materials
    (c) can waive specific requirements in the Code and allow alternate methods and material if safety is maintained
    (d) all of these

196. Installations of communications equipment under the exclusive control of communications utilities located outdoors or in building spaces used exclusively for such installations _____ covered by the Code.
    (a) are (b) are sometimes (c) are not (d) might be

197. When breaks occur in dwelling-unit kitchen counter top spaces for ranges, refrigerators, sinks, etc., each counter top surface is considered a separate counter space for determining receptacle placement.
    (a) True (b) False

198. A large single panel, frame, or assembly of panels on which are mounted, on the face or back, or both, switches, overcurrent and other protective devices, buses, and usually instruments is a _____.
    (a) switchboard (b) panel box (c) switch box (d) panelboard

199. _____ branch circuits supply energy to one or more outlets to which appliances are to be connected.
    (a) General purpose (b) Multiwire (c) Individual (d) Appliance

200. • Working space shall not be used for _____.
    (a) storage (b) raceways (c) lighting (d) accessibility

Unit 9

One-Family Dwelling-Unit Load Calculations

OBJECTIVES

After reading this unit, the student should be able to briefly explain the following concepts:

- Air conditioning versus heat
- Appliance (small) circuits
- Appliance demand load
- Clothes dryer demand load
- Cooking equipment calculations
- Dwelling-unit calculations – optional method
- Dwelling-unit calculations – standard method
- Laundry circuit
- Lighting and receptacles

After reading this unit, the student should be able to briefly explain the following terms:

- General lighting
- General–use receptacles
- Optional method
- Rounding
- Standard method
- Unbalanced demand load
- Voltages

PART A - GENERAL REQUIREMENTS

9–1 GENERAL REQUIREMENTS

Article 220 provides the requirement for residential branch circuit, feeders, and service calculations. Other applicable articles are Branch Circuits – 210, Feeders – 215, Services – 230, Overcurrent Protection – 240, Wiring Methods – 300, Conductors – 310, Appliances– 422, Electric Space Heating Equipment – 424, Motors – 430, and Air Conditioning – 440.

9–2 VOLTAGES [220–2]

Unless other voltages are specified, branch-circuit, feeder, and service loads shall be computed at nominal system voltages of 120, 120/240, 208Y/120, or 240, Fig. 9–1.

9–3 FRACTION OF AN AMPERE

There are no specific NEC rules for rounding when a calculation results in a fraction of an ampere, but Chapter 9, Part B Examples contains the following note; "except where the computations result in a major fraction of an ampere (.5 or larger), such fractions may be dropped."

279

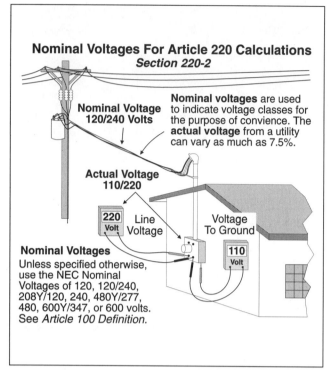

Figure 9-1
Nominal Voltages

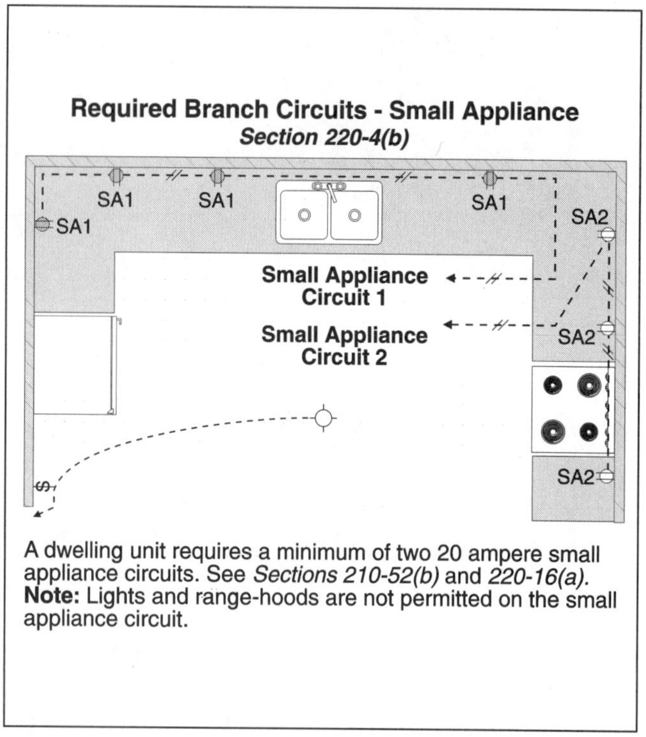

Figure 9-2
Required Branch Circuits – Small Appliances

9–4 APPLIANCE (SMALL) CIRCUITS [220–4(b)]

A minimum of two 20-ampere *small appliance branch circuits* is required for receptacle outlets in the kitchen, dining room, breakfast room, pantry, or similar dining areas [220–4(b)]. In general no other receptacle or lighting outlet can be connected to these 20-ampere small appliance branch circuits [210–52(b) Exceptions], Fig. 9–2.

Feeder And Service

When sizing the feeder or service, each dwelling-unit shall have two 20-ampere small appliance branch circuits with a feeder load of 1,500 volt-amperes for each circuit [220–16(a)].

Other Related Code Sections

Areas that the small appliance circuits supply [210–52(b)] and
Receptacle outlets required for kitchen counter tops [210–52(c)].
Fifteen- or 20-ampere rated receptacles can be used on 20-ampere circuit [210-21(b)(3) and 220–16(a)].

9–5 COOKING EQUIPMENT – BRANCH CIRCUIT [Table 220–19, Note 4]

To determine the branch circuit demand load for cooking equipment, we must comply with the requirements of Table 220–19, Note 4. The branch circuit demand load for one range shall be according to the demand loads listed in Table 220–19, but a minimum 40-ampere circuit is required for 8.75 kW and larger ranges [210–19(b)].

❏ **Less than 12 kW, Column *A* Example**

What is the branch circuit demand load (in amperes) for one 9-kW range, Fig. 9–3?

(a) 21 amperes   (b) 27 amperes   (c) 38 amperes   (d) 33 amperes

• Answer: (d) 33 amperes, 8 kW

The demand load for one range in Table 220–19 Column A is 8 kW. This can be converted to amperes by dividing the power by the voltage, $I = \dfrac{P}{E}$

$$I = \dfrac{8 \text{ kW} \times 1{,}000}{240 \text{ volts}*}$$

* Assume 120/240 volts for all calculations unless the question gives a specific voltage and system.

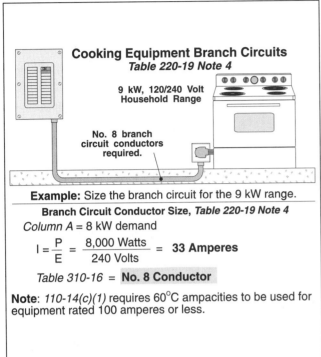

Figure 9-3

Branch Circuit Column "A" Example

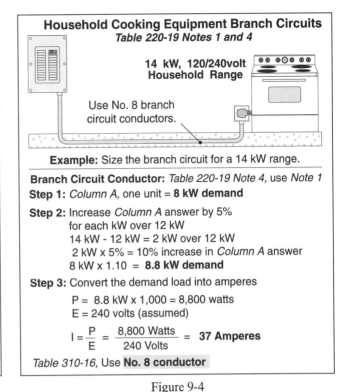

Figure 9-4

Branch Circuit Note 1 Example

❑ More Than 12 kW, Note 1 Example

What is the branch circuit load (in amperes) for one 14-kW range, Fig. 9–4?

(a) 33 amperes (b) 37 amperes (c) 50 amperes (d) 58 amperes

- Answer: (b) 37 amperes

Step 1: → Since the range exceeds 12 kW, we must comply with Note 1 of Table 220–19. The first step is to determine the demand load as listed in Column A of Table 220–19 for one unit, = 8 kW

Step 2: → We must increase the Column A value (8 kW) by 5 percent for each kW that the average range (in this case 14 kW) exceeds 12 kW. This results is an increase of the Column A value (8 kW) by 10 percent.
8 kW × 1.1 = 8.8 kW or (8,000 watts + 800 watts)

Step 3: → Convert the demand load in kW to amperes.

$$I = \frac{P}{E}, I = \frac{8,800 \text{ VA}}{240 \text{ volts}} = 36.7 \text{ amperes}$$

One Wall-Mounted Oven Or One Counter-Mounted Cooking Unit [220–19 Note 4]

The branch circuit demand load for one wall-mounted oven or one counter-mounted cooking unit shall be the nameplate rating.

❑ One Oven Example

What is the branch circuit load (in amperes) for one 6-kW wall-mounted oven, Fig. 9–5?

(a) 20 amperes (b) 25 amperes (c) 30 amperes (d) 40 amperes

- Answer: (b) 25 amperes

$$\text{Nameplate} = 6,000 \text{ VA} \quad I = \frac{VA}{E}, I = \frac{6,000 \text{ VA}}{240 \text{ volts}} = 25 \text{ amperes}$$

❑ One Counter-Mounted Cooking Unit Example

What is the branch circuit load (in amperes) for one 4.8-kW counter-mounted cooking unit?

(a) 15 amperes (b) 20 amperes (c) 25 amperes (c) 30 amperes

- Answer: (b) 20 amperes,

$$\text{Nameplate} = 4,800 \text{ VA}, I = \frac{VA}{E}, I = \frac{4,800 \text{ VA}}{240 \text{ volts}} = 20 \text{ amperes}$$

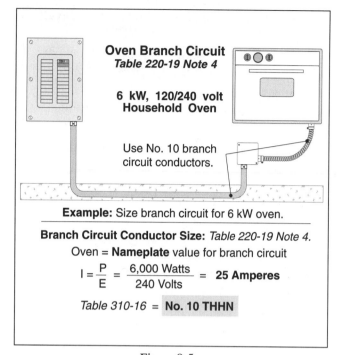

Figure 9-5
Branch Circuit One Oven Example

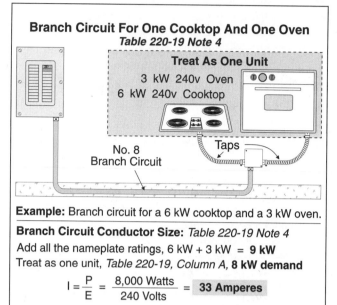

Figure 9-6
Branch Circuit One Cook-top And One Oven Example

One Counter-mounted Cooking Unit And Up To Two Ovens [220–19 Note 4]

To calculate the load for one counter-mounted cooking unit and up to two wall-mounted ovens, complete the following steps:

Step 1: → Total Load. Add the nameplate ratings of the cooking appliances and treat this total as one range.
Step 2: → Table 220-19 Demand. Determine the demand load for one unit from Table 220–19, Column A.
Step 3: → Net Computed Load. If the total nameplate rating exceeds 12 kW, increase Column A (8 kW) 5 percent for each kW, or major fraction (.5 kW), that the combined rating exceeds 12 kW.

❏ **Cook-Top And One Oven Example**

What is the branch circuit load (in amperes) for one 6-kW counter-mounted cooking unit and one 3-kW wall-mounted oven, Fig. 9–6?

(a) 25 amperes (b) 38 amperes (c) 33 amperes (d) 42 amperes

• Answer: (c) 33 amperes

Step 1: → Total connected load. 6 kW + 3 kW = 9 kW.
Step 2: → Demand load for one range, Column A of Table 220–19, 8 kW.

$$I = \frac{P}{E}, I = \frac{8,000 \text{ VA}}{240 \text{ volts}} = 33.3 \text{ amperes}$$

❏ **Cook-Top And Two Ovens Example**

What is the branch circuit load (in amperes) for one 6 kW Counter-mounted cooking unit and two 4 kW wall-mounted ovens, Fig. 9–7?

(a) 22 amperes (b) 27 amperes (c) 33 amperes (d) 37 amperes

• Answer: (d) 37 amperes

Step 1: → The total connected load:
6 kW + 4 kW + 4 kW = 14 kW
Step 2: → Demand load for one range, Column A or Table 220–19, = 8 kW
Step 3: → Since the combined total (14 kW) exceeds 12 kW, we must increase Column A (8 kW) 5 percent for each kW, or major fraction (.5 kW), that the range exceeds 12 kW. 14 kW exceeds 12 kW by 2 kW, which results in a 10 percent increase of the Column A value. 8 kW × 1.1 = 8.8 kW.

$$I = \frac{P}{E}, I = \frac{8,800 \text{ VA}}{240 \text{ volts}} = 36.7 \text{ amperes}$$

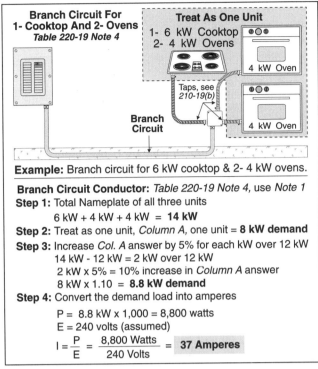

Figure 9-7

Branch Circuit, One Cook-Top And Two Ovens Example

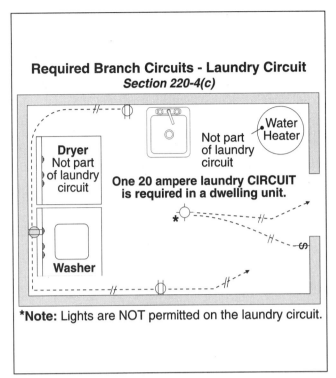

Figure 9-8

Laundry Room Circuit Requirement

9–6 LAUNDRY RECEPTACLE(S) CIRCUIT [220–4(c)]

One 20-ampere branch circuit for the laundry receptacle outlet (or outlets) is required and the laundry circuit cannot serve any other outlet, such as the laundry room lights [210–52(f)]. The NEC does not require a separate circuit for the washing machine but does require a separate circuit for the laundry room receptacle or receptacles. If the washing machine receptacle(s) are located in the garage or basement, GFCI protection might be required, see Section 210–8(a)(2) and (4) and their exceptions, Fig. 9–8.

Feeder and Service
Each dwelling-unit shall have a feeder load consisting of 1,500 volt-amperes for the 20-ampere laundry circuit [220-16(b)].

Other related Code rules
A laundry outlet must be within 6 feet of washing machine [210–50(c)].
Laundry area receptacle outlet required [210–52(f)].

9–7 LIGHTING AND RECEPTACLES

General Lighting Volt-Amperes Load [220–3(b)]
The NEC requires a minimum 3 volt-amperes per square-foot [Table 220–3(b)] for the *general lighting* and general-use receptacles. The dimensions for determining the area shall be computed from the outside of the building and shall not include open porches, garages, or spaces not adaptable for future use, Fig. 9–9.

Note. The 3 volt-ampere-per-square-foot rule for general lighting includes all 15- and 20-ampere general-use receptacles; but, it does not include the appliance or laundry circuit receptacles. See the Note at the Bottom of Table 220–3(b) for more details.

☐ General Lighting VA Load Example
What is the general lighting and receptacle load for a 2,100 square-foot dwelling-unit that has thirty four convenience receptacles and twelve lighting fixtures rated 100 watts each, Fig. 9–10?

(a) 2,100 VA   (b) 4,200 VA   (c) 6,300 VA   (d) 8,400 VA

- Answer: (c) 6,300 VA.

    2,100 square feet × 3 VA = 6,300 VA. No additional load is required for general-use receptacles and lighting outlets, see Note to Table 220–3(b).

284 Unit 9 One-Family Load Calculations

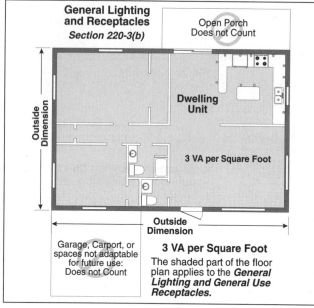

Figure. 9-9
General Lighting Requirement

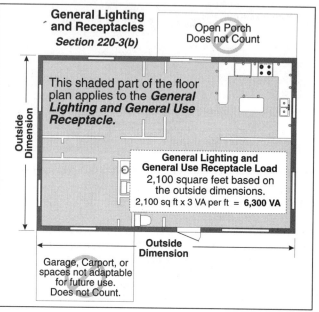

Figure. 9-10
General Lighting Example

Number Of Circuits Required [Chapter 9, Example No. 1(a)]

The number of branch circuits required for general lighting and receptacles shall be determined from the general lighting load and the rating of the circuits [220–4(a)].

To determine the number of branch circuits for general lighting and receptacles, follow these steps:

Step 1: → Determine the general lighting VA load:
Living area square footage × 3 VA.

Step 2: → Determine the general lighting ampere load:
Amperes = VA/E.

Step 3: → Determine the number of branch circuits:
$$\frac{\text{General Lighting Amperes (Step 2)}}{\text{Circuit Amperes (100\%)}}$$

❑ **Number Of 15 Ampere Circuits Example**

How many 15-ampere circuits are required for a 2,100 square foot dwelling-unit, Fig. 9–11?

(a) 2 circuits   (b) 3 circuits
(c) 4 circuits   (d) 5 circuits

• Answer: (c) 4

Step 1: → General lighting VA = 2,100 square feet × 3 VA = 6,300 VA

Step 2: → General lighting ampere:
$I = \frac{VA}{E}$, $I = \frac{6{,}300 \text{ VA}}{120 \text{ volts*}}$
I = 53 amperes

*Use 120 volts single-phase unless specified otherwise [220–2].

Step 3 → Determine the number of circuits:
$$\frac{\text{General Lighting Amperes}}{\text{Circuit Amperes}}$$
Number of circuits = $\frac{53 \text{ amperes}}{15 \text{ amperes}}$
Number of circuits = 3.53 or 4

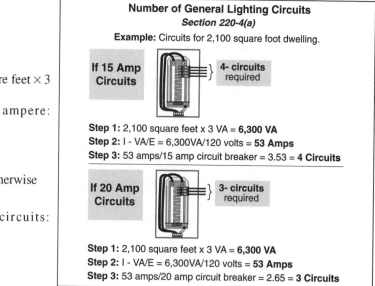

Figure 9-11
Number of General Lighting Circuits Example

Chapter 2 NEC Calculations And Code Questions

❏ **Number Of 20-Ampere Circuits Example**
How many 20-ampere circuits are required for a 2,100 square foot home?

(a) 2 circuits (b) 3 circuits (c) 4 circuits (d) 5 circuits

- Answer: (b) 3

Step 1: → General lighting VA = 2,100 square foot × VA = 6,300 VA.

Step 2: → General lighting amperes: $I = \frac{VA}{E}$, $I = \frac{6{,}300 \text{ VA}}{120 \text{ volts}}$, I = 53 amperes

Step 3: → Determine the number of circuits: $\frac{\text{General Lighting Amperes}}{\text{Circuit Amperes}}$

Number of circuits = $\frac{53 \text{ amperes}}{20 \text{ amperes}}$ Number of circuits, = 2.65 or 3 circuits

PART B - *STANDARD METHOD – FEEDER/SERVICE LOAD CALCULATIONS*

9–8 *DWELLING UNIT FEEDER/SERVICE LOAD CALCULATIONS (Part B Of Article 240)*

When determining a dwelling-unit feeder, or service using the standard method, the following steps should be used.

Step 1: → **General Lighting And Receptacles, Small Appliance And Laundry Circuits [220–11].** The NEC recognizes that the general lighting and receptacles, small appliance, and laundry circuits will not all be on, or loaded, at the same time and permits a demand factor to be applied [220–16]. To determine the feeder demand load for these loads, follow these steps.

(a) **General Lighting.** Determine the total connected load for general lighting and receptacles (3 VA per square foot), two small appliance circuits (3,000 VA), and one laundry circuit (1,500 VA).

(b) **Demand Factor.** Apply Table 220–11 demand factors to the total connected load (Step 1).
First 3,000 VA @ 100 percent demand
Remainder @ 35 percent demand

Step 2: → **Air Conditioning versus Heat.** Because the air conditioning and heating loads are not on at the same time, it is permissible to omit the smaller of the two loads [220–21].
The A/C demand load is sized at 125 percent of the largest A/C VA, plus the sum of the other A/C VA's [440–34]. Fixed electric space-heating demand loads are calculated at 100 percent of the total heating load [220–15].

Step 3: → **Appliances [220–17]**
A demand factor of 75 percent is permitted to be applied when four or more appliances fastened in place, such as, dishwasher, disposal, trash compactor, water heater, etc. are on the feeder. This does not apply to motors [220–14], space-heating equipment [220–15], clothes dryers [220–18], cooking appliances [220–19], or air-conditioning equipment [440–34].

Step 4: → **Clothes Dryer [220–18]**
The feeder or service demand load for electric clothes dryers located in a dwelling-unit shall not be less than: (1) 5000 watts or (2) the nameplate rating if greater than 5,000 watts. A feeder or service dryer load is not required if the dwelling-unit does not contain an electric dryer!

Step 5: → **Cooking Equipment [220–19]**
Household cooking appliances rated over 1¾ kW can have the feeder and service loads calculated according to the demand factors of Section 220–19, Table and Notes 1, 2, and 3.

Step 6: → **Feeder And Service Conductor Size**
Four Hundred Amperes And Less. The feeder or service conductors are sized according to Note 3 of Table 310–16 for 3-wire, single-phase, 120/240-volt systems up to 400 amperes and the grounded conductor is sized to the maximum unbalanced load [220–22] using Table 310–16.
Over Four Hundred Amperes. The ungrounded and grounded (neutral) conductors are sized according to Table 310–16.

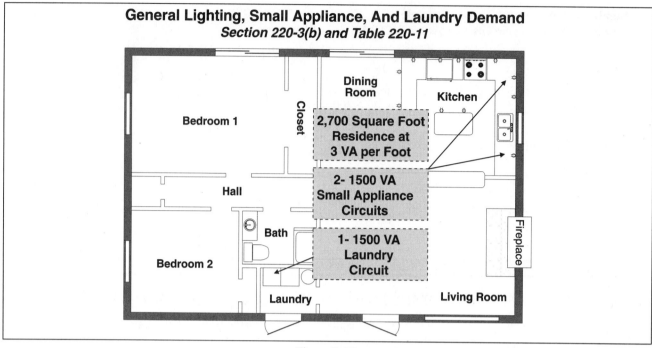

Figure 9-12

General Lighting, Small Appliance and Laundry Demand Example

9–9 DWELLING-UNIT FEEDER/SERVICE CALCULATIONS EXAMPLES

Step 1: General Lighting, Small Appliance And Laundry Demand [220–11]

➤ **General Lighting Question 1**

What is the general lighting, small appliance and laundry demand load for a 2,700 square foot dwelling-unit, Fig. 9–12?

(a) 8,100 VA   (b) 12,600 VA   (c) 2,700 VA   (d) 6,360 VA

• Answer: (d) 6,360 VA

General Lighting and Receptacles	2,700 square feet × 3 VA =	8,100 VA
Small Appliance Circuits	1,500 VA × 2 =	3,000 VA
Laundry Circuit	1,500 VA × 1 =	+ 1,500 VA
Total Connected Load		12,600 VA

Demand Factors:

Total Connected Load   12,600 VA
First 3,000 VA @ 100%   –3,000 VA × 1.00 =   3,000 VA
Remainder @ 35%   9,600 VA × .35 =   + 3,360 VA
General lighting, small appliance, and laundry demand load =   6,360 VA

➤ **General Lighting Question 2**

What is the general lighting, small appliance, and laundry demand load for a 6,540 square foot dwelling-unit?

(a) 8,100 VA   (b) 12,600 VA   (c) 2,700 VA   (d) 10,392 VA

• Answer: (d) 10,392 VA

General Lighting and Receptacles	6,540 square feet × 3 VA =	19,620 VA
Small Appliance Circuits	1,500 VA × 2 =	3,000 VA
Laundry Circuit	1,500 VA × 1 =	+ 1,500 VA
Total connected load		24,120 VA

Demand Factors:

Total Connected Load   24,120 VA
First 3,000 VA @ 100   –3,000 VA × 1.00 =   3,000 VA
Remainder @ 35%   21,120 VA × .35 =   + 7,392 VA
General lighting, small appliance, and laundry demand load =   10,392 VA

Step 2: Air Conditioning Versus Heat [220–15]

When determining the cooling versus heating load, we must compare the air conditioning load at 125 percent against the the heating load of 100 percent. Section 220–15 only refers to the heating load and Section 440–32 contains the requirement of the air conditioning load. We are permitted to omit the smaller of the two loads according to Section 220–21.

→ **Air Conditioning versus Heat Question 1**

What is the service or feeder demand load for a 4-horsepower 230-volt A/C unit and a 10-kW electric space heater?

(a) 10,000 VA (b) 3,910 VA
(c) 6,440 VA (d) 10,350 VA

- Answer: (a) 10,000 VA heat is greater than the 6,469 VA air conditioning

  A/C: [440–34] Since a 4-horsepower 230 volt motor is not listed in Table 430–148, we must determine the approximate VA rating by averaging the values for a 3-HP and 5-HP motor.

  3-HP = 230 volts × 17 amperes* = 3,910 VA
  5-HP = 230 volts × 28 amperes* = + 6,440 VA
  Total VA for 3-HP + 5-HP (8-HP) = 10,350 VA
  VA for 4-HP = 10,350 VA (8-HP)/2 = 5,175 VA × 1.25 = 6,469 VA, omit [220–21]

→ **Air Conditioning versus Heat Question 2**

What is the service demand load for a 5-horsepower 230-volt A/C versus three 3-kW baseboard heaters, Fig. 9–13?

(a) 6,400 VA (b) 3,000 VA (c) 8,050 VA (d) 9,000 VA

- Answer: (d) 9,000 VA is greater than the 8,050 VA air conditioning

  A/C [440–34] 230 volts × 28 amperes* = 6,440 × 1.25 = 8,050 VA, omit [220–21]

  Heat [220–15] 3,000 VA × 3 units = 9,000 VA, Table 430-148

  * Table 430–148

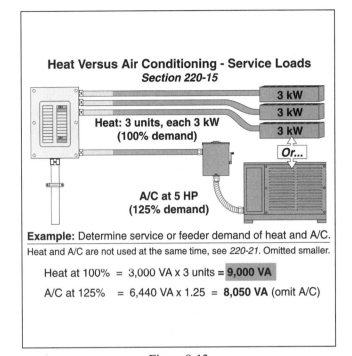

Figure 9-13
Air Conditioning versus Heat Example

Step 3: Appliance Demand Load [220–17]

→ **Appliance Question 1**

What is the demand load for a 940 VA disposal, a 1,250 VA dishwasher, and a 4,500 VA water heater?

(a) 5,018 VA (b) 6,690 VA (c) 8,363 VA (d) 6,272 VA

- Answer: (b) 6,690 VA, no demand factor for three units

  Disposal   940 VA
  Dishwasher   1,250 VA
  Water Heater + 4,500 VA
                6,690 VA   at 100%

→ **Appliance Question 2**

What is the demand load for a 940 VA disposal, a 1,250 VA dishwasher, a 1,100 VA trash compactor, and a 4,500 VA water heater, Fig. 9–14?

(a) 7,790 VA (b) 5,843 VA (c) 7,303 VA (d) 9,738 VA

- Answer: (b) 5,843 VA

  Disposal   940 VA
  Dishwasher   1,250 VA
  Trash Compactor 1,100 VA
  Water Heater + 4,500 VA
                  7,790 VA   × .75 = 5,843 VA

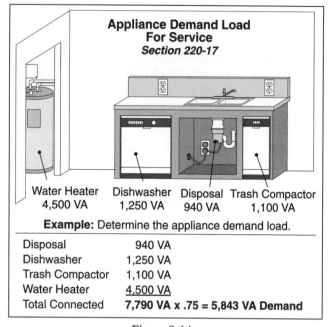

Figure 9-14
Appliance Demand Load Example

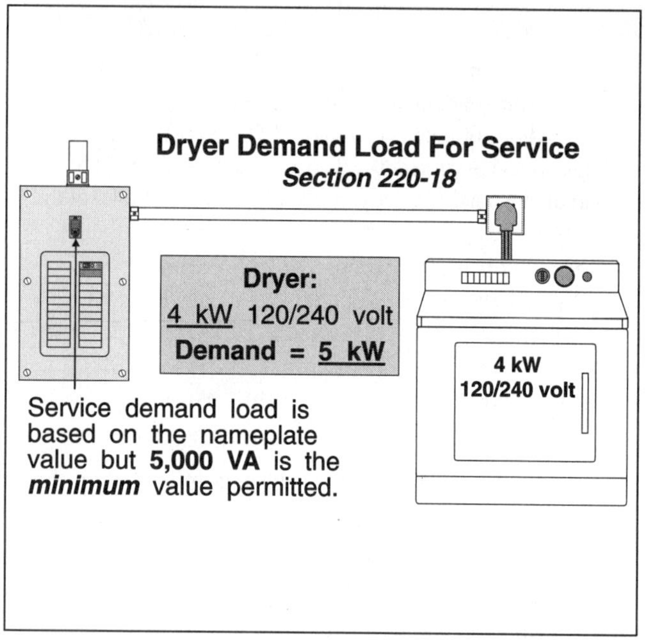

Figure 9-15
Dryer Demand Load Example

Step 4: Dryer Demand Load [220–18]

→ Dryer Question 1
What is the service and feeder demand load for a 4-kW dryer, Fig. 9–15?

(a) 4,000 VA (b) 3,000 VA (c) 5,000 VA (d) 5,500 VA

• Answer: (c) 5,000 VA, The dryer load must not be less than 5,000 VA.

→ Dryer Question 2
What is the service and feeder demand load for a 5.5-kW dryer?

(a) 4,000 VA (b) 3,000 VA (c) 5,000 VA (d) 5,500 VA

• Answer: (c) 5,500 VA, The dryer load must not be less than the nameplate rating if greater than 5 kW.

Step 5: Cooking Equipment Demand Load [220–19]

→ Column B Question 3
What is the service and feeder demand load for two 3-kW cooking appliances in a dwelling-unit?

(a) 3 kW (b) 4.8 kW (c) 4.5 kW (d) 3.9 kW

• Answer: (c) 4.5 kW, Table 220–19, Column B
3 kW × 2 units = 6 kW × 0.75 = 4.5 kW

→ Column C Question 4
What is the service and feeder demand load for one 6-kW cooking appliance in a dwelling-unit?

(a) 6 kW (b) 4.8 kW (c) 4.5 kW (d) 3.9 kW

• Answer: (b) 4.8 kW, Table 220–19, Column C
6 kW × 0.8 = 4.8 kW

→ Column B And C Question 5
What is the service and feeder demand load for two 3-kW ovens and one 6-kW cook-top in a dwelling-unit, Fig. 9–16?

(a) 6 kW (b) 4.8 kW (c) 4.5 kW (d) 9.3 kW

• Answer: (d) 9.3 kW, Table 220–19, Column B and C

Column C demand 6 kW × 0.8 =   4.8 kW
Column B demand (3 kW × 2) = 6 kW × 0.75 = + 4.5 kW
Total demand   9.3 kW

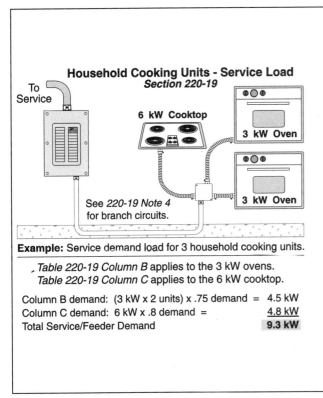

Figure 9-16
Cooking Equipment Demand Load Example

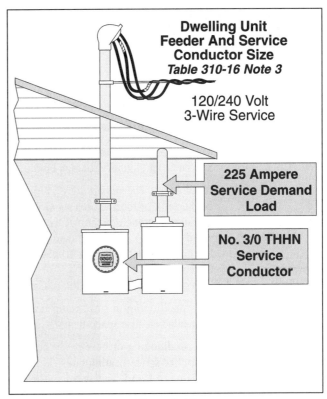

Figure 9-17
Feeder and Service Conductor Size – Note 3 of Table 310–16

→ **Column A Question 1**

What is the service and feeder demand load for an 11.5-kW range in a dwelling-unit?

(a) 11.5 kW (b) 8 kW (c) 9.2 kW (d) 6 kW

- Answer: (b) 8 kW, Table 220–19, Column A

→ **Note 1 Question 2**

What is the service and feeder demand load for a 13.6-kW range in a dwelling-unit?

(a) 8.8 kW (b) 8 kW (c) 9.2 kW (d) 6 kW

- Answer: (a) 8.8 kW,

    8 kW × 1.1* = 8.8 kW

* Column A value (8 kW) must be increased 5 percent for each kW or major fraction of a kW (.5 kW or larger) over 12 kW [Table 220–19, Note 1].

Step 6: Service Conductor Size [Note 3 of Table 310–16]

→ **Service Conductor Question 1**

What size THHN feeder or service conductor (120/240 volt, single-phase) is required for a 225 ampere service demand load, Fig. 9–17?

(a) No. 0 (b) No. 00 (c) No. 000 (d) No. 0000

- Answer: (c) No. 3/0

→ **Service Conductor Question 2**

What size THHN feeder or service conductor (208Y/120-volt, single-phase) is required for a 225-ampere service demand load?

(a) No. 1/0 (b) No. 2/0 (c) No. 3/0 (d) No. 4/0

- Answer: (d) 4/0, Table 310–16, Note 3, does not apply to 208Y/120-volt systems because the grounded conductor will carry current even if the loads are balanced.

PART C - OPTIONAL METHOD – FEEDER/SERVICE LOAD CALCULATIONS

9–10 DWELLING UNIT OPTIONAL FEEDER/SERVICE CALCULATIONS [220–30]

Instead of sizing the dwelling-unit feeder and/or service conductors according to the standard method (Part B of Article 220), an optional method [220–30] can be used. The following Steps can be used to determine the demand load:

Step 1: → **Total Connected Load.** To determine the total connected load, add the following loads:

(a) **General Lighting And Receptacles:** 3 volt-amperes per square-foot.

(b) **Small Appliance And Laundry Branch Circuits**: 1,500 volt-amperes for each 20 ampere small appliance and laundry branch circuit.

(c) **Appliances:** The *nameplate* VA rating of all appliances and motors that are fastened in place (permanently connected), or located on a specific circuit. Be sure to use the range and dryer nameplate rating! This does not include the A/C or heating loads.

Step 2: → **Demand Factor.** A 40 percent demand factor is permitted to be applied to that portion of the total connected load in excess of 10 kW. First 10 kW of total connected load (Step 1) is at 100 percent and remainder is at 40 percent.

Step 3: → **A/C versus Heat.** Determine the A/C verses Heat demand load:

(a) Air conditioning or heat-pump compressors at 100 percent vs. 65 percent of three or less separately controlled electric space-heating units, or

(b) Air conditioning or heat-pump compressors at 100 percent vs. 40 percent of four or more separately controlled space-heating units.

Step 4: → **Total Demand Load.** To determine the total demand load, add the demand load from Step 2 and the largest of Step 3.

I = VA/E to be used to convert the total demand VA load to amperes.

Step 5: → **Conductor Size**

Four Hundred Amperes And Less. The ungrounded conductors are permitted to be sized to Note 3 of Table 310–16 for 120/240 volt single-phase systems up to 400 amperes. The grounded neutral conductor must have an ampacity to carry the unbalanced load, sized to Table 310-16.

Over Four Hundred Amperes. The ungrounded and grounded neutral conductors are sized according to Table 310–16.

9–11 DWELLING UNIT OPTIONAL CALCULATION EXAMPLES

☐ **Optional Load Calculation Example No. 1**

What size service conductor is required for a 1,500 square foot dwelling-unit containing the following loads:

Dishwasher at 1,200 VA, 120 volt
Disposal at 900 VA, 120 volt
Cook top at 6,000 VA, 120/240 volt
A/C at 5 horsepower, 240 volt

Water heater at 4,500 VA, 240 volt
Dryer at 4,000 VA, 240 volt
Two-ovens each at 3,000 VA, 120/240 volt
Heat Strip at 10 kW, 240 volt

(a) No. 5   (b) No. 4   (c) No. 3   (d) No. 2

• Answer: (c) No. 3, 105 amperes – Note 3 of Table 310–16.

STEP 1: Total Connected Load

			Totals
(a) General Lighting 1,500 square feet × 3 VA =	4,500 VA		(a) 4,500 VA
(b) Small Appliance Circuits 1,500 VA × 2	3,000 VA		
Laundry Circuit	+ 1,500 VA		
	4,500 VA	=	(b) 4,500 VA
(c) Appliances (nameplate)			
Dishwasher	1,200 VA		
Water Heater	4,500 VA		
Disposal	900 VA		
Dryer*	4,000 VA		
Cook-top**	6,000 VA		
Ovens 3,000 VA × 2 units	+ 6,000 VA		
	22,600 VA	=	(c) 22,600 VA

Chapter 2 NEC Calculations And Code Questions — Unit 9 One-Family Load Calculations

* The dryer load is 4,000 VA, not 5,000 VA! Section 220–18 (Part B of Article) does not apply to Section 220–30 (Part C).
** Cooking equipment is also calculated at nameplate value.

STEP 2: The Demand Load

(a) General Lighting (and receptacles)	4,500 VA
(b) Small Appliance and Laundry Circuits	4,500 VA
(c) Appliance nameplate ratings	+ 22,600 VA
Total Connected Load	31,600 VA

Demand Factor

Total Connected Load	31,600 VA		
First 10,000 @ 100%	– 10,000 VA	× 1.00 =	10,000 VA
Remainder @ 40%	21,600 VA	× 0.40 =	+ 8,640 VA
Demand load			18,640 VA

STEP 3: Air-Conditioning Versus Heat

Air-conditioning or heat-pump compressors at 100 percent versus 65 percent of three or less separately controlled electric space-heating units.

Air Conditioner 230* volts × 28 amperes** = 6,440 VA, omit, Heat Strip 10,000 VA × .65 = 6,500 VA

* Some exams use 240 volts, ** Table 430–148

STEP 4: Total Demand Load

Total of Step 2 = Other Loads	18,640 VA
Total of Step 3 = Heat Load	+ 6,500 VA
Total Demand Load =	25,140 VA

STEP 5: Service Conductors [Table 310–16, Note 3]

$$I = \frac{VA}{E}, \quad I = \frac{25{,}140 \text{ VA}}{240 \text{ volts}} = 105 \text{ amperes}$$

The feeder and service conductor is sized to 110 amperes, Note 3 of Table 310–16 = No. 3 AWG.

☐ **Optional Load Claculation Example No. 2**

What size service conductor is required for a 2,330 square-foot residence with a 300 square-foot porch and a 150 square-foot carport that contains the following:

Receptacles, 15 extra
Dishwasher @ 1.5 kVA, 120 volt
Trash compactor @1.5 kVA, 120 volt
Range @ 14 kW, 120/240 volt
Air conditioner @ 6 kVA, 240 volt

Recessed Fixtures, 10 – 100 watts (high-hats)
Disposal @ 1 kVA 120 volt
Water heater @ 6 kW, 240 volt
Dryer @ 4.5 kVA 120/240 volt
Baseboard heat, four each @ 2.5 kW, 240 volt

(a) No. 5   (b) No. 4   (c) No. 3   (d) No. 2

• Answer: (d) No. 2, Note 3 of Table 310–16

STEP 1: Total Connected Load

		Totals
(a) General lighting (and receptacles): 2,330 square feet* × 3 VA = 6,990 VA		(a) 6,990 VA
(b) Small appliance and laundry branch circuits:		
Small Appliance Circuit 1,500 VA × 2 circuits =	3,000 VA	
Laundry Circuit	+1,500 VA	
	4,500 VA =	(b) 4,500 VA
(c) Appliance Nameplate VA:		
Dishwasher	1,500 VA	
Disposal	1,000 VA	
Trash Compactor	1,500 VA	
Water Heater	6,000 VA	
Range**	14,000 VA	
Dryer***	+ 4,500 VA	
	28,500 VA =	(c) 28,500 VA

*The open porch and carport is not counted [220–3(b)]. The 15 receptacles and the 10 recessed fixtures are included in the 3 VA per square-foot general lighting and receptacle load [Table Notes to Table 220–3(b)].

  **All appliances at nameplate rating

  *** The dryer load is 4,500 VA, NOT 5,000 VA! Be careful, Section 220–18 (Part B of Article 220) does not apply to Section 220–30 (Part C).

STEP 2: The Demand Load

(a) General Lighting (and receptacles)	6,990 VA		
(b) Small Appliance and Laundry Circuit	4,500 VA		
(c) Appliance nameplate ratings	+ 28,500 VA		
	39,990 VA		
First 10,000 @ 100% =	− 10,000 VA	× 1.0 =	10,000 VA
Remainder @ 40% =	29,990 VA	× .40 =	+ 11,996 VA
Demand load			21,996 VA

STEP 3: Air Conditioning Versus Heat

Air conditioning or heat-pump compressors at 100 percent vs. 40 percent of four or more separately controlled electric space-heating units.

Air Conditioner 230* volts × 28 amperes** = 6,440 VA
Heat Strip 10,000 VA × .40 = 4,000 VA* omit 220–21
* Some exams use 240 volts, ** Table 430–148

STEP 4: The Total Demand Load

Step 2 = Other Loads	21,996 VA
Step 3 = A/C Load	+ 6,440 VA
Total Demand Load	28,436 VA

STEP 5: Feeder And Service Conductors [Table 310–16, Note 3]

$$I = \frac{VA}{E}, \quad I = \frac{28,436 \text{ VA}}{240 \text{ volts}} = 118 \text{ amperes}$$

The feeder and service conductor size's are sized to 125 amperes, Note 3 of Table 310–16 = No. 2 AWG.

9–12 NEUTRAL CALCULATIONS – GENERAL [220–22]

The feeder and service neutral load is the maximum *unbalanced demand load* between the grounded conductor (neutral) and any one ungrounded conductor as determined by Article 220 Part B (standard calculations). This means that since 240 volt loads are not connected to the neutral conductor, they are not considered for sizing the neutral conductor.

❑ **Neutral Not Over 200 Amperes Example**

What size 120/240-volt single-phase ungrounded and grounded conductor is required for a 375-ampere two-family dwelling-unit demand load, of which 175 amperes consist of 240 volt loads, Fig. 9–18?

(a) Two 400 THHN and 1– 350 THHN
(b) Two 350 THHN and 1– 350 THHN
(c) Two 500 THHN and 1– 500 THHN
(d) Two 400 THHN and 1– No. 3/0 THHN

• Answer: (d) 400 kcmil THHN, and No. 3/0 THHN

The ungrounded conductor is sized at 375 amperes (400 kcmil THHN) according to Note 3 of Table 310-16. The grounded conductor (neutral) is sized to carry a maximum unbalanced load of 200 amperes (375 amperes less 175 amperes – 240 volt loads).

Cooking Appliance Neutral Load [220–22]

The feeder or service cooking appliance neutral load for household cooking appliances; such as, electric ranges, wall-mounted ovens, or counter-mounted cooking units, shall be calculated at 70 percent of the demand load as determined by Section 220–19.

❑ **Range Neutral Example**

What is the neutral load (in amperes) for one 14-kW range?

(a) 33 amperes (b) 26 amperes
(c) 45 amperes (d) 18 amperes

• Answer: (b) 26 amperes
  36.7 amperes × 0.7 = 25.69 amperes

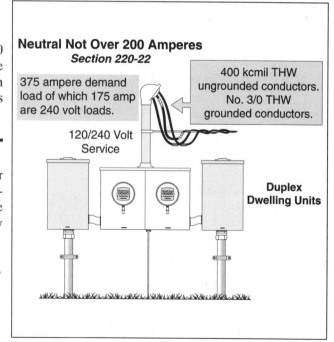

Figure. 9-18

Neutral Conductor Size Example

Step 1: → Since the range exceeds 12 kW, we must comply with Note 1 of Table 220–19. The first step is to determine the demand load as listed in Column A of Table 220–19 for one unit, = 8 kW

Step 2: → We must increase the Column A value (8 kW) by 5 percent for each kW that the average range (in this case 14 kW) exceeds 12 kW. This results is an increase of the Column A value (8 kW) by 10 percent.
8 kW x 1.1 = 8.8 kW or (8,000 watts + 800 watts)

Step 3: → Convert the demand load in kW to amperes.
$$I = \frac{P}{E}, I = \frac{8,800 \text{ VA}}{240 \text{ volts}} = 36.7 \text{ amperes} \times 0.7 = 26 \text{ amperes}$$

Dryer Neutral Load [220–22]

The feeder and service dryer neutral demand load for electric clothes dryers shall be calculated at 70 percent of the demand load as determined by Section 220–18.

❏ **Dryer Over 5-kW Example**
What is the neutral load (in amperes) for one 5.5-kVA dryer?
(a) 4 kW   (b) 5 kW   (c) 6 kW   (d) None of these

• Answer: (a) 4 kW, 5.5 kVA × 0.7 = 3.85 kW

Unit 9 – One-Family Dwelling-Unit Load Calculations Summary Questions

Part A – General Requirements

9–2 Voltages [220–2]

1. Unless other voltages are specified, branch-circuit, feeder and service loads shall be computed at a nominal system voltages of _____.
   (a) 120/240
   (b) 120/240
   (c) 208Y/120
   (d) any of these

9–3 Fraction Of An Ampere

2. There are no NEC rules for rounding when a calculation results in a fraction of an ampere, but it does contains the following note, except where the computations result in a _____ of an ampere or larger, such fractions may be dropped.
   (a) .05
   (b) .5
   (c) .49
   (d) .51

9–4 Appliance (Small) Circuits [220–4(b)]

3. A minimum of _____ 20-ampere small appliance branch circuits are required for receptacle outlets in the kitchen, dining room, breakfast room, pantry, or similar areas of dining.
   (a) one
   (b) two
   (c) three
   (d) none of these

4. When sizing the feeder or service, each dwelling-unit shall have a minimum feeder load of _____ VA for the two small appliance branch circuits.
   (a) 1,500
   (b) 2,000
   (c) 3,000
   (d) 4,500

9–5 Cooking Equipment – Branch Circuit [Table 220–19, Note 4]

Table 220–19 Column A

5. What is the branch circuit demand load (in amperes) for one 11-kW range?
   (a) 21 amperes
   (b) 27 amperes
   (c) 38 amperes
   (d) 33 amperes

6. What is the branch circuit load (in amperes) for one 13.5-kW range?
   (a) 33 amperes
   (b) 37 amperes
   (c) 50 amperes
   (d) 58 amperes

Wall Mounted Oven And/Or Counter-Mounted Cooking Unit

7. • What is the branch circuit load (in amperes) for one 6-kW wall-mounted oven?
   (a) 20 amperes
   (b) 25 amperes
   (c) 30 amperes
   (d) 40 ampere

Chapter 2 NEC Calculations And Code Questions   Unit 9 One-Family Load Calculations 295

8. • What is the branch circuit load (in amperes) for one 4.8-kW counter-mounted cooking unit?
   (a) 15 amperes   (b) 20 amperes
   (c) 25 amperes   (d) 30 amperes

9. What is the branch circuit load (in amperes) for one 6-kW counter-mounted cooking unit and one 3-kW wall-mounted oven?
   (a) 25 amperes   (b) 38 amperes
   (c) 33 amperes   (d) 42 amperes

10. What is the branch circuit load (in amperes) for one 6-kW counter-mounted cooking unit and two 4-kW wall-mounted ovens?
    (a) 22 amperes   (b) 27 amperes
    (c) 33 amperes   (d) 37 amperes

9–6 Laundry Receptacle(s) Circuit [220–4(c)]

11. The NEC does not require a separate circuit for the washing machine, but requires is a separate circuit for the laundry room receptacle or receptacles, of which one can be for the washing machine.
    (a) True   (b) False

12. Each dwelling-unit shall have a feeder load consisting of _____ VA volt-ampere for the 20 ampere laundry circuit.
    (a) 1,500   (b) 2,000
    (c) 3,000   (d) 4,500

9–7 Lighting And Receptacles

13. The NEC requires a minimum 3 volt-ampere per square foot [Table 220–3(b)] for the required general lighting [210–70(a)] and receptacles [210–52(a)]. The dimensions for determining the area shall be computed from the outside of the building and shall not include _____.
    (a) open porches
    (b) garages
    (c) spaces not adaptable for future use
    (d) all of these

14. The 3 volt-ampere per square foot for general lighting includes all 15- and 20-ampere general use receptacles, but not the appliance or laundry circuit receptacles.
    (a) True   (b) False

15. What is the general lighting load and receptacle load for a 2,100-square foot home that has fourteen extra convenience receptacles and nine extra recessed fixtures rated 100 watts each?
    (a) 2,100 VA
    (b) 4,200 VA
    (c) 6,300 VA
    (d) 8,400 VA

Number of General Lighting Circuits [Chapter 9, Example No. 1(a)]

16. The number of branch circuits required for general lighting and receptacles shall be determined from the general lighting load and the rating of the circuits.
    (a) True   (b) False

17. How many 15-ampere general lighting circuits are required for a 2,340-square foot home?
    (a) two   (b) three
    (c) four   (d) five

18. How many 20-ampere general lighting circuits are required for a 2,100-square foot home?
    (a) two   (b) three
    (c) four   (d) five

19. How many 20 ampere circuits are required for the general lighting and receptacles for a 1,500-square foot dwelling-unit?
    (a) two   (b) three

(c) four (d) five

Part B – Standard Method – Load Calculation Questions

9–8 Dwelling Unit Load Calculations (Part B Of Article 240)

20. The NEC recognizes that the general lighting and receptacles, small appliance and laundry circuits will not all be on or loaded at the same time and permits a _____ to be applied to the total of these loads.
    (a) demand factor
    (b) adjustment factor
    (c) correction factor
    (d) correction

21. Because the A/C and heating loads are not on at the same time (simultaneously), it is permissible to omit the smaller of the two loads when determining the A/C versus heat demand load.
    (a) True (b) False

22. A load demand factor of 75 percent is permitted for _____ or more appliances fastened in place, such as dishwasher, disposal, trash compactor, water heater, etc.
    (a) one (b) two
    (c) three (d) four

23. The feeder or service demand load for electric clothes dryers located in dwelling-units shall not be less than _____.
    (a) 5000 watts
    (b) nameplate rating
    (c) a or b
    (d) the greater of a or b

24. • A feeder or service dryer load is required, even if the dwelling-unit does not contain a electric dryer.
    (a) True (b) False

25. Household cooking appliances rated 1¾-kW can have the feeder and service loads calculated according to the demand factors of Section 220–19, Table, and Notes.
    (a) True (b) False

26. Dwelling-unit feeder and service ungrounded conductors are sized according to Note 3 of Table 310–16 for _____.
    (a) 3-wire single phase 120/240 volt systems up to 400 amperes
    (b) 3-wire single phase 120/208 volt systems up to 400 amperes
    (c) 4-wire single phase 120/240 volt systems up to 400 amperes
    (d) none of these

27. What is the general lighting load for a 2,700-square foot dwelling-unit?
    (a) 8,100 VA (b) 12,600 VA
    (c) 2,700 VA (d) 6,840 VA

28. What is the total connected load for general lighting and receptacles, small appliance and laundry circuits for a 6,540-square foot dwelling-unit?
    (a) 8,100 VA (b) 12,600 VA
    (c) 2,700 VA (d) 24,120 VA

29. What is the feeder or service demand load for a 5-horsepower 230-volt A/C and three 3-kW baseboard heaters?
    (a) 6,400 VA (b) 3,000 VA
    (c) 8,050 VA (d) 9,000 VA

30. What is the feeder or service demand load for a 4-horsepower 230-volt A/C and a 10-kW electric space heater?
    (a) 10,000 VA (b) 3,910 VA
    (c) 6,440 VA (d) 10,350 VA

31. What is the feeder or service demand load for a 940 VA disposal, 1,250 VA dishwasher and a 4,500 VA water heater?
    (a) 5,018 VA (b) 6,690 VA

(c) 8,363 VA (d) 6,272 VA

32. • What is the feeder or service demand load for a 940 VA disposal, 1,250 VA dishwasher, 1,100 VA trash compactor and a 4,500 VA water heater?
    (a) 7,790 VA (b) 5,843 VA
    (c) 7,303 VA (d) 9,738 VA

33. What is the feeder or service demand load for a 4-kW dryer?
    (a) 4,000 VA (b) 3,000 VA
    (c) 5,000 VA (d) 5,500 VA

34. What is the feeder or service demand load for a 5.5-kW dryer?
    (a) 4,000 VA (b) 3,000 VA
    (c) 5,000 VA (d) 5,500 VA

35. • What is the feeder or service demand load for two 3-kW cooking appliances in a dwelling-unit?
    (a) 3 kW (b) 4.8 kW
    (c) 4.5 kW (d) 3.9 kW

36. What is the feeder or service demand load for one 6-kW cooking appliance in a dwelling-unit?
    (a) 6 kW (b) 4.8 kW
    (c) 4.5 kW (d) 3.9 kW

37. What is the feeder or service demand load for one 6-kW and two 3-kW cooking appliances in a dwelling-unit?
    (a) 6 kW (b) 4.8 kW
    (c) 4.5 kW (d) 9.3 kW

38. What is the feeder or service demand load for an 11.5-kW range in a dwelling-unit?
    (a) 11.5 kW (b) 8 kW
    (c) 9.2 kW (d) 6 kW

39. What is the feeder or service demand load for a 13.6-kW range in a dwelling-unit?
    (a) 8.8 kW (b) 8 kW
    (c) 9.2 kW (d) 6 kW

40. What size 120/240-volt single-phase service or feeder THHN conductors are required for a dwelling-unit that has a 190-ampere service demand load?
    (a) No. 1/0 (b) No. 2/0
    (c) No. 3/0 (d) No. 4/0

41. • What is the net computed demand load for the general lighting, small appliance and laundry circuits for a 1,500-square foot dwelling-unit?
    (a) 4,500 VA (b) 9,000 VA
    (c) 5,100 VA (d) none of these

42. What is the feeder or service demand load for a 3-horsepower, 230-volt air conditioning with a $\frac{1}{8}$-horsepower blower versus 4-kW heat?
    (a) 3,910 VA (b) 5,222 VA
    (c) 4,000 VA (d) None of these

43. What is the feeder or service demand load a dwelling-unit that has one water heater 4,000 VA and one dishwasher 1,500 VA?
    (a) 4 kW (b) 5.5 kW
    (c) 4.13 kW (d) none of these

44. What is the net computed load for a 4.5-kW dryer?
    (a) 4.5 kW (b) 5 kW
    (c) 5.5 kW (d) none of these

45. What is the total net computed demand load for a 14-kW range?
    (a) 8 kW (b) 8.4 kW

(c) 8.8 kW (d) 14 kW

46. If the total demand load is 30-kW for a 120/240-volt dwelling-unit, what is the service and feeder conductor size?
    (a) No. 4 (b) No. 3
    (c) No. 2 (d) No. 1

47. If the service raceway contains No. 2 service conductors, what size bonding jumper is required for the service raceway?
    (a) No. 8 (b) No. 6
    (c) No. 4 (d) No. 3

48. If the service contains No. 2 conductors, what is the minimum size grounding electrode conductor required?
    (a) No. 8 (b) No. 6
    (c) No. 4 (d) No. 3

Part C - Optional Method – Load Calculation Questions

9–10 Dwelling Unit Optional Feeder/Service Calculations (220–30)

49. When sizing the feeder or service conductor according to the optional method we must:
    (a) Determine the total connected load of the general lighting and receptacles, small appliance and laundry branch circuits, and the nameplate VA rating of all appliances and motors.
    (b) Determine the demand load by applying the following demand factor to the total connected load. First 10 kVA 100 percent, remainder 40 percent.
    (c) Determine the A/C versus Heat demand load.
    (d) All of these.

Use this information for Question 50.
A 1,500-square foot dwelling-unit containing the following loads:

Dishwasher at 1,500 VA   Water heater at 5,000 VA
Disposal at 1,000 VA   Dryer at 5,500 VA
Cooktop at 6,000 VA   Two ovens each at 3,000 VA
A/C at 3 horsepower   Heat strip at 10 kW

50. • Using the optional method, what size aluminum conductor is required for the service for the described loads?
    (a) No. 1   (b) No. 1/0
    (c) No. 2/0 (d) No. 3/0

A 2,330-square foot residence with a 300-square foot porch and 150-square foot carport contains the following:

15 extra general-use receptacles   30 – 100 watt Recessed fixtures (high hats)
Dishwasher at 1.5 kVA, 120 volt   Disposal at 1 kVA, 120 volt
Trash compactor at 1.5 kVA, 120 volt   Water heater at 6 kW, 240 volt
Range at 14 kW, 120/240 volt   Dryer at 4.5 kW 120/240 volt
Air conditioner at 5 HP, 240 volt   Baseboard heat four each at 2.5 kW, 240 volt

51. Using the optional method, what size conductor is required for the service, using the described loads?
    (a) No. 5 (b) No. 4
    (c) No. 3 (d) No. 2

9–12 Neutral Calculations General [220–22]

52. The feeder and service neutral load is the maximum unbalanced demand load between the grounded conductor (neutral) and any one ungrounded conductor as determined by Article 220 Part B. Since 240 volt loads cannot be connected to the neutral conductor, 240 volt loads are not considered for sizing the feeder neutral conductor.
    (a) True (b) False

Chapter 2   NEC Calculations And Code Questions   Unit 9   One-Family Load Calculations   299

53. • What size single-phase 3-wire feeder is required for a for a 475 ampere demand load of which 275 amperes consists of 240 volt loads?
(a) 2– 750 THHN and 1– 350 THHN
(b) 2– 500 THHN and 1– 350 THHN
(c) 2– 750 THHN and 1– 500 THHN
(d) 2– 750 THHN and 1– No. 3/0 THHN

54. The feeder or service neutral load for household cooking appliances such as electric ranges, wall-mounted ovens, or counter-mounted cooking units, shall be calculated at ____ percent of the demand load as determined by Section 220–19.
(a) 50   (b) 60
(c) 70   (d) 80

55. What is the dwelling-unit service neutral load for one 9-kW ranges?
(a) 5.6 kW   (b) 6.5 kW
(c) 3.5 kW   (d) 12 kW

56. The service neutral demand load for household electric clothes dryers shall be calculated at ____ percent of the demand load as determined by Section 220–18.
(a) 50   (b) 60
(c) 70   (d) 80

57. What is the feeder neutral load for one 6-kW household dryers?
(a) 4.2 kW   (b) 4.7 kW
(c) 5.4 kW   (d) 6.6 kW

Challenge Questions

9–5 Cooking Equipment – Branch Circuit [Table 220–19 Note 4]

58. The branch-circuit demand load for one 18-kW range is _____ kW.
(a) 12   (b) 8
(c) 10.4   (d) 18

59. A dwelling-unit kitchen will have the following appliances: One 9-kW cooktop and one wall-mounted oven rated 5.3-kW. The branch-circuit demand load for these appliances will be _____ kW.
(a) 14.3   (b) 12
(c) 8.8   (d) 8

9–8 Standard Method – Load Calculation Questions

General Lighting [Table 220-11]

60. An apartment building contains twenty units, and each unit has 840-square feet. What is the general lighting feeder demand load for each dwelling-unit? Note: The laundry facilities are provided on the premises for all tenants and no laundry circuit is required in each unit, see 210–52(f) Exception No. 1.
(a) 3,520 VA   (b) 3,882 VA
(c) 4,220 VA   (d) 6,300 VA

61. An 1,800-square foot residence has a 300-square foot open porch and a 450-square foot carport with the following loads: 45 general use receptacles; 19 high-hats; and six 150 VA flood lights. What is the demand load for the general lighting, receptacles, small appliance circuits, and laundry circuits?
(a) 5,400 VA
(b) 7,900 VA
(c) 5,415 VA
(d) 6,600 VA

62. How many 15-ampere, 120-volt branch circuits are required for general use receptacle and lighting for a 1,800-square foot dwelling-unit?
    (a) 1 circuit   (b) 2 circuits
    (c) 3 circuits   (d) 4 circuits

63. How many 20-ampere, 120-volt branch circuits are required for general use receptacle and lighting for a 2,800-square foot dwelling-unit?
    (a) 2 circuit   (b) 6 circuits
    (c) 7 circuits   (d) 4 circuits

Appliance Demand Factors [Section 220–17]

64. A dwelling-unit contains the following: A 1.2-kW washing machine; a 4-kW water heater; a 1.2-kW dishwasher; and a 1.5-kW trash compactor. The appliance demand load added to the service is _____.
    (a) 5.9 kW   (b) 6.7 kW
    (c) 7.7 kW   (d) 8.8 kW

65. • A dwelling-unit contains the following: A water heater (4-kW); a dishwasher (½-horsepower); a dryer; a pool pump (¾-horsepower); a cooktop (6-kW); an oven (6-kW); A/C (4-horsepower); and heat (6-kW). What is the appliance demand load for the dwelling-unit?
    (a) 6.7 kW   (b) 4 kW
    (c) 9 kW   (d) 11 kW

Dryer Calculation [Section 220–18]

66. A dwelling-unit contains a 5.5-kW electric clothes dryer. What is the feeder demand load for the dryer?
    (a) 3.38 kW   (b) 4.5 kW
    (c) 5 kW   (d) 5.5 kW

Cooking Equipment Demand Load [220–19]
General Calculations

67. The feeder load for twelve 1¾-kW cooktops is _____.
    (a) 21 kW   (b) 10.5 kW
    (c) 13.5 kW   (d) 12.5 kW

68. The minimum neutral for a branch-circuit to an 8¾-kW household range is _____. See Article 210.
    (a) No. 10   (b) No. 12
    (c) No. 8   (d) No. 6

69. What is the maximum dwelling-unit feeder or service demand load for fifteen 8-kW cooking units?
    (a) 38.4 kW   (b) 30 kW
    (c) 110 kW   (d) 27 kW

Note 2 of Table 220–19

70. What is the feeder demand load for five 10-kW, five 14-kW, and five 16-kW household ranges?
    (a) 70 kW   (b) 23 kW
    (c) 14 kW   (d) 33 kW

Note 3 to Table 220–19

71. A dwelling-unit has one 6-kW cooktop and one 6-kW oven. What is the minimum feeder demand load for the cooking appliances?
    (a) 13 kW   (b) 8.8 kW
    (c) 7.8 kW   (d) 8.2 kW

Chapter 2 NEC Calculations And Code Questions Unit 9 One-Family Load Calculations 301

72. The maximum feeder demand load for an 8½-kW range is _____.
(a) 8.5 kW (b) 8 kW
(c) 6.3 kW (d) 12 kW

73. Given: Five 5-kW ranges; two 4-kW ovens; and four 7-kW cooking units. The minimum feeder demand load is _____.
(a) 61 kW (b) 22 kW
(c) 19.5 kW (d) 18 kW

74. Given: Two 3-kW wall mounted ovens and one 6-kW Cooktop. The feeder demand load would be _____.
(a) 9.3 kW (b) 11 kW
(c) 8.4 kW (d) 9.6 kW

9–9 Service Questions

75. • Both units of a duplex apartment require a 100 ampere main, the resulting 200 ampere service would require _____ THHN.
(a) No. 1/0 (b) No. 2/0
(c) No. 3/0 (d) No. 4

76. After all demand factors have been taken into consideration, the demand load for a dwelling-unit is 21,560 VA. The minimum service size for this residence would be _____ if the optional method of service calculations were used. Be sure to read 220–30 carefully!
(a) 90 amperes (b) 100 amperes
(c) 110 amperes (d) 125 amperes

77. After all demand factors have been taken into consideration, the net computed load for a service is 24,221 VA. The minimum service size is _____.
(a) 150 amperes (b) 175 amperes
(c) 125 amperes (d) 110 amperes

9–10 Optional Method – Load Calculation Questions [220–30]

78. Using the optional calculations method, the demand load for a 6-kW central space heating and a 4-kW air-conditioning unit would be _____.
(a) 9,000 watts (b) 3,900 watts
(c) 4,000 watts (d) 5,000 watts

79. The total connected load of a dwelling-unit is 25 kVA, not including heat or air conditioning. If the heat is separately controlled in five rooms (10 kW), and the air conditioning is 6 kW, then the total service demand load would be _____ if the optional method is used. Note: Don't forget to apply optional method demand factors.
(a) 22 kVA (b) 29 kVA
(c) 35 kVA (d) 36 kVA

80. Using the optional method, the service size would be _____ amperes for the following: 1,200-square feet first floor, plus 600-square feet on the second floor, and a 200-square foot open porch; a 4-kW, 230 volt water heater; a ½-horsepower, 115 volt dishwasher; a 4-kW clothes dryer; 4-horsepower A/C, 6-kW heat strips; a ¾-horsepower 115 volt pool pump; one 6-kW oven; one 6-kW Cooktop; 20 receptacles; and 10 lights. The service source voltage is 230/115 volt single phase.
(a) 150 (b) 175
(c) 125 (d) 110

81. An 1,800-square foot residence contains the following: a 300-square foot porch; a 150-square foot carport; a water heater of 4-kW; 10-kW heat separated in five rooms; a 1.5-kW dishwasher; a 6-kW range; a 4.5-kW dryer; two 3-kW ovens; a 6-kW air conditioner; 35 general use receptacles; and sixteen 75 watt lighting fixtures. The service size for the loads would be _____ amperes. The optional method of service calculations were used.
(a) 175 (b) 110
(c) 125 (d) 150

82. A dwelling-unit has 1200-square feet on the first floor, 600-square feet upstairs (unfinished but adaptable for future use), and a 200-square foot open porch with the following loads: a pool pump ¾-horsepower; range – 13.9-kW; dishwasher – 1.2-kW; water heater – 4-kW; dryer – 4-kW; A/C 5 horsepower; and Heat 6-kW heat. Using the optional calculation, what is the demand load for the service?
    (a) No. 4   (b) No. 3
    (c) No. 2   (d) No. 1

NEC Questions
Random Sequence, Section 90–5 Through Section 511–2

83. Concrete, brick, or tile walls shall be considered as _____ , as it applies to working space requirements..
    (a) inconsequential   (b) in the way
    (c) grounded   (d) none of these

84. Equipment termination provisions shall be permitted to be used with the higher rated conductors at the ampacity of the higher rated conductors, provided the equipment is _____ and identified for use with the higher rated conductors.
    (a) listed   (b) tested
    (c) installed   (d) rated

85. An outlet intended for the direct connection of a lampholder, a lighting fixture, or a pendant cord terminating in a lampholder is a(n) _____.
    (a) outlet   (b) receptacle outlet
    (c) lighting outlet   (d) general purpose outlet

86. Some spray cleaning and lubricating compounds contain chemicals that cause severe deteriorating reactions with plastics.
    (a) True   (b) False

87. The correct word used to define wiring which is not concealed is _____.
    (a) open   (b) uncovered
    (c) exposed   (d) bare

88. For other than dwelling-units, a wall switched lighting outlet is required near equipment requiring servicing in attics or underfloor spaces and the switch must be located at the point of entrance to the attic or underfloor space.
    (a) True   (b) False

89. Adequate illumination shall be provided for all working spaces about _____ rated over 600 volts.
    (a) raceways   (b) electrical equipment
    (c) busways   (d) lighting

90. On circuits of 600 volts or less, overhead spans up to 50 feet in length shall have conductors not smaller than _____.
    (a) No. 14   (b) No. 12
    (c) No. 6   (d) No. 10

91. Working space with _____ entrance(s) provided shall be so located that the edge of the entrance nearest the equipment is the minimum clear distance given in Table 110–16(a) away from such equipment.
    (a) 1   (b) 2
    (c) 3   (d) 4

92. Separately installed pressure connectors shall be used with conductors at the _____ not exceeding the ampacity at the listed and identified temperature rating of the connector.
    (a) voltages   (b) temperatures
    (c) listings   (d) ampacities

93. Connection by means of wire binding screws or studs and nuts having upturned lugs or the equivalent shall be permitted for _____ or smaller conductors.
    (a) No. 10   (b) No. 8
    (c) No. 6   (d) none of these

94. Where switches, cutouts, or other equipment operating at 600 volts, nominal, or less, are installed in a room or enclosure where there are exposed live parts or exposed wiring operating at over 600 volts, nominal, the high-voltage equipment shall be effectively separated from the space occupied by _____ by a suitable partition, fence, or screen.
    (a) the access area
    (b) the low-voltage equipment
    (c) unauthorized persons
    (d) motor-control equipment

95. A form of general-use switch so constructed that it can be installed in flush device boxes or on outlet box covers, or otherwise used in conjunction with wiring systems recognized by the Code is a _____ switch.
    (a) transfer
    (b) motor-circuit
    (c) general-use snap
    (d) bypass isolation

96. What is the maximum cord-and plug connected load permitted on a 15 ampere rated multioutlet receptacle that is supplied by a 20-ampere rated circuit.
    (a) 12
    (b) 16
    (c) 20
    (d) 24

97. The minimum ampacity for 120/240-volt service entrance conductors is _____ amperes.
    (a) 15
    (b) 30
    (c) 60
    (d) none of these

98. Three hundred volt cartridge fuses and fuse holders are not permitted on circuits exceeding 300 volts_____.
    (a) between conductors
    (b) to ground
    (c) of the circuit
    (d) a or c

99. Where raceway-type service masts are used, all raceway fittings shall be _____ for use with service masts.
    (a) identified
    (b) approved
    (c) heavy-duty
    (d) none of these

100. The added ground-fault protective equipment at the wye service equipment will make it necessary to _____ the overall wiring system for proper selective overcurrent protection coordination.
    (a) redesign
    (b) review
    (c) inspect
    (d) none of these

101. Where used as switches in 120-volt and 277-volt fluorescent lighting circuits, circuit breakers shall be marked _____.
    (a) UL
    (b) SWD
    (c) Amps
    (d) VA

102. For installations other than one-circuit, two-circuit, or one-family dwellings, the service disconnecting means shall have a rating of not less than _____ amperes.
    (a) 60
    (b) 80
    (c) 100
    (d) 125

103. Overcurrent protection devices are not permitted to be located _____.
    (a) where exposed to physical damage
    (b) near easily ignitable materials, such as clothes closets
    (c) in dwelling-unit bathrooms
    (d) all of these

104. The minimum point of attachment of the service-drop conductors to a building shall in no case be less than _____ feet.
    (a) 8
    (b) 10
    (c) 12
    (d) 15

105. Bonding shall be provided where necessary to _____.
    (a) ensure electrical continuity and the capacity to conduct safely any fault current likely to be imposed
    (b) provide an equipment grounding means for all nonservice related, general use equipment
    (c) please the authority having jurisdiction
    (d) all of these

106. • The service disconnecting means shall plainly indicate whether it is in the _____ position.
   (a) open or closed (b) on or off
   (c) up or down (d) correct

107. The equipment bonding jumper can be installed on the outside of a raceway providing the length of the run is not more than _____ inches and the bonding jumper is routed with the raceway.
   (a) 12 (b) 24
   (c) 36 (d) 72

108. • For each farm building or load supplied by _____ or more branch circuits, the load for feeders, service-entrance conductors, and service equipment shall be computed in accordance with demand factors not less than indicated in Table 220–40.
   (a) one (b) two
   (c) three (d) four

109. The minimum feeder allowance for show window lighting expressed in volt-amps per linear foot shall be _____ VA.
   (a) 400 (b) 200
   (c) 300 (d) 180

110. Overhead conductors for festoon lighting shall not be smaller than No. _____.
   (a) 10 (b) 12
   (c) 14 (d) 18

111. Fuses are required to be marked with _____.
   (a) ampere and voltage rating
   (b) interrupting rating where other than 10,000 amperes
   (c) the name or trademark of the manufacturer
   (d) all of these

112. For a dwelling unit having the total connected load served by a single 3-wire, 120/240-volt or 208Y/120-volt set of service-entrance or feeder conductors with an ampacity of _____ or greater, it shall be permissible to compute the feeder and service loads in accordance with Table 220–30 instead of the method specified in Part B of article 220.
   (a) 100 (b) 125
   (c) 150 (d) 175

113. Individual open conductors and cables other than service-entrance cables shall not be installed within _____ feet of grade level or where exposed to physical damage.
   (a) 8 (b) 10
   (c) 12 (d) 15

114. For large capacity multibuilding industrial installations under single management, where it is assured that the disconnecting can be accomplished by establishing and maintaining safe _____, the disconnecting means shall be permitted to be located elsewhere on the premises.
   (a) procedures
   (b) switching procedures
   (c) protection for persons
   (d) all of these

115. • Service conductors shall not be spliced.
   (a) True (b) False

116. • Article 220 applies to the load calculations for sizing _____.
   (a) branch circuits (b) feeders
   (c) services (d) all of these

117. For the purpose of lighting outlets in dwelling units, a vehicle door in a garage is considered as an outdoor entrance.
   (a) True (b) False

118. When sizing a feeder, the appliance loads in dwelling units can apply a demand factor of 75 percent of nameplate ratings for _____ or more appliances fastened in place on the same feeder.
    (a) two
    (b) three
    (c) four
    (d) five

119. The sum of the ratings of the circuit breakers or fuses shall be permitted to exceed the _____ of the service conductors, provided the calculated load in accordance with Article 220 does not exceed the ampacity of the service conductors.
    (a) ampacity
    (b) voltage
    (c) current
    (d) none of these

120. The general rule for mounting enclosures for overcurrent devices is that the enclosure must be mounted in a _____ position.
    (a) vertical
    (b) horizontal
    (c) vertical or horizontal
    (d) there are no requirements

121. Feeder ground-fault protection of equipment shall not be required where ground-fault protection of equipment is provided on the _____ side of the feeder.
    (a) load
    (b) supply
    (c) service
    (d) none of these

122. • The connected load to which the demand factors of Table 220–32 apply shall include the _____ rating of all appliances that are fastened in place, permanently connected or located to be on a specific circuit; ranges, wall-mounted ovens, counter-mounted cooking units, clothes dryers, water heaters, and space heaters.
    (a) calculated
    (b) nameplate
    (c) circuit
    (d) overcurrent protection

123. Where continuous and noncontinuous loads are present when calculating feeder conductors, the rating of the overcurrent device shall not be less than _____.
    (a) 125 percent of the noncontinuous loads plus 100 percent of the continuous loads
    (b) 100 percent of the noncontinuous loads plus 125 percent of the continuous loads
    (c) 80 percent of the branch circuit rating
    (d) 125 percent of all loads involved

124. For services exceeding 600 volts, nominal, cable tray systems shall be permitted to support cables identified as _____ conductors.
    (a) service-entrance
    (b) overcurrent protection
    (c) two-wire circuits
    (d) three-wire circuits

125. Where the load is computed on a volt-amperes-per-square-foot basis, the load shall be evenly proportioned among multioutlet branch circuits within the _____.
    (a) premises
    (b) branch circuits
    (c) panelboard(s)
    (d) dwelling unit

126. Where installed in a metal raceway all conductors of all feeders using a common neutral shall be _____.
    (a) insulated for 600 volts
    (b) enclosed within the same raceway
    (c) shielded
    (d) none of these

127. To prevent water from entering service equipment, service entrance conductors must be _____.
    (a) connected to service drop conductors below the level of the service head
    (b) have drip loops formed on the service entrance conductors
    (c) either a or b
    (d) both a and b

128. • A _____ counter top is measured from the connecting edge, for the purpose of determining the placement of receptacles.
    (a) kitchen
    (b) usable
    (c) peninsular
    (d) cooking

129. • Communication conductors below power conductors rated over 300 volts supported on poles shall provide a horizontal climbing space not less than _____ inches.
    (a) 6
    (b) 10

(c) 20 (d) 30

130. • Where the service disconnecting means is mounted on a switchboard having exposed busbars on the back, a raceway shall be permitted to terminate at a _____.
(a) connector (b) box
(c) bushing (d) cabinet

131. • Service conductors between the street main and the first point of connection to the service entrance run underground are known as the _____.
(a) utility service (b) service lateral
(c) service drop (d) main service conductors

132. • A synthetic nonflammable insulating media which, when decomposed by electric arcs, produces predominantly nonflammable gaseous mixtures is known as _____.
(a) oil (b) geritol
(c) askarel (d) phenol

133. Equipment or materials to which a symbol or other identifying mark acceptable to the authority having jurisdiction has been attached is known as _____.
(a) listed (b) labeled
(c) approved (d) rated

134. Overcurrent devices shall be readily accessible except for _____ as provided in Section 364–12.
(a) feeder taps (b) cable bus (c) busways (d) wireways

135. Explanatory material in the National Electrical Code appears in the form of _____.
(a) footnotes (b) fine print notes
(c) obelisks and asterisks (d) red print

136. • The connection between the grounded circuit conductor and the equipment grounding conductor at the service is the _____ jumper.
(a) main bonding (b) bonding
(c) equipment bonding (d) circuit bonding

137. A single panel or group of panel units designed for assembly in the form of a single panel is called a _____.
(a) switchboard (b) disconnect
(c) panelboard (d) main

138. In dwelling-units, the voltage between conductors shall not exceed 120 volts, nominal, between conductors that supply the terminals of _____.
(a) lighting fixtures
(b) cord- and plug-connected loads of less than 1,440 VA, nominal
(c) cord- and plug-connected loads of more than 1,440 VA, nominal
(d) a and b

139. A grounding electrode conductor shall be used to connect the equipment grounding conductors, the service-equipment enclosures, and, where the system is grounded, the grounded service conductor to the _____ electrode.
(a) ground (b) grounded
(c) grounding (d) none of these

140. • A separately derived system is a wiring system where power is received from _____.
(a) transformers (b) generators
(c) converter windings (d) all of these

141. A(n) _____ shall be of such design that any alteration of its trip point (calibration) or the time required for its operation will require dismantling of the device or breaking of a seal for other than intended adjustments.
(a) Type S fuse (b) Edison base fuse
(c) circuit breaker (d) fuseholder

142. The minimum headroom of working spaces about motor control centers shall be _____ feet.
(a) 3½ (b) 5

(c) 6⅓   (d) 6½

143. Ground-fault protection of equipment shall be provided for solidly grounded wye electrical services of more than 150 volts to ground, but not exceeding 600 volts phase-to-phase for each service disconnecting means rated _____ amperes or more.
(a) 1000   (b) 1500
(c) 2000   (d) 2500

144. In a multiple-occupancy building, each occupant shall have access to the occupant's service _____.
(a) conductor   (b) disconnecting means
(c) cable   (d) overcurrent protection

145. • The 3 volt-ampere per square foot for dwelling-unit general lighting includes all 15 and 20 ampere general use receptacles. The floor area shall _____ open porches, garages, or unused or unfinished spaces, not adaptable for future use.
(a) include   (b) not include

146. Service cables, shall be formed in a gooseneck and taped and painted or taped with a self-sealing, weather-resistant _____.
(a) cover   (b) protected
(c) thermoplastic   (d) none of these

147. Hallways in dwelling-units that are _____ feet long, or longer, require a receptacle outlet.
(a) 12   (b) 10
(c) 8   (d) 15

148. A plug fuse of the Edison base has a maximum rating of _____ amperes.
(a) 20   (b) 30
(c) 40   (d) 50

149. Where individual open conductors enter a building or other structure, they shall enter through roof bushings or through the wall in an upward slant through individual, noncombustible, nonabsorbent insulating _____.
(a) tubes   (b) raceways
(c) sleeves   (d) bushels

150. • A two-family dwelling unit, such as a duplex apartment, is considered a one-family dwelling.
(a) True   (b) False

151. The one or more additional service disconnecting means for fire pumps or for emergency, legally required standby, or optional standby services permitted by Section 230–2 shall be installed sufficiently remote from the one to six service disconnecting means for normal service to minimize the possibility of _____ interruption of supply.
(a) accidental   (b) intermittent
(c) simultaneous   (d) prolonged

152. No _____ shall be attached to any terminal or lead so as to reverse designated polarity.
(a) grounded conductor   (b) grounding conductor
(c) ungrounded conductor   (d) grounding connector

153. Service conductors shall be connected to the disconnecting means by _____ or other approved means.
I. pressure connectors   II. clamps   III. solder
(a) I only   (b) II only
(c) III only   (d) I or II only

154. Overhead spans of open conductors and open multiconductor cables of not over 600 volts, nominal, shall conform to _____ feet above finished grade, sidewalks, or from any platform or projection from which they might be reached where the supply conductors are limited to 150 volts to ground and accessible to pedestrians only.
(a) 10   (b) 12
(c) 15   (d) 18

155. A single building or other structure sufficiently large to make two or more services necessary is permitted by _____.
(a) architects   (b) special permission

(c) written authorization (d) master electricians

156. Where outdoor lampholders have terminals of a type that puncture the _____ and make contact with the conductors, they shall be attached only to conductors of the stranded type.
(a) exterior of the building (b) paint
(c) insulation (d) none of these

157. A signaling circuit is any electric circuit that energizes signaling equipment.
(a) True (b) False

158. Circuits used only for the operation of fire alarm, other protective signaling systems, or the supply to fire pump equipment shall be permitted to be connected on the _____ of the service overcurrent device where separately provided with overcurrent protection.
(a) base (b) load side
(c) supply side (d) top

159. Identification of the grounded conductor shall be by a metal or metal coating substantially white in color or the word _____ located adjacent to the identified terminal.
(a) grounded (b) white
(c) connected (d) danger

160. The minimum size service drop conductor permitted by the Code is No. _____ copper and No. _____ aluminum.
I. 6 cu   II. 8 cu   III. 6 al   IV. 8 al
(a) II and III (b) I and IV
(c) I and III (d) II and IV

161. Where oil switches or air, oil, vacuum, or sulfur hexafluoride circuit breakers constitute the service disconnecting means (for services exceeding 600 volts, nominal), an air-break isolating switch shall be installed on the supply side of the disconnecting means and all associated service equipment, except where such equipment is mounted on removable truck panels or metal-enclosed switch gear units, which cannot be opened unless the circuit is disconnected, and which, when removed from the normal operating position, automatically disconnect the circuit breaker or switch from all energized parts.
(a) True (b) False

162. • Where necessary to prevent tampering, an automatic overcurrent device protecting service conductors supplying only a specific load, such as a water heater, shall be permitted to be _____ where located so as to be accessible.
(a) locked (b) sealed
(c) either a or b (d) neither a nor b

163. Receptacles outlets in dwelling units are part of the _____ VA per square foot general lighting load.
(a) 3 (b) 5
(c) 2 (d) 4

164. Where fixed multioutlet assemblies are employed, each _____ feet or fraction thereof of each separate and continuous length shall be considered as one outlet of not less than 180 volt-amperes capacity.
(a) 5 (b) 5½
(c) 6 (d) 6½

165. Where the voltage between conductors does not exceed 300 volts, and the roof has a slope of not less than 4 inches in 12 inches, a reduction in clearance to _____ feet shall be permitted.
(a) 1½ (b) 2
(c) 2½ (d) 3

166. Each service disconnecting means shall be suitable for _____.
(a) hazardous locations (b) disconnecting the service
(c) protection of persons (d) the prevailing conditions

167. In _____ rooms, other than kitchens and bathrooms of dwelling units, one or more receptacles controlled by a wall switch shall be permitted in lieu of lighting outlets.
(a) habitable (b) finished

168. The standard ampere ratings for fuses and fixed trip circuit breakers are listed in Chapter 2 of the NEC, additional standard ratings for fuses shall be _____ ampere(s).
    (a) 1 (b) 6
    (c) 601 (d) all of these

169. Where the work space is doubled, _____ entrance(s) to the working space is/are required.
    (a) 1 (b) 2
    (c) 3 (d) 4

170. Where nails or screws are likely to penetrate nonmetallic-sheathed cable, or electrical nonmetallic tubing, a steel sleeve, steel plate, or steel clip not less than _____ inch in thickness shall be used to protect the cable or tubing.
    (a) 1/16 (b) 1/8
    (c) 1/4 (d) 1/2

171. In general, areas where _____ are handled and stored may present severe corrosive conditions, particularly when wet or damp.
    (a) laboratory chemicals and acids
    (b) acids and alkali chemicals
    (c) acids and water
    (d) chemicals and water

172. • Conductors on poles shall have a separation of not less than _____ where not placed on racks or brackets.
    (a) 3 inches (b) 6 inches
    (c) 1 foot (d) 1½ feet

173. Where two to six service disconnecting means in separate enclosures are grouped at one location and supply separate loads from one service drop or lateral, _____ set(s) of service-entrance conductors shall be permitted to supply each or several such service equipment enclosures.
    (a) one (b) two
    (c) three (d) four

174. When normally enclosed live parts are exposed for inspection or servicing, the working space, if in a passageway or general open space, shall be suitably _____.
    (a) accessible (b) guarded
    (c) open (d) enclosed

175. There are four principal determinants of conductor operating temperature. One of those is _____ generated internally in the conductor as the result of load current flow.
    (a) friction (b) magnetism
    (c) heat (d) none of these

176. The Code does not cover installations in ships, watercraft, railway rolling stock, aircraft, or automotive vehicles.
    (a) True (b) False

177. Where equipment or devices are installed in ducts or plenum chambers used to transport environmental air and illumination is necessary to facilitate maintenance and repair, enclosed _____-type fixtures shall be permitted.
    (a) screw (b) plug
    (c) gasketed (d) neon

178. When live parts of electric equipment are guarded by suitable permanent, substantial partitions or screens, any openings in such partitions or screens shall be so sized and located that persons are not likely to come into accidental contact with the live parts or to bring _____ into contact with them.
    (a) dust (b) conducting objects
    (c) wires (d) contaminating parts

179. As used in Section 110-16, a motor control center is an assembly of one or more enclosed sections having a common _____ and principally containing motor control units.
    (a) circuit breaker (b) raceway (c) service (d) power bus

180. A circuit breaker with a _____ voltage rating, e.g., 240 volt or 480 volt, may be applied in a circuit in which the nominal voltage between any two conductors does not exceed the circuit breaker's voltage rating; except that a two-pole circuit breaker is not suitable for protecting a three-phase corner-grounded delta circuit unless it is marked 1ϕ/3ϕ to indicate such suitability.
    (a) straight (b) slash (c) high (d) low

181. Where the service disconnecting means is power operable, the _____ shall be permitted to be connected ahead of the service disconnecting means if suitable overcurrent protection and disconnecting means are provided.
    (a) control circuit
    (b) panel
    (c) grounding conductor
    (d) none of these

182. In the event the Code requires new products, constructions, or materials that are not yet available at the time the Code is adopted, the _____ may permit the use of the products, constructions, or materials that comply with the most recent previous edition of this Code adopted by the jurisdiction.
    (a) architect
    (b) master electrician
    (c) authority having jurisdiction
    (d) supply house

183. _____ shall not be installed beneath openings through which material may be moved, such as openings in farm and commercial buildings, and shall not be installed where they will obstruct entrance to these building openings.
    (a) Overcurrent protection devices
    (b) Overhead service conductors
    (c) Grounding conductors
    (d) Wiring systems

184. Plug fuses of 15-ampere and lower rating shall be identified by a _____ configuration of the window, cap, or other prominent part to distinguish them from fuses of higher ampere ratings.
    (a) octagonal (b) rectangular (c) hexagonal (d) triangular

185. Overhead branch-circuit and feeder conductors shall not be installed _____ openings through which materials may be moved, such as openings in farm and commercial buildings, and shall not be installed where they will obstruct entrance to these building openings.
    (a) above
    (b) beneath
    (c) inside
    (d) outside

186. Circuit breakers rated at _____ amperes or less and _____ volts or less shall have the ampere rating molded, stamped, etched, or similarly marked into their handles or escutcheon areas.
    (a) 100, 600
    (b) 600, 100
    (c) 1000, 6000
    (d) 6000, 1000

187. In general, conductors, other than flexible cords and fixture wires, shall be protected against overcurrent in accordance with their _____ as specified in Section 310–15.
    (a) rating (b) markings (c) ampacities (d) listings

188. The work space shall not be less than _____ inches wide in front of the electric equipment.
    (a) 15
    (b) 20
    (c) 25
    (d) 30

189. In locations where _____ would be exposed to physical damage, enclosures or guards shall be so arranged and of such strength as to prevent physical damage.
    (a) motor control panels
    (b) electric equipment
    (c) generators
    (d) circuit breakers

190. The requirement for maintaining the vertical clearance 3 feet from the edge of the roof shall not apply to the final conductor span where the service drop is attached to _____.
    (a) a service pole
    (b) the side of a building
    (c) an antenna
    (d) the base of a building

191. Conductors for _____ lighting shall be of the rubber-covered or thermoplastic type.
    (a) indirect   (b) festoon   (c) show window   (d) temporary

192. Supplementary grounding electrodes are permitted to supplement the equipment grounding conductor, but they cannot be used as the sole equipment grounding conductor.
    (a) True   (b) False

193. Conductors of light and power systems may occupy the same enclosure _____.
    (a) if all the conductors are less than 600 volts and insulated for the maximum circuit voltage of any conductor within the enclosure
    (b) if the power system is over 600 volts and the light system is under 600 volts
    (c) if the power system is over 600 volts and the individual circuits are alternating current
    (d) in most instances without qualification

194. A unit or assembly of units or sections, and associated fittings, forming a rigid structural system used to support cables and raceways is known as a _____ system.
    (a) cablebus system   (b) busway   (c) wireway   (d) cable tray

195. A conductor used for open wiring, where sleeved, shall be carried through a _____.
    (a) separate sleeve or tube
    (b) weatherproof tube
    (c) tube of absorbent material
    (d) grounded metallic tube

196. The grounding electrode conductor shall be _____ and shall be installed in one continuous length without a splice or joint.
    I. solid   II. solid or stranded   III. insulated, covered or bare
    (a) I   (b) I and III   (c) II and III   (d) III

197. The smallest diameter pipe or conduit permitted for a made electrode shall not be less than _____ inch.
    (a) 1/2   (b) 3/4   (c) 1   (d) 5/8

198. A bare No. 4 copper conductor installed near the bottom of a concrete foundation or footing that is in direct contact with the earth, may be used as a grounding electrode when at least _____ feet in length.
    (a) 25   (b) 15   (c) 10   (d) 20

199. Table 310–74 provides ampacities of insulated triplexed or three single conductor aluminum cables in isolated conduit in air based on conductor temperature of 90°C (194°F) and ambient air temperature of 40°C (104°F). If the conductor size is 350 kcmil, and the voltage range is 2,001 to 5,000, then the ampacity is _____ amperes.
    (a) 305   (b) 310
    (c) 380   (d) 385

200. Where a single equipment grounding conductor is used for multiple circuits in the same raceway, the single equipment grounding conductor must be sized according to _____.
    (a) a combined rating of all the overcurrent protection devices
    (b) the largest overcurrent protection device of the multiple circuits
    (c) a combined rating of all the loads
    (d) any of these

CHAPTER 3
Advanced NEC Calculations And Code Questions

Scope of Chapter 3

UNIT 10   MULTIFAMILY DWELLING-UNIT LOAD CALCULATIONS

UNIT 11   COMMERCIAL LOAD CALCULATIONS

UNIT 12   DELTA/DELTA AND DELTA/WYE TRANSFORMER CALCULATIONS

Unit 10

Multifamily Dwelling-Unit Load Calculations

OBJECTIVES

After reading this unit, the student should be able to explain the following concepts:

- Air conditioning versus heat
- Appliance (small) circuits
- Appliance demand load
- Clothes dryer demand load
- Cooking equipment calculations
- Dwelling-unit calculations – Optional method
- Dwelling-unit calculations – Standard method
- Laundry circuit
- Lighting and receptacles

After reading this unit, the student should be able to explain the following terms:

- General lighting
- General–use receptacles
- Optional method
- Rounding
- Standard method
- Unbalanced demand load
- Voltages

10–1 MULTIFAMILY DWELLING-UNIT CALCULATIONS – STANDARD METHOD

When determining the ungrounded and grounded (neutral) conductor for multifamily dwelling units (Fig. 10–1), apply the following multifamily standard method steps:

Step 1: → **General Lighting And Receptacles, Small Appliance And Laundry Circuits [220–11].** The NEC recognizes that the general lighting and receptacles, small appliance, and laundry circuits will not all be on, or loaded, at the same time. The NEC permits the following demand factor to be applied to these loads [220–16].

(a) **Total Connected Load.** Determine the total connected general lighting and receptacles (3 VA), small appliance (3,000 VA), and the laundry (1,500 VA) circuit load of all dwelling units. The laundry load (1,500 VA) can be omitted if laundry facilities are provided on the premises [210–52(f) Exception 1].

(b) **Demand Factor.** Apply Table 220–11 demand factors to the total connected load (Step 1a).
First 3,000 VA at 100% demand
Next 117,000 VA at 35% demand
Remainder at 25% demand

Step 2: → **A/C versus Heat [220–21].** Because the A/C and heating loads are not on at the same time (simultaneously), it is permissible to omit the smaller of the two loads.
Air conditioning: The A/C demand load shall be calculated at 125 percent of the largest A/C motor VA, plus the sum of the other A/C motor VA's [440–34].
Heat: Fixed electric space-heating loads shall be computed at 100 percent of the total connected load [220–15].

315

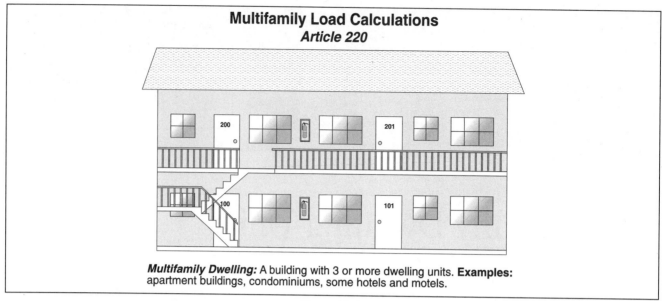

Figure 10-1

Multifamily Load Calculations

Step 3: → **Appliances [220–17].** A demand factor of 75 percent is permitted for four or more appliances fastened in place; such as a dishwasher, disposal, trash compactor, water heater, etc. This does not apply to space-heating equipment [220–15], clothes dryers [220–18], cooking appliances [220–19], or air-conditioning equipment [440–34].

Step 4: → **Clothes Dryers [220–18].** The feeder or service demand load for electric clothes dryers located in dwelling units shall not be less than: (1) 5,000 watts, or (2) the nameplate rating (whichever is greater) adjusted according to the demand factors listed in Table 220–18.

Note. A dryer load is not required if the dwelling unit does not contain an electric dryer. Laundry room dryers shall not have their loads calculated according to this method. This is covered in Unit 11.

Step 5: → **Cooking Equipment [220–19].** Household cooking appliances rated over $1\frac{3}{4}$ kW can have their feeder and service loads calculated according to the demand factors of Section 220–19, Table and Notes.

Step 6: → **Feeder And Service Conductor Size**

Four hundred amperes and less. The ungrounded conductors are sized according to Note 3 of Table 310–16 for 120/240-volt single-phase systems.

Over four hundred amperes. The ungrounded conductors are sized according to Table 310–16 to the calculated unbalanced demand load.

10–2 MULTIFAMILY DWELLING-UNITS CALCULATION EXAMPLES – STANDARD METHOD

Step 1. General Lighting, Small Appliance, And Laundry Demand [220–11]

→ **Question 1.**

What is the demand load for an apartment building that contains twenty units? Each apartment is 840 square feet.
Note. Laundry facilities are provided on the premises for all tenants [210-52(f)].

(a) 5,200 VA (b) 40,590 VA (c) 110,400 VA (d) none of these

• Answer: (b) 40,590 VA

General Lighting   840 square feet × 3 VA   2,520 VA
Small Appliance Circuits 1,500 VA × 2 circuits   3,000 VA
Laundry Circuit (none)   +   0 VA
Total connected load × units   5,520 VA × 20 = 110,400 VA

                                                        110,400 VA
First 3,000 VA at 100%   – 3,000 VA × 1.00 =   3,000 VA
Next 117,000 VA at 35%   107,400 VA × .35 = +37,590 VA
Total demand load =   40,590 VA

Chapter 2 Advanced NEC Calculations And Code Questions   Unit 10 Multifamily Load Calculations   317

➤ Question 2.
What is the general lighting net computed demand load for a twenty unit apartment? Each unit is 990 square feet.

(a) 74,700 VA (b) 149,400 VA (c) 51,300 VA (d) 105,600 VA

- Answer: (c) 51,300 VA

General lighting	990 square feet × 3 VA	2,970 VA
Small appliance	1,500 VA × 2 circuits	3,000 VA
Laundry circuit	1,500 VA × 1 circuits	+1,500 VA
Total connected load × units		7,470 VA × 20 = 149,400 VA

```
                                        149,400 VA
First 3,000 VA at 100%                  − 3,000 VA   × 1.00 =    3,000 VA
                                        146,400 VA
Next 117,000 VA at 35%                 −117,000 VA   × .35 =    40,950 VA
Remainder VA at 25%                      29,400 VA   × .25 =   + 7,350 VA
Total demand load =                                            51,300 VA
```

Step 2. Air Conditioning Versus Heat [220–15 and 440-34]

➤ Question 1.
What is the net computed A/C versus heat load for a forty unit multifamily building that has a 3-horsepower 230-volt A/C unit and two 3-kW baseboard heaters in each unit?

(a) 160 kW (b) 240 kW (c) 60 kW (d) 50 kW

- Answer: (b) 240 kW

    A/C [440-34] – 230 volts × 17 amperes* = 3,910 VA, 3,910 VA × 1.25 = 4,888 VA

    4,888 VA + (3,910 VA × 39 units) = 157,378 VA, Omit smaller than heat [220-21].

    Heat [220–15] 3,000 watts × 2 units = 6,000 watts, 6,000 × 40 units = 240,000 watts/1,000 = 240 kW

➤ Question 2.
What is the A/C versus heat demand load for a twenty-five-unit multifamily building that has a 3-horsepower 230 volt A/C unit and a 5-kW heat strip?

(a) 160 kVA (b) 125 kVA (c) 6 kVA (d) 5 kVA

- Answer: (b) 125 kVA

    Air Conditioning [220–14] – 230 volts × 17 amperes* = 3,910 VA, 3,910 VA × 1.25 = 4,888 VA

    4,888 VA + (3,910 × 24 units) = 98,728 VA, Omit smaller than heat [220-21].

    Heat [220–15] 5,000 watts × 25 units = 125,000 watts/1,000 = 125 kW

* Table 430–148

Step 3. Appliance Demand Load [220–17]

➤ Question 1.
What is the appliance demand load for a twenty-unit multifamily building that contains a 940 VA disposal, a 1,250 VA dishwasher, and a 4,500 VA water heater?

(a) 100 kVA (b) 134 kVA (c) 7 kVA (d) 5 kVA

- Answer: (a) 100 kVA

```
Disposal            940 VA
Dishwasher        1,250 VA
Water heater    + 4,500 VA
                  6,690 VA   × 20 units = 133,800 VA
Demand factor at 75%                          × .75
Total demand load =                       100,350 VA
```

Question 2.
What is the appliance demand load for a thirty-five-unit multifamily building that contains a 900 VA disposal, a 1,200 VA dishwasher, and a 5,000 VA water heater.

(a) 71 kVA (b) 142 kVA (c) 107 kVA (d) 186 kVA

- Answer: (d) 186 kVA

Disposal	900 VA
Dishwasher	1,200 VA
Water heater	+ 5,000 VA
Unit total connected load	7,100 VA × 35 Units = 248,500 VA
Demand factor at 75%	× .75
Total demand load =	186,375 VA/1,000 = 186 VA

Step 4. Dryer Demand Load [220–18]

Question 1.
A multifamily dwelling (twelve-unit building) contains a 4.5-kVA electric clothes dryer in each unit. What is the feeder and service dryer demand load for the building?

(a) 5 kVA (b) 27 kVA (c) 60 kVA (d) none of these

- Answer: (b) 27 kVA

  5 kVA* × 12 units = 60 kVA × .45 = 27 kVA

* The minimum load is 5 kVA, for standard calculations.

Question 2.
What is the demand load for twenty 5.25-kW dryers installed in dwelling-units of a multifamily building?

(a) 5 kVA (b) 27 kVA (c) 60 kVA (d) 37 kVA

- Answer: (d) 37 kVA

  5.25 kVA × 20 units = 105 kVA × .35 = 36.75 kVA

Step 5. Cooking Equipment Demand Load [220–19]

❏ Column A Example
What is the feeder and service demand load for five 9-kW ranges?

(a) 9 kW (b) 45 kW (c) 20 kW (d) none of these

- Answer: (c) 20 kW [Table 220–19, Column A]

❏ Note 1 Example
What is the feeder and service demand load for three ranges rated 16 kW each?

(a) 15 kW (b) 14 kW (c) 17 kW (d) 21 kW

- Answer: (c) 17 kW [220–19 Note 1] (closest answer)

  Step 1. → "Column A" demand load = 14 kW (3 units).
  Step 2. → The average range (16 kW) exceeds 12 kW by 4 kW, increase "Column A" demand load (14 kW) by 20 percent, 14 kW × 1.2 = 16.8 kW or 14 kW + 2.8 kW = 16.8 kW

❏ Note 2 Example.
What is the feeder and service demand load for three ranges rated 9-kW and three ranges rated 14-kW?

(a) 36 kW (b) 42 kW (c) 78 kW (d) 22 kW

- Answer: (d) 22 kW [220–19, Note: 2]

  Step 1: → Determine the total connected load.
  9 kW (use minimum 12 kW) 3 × 12 kW =   36 kW
  14 kW   3 × 14 kW = + 42 kW
                                         78 kW

  Step 2. → Determine the average range rating, 78 kW/6 units = 13 kW average rating.
  Step 3. → Demand load Table 220–19 "Column A", 6 ranges = 21 kW.
  Step 4. → The average range (13 kW) exceeds 12 kW by 1 kW. Increase "Column A" demand load (21 kW) by 5 percent, 21 kW × 1.05 = 22.05 kW, or 21 kW + 1.05 kW = 22.05 kW.

Note 3 Example No. 1

What is the feeder and service demand load for ten 3-kW ovens?

(a) 10 kW (b) 30 kW
(c) 15 kW (d) 20 kW

- Answer: (c) 15 kW [220–19, Column B]

  3 kW × 10 units = 30 kW × .49 = 14.70 kW demand load.

Note 3 Example No. 2

What is the feeder and service demand load for eight 6-kW cook tops?

(a) 10 kW (b) 17 kW
(c) 14.7 kW (d) 48 kW

- Answer: (b) 17 kW [220–19, Column C]

  6 kW × 8 units = 48 kW × .36 = 17.28 kW demand load.

Step 6. Service Conductor Size

→ **Question 1.**

What size aluminum service conductors are required for a 120/240-volt, single-phase multifamily building that has a total demand load of 93 kVA, Fig. 10–2?

(a) 300 THHN kcmil (b) 350 THHN kcmil
(c) 500 THHN kcmil (d) 600 THHN kcmil

- Answer: (d) 600 kcmil aluminum

  I = VA/E

  I = 93,000 VA/240 volts

  I = 388 amperes [Note 3 of Table 310–16]

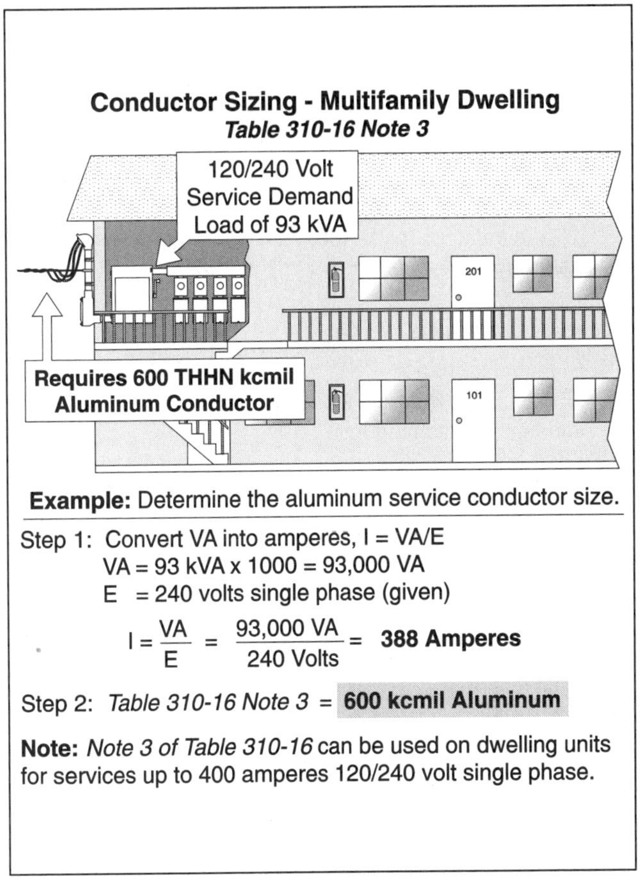

→ **Question 2.**

What size service conductors are required for a multifamily building that has a total demand load of 270 kVA for a 120/208-volt three-phase system?

Note. Service conductors are run in parallel.

(a) 2 – 300 THHN kcmil per phase (b) 2 – 350 THHN kcmil per phase
(c) 2 – 500 THHN kcmil per phase (d) 2 – 600 THHN kcmil per phase

- Answer: (c) 2 – 500 THHN kcmil per phase

  I = VA/(E × 1.732)

  I = 270,000 VA/(208 volts × 1.732)

  I = 750 amperes

  Amperes per parallel set = 750 amperes/2, = 375 amperes.

  500 THHN kcmil conductor has an ampacity of 380 amperes [Table 310–16 at 75ºC]. These parallel conductors rated 760 amperes (380 amperes × 2) are permitted to be protected by a 800 ampere protection device [240–3(b)].

Conductor Sizing - Multifamily Dwelling
Table 310-16 Note 3

120/240 Volt Service Demand Load of 93 kVA

Requires 600 THHN kcmil Aluminum Conductor

Example: Determine the aluminum service conductor size.

Step 1: Convert VA into amperes, I = VA/E
VA = 93 kVA × 1000 = 93,000 VA
E = 240 volts single phase (given)

$$I = \frac{VA}{E} = \frac{93,000\ VA}{240\ Volts} = 388\ Amperes$$

Step 2: Table 310-16 Note 3 = **600 kcmil Aluminum**

Note: *Note 3 of Table 310-16* can be used on dwelling units for services up to 400 amperes 120/240 volt single phase.

Figure 10-2
Conductor Sizing – Multifamily Dwelling

10–3 MULTIFAMILY DWELLING UNITS CALCULATION EXAMPLES – STANDARD METHOD

Multifamily Load Calculation Example

What is the demand load for an apartment building that contains twenty-five units? Each apartment is 1,000 square feet. Air conditioning 5-horsepower, heat 7.5-kW, dishwasher 1.2-kVA, disposal 1.5-kVA, water heater 4.5-kW, dryer 4.5-kW, and range 13.5-kW. System voltage 208/120-volt wye three-phase.

Step 1. General Lighting, Small Appliance, And Laundry Demand [220–11]

Twenty units each containing:
General lighting 1,000 square feet × 3 VA   3,0000 VA
Small appliance circuits 1,500 VA × 2   3,000 VA
Laundry circuit   + 1,500 VA
                                               7,500 VA

Total connected load, 7,500 VA × 25 units =   187,500 VA
First 3,000 VA at 100% =   − 3,000 VA × 1.00 =   3,000 VA
                                               184,000 VA

Next 117,000 VA at 35%   − 117,000 VA × .35 =   40,950 VA
Remainder at 25%   67,000 VA × .25 =   + 16,750 VA
Total general lighting, small appliance,   60,700 VA
and laundry demand load

Step 2. Air Conditioning Versus Heat [220–15 And 440–34]

Twenty five units each containing A/C 5-horsepower versus 7.5-kW heat:
A/C 5 horsepower single-phase [440-34], 208 volts × 30.8 amperes* = 6,406 VA
6,406 VA × 1.25 = 8,008 VA, 8,008 VA + (6,406 × 24 units) = 161,752 VA
Omit smaller than heat [220–21]
Total heat demand load [220–15] 7,500 VA × 25 units = 187,500 VA

Step 3. Appliance Demand Load [220–17]

Twenty five units each contining:
Disposal   1,200 VA
Dishwasher   1,500 VA
Water heater   + 4,500 VA
Total connected load   7,200 VA × 25 units = 180,000 VA
Demand factor at 75%   × .75
Total appliance demand load   135,000 VA

Step 4. Dryer Demand Load [220–18]

Twenty five 4.5-kW dryers:
Total connected load = 5 kVA* × 25 units = 125 kW
Demand factor for 25 unit = 32.5%
Total demand dryer load = 125 kW × .325 = 40.625 kW
* The minimum dryer load for standard load calculations is 5 kW

Step 5. Cooking Equipment Demand Load [220–19]

Twenty five 13.5-kW ranges:
Step 1. Column "A" demand load for 25 units = 40 kW.
Step 2. The average range (13.5 kW) exceeds 12 kW by 3.5 kW, increase Column "A" demand load (40 kW) by 20%.
     Range demand load = 40 kW × 1.2 = 48 kW or 40 kW + 8 kW = 48 kW

Step 6. Service Conductor Size [Note 3 Of Table 310–16]

Step 1. Total general lighting, small appliance, laundry demand load   60,700 VA
Step 2. Total heat demand load [220–15] 7,500 VA × 25   187,500 VA
Step 3. Total appliance demand load   135,000 VA
Step 4. Total demand dryer load   40,625 W
Step 5. Range demand load   + 48,000 W
Total demand load   471,825 VA
Service conductor amperes = VA/(E × 1.732)
I = 471,825 VA/(208 volts × 1.732), I = 1,310 amperes

10–4 MULTIFAMILY DWELLING-UNIT CALCULATIONS [220–32] – OPTIONAL METHOD

Instead of sizing the ungrounded conductors according to the standard method (Part B of Article 220), the optional method can be used for feeders and service conductors in multifamily dwelling units. Follow the following steps for determining the demand load.

Step 1: → **Determine the total connected load.** Add the following loads:

   (a) **General Lighting.** 3 volt-amperes per square-foot.

   (b) **Small appliance and laundry branch circuit.** 1,500 volt-amperes for each 20 ampere small appliance and laundry branch circuit.

   (c) **Appliances.** The *nameplate* VA rating of all appliances and motors fastened in place (permanently connected), but not the A/C or Heating Load. Be sure to use the range and dryer nameplate rating!

   (d) **A/C versus Heat.** Determine the largest of the A/C versus heat.
      (1) Air conditioning or heat-pump compressors at 100 percent.
      (2) Heat at 100 percent.

Step 2. → **Determine the demand load.** The net computed demand load is determined by applying the demand factor from Table 220–32 to the total connected load (Step 1). The net computed demand load (kVA) can be converted to amperes by:

$$\text{Single Phase } I = \frac{VA}{E} \qquad \text{Three Phase } I = \frac{VA}{(E \times 1.732)}$$

Step 3. → **Feeder and Service conductor**
   Four hundred amperes and less. The ungrounded conductors are sized according to Note 3 of Table 310–16 for 120/240-volt, single-phase systems up to 400 amperes.
   Over four hundred amperes. The ungrounded conductors are sized according to Table 310–16 based on the calculated demand load.

When do you use the standard method verses the optional method? For the purpose of exam preparation, always use the standard load calculations unless the question specifies the optional method. In the field, you will probably want to use the optional method because it provides for a smaller service.

10–5 MULTIFAMILY DWELLING-UNIT EXAMPLE QUESTIONS [220–32] – OPTIONAL METHOD

A multifamily building has twelve units, each is 1,500 square feet and contain the following:

25 extra receptacles	Dryer at 4.5 kVA
Washing machine at 1.2 kVA	Range at 14.4 kW
Dishwasher at 1.5 kVA	Water heater at 4 kW
Heat at 5 kVA	A/C 3-horsepower with 1/8-horsepower compressor fan

→ **General Lighting Demand Question**
Using the optional calculations, what is the net computed load for the building general lighting and general-use receptacles, small appliance, and laundry circuits?

(a) 45 kVA (b) 90 kVA (c) 108 kVA (d) 60 kVA

• Answer: (a) 45 kVA [Table 220–32]

(a) General Lighting, 1,500 square feet × 3 VA =   4,500 VA
(b) Two Small Appliance Circuits   3,000 VA
(c) Laundry Circuit   + 1,500 VA
                                                    9,000 VA

Total Demand Load for 12 units × 9,000 VA = 108,000 VA × .41 = 44,280 VA
Note. The washing machine is calculated as part of the 1,500 VA laundry circuit.

→ **Air Conditioning Versus Heat Question**
Using the optional calculations, what is the demand load for the building air conditioning versus heat?

(a) 52 kVA (b) 60 kVA (c) 30 kVA (d) 25 kVA

• Answer: (d) 25 kVA [220–32]

A/C = 3-horsepower, 230 volts × 17 amperes =   3,910 VA
1/8-horsepower, 230 volts × 1.45 amperes =   + 334 VA
                                                 4,244 VA

4,244 VA × 12 units × .41 = 20,880 VA *
Heat = 5,000 watts × 12 units = 60,000 VA × .41 = 24,600 watts.

* Omit the air conditioner load because it is smaller than the heat [220–21].

→ Appliance Demand Question

Using the optional calculations, what is the twenty unit building net demand load for the water heaters and dishwashers?

(a) 41 kVA (b) 53 kVA
(c) 33 kVA (d) 27 kVA

• Answer: (d) 27 kVA [220–32]

Water heater   4,000 VA
Dishwasher   + 1,500 VA
               5,500 VA

Total Appliance Demand Load for the 12 units,
5,500 VA × 12 units = 66,000 VA × .41 = 27,060 VA

Note. The washing machine is calculated as part of the laundry circuit.

Dryer Demand Load Question

Using the optional calculations, what is the twenty unit building net computed load for the 4.5-kVA dryer?

(a) 60 kVA (b) 25 kVA
(c) 55 kVA (d) 22 kVA

• Answer: (d) 22 kVA

4.5 kW* × 12 units × .41 = 22.14 kVA

* Be sure to use the nameplate rating, not 5,000 VA.

Range Demand Load Question

Using optional calculations, what is the twenty unit building demand load for the 14.4-kW ranges?

(a) 70 kW (b) 40 kW (c) 170 kW (d) 105 kW

• Answer: (a) 70 kW

14.4 kW × 12 units × .41 = 70.85 kW

Service Conductor Size Question

If the total demand load is 189 kVA, what is the service conductor size? Service is 208Y/120 volt wye three-phase, Fig. 10–3.

(a) 600 amperes (b) 800 amperes (c) 1,000 amperes (d) 1,200 amperes

• Answer: (a) 600 amperes

$$I = \frac{VA}{(E \times 1.732)}$$

$$I = \frac{189,00 \text{ VA}}{(208 \text{ volts} \times 1.732)} = 525 \text{ amperes}$$

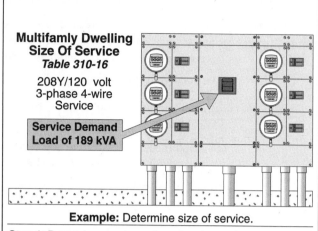

Figure 10-3

Multifamily Dwelling – Size Of Service

Unit 10 – Multifamily Dwelling-Unit Load Calculations Summary Questions

10–1 Multifamily Dwelling Calculations – Standard Method

1. • An apartment building contains twenty dwelling-units, 840 square feet each. What is the general lighting feeder demand load for the building? *Note.* Laundry facilities are provided on the premises for all tenants.
   (a) 5,200 VA (b) 40,590 VA (c) 110,400 VA (d) None of these

2. An apartment building contains twenty dwelling-units, 990 square feet each. What is the general lighting net computed demand load for the building?
   (a) 74,700 VA (b) 149,400 VA (c) 51,300 VA (d) 105,600 VA

3. A forty unit multifamily building has a 3-horsepower, 230-volt A/C and two 3-kW baseboard heaters in each unit. What is the A/C versus heat net computed demand load?
   (a) 160 kW (b) 240 kW (c) 60 kW (d) 50 kW

4. A twenty-five unit multifamily building has a 3-horsepower, 230-volt A/C and a 5-kW heat strip. What is the A/C versus heat net computed demand load?
   (a) 160 kW (b) 125 kW (c) 6 kW (d) 5 kW

5. A sixteen unit multifamily building contains a 940 VA disposal, a 1,250 VA dishwasher, and a 4,500 VA water heater. What is the service demand load for these appliances?
   (a) 100 kVA (b) 134 kVA (c) 80 kVA (d) 5 kVA

6. An apartment building contains twenty-eight dwelling-units, each unit contains a 900 VA disposal, 1,200 VA dishwasher and a 5,000 VA water heater. What is the feeder and service demand load for the appliances?
   (a) 149 kVA (b) 142 kVA (c) 107 kVA (d) 186 kVA

7. A multifamily dwelling (forty unit building) contains a 4.5-kW electric clothes dryer in each unit. What is the feeder and service demand load for all the dryers?
   (a) 50 kW (b) 27 kW (c) 60 kW (d) none of these

8. What is the demand load for ten 5.25-kW dryers installed in dwelling-units of a multifamily building?
   (a) 26 kW (b) 37 kW (c) 60 kW (d) 37 kW

9. What is the demand load for twelve 3.25-kW ovens?
   (a) 10 kW (b) 18 kW (c) 15 kW (d) 20 kW

10. What is the demand load for eight 7-kW cooktops?
    (a) 20 kW (b) 17 kW (c) 14.7 kW (d) 48 kW

11. • What is the demand load for five 12.4-kW ranges?
    (a) 9 kW (b) 45 kW (c) 20 kW (d) None of these

12. What is the demand load for three ranges rated 15.5-kW?
    (a) 15 kW (b) 14 kW (c) 17 kW (d) 21 kW

13. What is the feeder and service demand load for 3 ranges rated 11-kW and 3 ranges rated 14-kW?
    (a) 36 kW (b) 42 kW (c) 78 kW (d) 22 kW

14. • What size aluminum service conductors are required for a 115/230-volt, single-phase multifamily building that has a total demand load of 90-kW?
    (a) 500 THHN kcmil (b) 600 THHN kcmil (c) 700 THHN kcmil (d) 800 THHN kcmil

15. What size service conductors are required for a multifamily building that has a total demand load of 260-kW for a 120/208-volt three phase system?
    (a) 2 – 300 THHN kcmil
    (b) 2 – 350 THHN kcmil
    (c) 2 – 500 THHN kcmil
    (d) 2 – 600 THHN kcmil

For The Next Five Questions
A multifamily building has 12 units. Each is 1500 square feet and contains the following:
System voltage 120/240-volt, single-phase.

25 extra receptacles	Dryer at 4.5-kW
Washing machine at 1.2 kVA	Range at 14.45-kW
Dishwasher at 1.5 kVA	Water heater at 4-kW
Heat at 5-kW	A/C 3 HP with ⅛-horsepower compressor fan

16. • What is the building's net computed load for the general lighting, small appliance and laundry in VA?
    (a) 40 kVA
    (b) 108 kVA
    (c) 105 kVA
    (d) 90 kVA

17. What is the building's demand load for 3-horsepower air conditioning with a ⅛-horsepower compressor fan versus 5-kW heat in VA?
    (a) 52 kVA
    (b) 60 kVA
    (c) 30 kVA
    (d) 105 kVA

18. What is the building demand load for the appliances in VA?
    (a) 40 kVA
    (b) 55 kVA
    (c) 43 kVA
    (d) 50 kVA

19. What is the building demand load for the 4.5-kW dryer?
    (a) 60 kVA
    (b) 27 kVA
    (c) 55 kVA
    (d) 101 kVA

20. What is the building demand load for the 14.45-kW range?
    (a) 27 kW
    (b) 35 kW
    (c) 168 kW
    (d) 30 kW

21. If the total demand load of a multifamily dwelling-unit is 206 kVA, what is the service and feeder conductor size? The system voltage is 120/240, single-phase.
    (a) 600 amperes
    (b) 800 amperes
    (c) 1,000 amperes
    (d) 1,200 amperes

22. • If a service contains three sets of parallel 400 kcmil conductors, what size bonding jumper is required for each service raceway?
    (a) No. 1
    (b) No. 1/0
    (c) No. 2/0
    (d) No. 3/0

23. • If 400 kcmil service conductors are in parallel in three raceways, what is the minimum size grounding electrode conductor required?
    (a) No. 1/0
    (b) No. 2/0
    (c) No. 3/0
    (d) No. 4/0

10–3 Multifamily Dwelling Calculations [220–32] – Optional Method

24. • When determining the service (using optional calculations) for a multifamily dwelling, the total connected load shall have the demand factors of Table 220–32 applied. When determining the total connected load, the largest of the _____ shall be used.
    (a) 100 percent of the air conditioning
    (b) 125 percent of the air conditioning
    (c) 100 percent of the heat
    (d) a or c

25. A building has sixty units. Each unit is 1,500 square feet. Using the optional dwelling-unit calculations, what is the net computed load for the building general lighting and general use receptacles, small appliance and laundry circuits?
    (a) 145 kVA
    (b) 190 kVA
    (c) 108 kVA
    (d) 130 kVA

26. A 60 unit multifamily dwelling has a 3-horsepower air conditioner with a ⅛-horsepower blower and 5-kW heat. Using the optional method, what is the A/C versus heat demand load?
    (a) 50 kVA
    (b) 61 kVA
    (c) 30 kVA
    (d) 72 kVA

27. Using the optional method for dwelling-unit calculations, what is a sixty unit multifamily building net demand load for a 4-kW water heater and a 1.5-kVA dishwasher?
    (a) 80 kVA (b) 50 kVA (c) 30 kVA (d) 60 kVA

28. Each unit of a sixty unit apartment building has a 4-kW dryer. Using the optional method, the demand load that would be added to the service is _____ kW.
    (a) 75 (b) 240 (c) 72 (d) 58

29. Using the optional method for dwelling-unit calculations, what is the sixty unit multifamily building net computed load for a 4.5-kW dryer?
    (a) 65 kW (b) 25 kW (c) 55 kW (d) 75 kW

30. Using the optional method for dwelling-unit calculations, what is a sixty unit multifamily building demand load for the 14-kW range?
    (a) 150 kW (b) 50 kW (c) 100 kW (d) 200 kW

31. If the total demand load is 270 kVA, what is the service and feeder conductor size? The service is 208/120-volt wye three phase.
    (a) 600 amperes (b) 800 amperes (c) 1,000 amperes (d) 1,200 amperes

32. Each unit of a twenty unit multifamily dwelling has 900 square feet of living space. The air conditioning is 5-horsepower, heat 5-kW, water heater 5-kW, and range 14-kW. The service for this apartment building is approximately _____ kVA if the optional method calculation is used.
    (a) 200 (b) 250 (c) 280 (d) 320

33. • If the service conductors are in parallel in two raceways (500 THHN kcmil), what size bonding jumper is required for each service raceway?
    (a) No. 2 (b) No. 1 (c) No. 1/0 (d) No. 2/0

34. • If the service conductors are in parallel in two raceways (500 THHN kcmil), what is the minimum size grounding electrode conductor required?
    (a) No. 1/0 (b) No. 2/0 (c) No. 3/0 (d) No. 4/0

Challenge Questions

General Lighting and Receptacle Calculations [Table 220–11]

35. Each dwelling-unit of a twenty unit multifamily building has 900 square feet of living space. What is the general lighting load for the multifamily dwelling-unit apartment building?
    (a) 45 kVA (b) 60 kVA (c) 37 kVA (d) 54 kVA

36. An apartment building contains twenty dwelling-units and each dwelling-unit has 840 square feet of living space. What is the general lighting feeder demand load for the multifamily building, if laundry facilities are provided on the premises for all tenants and no laundry circuit is installed in each unit?
    (a) 35 kVA (b) 41 kVA (c) 45 kVA (d) 63 kVA

Appliance Demand Factors [Section 220–17].

37. An apartment building contains twenty dwelling-units. Each unit contains a 900 VA disposal, a 1,200 VA dishwasher, and a 5,000 VA water heater. What is the feeder demand load for the appliances in this building?
    (a) 106,500 VA (b) 117,100 VA (c) 137,000 VA (d) 60,000

Dryer Calculation [Section 220–18]

38. The nameplate rating for each household dryer in a ten unit apartment building is 4-kW. This would add _____ kW to the service size.
    (a) 20 (b) 25 (c) 40 (d) 50

Ranges – Note 1 of Table 220–19

39. • The demand load for thirty 15.8-kW household ranges is _____ kW.
    (a) 31 (b) 47 (c) 54 (d) 33

Ranges – Note 2 of Table 220–19

40. What is the feeder demand load for five 10-kW, five 14-kW, and five 16-kW household ranges?
    (a) 210 kW (b) 30 kW (c) 14 kW (d) 33 kW

41. What is the kW to be added to service loads for ten 12-kW, eight 14-kW, and two 9-kW household ranges?
    (a) 33 kW (b) 35 kW (c) 36.75 kW (d) 29.35 kW

Ranges – Note 3 to Table 220–19

42. What is the minimum demand load for five 5-kW cooktops, two 4-kW ovens, and four 7-kW ranges?
    (a) 15.5 kW (b) 8.8 kW (c) 19.5 kW (d) 18.2 kW

43. • What is the maximum dwelling-unit feeder or service demand load for fifteen 8-kW cooking units?
    (a) 38.4 kW
    (b) 30 kW
    (c) 120 kW
    (d) none of these

Ranges – Note 5 of Table 220–19

44. A school has twenty 10-kW ranges installed in the home economics class. The minimum load this would add to the service is _____ kW.
    (a) 35 (b) 44.8 (c) 56 (d) 160

10–12 Neutral Calculation [220–22]

45. The service neutral demand load for household electric clothes dryers shall be calculated at _____ percent of the demand load as determined by Section 220–18.
    (a) 50 (b) 60 (c) 70 (d) 80

46. The feeder neutral demand for fifteen 8-kW cooking units would be _____ kW.
    (a) 38.4 (b) 30 (c) 120 (d) 21

47. What is the feeder neutral load (in amperes) for fifteen 6-kW household dryers?
    (a) 105 amperes
    (b) 125 amperes
    (c) 150 amperes
    (d) none of these

48. A twelve unit multifamily dwelling contains a 4 kVA electric clothes dryer in each unit. What is the feeder or service neutral demand load (in amperes)?
    (a) 115 amperes
    (b) 225 amperes
    (c) 80 amperes
    (d) 55 amperes

49. • The feeder and service neutral demand load can be reduced 70 percent for that portion of the unbalanced load over 200 amperes. This applies to _____ systems.
    (a) 3-wire single phase 120/240-volt
    (b) 3-wire single phase 120/208-volt
    (c) 4-wire three phase 120/208-volt
    (d) a and c

50. • What is the dwelling-unit service neutral load (in amperes) for ten 9-kW ranges?
    (a) 40 amperes
    (b) 55 amperes
    (c) 75 amperes
    (d) 105 amperes

NEC Questions

Random Sequence, Article 100 Through Section 328–12(a)

51. Grounding electrode conductor connections to a concrete-encased or buried grounding electrode are required to be readily accessible.
    (a) True   (b) False

52. The terminal for the connection of the equipment grounding conductor shall be identified by a green-colored, _____.
    I. not readily removable terminal screw with a hexagonal head
    II. hexagonal, not readily removable terminal nut
    III. pressure wire connector
    (a) I only   (b) I and III   (c) I or II   (d) I, II, or III

53. • Grounding electrode conductor fittings shall be protected from physical damage by being enclosed in _____.
    (a) metal   (b) wood
    (c) the equivalent of a or b   (d) none of these

54. Grounding and bonding conductors shall not be connected by _____.
    (a) pressure connections   (b) solder   (c) lugs   (d) approved clamps

55. There is no time limit for temporary electrical power and lighting except that it must be removed upon completion for which of the following?
    (a) Construction or remodeling
    (b) Maintenance or repair
    (c) Demolition of buildings
    (d) All of these

56. Rigid metal conduit that is direct buried outdoors must have at least _____ inches of cover.
    (a) 6   (b) 12   (c) 18   (d) 24

57. Flexible cords and cables used for temporary wiring shall be protected _____.
    (a) from accidental damage   (b) where passing through doors
    (c) from sharp corners and projections   (d) all of these

58. For temporary wiring over 600 volts, nominal, _____ shall be provided to prevent access of other than authorized and qualified personnel.
    I. fencing   II. barriers   III. signs
    (a) I only   (b) II only   (c) III only   (d) I or II

59. Open wiring on insulators is a(n) _____ wiring method using cleats, knobs, tubes, and flexible tubing for the protection and support of single insulated conductors run in or on buildings, and not concealed by the building structure.
    (a) temporary   (b) acceptable   (c) enclosed   (d) exposed

60. All conductors of a circuit are required to be _____.
    (a) in the same raceway
    (b) close proximity in the same trench
    (c) the same size
    (d) a and b

61. Fittings shall be used only with the specific wiring methods for which they are designed and listed.
    (a) True   (b) False

62. Where underground conductors and cables emerge from underground, they shall be protected to a point _____ feet above finished grade. In no case shall the protection be required to exceed 18 inches below grade.
    (a) 3   (b) 6   (c) 8   (d) 10

63. _____ is not a magnetic metal and there will be no heating due to inductive hysteresis heating.
    (a) Steel   (b) Iron   (c) Aluminum   (d) All of these

64. Use of FCC systems in damp locations shall be _____.
    (a) restricted
    (b) permitted
    (c) permitted provided the system is encased in concrete
    (d) approved by special permission

65. Surge arresters shall be _____.
    (a) permitted to be located indoors
    (b) permitted to be located outdoors
    (c) made inaccessible to unqualified persons
    (d) all of these

66. Where knob-and-tube conductors pass through wood cross members in plastered partitions, conductors shall be protected by noncombustible, nonabsorbent, insulating tubes extending not less then _____ inches beyond the wood member.
    (a) 2 (b) 3 (c) 4 (d) 6

67. Exposed noncurrent carrying metal parts likely to become energized must be grounded where _____.
    (a) within 8 feet vertically or 5 feet horizontally of ground or grounded objects
    (b) located in wet or damp locations
    (c) in electrical contact with metal
    (d) all of these

68. Equipment grounding conductors shall be the same size as the circuit conductors for _____ ampere circuits.
    (a) 15 (b) 20 (c) 30 (d) all of these

69. Note 8 (a) derating factors shall not apply to conductors in nipples having a length not exceeding _____ inches.
    (a) 12 (b) 24 (c) 36 (d) 48

70. Receptacles (15- and 20-ampere 125 volt) used by construction site personnel, which are not part of the permanent wiring of the building, are required to be protected by _____.
    (a) ground fault circuit interrupters
    (b) suitable guards
    (c) inverse time breakers
    (d) instantaneous trip breakers

71. Supports for concealed knob-and-tube wiring shall be installed within _____ inches of each side of each tap or splice, and at intervals not exceeding _____ feet.
    (a) 3, 2 (b) 2, 3 (c) 6, 4½ (d) 4½, 6

72. A load is considered to be continuous if it is expected to continue for _____ hour(s).
    (a) ½ (b) 1 (c) 2 (d) 3

73. It is the intent of Section 318–6 to permit discontinuous segments and termination of cable tray installations where the system provides for the support of cables in accordance with their corresponding articles and where adequate _____ is provided in the cable tray system design.
    (a) bonding (b) grounding (c) jumping (d) safety

74. Concealed knob-and-tube wiring shall be permitted to be used only for extensions of existing installations and elsewhere only by special permission as in the hollow spaces of _____.
    (a) walls (b) ceilings (c) a and b (d) a nor b

75. Equipment and devices shall be permitted within ducts or plenum chambers used to transport environmental air only if necessary for their direct action upon, or sensing of, the _____.
    (a) contained air (b) air quality (c) air temperature (d) none of these

76. Open conductors entering or leaving locations subject to dampness, wetness, or corrosive vapors shall have _____ formed on them and shall then pass upward and inward from the outside of the buildings, or from the damp, wet, or corrosive location, through noncombustible, nonabsorbent insulating tubes.
    (a) weather heads (b) drip loops (c) identification (d) blisters

77. A fuse or an overcurrent trip unit of a circuit breaker shall be connected in series with each ungrounded _____.
    (a) device   (b) conductor   (c) branch-circuit   (d) all of these

78. Individually covered or insulated _____ conductors shall have a continuous outer finish that is either green, or green with one or more yellow stripes.
    (a) grounded   (b) grounding   (c) neutral   (d) load

79. Line and ground connecting conductors for a surge arrester shall not be smaller than No. _____.
    (a) 14   (b) 12   (c) 10   (d) 8

80. Grounding electrodes that consist of rods, require a minimum of _____ feet is in contact with the soil.
    (a) 10   (b) 8   (c) 6   (d) 12

81. Steel or aluminum cable tray systems shall be permitted to be used as equipment grounding conductors provided four requirements are met. One of those requirements is that the cable tray sections and fittings shall be identified for _____ purposes.
    (a) grounding   (b) special   (c) industrial   (d) all

82. For the ampacity of cables, rated 2,000 volts or less, in cable trays, the derating factors of Article 310, Note 8(a) of the Notes to Ampacity Tables of 0 to 2,000 Volts shall apply only to multiconductor cables with more than _____ current-carrying conductors.
    (a) one   (b) two   (c) three   (d) four

83. Where solid bottom cable trays contain multiconductor power or lighting cables, or any mixture of multiconductor power, lighting, control, and signal cables, the maximum number of _____ shall conform to three requirements.
    (a) cables   (b) conductors   (c) connections   (d) splices

84. Where ladder or ventilated trough cable trays contain single conductor cables (rated 2,000 volts or less), the maximum number of _____ conductors shall conform to four requirements.
    (a) single   (b) double   (c) triple   (d) quadruple

85. The messenger shall be supported at dead ends and at intermediate locations so as to eliminate _____ on the conductors.
    (a) static   (b) magnetism   (c) tension   (d) induction

86. The top shield installed over all floor-mounted Type FCC cable shall completely _____ all cable runs, corners, connectors, and ends.
    (a) cover   (b) encase   (c) protect   (d) none of these

87. Table 310–71 provides ampacities of an insulated three-conductor copper cable isolated in air, based on conductor temperature of 90ºC (194ºF) and ambient air temperature of 40ºC (104ºF). If the conductor size is 4/0 AWG, and the voltage range is 2,001 to 5,000, then the ampacity is _____ amperes.
    (a) 250   (b) 285   (c) 320   (d) 325

Random Sequence, Section 300–4(a) Through Section 511–2

88. Cables laid in wood notches require protection against nails or screws by a steel plate at least _____ inch thick.
    (a) ½   (b) ¼   (c) 1/16   (d) ¾

89. Cable _____ made and insulated by approved methods shall be permitted to be located within a cable tray provided they are accessible and do not project above the side rails.
    (a) connections   (b) jumpers   (c) splices   (d) conductors

90. Where buildings exceed three stories or 50 feet in height, overhead lines shall be arranged, where practicable, so that a clear space (or zone) at least 6 feet wide will be left either adjacent to the buildings or beginning not over 8 feet from them to facilitate the _____ when necessary for fire fighting.
    (a) safety of persons (b) entrance of water (c) raising of ladders   (d) all of these

91. Parking garages used for parking or storage and where no repair work is done except for exchange of parts and routine maintenance requiring no use of electrical equipment, open flame, welding, or the use of volatile flammable liquids are not classified as hazardous locations.
    (a) True   (b) False

92. Lighting track fittings shall be permitted to be equipped with general-purpose receptacles.
    (a) True   (b) False

93. In a concealed flexible metal conduit installation, _____ connectors shall not be used.
    (a) straight   (b) angle   (c) grounding type   (d) none of these

94. Electrical nonmetallic tubing must be securely fastened in place every _____ inches.
    (a) 12   (b) 18   (c) 24   (d) 36

95. Rigid metal conduit shall be permitted to be installed in concrete, in direct contact with the earth, or in areas subject to severe corrosive influences where protected by _____ and judged suitable for the condition.
    (a) ceramic   (b) corrosion protection
    (c) PVC   (d) orangeburg

96. Flat cable assembly shall consist of _____ conductors.
    (a) 2   (b) 3   (c) 4   (d) any of these

97. Where electrical nonmetallic tubing is installed concealed in walls, floors, and ceilings of buildings exceeding three floors above grade, a barrier must have a minimum _____ minute finish rating as listed for fire rated assemblies.
    (a) 5   (b) 10   (c) 15   (d) 30

98. UF cable is permitted for _____.
    (a) interior use   (b) wet locations (c) direct burial   (d) all of these

99. Rigid metal conduit _____ used in or under cinder fill where subject to permanent moisture.
    (a) shall be   (b) shall not be   (c) should be   (d) is the only conduit

100. The National Electrical Code permits ENT to be installed in _____ if used with fittings listed for the purpose.
    (a) wet locations indoors   (b) in a concrete slab on or below grade
    (c) direct earth burial   (d) a and b

101. Rigid nonmetallic conduit shall be securely fastened within _____ inches of each box.
    (a) 6   (b) 24   (c) 12   (d) 36

102. Nonmetallic extensions shall be secured in place by approved means at intervals not exceeding _____ inches.
    (a) 6   (b) 8   (c) 10   (d) 16

103. How many No. 12 XHHW conductors, not counting a bare ground wire, are allowed in a 3/8 inch flexible metal conduit (maximum of six feet) with outside fittings?
    (a) 4   (b) 3   (c) 2   (d) 5

104. Metal poles used for the support fixtures must be bonded to a(n) _____.
    (a) grounding electrode
    (b) grounded conductor
    (c) equipment grounding conductor
    (d) any of these

105. Nonmetallic sheath cable must closely follow the surface of the building finish or running boards when run exposed.
    (a) True   (b) False

106. Type MI cable shall be securely supported at intervals not exceeding _____ feet.
    (a) 3   (b) 3½   (c) 5   (d) 6

107. Materials such as straps, bolts, screws, etc., associated with the installation of intermediate metal conduit in concrete, direct contact with the earth are required to be _____.
    (a) weatherproof   (b) weathertight   (c) corrosion resistant (d) none of these

108. • Electrical metallic tubing shall not be threaded.
    (a) True   (b) False

109. Rigid nonmetallic conduit shall not be used:
    I. In hazardous (classified) locations.
    II. For the support of fixtures or other equipment.
    III. Where subject to physical damage unless identified for such use.
    IV. Where subject to ambient temperatures exceeding those for which the conduit is approved.
    (a) I and II (b) II and IV (c) I, II and IV (d) All of these

110. Two conductor nonmetallic sheath cables cannot be stapled on edge.
    (a) True (b) False

111. A run of intermediate metal conduit shall not contain more than the equivalent of _____ quarter bends including the offsets located immediately at the outlet or fitting.
    (a) 1 (b) 2 (c) 3 (d) 4

112. The ampacity of NM and NMC cable shall be that of _____ conductors.
    I. 90ºC II. 75°C III. 60ºC
    (a) I only (b) II only (c) III only (d) I, II, or III

113. In general, the voltage limitation between conductors in a surface metal raceway shall not exceed _____ volts unless the metal has a thickness listed for higher voltage.
    (a) 300 (b) 150 (c) 600 (d) 1,000

114. It shall be permissible to extend a metal multioutlet assembly through (not run within) dry partitions, if arrangements are made for removing the cap or cover on all exposed portions and no outlet is located within the partitions.
    (a) True (b) False

115. Water heaters having a capacity of _____ gallons or less shall have a branch circuit rating not less than 125% of the rating of the water heater.
    (a) 60 (b) 75 (c) 90 (d) 120

116. Rigid nonmetallic conduit and fittings used above ground shall be resistant to _____.
    (a) moisture and chemical atmospheres
    (b) low temperatures and sunlight
    (c) distortion from heat
    (d) all of these

117. Lighting track shall not be installed less than _____ feet above the finished floor except where protected from physical damage or track operating at less than 30 volts RMS open-circuit voltage.
    (a) 4½ (b) 5 (c) 5½ (d) 6

118. One inch intermediate metal conduit must be supported every _____ feet.
    (a) 8 (b) 10 (c) 12 (d) 14

119. Electrical metallic tubing shall not be used in cinder concrete or cinder fill where subject to permanent moisture unless protected on all sides by a layer of noncinder concrete at least _____ inches thick or unless the tubing is at least _____ inches under the fill.
    (a) 2, 18 (b) 2½, 18 (c) 2, 12 (d) 2½, 12

120. Bends in Type MI cable shall be so made as not to _____ the cable.
    (a) damage (b) shorten (c) both a and b (d) none of these

121. For cellular concrete floor raceways, junction boxes shall be _____ the floor grade and sealed against the free entrance of water or concrete.
    (a) leveled to (b) above (c) below (d) perpendicular to

122. Nonmetallic underground conduit with conductors shall not be used _____.
    (a) in exposed locations
    (b) inside buildings
    (c) in hazardous (classified) locations
    (d) all of these

123. An autotransformer which is used to raise the voltage to more than 300 volts, as part of a ballast for supplying lighting units, shall be supplied by a(n) _____ system.
    (a) system (b) grounded (c) listed (d) identified

124. Exposed conductive parts of lighting fixtures shall be _____.
    (a) grounded (b) painted (c) bonded (d) a and b

125. When an outlet from an underfloor raceway is discontinued, the circuit conductors supplying the outlet _____.
    (a) may be spliced
    (b) may be reinsulated
    (c) may be handled like abandoned outlets on loop wiring
    (d) shall be removed from the raceway

126. Ballasts for fluorescent electric discharge lighting installed indoors must have _____ protection.
    (a) short circuit (b) overcurrent (c) integral thermal (d) none of these

127. Type HPD cord is permitted for _____ usage.
    (a) not hard (b) hard (c) extra hard (d) all of these

128. Cords shall be protected against overcurrent according to Section _____.
    (a) 240–2 (b) 240–4 (c) 240–6 (d) 240–8

129. No conductor larger than a No. _____ shall be installed in a cellular metal floor raceway.
    (a) 1/0 (b) 4/0 (c) 250 (d) no restriction

130. Panelboards supplied by a 4-wire delta 3-phase system, shall have the high leg conductor connected to the _____ phase.
    (a) A (b) B (c) C (d) any of these

131. Lampholders installed in wet or damp locations shall be of the _____ type.
    (a) waterproof (b) weatherproof (c) moistureproof (d) moisture resistant

132. The individual conductors in a cablebus shall be supported at intervals not greater than _____ feet for vertical runs.
    (a) ½ (b) 1 (c) 1½ (c) 2

133. Flat cable assemblies shall not be installed _____.
    (a) where subject to corrosive vapors unless suitable for the application
    (b) in hoistways
    (c) in any hazardous (classified) location
    (d) all of these

134. TC cable shall be permitted to be used _____.
    (a) for power and lighting circuits
    (b) in cable trays in hazardous locations
    (c) in Class 1 control circuits
    (d) all of these

135. Receptacles installed outdoors, or in other wet locations, must be _____ when the attachment plug cap is inserted.
    (a) weatherproof (b) weathertight (c) rainproof (d) raintight

Random Sequence, Article 100 Through Chapter 9

136. Central heating equipment, other than fixed electrical space heating equipment, shall be required to be supplied by a(n) _____ branch circuit.
    (a) multiwire (b) individual
    (c) multipurpose (d) small appliance branch circuit

137. All extensions from flat cable assemblies shall be made by approved wiring methods, within the _____, installed at either end of the flat cable assembly runs.
    (a) end-caps (b) junction boxes
    (c) surface metal raceway (d) underfloor metal raceway

138. Enclosures for switches or circuit breakers shall not be used as _____.
    (a) junction boxes (b) auxiliary gutters (c) raceways (d) all of these

139. Power-limited fire alarm circuit cables installed within buildings in an air-handling space shall be of Type _____.
    (a) FPL (b) CL3P (c) OFNP (d) FPLP

140. For wall-mounted ovens and counter-mounted cooking units, a separable connector or a plug and receptacle combination in the supply line to the oven or cooking unit shall be approved for the _____ of the space in which it is located.
    (a) size (b) temperature (c) conditions (d) use

141. • Personnel doors for transformer vaults shall _____ and be equipped with panic bars, pressure plates, or other devices that are normally latched but open under simple pressure.
    (a) be clearly identified (b) swing out (c) a and b (d) a or b

142. For phase converters, the ampacity of the single-phase supply conductors shall not be less than _____ percent of the phase converter nameplate single-phase input full-load amperes.
    (a) 75 (b) 100 (c) 125 (d) 150

143. When installing duct heaters, sufficient clearance shall be maintained to permit replacement and adjustment of controls and heating elements.
    (a) True (b) False

144. Resistors and reactors (over 600 volts, nominal) shall be isolated by _____ to protect personnel from accidental contact with energized parts.
    (a) an enclosure (b) elevation (c) a or b (d) a and b

145. Capacitors shall be _____ so that persons cannot come into accidental contact or bring conducting materials into accidental contact with exposed energized parts, terminals, or busses associated with them.
    (a) enclosed (b) located (c) guarded (d) any of these

146. The motor branch-circuit short-circuit and ground-fault protective device shall be capable of carrying the _____ current of the motor.
    (a) varying (b) starting (c) running (d) continuous

147. The branch circuit conductor and overcurrent protection device for fixed electric space heating equipment loads shall not be smaller than _____ percent of the total load.
    (a) 80 (b) 100 (c) 125 (d) 150

148. • Conductors, including splices and taps, shall not fill the auxiliary gutter to more than _____ percent of its area.
    (a) 30 (b) 40 (c) 60 (d) 75

149. Type UF cable shall not be used _____.
    (a) in any hazardous (classified) location
    (b) embedded in poured cement, concrete, or aggregate
    (c) where exposed to direct rays of the sun, unless identified as sunlight-resistant
    (d) all of these

150. Preassembled cable in nonmetallic conduit larger than _____ inch(es) electrical trade size shall not be used.
    (a) 1 (b) 2 (c) 3 (d) 4

151. Each receptacle of dc plugging boxes shall be rated at not _____ amperes when used on a stage or set of a motion picture studio.
    (a) more than 30 (b) less than 20 (c) less than 30 (d) more than 20

152. A generator set for a required standby system shall _____.
    (a) have means for automatically starting the prime movers
    (b) have two hours full-demand fuel supply if an internal combustion engine
    (c) not be solely dependent on public utility gas system
    (d) all of these

153. Each arc welder shall have overcurrent protection rated or set at not more than _____ percent of the rated primary current of the welder.
    (a) 100 (b) 125 (c) 150 (d) 200

154. All receptacles in dressing rooms of theaters shall be _____.
    (a) controlled by wall switches with pilot lights
    (b) controlled by wall switches located within the room
    (c) limited to a 30 ampere branch circuit
    (d) a and b

155. • Emergency lighting, emergency power, or both in a building or group of buildings will be available within the time period required for the application, but not to exceed _____ seconds.
    (a) 5 (b) 10 (c) 30 (d) 60

156. Where used for sound-recording and reproduction, conductors in wireways and auxiliary gutters shall not fill the raceway to more than _____ percent.
    (a) 20 (b) 58 (c) 50 (d) 75

157. Each commercial building and occupancy with ground floor access for pedestrians shall have at least one outside sign outlet. The outlet shall be supplied by a _____ ampere branch circuit that supplies no other loads.
    (a) 15 (b) 20 (c) either a or b (d) neither a nor b

158. The pilot light provided within the portable stage switchboard enclosure shall have overcurrent protection rated or set an not over _____ amperes.
    (a) 10 (b) 15 (c) 20 (d) 30

159. A switch for the control of parking lights in a theater may be permitted to be installed inside the projection booth.
    (a) True (b) False

160. When calculating the ampacity of conductors that supply a motor generator arc welder the multiplier or demand factor will be _____ when the welder has a one hour duty cycle.
    (a) .78 (b) 1.00 (c) .89 (d) .80

161. A disconnecting means for X-ray equipment shall have adequate capacity for at least _____.
    (a) 50 percent of the input required for the momentary rating of equipment
    (b) 100 percent of the input required for the momentary rating of equipment
    (c) 50 percent of the input requirement for the long time rating
    (d) 125 percent of the input requirement

162. Low-voltage equipment in anesthetic areas that are frequently in contact with the bodies of persons or have exposed current carrying elements shall _____.
    (a) operate on an electrical potential of 10 volts or less
    (b) be moisture resistant
    (c) be intrinsically safe or double-insulated
    (d) all of these

163. Receptacles located within _____ feet of the inside walls of a pool shall be protected by a ground-fault circuit-interrupter.
    (a) 8 (b) 10 (c) 15 (d) 20

164. A generator output circuit shall be isolated from ground, except where the capacitive coupling inherent in the generator causes the generator terminals to have voltages from terminal to ground that are equal.
    (a) True (b) False

165. Fire alarm signaling systems include _____.
    (a) fire alarm
    (b) guard tour
    (c) sprinkler water flow
    (d) all of these

166. When considering the number of conductors permitted when calculating conduit or tubing fill, the equipment grounding or bonding conductors, when installed shall _____.
(a) not be counted
(b) have the actual dimensions used
(c) not be counted if in a nipple
(d) not be counted if for a wye 3-phase balanced load

167. • The transmitter enclosure of a radio or TV station shall have interlocks to disconnect the power to the transmitter in the enclosure access door if the voltage between conductors is over _____ volts.
(a) 150 (b) 250 (c) 350 (d) 480

168. The minimum size of a receiving station outdoor antenna conductor, where the span is 75 feet shall be at least No. _____ if a copper-clad steel conductor is used.
(a) 10 (b) 12 (c) 14 (d) 17

169. Stainless steel, copper, or copper alloy clamps approved for direct burial use can be used to connect the bonding conductor to the common bonding grid of a swimming pool.
(a) True (b) False

170. Since Class 3 control circuits permit higher allowable levels of voltage and current, additional _____ are specified to provide protection against the electric shock hazard that could be encountered.
(a) circuits (b) safeguards (c) conditions (d) requirements

171. The feeder for six 20-ampere receptacles supplying shore power shall be calculated at _____ percent of the sum of the rating of the receptacles.
(a) 70 (b) 80 (c) 90 (d) 100

172. Examples of "Places of Assembly" could include (but are not limited to) _____.
I. restaurants   II. conference rooms   III. pool rooms
(a) I and II (b) I and III (c) II and III (d) I, II, and III

173. The minimum cover required for a direct buried cable for a 23-kV cable is _____ inches.
(a) 18 (b) 24 (c) 36 (d) 42

174. The authority having jurisdiction may judge a location utilized for _____ as a "Non-hazardous Location" providing the conditions of the Code are met.
(a) drying or curing (b) dipping and coating
(c) spraying operations (d) none of these

175. Lighting fixture or ceiling fans mounted less than 12 feet above the water level cannot be installed within _____ feet of an outdoor pool, fountain, or spa.
(a) 3 (b) 5 (c) 10 (d) 8

176. High-voltage conductors in tunnels shall be installed in _____.
(a) rigid conduit (b) MC cable
(c) other metal raceways (d) any of these

177. A storable swimming pool is capable of holding water to a maximum height of _____ inches.
(a) 24 (b) 36 (c) 42 (d) 48

178. In a residential mobile home, at least _____ inches of free conductor shall be left at each box except where conductors are intended to loop without joints.
(a) 4 (b) 6 (c) 8 (d) none of these

179. Where overcurrent protection is provided as part of the industrial machine, the machine shall be marked _____.
(a) overcurrent protection provided at machine supply terminals
(b) this unit contains overcurrent protection
(c) fuses or circuit breaker enclosed
(d) with the overcurrent protection category and type

180. Service equipment for floating docks or marinas must be located _____ the floating structure.
    (a) next to (b) on (c) in (d) all of these

181. Ground-fault protection in a mobile home is required for _____.
    (a) outlets installed outdoors
    (b) receptacles within 6 feet of a lavatory
    (c) a receptacle within a bathroom light fixture
    (d) all of these

182. Communication wires and cables shall be separated at least 2 inches from conductors of _____ circuits.
    I. power   II. lighting   III. Class 1
    (a) I only (b) II only (c) III only (d) I, II and III

183. The National Electrical Code specifies wiring methods for prefabricated buildings (manufactured buildings).
    (a) True (b) False

184. Each unit of a data processing system for use on a branch circuit shall have a nameplate that includes the _____.
    (a) rating in volts
    (b) operating frequency
    (c) total load in amperes
    (d) all of these

185. Duty on escalator motors shall be classed as _____.
    (a) full time (b) continuous (c) various (d) long term

186. Storage batteries used with sound recording equipment shall have leads that are _____ covered.
    I. lead II. rubber III. thermoplastic IV. mineral
    (a) I or III (b) I or IV (c) II or III (d) II or IV

187. Forming shells for wet-niche lighting fixtures must be installed with the top level of the fixture lens at least _____ inches below the normal water level of the pool or spa.
    (a) 6 (b) 12 (c) 18 (d) 24

188. The alternate source for emergency systems _____ be required to have ground-fault protection of equipment.
    (a) shall (b) shall not

189. Where internal combustion engines are used as the prime-mover for an emergency system or legally required stand-by system, an on-site fuel supply shall be provided with an on-premise fuel supply for not less than _____ hours full demand.
    (a) 2 (b) 3 (c) 4½ (d) 5

190. A GFCI protected 125-volt convenient receptacle must be located within _____ feet, but not less than 5 feet from the inside wall of a indoor spa or hot tub.
    (a) 3 (b) 5 (c) 10 (d) 20

191. Extensions from wireways are not permitted.
    (a) True (b) False

192. Round boxes shall not be used with any wiring method connector where a locknut or bushing is connected to the side of the box.
    (a) True (b) False

193. Where practical, dissimilar metals in contact anywhere in the system shall be avoided to prevent _____.
    (a) corrosion (b) galvanic action (c) shorts (d) none of these

194. The ampacity of the phase conductors from the generator terminals to the first overcurrent device shall not be less than _____ percent of the nameplate rating of the generator.
    (a) 75 (b) 115 (c) 125 (d) 140

195. Where threadless couplings and connectors used with rigid metal conduit are installed in wet locations, they shall be of the _____ type.
    (a) raintight

(b) wet and damp location
(c) nonabsorbent
(d) weatherproof

196. Open motors having commutators shall be located or protected so that sparks cannot reach adjacent combustible material, but this shall not prohibit the installation of these motors _____.
(a) on wooden floors
(b) over combustible fiber
(c) under combustible material
(d) none of these

197. The conductors, including splices and taps in a metal surface raceway shall not fill the raceway to more than _____ percent of its area at that point.
(a) 75   (b) 40   (c) 38   (d) 53

198. 1 inch rigid nonmetallic conduit must be secured every _____ feet.
(a) 2   (b) 3   (c) 4   (d) 6

199. Flexible metal conduit shall be secured _____.
(a) at intervals not exceeding 4½ feet
(b) within 12 inches on each side of a box where fished
(c) where fished
(d) at lengths not exceeding 3 feet at motor terminals

200. All theater fixed stage switchboards that are not completely enclosed dead front and dead rear or recessed into a wall, shall be provided with a metal hood extending the full length of the board to protect all equipment from falling objects.
(a) True   (b) False

Unit 11

Commercial Load Calculations

OBJECTIVES

After reading this unit, the student should be able to briefly explain the following concepts:

Part A – General
Conductor overcurrent protection [240–3]
Conductor ampacity
Fraction of an ampere
General requirements
Part B – Loads
Air conditioning
Dryers
Electric heat

Kitchen equipment
Laundry circuit
Lighting demand factors
Lighting without demand factors
Lighting miscellaneous
Multioutlet assembly [220–3(c) Exception No. 1]
Receptacles [220–3(b)(7)]
Banks and offices general lighting and receptacles

Signs [600–6(c)]
Part C – Load Calculations
Marina [555–5]
Mobile home park [550–22]
Recreational vehicle park [551–73]
Restaurant – Optional method Section 220–36
School – Optional method Section 220–34

After reading this unit, the student should be able to briefly explain the following terms:

General lighting
General lighting demand factors
Nonlinear loads
VA rating

Ampacity
Continuous load
Next size up protection device
Overcurrent protection

Rounding
Standard ampere ratings for overcurrent protection devices
Voltages

PART A – GENERAL

11–1 GENERAL REQUIREMENTS

Article 220 provides the requirement for branch circuits, feeders and services. In addition to this Article, other Articles are applicable such as Branch Circuits–210, Feeders–215, Services–230, Overcurrent Protection–240, Wiring Methods–300, Conductors–310, Appliances–422, Electric Space-Heating Equipment–424, Motors–430, and Air-Conditioning–440.

11–2 CONDUCTOR AMPACITY [ARTICLE 100]

The ampacity of a conductor is the rating, in amperes, that a conductor can carry continuously without exceeding its insulation temperature rating [NEC Definition – Article 100]. The allowable ampacities as listed in Table 310–16 are affected by ambient temperature, conductor insulation, and conductor bunching or bundling [310–10], Fig. 11–1.

Continuous Loads
Conductors are sized at 125 percent of the continuous load before any derating factor and the overcurrent protection devices are sized at 125 percent of the continuous loads [210–22(c), 220–3(a), 220–10(b), and 384–16(c)].

Chapter 3 Advanced NEC Calculations And Code Questions

Unit 11 Commercial Load Calculations

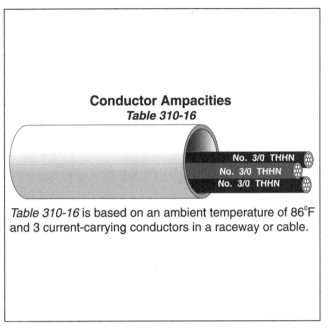

Figure 11-1
Conductor Ampacities – Table 310–16

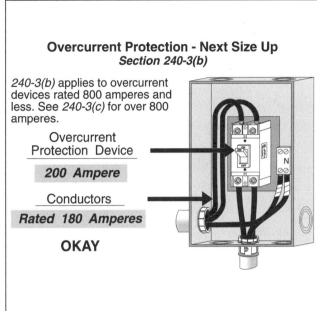

Figure 11-2
Overcurrent Protection – Next Size Up

11–3 CONDUCTOR OVERCURRENT PROTECTION [240–3]

The purpose of overcurrent protection devices is to protect conductors and equipment against excessive or dangerous temperatures [240–1 FPN]. There are many rules in the National Electrical Code for conductor protection and there are many different installation applications where the general rule of protecting the conductor at its ampacity does not apply. Examples would be motor circuits, air-conditioning, tap conductors, etc. See Section 240–3 for specific rules on conductor overcurrent protection.

Next Size Up Okay [240–3(b)]

If the ampacity of a conductor does not correspond with the standard ampere rating of a fuse or circuit breaker, as listed in Section 240–6(a), the next size up protection device is permitted. This practice only applies if the conductors do not supply multioutlet receptacles and if the next size up overcurrent protection device does not exceed 800 amperes [Note 9 of Table 310–16], Fig. 11–2.

Standard Size Overcurrent Devices [240–6(a)]

The following is a list of some of the standard ampere ratings for fuses and inverse time circuit breakers: 15, 20, 25, 30, 35, 40, 45, 50, 60, 70, 80, 90, 100, 110, 125, 150, 175, 200, 225, 250, 300, 350, 400, 500, 600, 800, 1,000, 1,200, and 1,600 amperes.

11–4 VOLTAGES [220–2]

Unless other voltages are specified, branch-circuit, feeder, and service loads shall be computed at a nominal system voltage of 120, 120/240, 208Y/120, 240, 480Y/277, 480 volts, or 600Y/347, Fig. 11–3.

11–5 FRACTION OF AN AMPERE

There is no specific NEC rule for rounding fractions of an ampere but Chapter 9, Part B, Examples, contains the following note, "except where the computations result in a major fraction of an ampere (0.5 or larger), such fractions may be dropped." So for all practical purposes, we should follow this practice.

PART B – LOADS

11–6 AIR CONDITIONING

Branch Circuit [440–22(a) And 440–32]

Branch circuit conductors that supply air-conditioning equipment must be sized no less than 125 percent of the air conditioner rating [440–32], and the protection is sized between 175 percent and up to 225 percent of the air-conditioning rating.

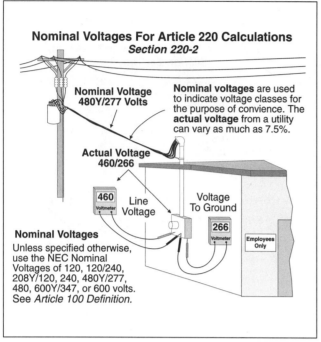

Figure 11-3
Nominal Voltage

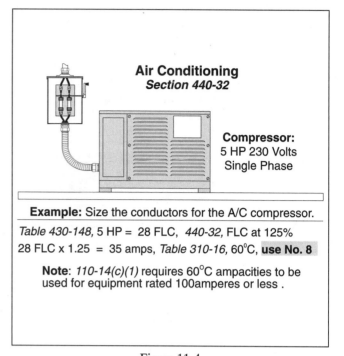

Figure 11-4
Air Conditioning – Branch Circuit Sizing Example

❑ **Air Conditioning Branch Circuit Example**

What size branch circuit is required for a 5-horsepower, 230 volt air conditioner that has a current rating of 28 amperes, Fig. 11–4?

(a) No. 10   (b) No. 8   (c) No. 14   (d) none of these

• Answer: (b) No. 8

Branch Circuit Conductor: The ampere rating of a single-phase 5-horsepower motor is 28 amperes. Always assume single-phase unless the question specifies three-phase. The branch circuit conductors are sized no less than 125 percent of 28 amperes, 28 amperes × 1.25 = 35 amperes, which requires a No. 8 at 60ºC [Table 310–16 and 110–14(c)].

Branch Circuit Protection: The branch circuit protection is sized from 175 percent up to 225 percent of the ampere rating, 28 amperes × 1.75 = 49 amperes up to 28 amperes × 2.25 = 63 amperes.

Feeder or Service Conductors [440–34]

Feeder circuit conductors that supply air-conditioning equipment must be sized no less than 125 percent of the largest air conditioner VA rating plus 100 percent of the other air conditioners' VA ratings.

VA Rating – The VA rating of an air conditioner is determined by multiplying the voltage rating of the unit by its ampere rating. For the purpose of this book, we will determine the unit ampere rating by using motor full-load current ratings as listed in Article 430, Table 430–148, and Table 430–150 (three-phase).

❑ **Air Conditioning Feeder/Service Conductor Example**

What is the demand load required for the air-conditioning equipment of a twelve-unit office building where each unit contains one air conditioner rated 5-horsepower, 230 volts?

(a) 77 kVA   (b) 79 kVA   (c) 45 kVA   (d) none of these

• Answer: (b) 79 kVA

The VA of a single-phase 5-horsepower motor rated 230 volts is calculated as volts times amperes, 230 volts × 28 amperes = 6,440 VA. The feeder demand load for air-conditioning is calculated at no less than 125 percent of the largest air conditioner plus 100 percent of all other units. [(6,440 VA × 1.25) + (6,440 VA × 11 units)] = 78,890 VA/1,000 = 79 kVA.

Note. Section 220–21 of the NEC permits the smaller of the air-conditioning or heat load to be omitted for calculation purposes.

11-7 DRYERS

The branch circuit conductor and overcurrent protection device for commercial dryers is sized to the appliance nameplate rating. The feeder demand load for dryers is calculated at 100 percent of the appliance rating. Section 220-18 demand factors do not apply to commercial dryers.

☐ **Dryer Branch Circuit Example**

What size branch circuit conductor and overcurrent protection is required for a 7-kW dryer rated 240 volts when the dryer is located in the laundry room of a multifamily dwelling, Figure 11-5?

(a) No. 12 with a 20-ampere breaker

(b) No. 10 with a 20-ampere breaker

(c) No. 12 with a 30-ampere breaker

(d) No. 10 with a 30-ampere breaker

- Answer: (d) No. 10 with a 30-ampere breaker.
  The ampere rating of the dryer is: I = VA/E
  I = 7,000 VA/240 volts = 29 amperes.

The ampacity of the conductor and overcurrent device must not be less than 29 amperes [240-3]

Table 310-16, No. 10 conductor at 60°C is rated 30-amperes protected with a 30-ampere protection device [240-6(a)].

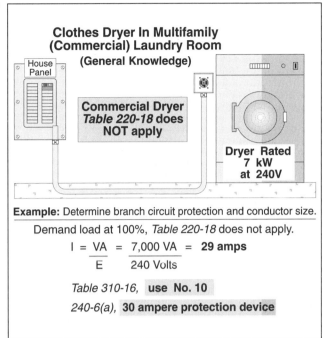

Figure 11-5
Clothes Dryer – Branch Circuit Sizing Example

☐ **Dryer Feeder Example**

What is the service demand load for ten 7-kW dryers located in a laundry room?

(a) 70 kVA (b) 52.5 kVA (c) 35 kVA (d) none of these

- Answer: (a) 70 kVA
  The NEC does not permit a demand factor for commercial dryers, therefore the dryer demand load must be calculated at 100 percent; 7 kW × 10 units = 70 kVA.

Note. If the dryers are on continuously, the conductor and protection device must be sized at 125 percent of the load [210-22(c), 220-3(a), 220-10(b), and 384-16(c)].

11-8 ELECTRIC HEAT

Branch Circuit Sizing [424-3(b)]

Branch circuit conductors and the overcurrent protection device for electric heating shall be sized not be less than 125 percent of the total heating load, including blower motors.

☐ **Heating Branch Circuit Example**

What size conductor and protection is required for a three-phase 240-volt, 15-kW heat strip with a 5.4-ampere blower motor, Fig. 11-6?

(a) No. 10 with 30-ampere protection

(b) No. 6 with 50-ampere protection

(c) No. 8 with 40-ampere protection

(d) No. 6 with 60-ampere protection

- Answer: (d) No. 6 with 60-ampere protection [424-3(b)]
  I = VA/(E × $\sqrt{3}$), I = 15,000 VA/(240 volts × 1.732) = 36 amperes. The conductors and protection must not be less than 125 percent of [36 amperes (heat) + 5.4 amperes (motor)] = 41.4 amperes × 1.25 = 51.75 amperes. The conductors are sized to the 60°C column ampacities of Table 310-16 [110-14(c)(1)], which is a No. 6 rated 55 amperes. The overcurrent protection device must be sized no less than 52 amperes, which is a 60-ampere device [240-6(a)].

Heating Feeder/Service Demand Load [220-15]

The feeder and service demand load for electric heating equipment is calculated at 100 percent of the total heating load.

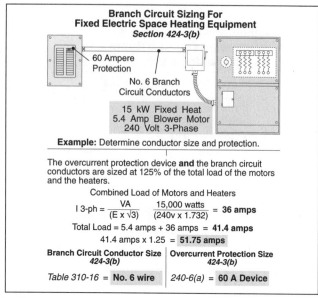

Figure 11-6

Fixed Electric Heat – Branch Circuit Sizing Example

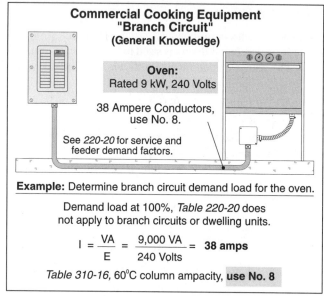

Figure 11-7

Commercial Cooking Equipment – Branch Circuit Sizing Example

❑ **Heating Feeder/ Service Demand Load Example**

What is the feeder and service demand load for a building that has seven three-phase 208 volt, 10-kW heat strips with a 5.4-ampere blower motor (1,945 VA) for each unit?

(a) 84 kVA (b) 53 kVA (c) 129 kVA (d) 154 kVA

• Answer: (a) 84 kVA

10,000 VA + 1,945 VA = 11,945 VA × 7 units = 83,615 VA/1,000 = 83.615 kVA.

Note. Section 220–21 of the NEC permits the smaller of the air-conditioning or heat load to be omitted for feeder and service calculations.

11–9 KITCHEN EQUIPMENT

Branch Circuit

Branch circuit conductors and overcurrent protection for commercial kitchen equipment are sized according to the appliance nameplate rating.

❑ **Kitchen Equipment Branch Circuit Example No. 1**

What is the branch circuit demand load (in amperes) for one 9-kW oven rated 240 volts, Fig. 11–7?

(a) 38 amperes (b) 27 amperes (c) 32 amperes (d) 33 amperes

• Answer: (a) 38 amperes

I = VA/E

I = 9,000 VA/240 volts = 38 amperes

❑ **Kitchen Equipment Branch Circuit Example No. 2**

What is the branch circuit load for one 14.47-kW range rated 208 volts, three-phase?

(a) 60 amperes (b) 40 amperes (c) 50 amperes (d) 30 amperes

• Answer: (b) 40 amperes

I = VA/(E × 1.732)

I = 14,470 VA/(208 volts × 1.732) = 40 amperes

Kitchen Equipment Feeder/Service Demand Load [220–20]

The service demand load for thermostatic control or intermittent use commercial kitchen equipment is determined by applying the demand factors from Table 220–20, to the total connected kitchen equipment load. *The feeder or service demand load cannot be less than the two largest appliances!*

Note. The demand factors of Table 220–20 do not apply to space heating, ventilating, or air–conditioning equipment.

❑ **Kitchen Equipment Feeder/Service Example No. 1**
What is the demand load for the following kitchen equipment loads, Fig. 11–8?

Water heater	5.0 kW	Booster heater	7.5 kW
Mixer	3.0 kW	Oven	5.0 kW
Dishwasher	1.5 kW	Disposal	1.0 kW

(a) 15 kW   (b) 23 kW
(c) 12.5 kW   (d) none of these

• Answer: (a) 15 kW

Water heater	5 kW
Booster heater	7.5 kW
Mixer	3 kW
Oven	5 kW
Dishwasher	1.5 kW
Disposal	+ 1 kW
Total connected	23 kW

The demand factor for six loads is 65 percent

23 kW × .65 = 14.95 kW.

The demand load cannot be less than the two largest appliances; 5 kW + 7.5 kW = 12.5 kW.

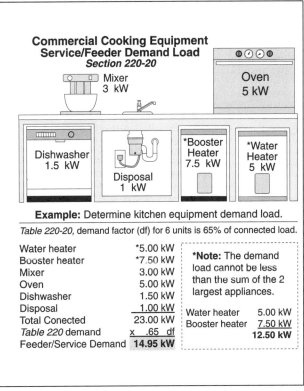

Figure 11-8

Commercial Cooking Equipment – Service Demand Load Example

❑ **Kitchen Equipment Feeder/Service Example No. 2**
What is the demand load for the following kitchen equipment loads?

Water heater	10 kW
Booster heater	15 kW
Mixer	4 kW
Oven	6 kW
Dishwasher	1.5 kW
Disposal	1 kW

(a) 24.4 kW   (b) 38.2 kW   (c) 25.0 kW   (d) 18.9

• Answer: (c) 25 kW*

Water heater	10 kW
Booster heater	15 kW
Mixer	4 kW
Oven	6 kW
Dishwasher	1.5 kW
Disposal	+ 1 kW
Total connected	37.5 kW

The demand factor for six appliances is 65 percent; 37.5 kW × .65 = 24.4 kW

* The demand load cannot be less than the two largest appliances; 10 kW + 15 kW = 25 kW

11–10 LAUNDRY EQUIPMENT

Laundry equipment circuits are sized to the appliance nameplate rating. For exam purposes, it is generally accepted that a laundry circuit is not considered a continuous load and assume all commercial laundry circuits to be rated 1,500 VA unless noted otherwise in the question.

❑ **Laundry Equipment Example**
What is the demand load for ten washing machines located in a laundry room, Fig. 11–9?

(a) 1,500 VA   (b) 15,000 VA   (c) 1,125 VA   (d) none of these

• Answer: (b) 15,000 VA

1,500 VA × 10 units = 15,000 VA

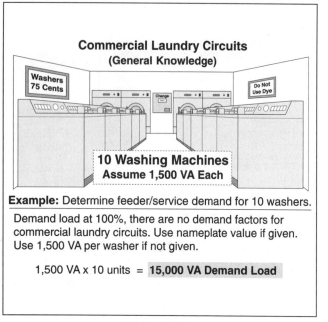

Figure 11-9
Commercial Laundry Service Demand Load Example

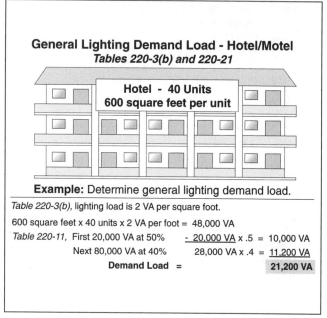

Figure 11-10
General Lighting Demand Factor – Hotel Example

11–11 LIGHTING – DEMAND FACTORS [TABLE 220–3(b) and 220–11]

The NEC requires a minimum service load per square foot for general lighting depending on the type of occupancy [Table 220–3(b)]. For the guest rooms of hotels and motels, hospitals, and storage warehouses, the general lighting demand factors of Table 220–11 can be applied to the general lighting load.

Hotel or Motel Guest Rooms – General Lighting

The general lighting demand load of 2-VA per square foot [Table 220–3(b)] for the guest rooms of hotels and motels is permitted to be reduced according to the demand factors listed in Table 220–11:

General Lighting Demand Factors

First 20,000 VA at 50 percent demand factor
Next 80,000 VA at 40 percent demand factor
Remainder VA at 30 percent demand factor

❑ **Hotel General Lighting Demand Example**

What is the general lighting demand load for a forty room hotel? Each unit contains 600 square feet of living area, Fig. 11–10.

(a) 48 kVA (b) 24 kVA (c) 20 kVA (d) 21 kVA

• Answer: (d) 21 kVA [Table 220–3(b) and Table 220–11]

```
40 units × 600 square feet =    24,000 square feet × 2 VA =   48,000 VA
First 20,000 VA at 50 %                          − 20,000 VA  × .5 =    10,000 VA
Next 80,000 VA at 40 %                             28,000 VA  × .4 =  + 11,200 VA
Total demand load                                                      21,200 VA
```

11–12 LIGHTING WITHOUT DEMAND FACTORS [TABLE 220–3(b) and 220–10(b)].

The general lighting load for commercial occupancies other than guest rooms of motels and hotels, hospitals, and storage warehouses is assumed continuous and shall be calculated at 125 percent [220–10(b)] of the general lighting load as listed in Table 220–3(b).

❑ **Store General Lighting Example**

What is the general lighting load load for a 21,000 square foot store, Fig. 11–11?

(a) 40 kVA (b) 63 kVA (c) 79 kVA (d) 81 kVA

• Answer: (c) 79 kVA

21,000 square feet × 3 VA × 1.25 = 78,750 VA

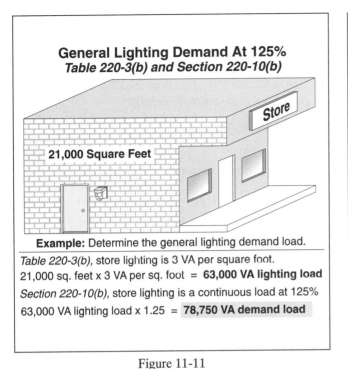

Figure 11-11
General Lighting – No Demand Factor – Store Example

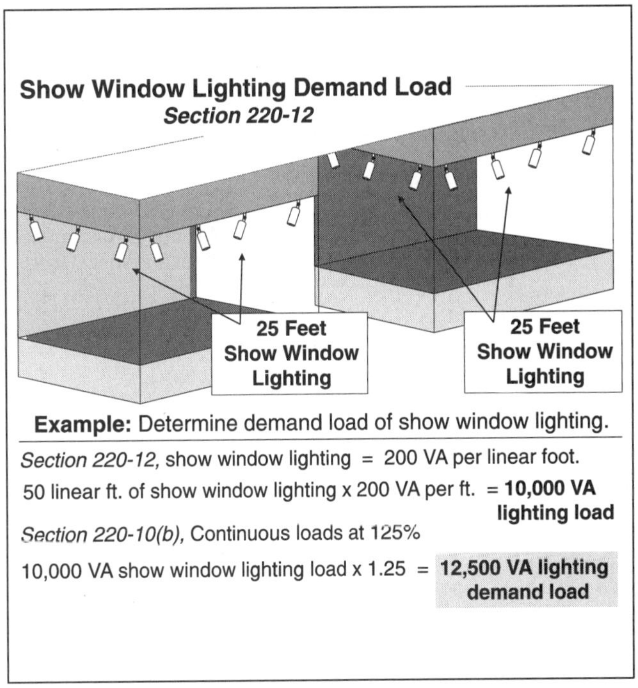

Figure 11-12
Show Window Demand Load Example

❑ **Club General Lighting Example**
What is the general lighting load for a 4,700 square foot dance club?

(a) 4,700 VA (b) 9,400 VA (c) 11,750 VA (d) 250 kVA

• Answer: (c) 11,750 VA

4,700 square feet × 2 VA × 1.25 = 11,750 VA

❑ **School General Lighting Example**
What is the general lighting load for a 125,000 square foot school?

(a) 125 kVA (b) 375 kVA (c) 475 kVA (d) 550 kVA

• Answer: (c) 475 kVA

125,000 square feet × 3 VA × 1.25 = 468,750 VA

11–13 LIGHTING – MISCELLANEOUS

Show-Window Lighting [220–12]

The demand load for each linear foot of show-window lighting shall be calculated at 200 VA per foot. Show-window lighting is assumed to be a continuous load; see Example 3 in the back of the NEC, also Section 220–3(c) Exception No. 3 for the requirements for show-window branch circuits.

❑ **Show-Window Load Example**
What is the demand load in kVA for 50 feet of show-window lighting, Fig. 11–12?

(a) 6 kVA (b) 7.5 kVA (c) 9 kVA (d) 12.5 kVA

• Answer: (d) 12.5 kVA

50 feet × 200 VA = 10,000 VA × 1.25 = 12,500 VA

11–14 MULTIOUTLET RECEPTACLE ASSEMBLY [220–3(c) EXCEPTION No. 1]

Each 5 feet, or fraction of a foot, of multioutlet receptacle assembly shall be considered to be 180 VA for service calculations. When a multioutlet receptacle assembly is expected to have a number of appliances used simultaneously, each foot, or fraction of a foot, shall be considered as 180 VA for service calculations. A multioutlet receptacle assembly is not generally considered to be a continuous load.

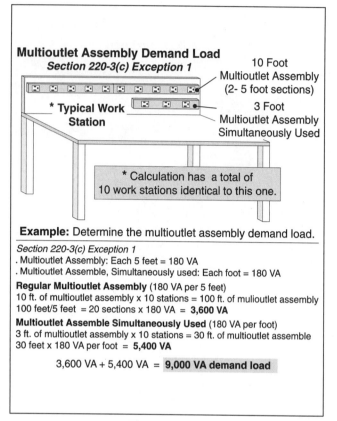

Figure 11-13
Multioutlet Assembly Demand Load Example

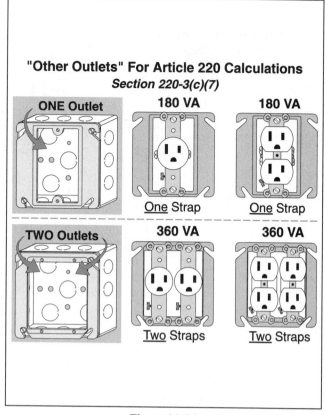

Figure 11-14
Receptacle Outlet – 180 VA

❏ **Multioutlet Receptacle Assembly Example**

What is the demand load for ten work stations that have 10 feet of multioutlet receptacle assembly and 3 feet of multioutlet receptacle assembly simultaneously used, Fig. 11–13?

(a) 5 kVA (b) 6 kVA (c) 7 kVA (d) 9 kVA

• Answer: (d) 9 kVA

100 feet/ 5 = 20 sections at 180 VA =   3,600 VA
30 feet/ 1 = 30 sections at 180 VA =   5,400 VA
                                          9,000 VA /1,000 = 9 kVA

11–15 RECEPTACLES VA LOAD [220–3(c)(7) and 220–13]

Receptacle VA Load

The minimum load for each commercial or industrial general-use receptacle outlet shall be 180 volt-amperes. Receptacles are generally not considered to be a continuous load, Fig. 11–14.

Number Of Receptacles Permitted On A Circuit

The maximum number of receptacle outlets permitted on a commercial or industrial circuit is dependent on the circuit ampacity. The number of receptacles per circuit is calculated by dividing the VA rating of the circuit by 180 VA for each receptacle strap.

❏ **Receptacles Per Circuit Example**

How many receptacle outlets are permitted on a 15-ampere, 120-volt circuit, Fig. 11–15?

(a) 10 (b) 13 (c) 15 (d) 20

• Answer: (a) 10 Receptacles [220–3(c)(7)]

The total circuit VA load for a 15-ampere circuit is 120 volts × 15 amperes = 1,800 VA.

The number of receptacle outlets per circuit = 1,800 VA/180 VA = 10 receptacles.

Note. Fifteen ampere circuits are permitted for commercial and industrial occupancies according to the National Electrical Code, but some local codes require a minimum 20-ampere rating for commercial and industrial circuits [310–5].

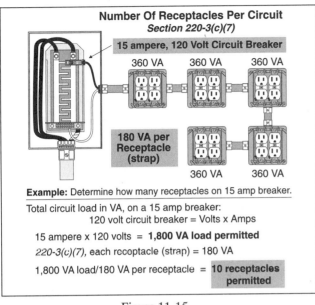

Figure 11-15

Number Of Receptacles Per Circuit Example

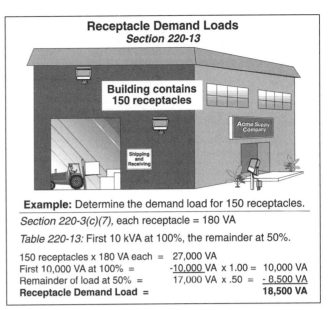

Figure 11-16

Receptacle Service Demand Load Example

Receptacle Service Demand Load [220–13]

The feeder and service demand load for commercial receptacles is calculated at 180 volt-amperes per receptacle strap [220–3(c)(7)]. The demand factors of Table 220–13 can be used for that portion of the receptacle load in excess of 10 kVA. Receptacle loads are generally not considered to be a continuous load.

Table 220–13 Receptacle Demand Factors:

First 10 kVA at 100 percent demand factor

Remainder kVA at 50 percent demand factor

❑ **Receptacle Service Demand Load Example**

What is the service demand load for one hundred and fifty, 20-ampere, 120 volt general-use receptacles in a commercial building, Fig. 11–16?

(a) 27 kVA   (b) 14 kVA   (c) 4 kVA   (d) none of these

• Answer: (d) none of these [220–13]

Total receptacle load	150 receptacles × 180 VA =	27,000 VA
First 10 kVA at 100%		− 10,000 VA × 1.00 = 10,000 VA
Remainder at 50%		17,000 VA × .50 = + 8,500 VA
Total receptacle demand load		18,500 VA/1,000 = 18.5 kVA

11–16 BANKS AND OFFICES GENERAL LIGHTING AND RECEPTACLES

Some testing agencies include the receptacle demand load to be part of the general lighting load for banks and offices. If that is the case, the general lighting demand load for banks and offices would be calculated as 3.5-VA per square foot times 125 percent for continuous lighting load, plus the receptacle demand load after applying Table 220–13 demand factors.

Receptacle Demand [Table 220–13]. The receptacle demand load is calculated at 180 volt-amperes for each receptacle outlet [220–3(c)(7)] if the number of receptacles are known, or one (1) VA per square foot if the number of receptacles are unknown.

❑ **Bank General Lighting And Receptacle Example**

What is the general lighting demand load (including receptacles) for an 18,000 square foot bank? The number of receptacles are unknown, Fig. 11–17.

(a) 68 kVA   (b) 110 kVA   (c) 84 kVA   (d) 93 kVA

• Answer: (d) 93 kVA

Lighting demand	78,750 VA*
Receptacle demand	+ 14,000 VA*
Total demand load	92,750 VA/1,000 = 92.75 kVA

* See next page for details.

Bank - General Lighting & Receptacle Demand Load
*Section 220-3(b) Note ***

Bank:
18,000 Square Feet
Number of Receptacles, Unknown

Example: Determine demand load for lighting and receptacles.

Table 220-3(b), Lighting load for a bank is 3 1/2 VA per square foot
18,000 square feet x 3 1/2 VA per foot = **63,000 VA lighting load**

Section 220-10(b), continuous loads at 125%
63,000 VA lighting load x 1.25 = **78,750 VA lighting demand load,** *but...*

...Table 220-3(b) **Note** requires an additional unit load of 1 VA per square foot when the number of receptacles is unknown.
18,000 square feet x 1 VA per foot = 18,000 VA receptacle load
Table 220-13, First 10,000 VA at 100% = 10,000 VA x 1 = 10,000 VA
Remainder at 50%   8,000 VA x .5 = 4,000 VA
Receptacle Demand Load = 14,000 VA

78,750 VA lighting + 14,000 VA receptacle = **92,750 VA demand load**

Figure 11-17
General Lighting Demand Load – Bank Example

Signs - Commercial Buildings, Pedestrian Access
Section 220-3(a) and 600-5

Section 600-5(a): Load for exterior outling lighting is 1,200 VA.

Commercial Building With Pedestrian Access

Example: Determine service demand load for exterior sign.

600-5(a) requires a 20 ampere branch circuit for an exterior sign.
600-5(b) states that the sign is computed at 1,200 VA.
220-10(b), continuous loads on feeders are calculated at 125%.

1,200 VA sign outlet x 1.25 = **1,500 VA demand load**

Figure 11-18
Sign Demand Load Example

General Lighting Load: 18,000 square feet × 3.5 VA × 1.25 = 78,750 VA

Receptacle Load: 1 VA per square foot and apply Table 220–13 demand factors.

18,000 square feet × 1 VA =	18,000 VA	
First 10 kVA at 100%	−10,000 VA × 1.00 =	10,000 VA
Remainder at 50%	8,000 VA × .50 =	+ 4,000 VA
Receptacle demand load		14,000 VA

❑ **Office General Lighting And Receptacle Example**

What is the general lighting demand load for a 28,000 square foot office building that has 150 receptacles?

(a) 111 kVA   (b) 110 kVA   (c) 128 kVA   (d) 141 kVA

• Answer: (d) 141 kVA

General lighting   122,500 VA
Receptacles   + 18,500 VA
Total load   141,000 VA/1,000 = 141 kVA

General Lighting Load: 28,000 square feet × 3.5 VA × 1.25 = 122,500 VA.
Receptacle Load:

150 receptacles × 180 VA =	27,000 VA	
First 10, 000 VA at 100%	− 10,000 VA × 1.00 =	10,000 VA
Remainder at 50%	17,000 VA × .50 =	+ 8,500 VA
Receptacle demand load		18,500 VA

11–17 SIGNS [220-3(c)(6) AND 600–5(b)]

The NEC requires each commercial occupancy that is accessible to pedestrians to be provided with at least one 20-ampere branch circuit for a sign [600–5(b)(1)]. The load for the required exterior signs or outline lighting shall be a minimum of 1,200 VA [220–3(c)(6) and 600-5(b)(3)]. A sign outlet is considered to be a continuous load and the feeder load must be sized at 125 percent ot the continuous load [210–22(c), 220–3(a), 220–10(b), and 384–16(c)].

❑ **Sign Demand Load Example**

What is the demand load for one electric sign, Fig. 11–18?

(a) 1,200 VA   (b) 1,500 VA   (c) 1,920 VA   (d) 2,400 VA

• Answer: (b) 1,500 VA

1,200 VA × 1.25 = 1,500 VA

11-18 NEUTRAL CALCULATIONS [220-22]

The neutral load is considered the maximum unbalanced demand load between the grounded (neutral) conductor and any one ungrounded (hot) conductor, as determined by the calculations in Article 220, Part B. This means that line-to-line loads are not considered when sizing the neutral conductor.

Reduction Over 200 Amperes

For balanced 3-wire, single-phase and 4-wire, 3-phase wye systems, the neutral demand load can be reduced 70 percent for that portion of the unbalanced load over 200 amperes.

Reduction Not Permitted

The neutral demand load shall not be permitted to be reduced for 3-wire, single-phase 208Y/120-, or 480Y/277-volt circuits consisting of two line wires and the common conductor (neutral) of a 4-wire, 3-phase wye system. This is because the common (neutral) conductor of a 3-wire circuit connected to a 4-wire, 3-phase wye system carries approximately the same current as the phase conductors; see Note 10(b) of Table 310–16. This can be proven with the following formula:

$$I_n = \sqrt{L_1^2 + L_2^2 - (L_1 \times L_2)}$$

L_1 = Current of Line$_1$, Line$_2$ = Current of Line$_2$

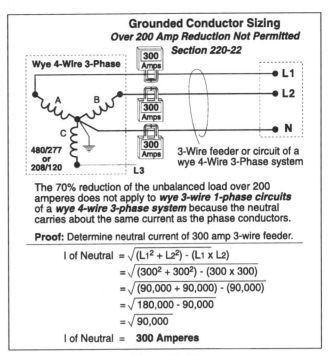

Figure 11-19

Grounded Conductor Not Permitted To Be Reduced Example

❏ **Three-Wire Wye Neutral Current Example**

What is the neutral current for a balanced 300-ampere 3-wire 208Y/120-volt feeder, Fig. 11–19?

(a) 100 amperes (b) 200 amperes (c) 300 amperes (d) none of these

• Answer: (c) 300 amperes

$$I_n = \sqrt{L_1^2 + L_2^2 - (L_1 \times L_2)}$$
$$I_n = \sqrt{(300^2 + 300^2) - (300 \times 300)}$$
$$I_n = \sqrt{180,000 - 90,000} \quad I_n = \sqrt{90,000} \quad I_n = 300 \text{ amperes}$$

Nonlinear Loads

The neutral demand load cannot be reduced for electric-discharge lighting, electronic ballasts, dimmers, controls, computers, laboratory test equipment, medical test equipment, recording studio equipment, or other *nonlinear loads*. This restriction only applies to circuits that are supplied from a 4-wire, wye-connected, 3-phase system, such as a 208Y/120- or 480Y/277-volt system. Nonlinear loads can cause triplen *harmonic currents* that add on the neutral conductor, which can require the neutral conductor to be larger than the ungrounded conductor load; see Section 220–22 FPN No. 2.

PART C – LOAD CALCULATIONS

MARINA [555–5]

The National Electrical Code permits a demand factor to apply to the receptacle outlets for boat slips at a marina. The demand factors of Table 555–5 are based on the number of receptacles. The receptacles must also be balanced between the lines to determine the number of receptacles on any given line.

❏ **Marina Receptacle Outlet Demand Example**

What size 120/240 volt, single-phase service is required for a marina that has twenty 20-ampere, 120 volt receptacles and twenty 30-ampere, 240 volt receptacles, Fig. 11–20?

(a) 200 amperes (b) 400 amperes (c) 600 amperes (d) 800 amperes

• Answer: (c) 600 amperes

Section 555–5 permits a demand factor according to the number of receptacles. The receptacles must be balanced to determine the number of receptacles on any given line. Ten 20-ampere, 120-volt receptacles are on line 1, ten 20-ampere, 120-volt receptacles are on line 2. Twenty 30-ampere, 240-volt receptacles are on line 1 and 2.

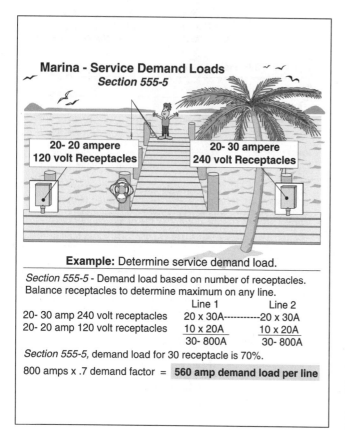

Figure 11-20
Marina Service Demand Load Example

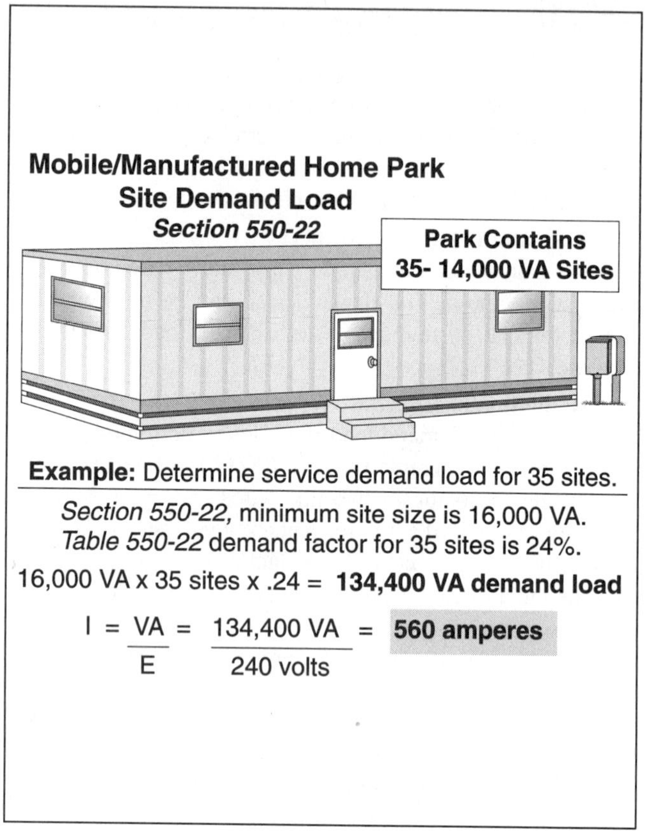

Figure 11-21
Mobile Home Park Service Demand Load Example

The total load on each line is [(10 receptacles × 20 amperes) + (20 receptacles × 30 ampere)] = 800 amperes per line. The demand factor for 30 receptacles (per line) is 70 percent; 800 amperes × 0.7 = 560 amperes per line.

MOBILE/MANUFACTURED HOME PARK [550–22]

The service demand load for a mobile/manufactured home park is sized according to the demand factors of Table 550–22 to the larger of; 16,000 VA for each mobile/manufactured home lot, or the calculated load for each mobile/manufactured home site according to Section 550–13.

☐ **Mobile/Maunfactured Home Park Example**

What is the demand load for a mobile/manufactured home park that has facilities for 35 sites? The system is 120/240-volts single-phase, Fig. 11–21.

(a) 400 amperes (b) 600 amperes (c) 800 amperes (d) 1,000 amperes

• Answer: (b) 600 amperes

16,000 VA × 35 sites × .24 = 134,400 VA
I = VA/E
I = 134,400 VA/240 volts = 560 amperes.

RECREATIONAL VEHICLE PARK [551–73]

Recreational vehicle parks are calculated according to the demand factor of Table 551–73. The total calculated load is based on:

2,400 VA for each 20-ampere supply facilities site,

3,600 VA for each 20- and 30-ampere supply facilities site, and

9,600 VA for each 50-ampere, 120/240-volt supply facilities site.

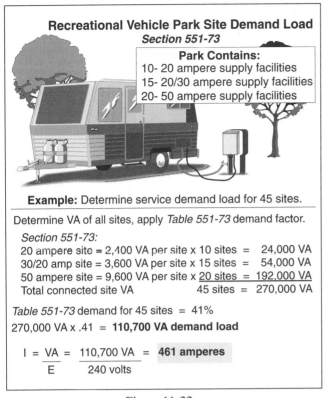

Figure 11-22
Recreational Vehicle Park Site Demand Load Example

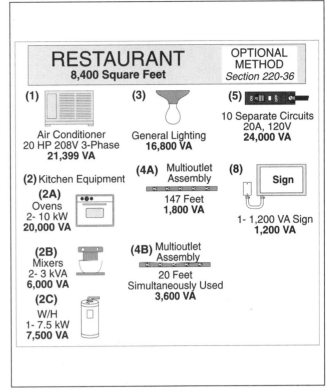

Figure 11-23
Restaurant Demand Load Optional Method – Example

☐ **Recreational Vehicle Park Example**

What is the demand load for a recreational vehicle park that has; ten 20-ampere supply facilities, fifteen 20- and 30-ampere supply facilities, and twenty 50-ampere supply facilities? The system is 120/240-volts single-phase, Fig. 11–22.

(a) 400 amperes (b) 500 amperes (c) 600 amperes (d) 700 amperes

- Answer: (b) 500 amperes

The service for the recreational vehicle park is sized according to the demand factors of Table 551–73

20-ampere supply facilities, 2,400 VA × 10 sites =	24,000 VA
20- and 30-ampere supply facilities, 3,600 VA × 15 sites =	54,000 VA
50-ampere supply facilities, 9,600 VA × 20 sites =	+ 192,000 VA
Total connected load	270,000 VA

Demand factor [Table 551–73] for 45 sites, 41%
Demand load = 270,000 VA × .41 = 110,700 VA
I = VA/E
I = 110,700 VA/240 volts = 461 amperes

RESTAURANT – OPTIONAL METHOD [220–36]

An optional method of calculating the demand service load for a restaurant is permitted. The following steps can be used to determine the service size.

Step 1: → Determine the total connected load. Add the nameplate rating of all loads at 100 percent and include both the air conditioner and heat load.

Step 2: → Apply the demand factors from from Table 220–36 to the total connected load.

All Electric Restaurant Demand Factors

0 - 250 kVA	*80%*
251- 280 kVA	*70%*
281 - 325 kVA	*60%*
Over 326 kVA	*50%*

Not All Electric Restaurant Demand Factors

0 - 250 kVA	*100%*
251- 280 kVA	*90%*
281 - 325 kVA	*80%*
326 - 375 kVA	*70%*
376 - 800 kVA	*65%*
Over 800 kVA	*50%*

☐ Restaurant Optional Method Example

Using the optional method, what is the service size for the loads listed in Figure 11–23?

Step 1: → Determine the total connected load.

(1) Air conditioning demand [Table 430–150 and 440–34]
VA = 208 Volts × 59.4 amperes × 1.732 =   21,399 VA

(2) Kitchen equipment
Ovens	10 kVA × 2 units =	20.0 kVA
Mixers	3 kVA × 2 units =	6.0 kVA
Water heaters	7.5 kVA × 1 unit =	+ 7.5 kVA
Total connected load		33.5 kVA

(3) Lighting [Table 220–3(b)]
General lighting, 8,400 square feet × 2 VA per square foot =   16,800 VA

(4) Multioutlet assembly [220–3(c) Exception]
147 feet/5 feet = thirty sections at 5-feet, 180 VA × 30 =   5,400 VA
Simultaneous used 180 VA × 20 feet =   3,600 VA

(5) Receptacles [220–3(c)(7)]
180 VA per receptacle × 40 receptacles =   7,200 VA

(6) Separate circuits (noncontinuous)
120 volts × 20 amperes = 2,400 VA × 10 circuits =   24,000 VA
Sign [220–3(c)(6)and 600-5] =   1,200 VA

(1) Air conditioning	21,399 VA
(2) Kitchen equipment	33,500 VA
(3) Lighting	16,800 VA
(5) Multioutlet assembly	5,400 VA
(6) Receptacles	7,200 VA
(7) Separate circuits	24,000 VA
(8) Sign	+ 1,200 VA
Total connected load	109,499 VA

Step 2: → Apply the demand factors from Table 220–36 to the total connected load.

All Electric – If the restaurant is all electric, a demand factor of 80 percent is permitted to apply to the first 250 kVA.

109,499 VA × 0.8 = 87,599 VA
I = VA/(E × √3)
I = 87,599 VA/(208 volts × 1.732) = 243 amperes

Not All Electric – If the restaurant is not all electric, then a demand factor of 100 percent applies to the first 250 kVA.

109,499 VA at 100% = 109,499 VA
I = VA/(E × √3)
I = 109,499 VA/(208 volts × 1.732) = 304 amperes

SCHOOL – OPTIONAL METHOD [220–34]

An optional method of calculating the demand service load for a school is permitted. The following steps can be used to determine the service size.

Step 1: → Determine the total connected load. Add the nameplate rating of all loads at 100 percent and select the larger of the air-conditioning versus heat load.

Step 2: → Determine the average VA per square foot by dividing the total connected load (Step 1) by the square feet of the building.

Step 3: → Determine the demand VA per square foot by applying the demand factors of Table 220–34 to the average VA per square foot (Step 2).

Step 4: → Determine the school net VA, by multiplying the demand VA per square foot (Step 3) by the square foot of the school building.

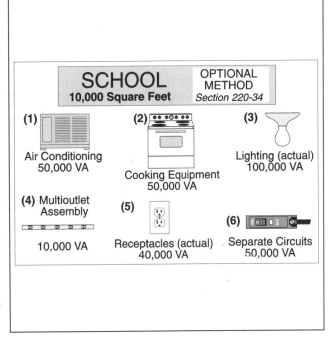

Figure 11-24

School Demand Load Optional Method – Example

❑ **School Optional Method Example**

What is the service size for the following loads? The system voltage is 208Y/120-volt three-phase, Fig. 11–24.

(1) Air conditioning   50,000 VA
(2) Cooking equipment   40,000 VA
(3) Lighting   100,000 VA
(4) Multioutlet assembly   10,000 VA
(5) Receptacles   40,000 VA
(6) Separate circuits   + 40,000 VA
                             280,000 VA

(a) 300 amperes   (b) 400 amperes   (c) 500 amperes   (d) 600 amperes

• Answer: (c) 500 amperes

Step 1: → Determine the average VA per square foot.
(Total Connected Load/Square Feet Area): 280,000 VA/ 10,000 square feet = 28 VA per square foot.

Step 2: → Apply the demand factors from Table 220–34 to the average VA per square foot.

Average VA per square foot	28 VA	
First 3 VA at 100%	– 3 VA × 1.00 =	3.00 VA
	25 VA	
Next 17 VA at 75%	– 17 VA × .75 =	12.75 VA
Remainder at 25%	8 VA × .25 =	+ 2.00 VA
Net VA per square foot		17.75 VA

Total demand load = 17.75 VA per square foot × 10,000 square feet = 175,000 VA

Service Size – I = VA/(E × $\sqrt{3}$)

I = 175,000 VA/(208 volts × 1.732) = 486 amperes

Service Demand Load Using the Standard Method

For exam preparation purposes, you are not expected to calculate the total demand load for a commercial building. However, you are expected to know how to determine the demand load for the individual loads (Steps 1 through 10 below). For your own personal knowledge, you can use the following steps to determine the total demand load for a commercial building for the purpose of sizing the service.

Part A – Determine The Demand Load For Each Type Of Load

Step 1: → Determine the general lighting load Table 220–3(b) and Table 220–11.
Step 2: → Determine the receptacle demand, load. [220–13].
Step 3: → Determine the appliance demand load at 100 percent.
Step 4: → Determine the demand load for show windows at 125 percent, [220–12].
Step 5: → Determine the demand load for multioutlet assembly, [220–3(c) Exception No. 1].
Step 6: → Determine the larger of :
Air Conditioner, largest 125 percent plus all others at 100 percent, [440–33] vs. heat at 100 percent, [220–15].

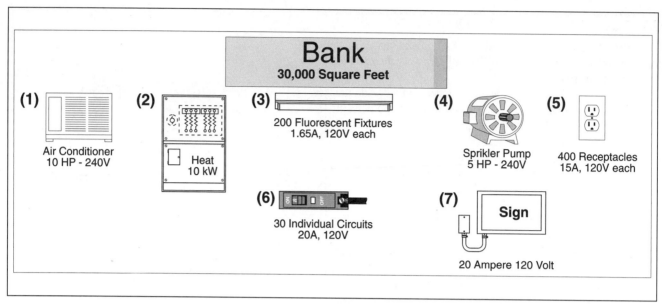

Figure 11-25
Bank Example

Step 7: → Determine the demand load for educational cooking equipment, [220–19 Note 5].
Step 8: → Determine the kitchen equipment demand load, [Table 220–20].
Step 9: → All other noncontinuous loads at 100 percent, [220–10(b)].
Step 10: → All other continuous loads at 125 percent, [220–10(b)].

Part B – Determine the Total Demand Load
Add up the individual demand loads of Steps 1 through 10.

Part C – Determine the Service Size
Divide the total demand load (Part B) by the system voltage.
Single-Phase = I = VA/E   Three-Phase = I = VA/(E × 1.732)

PART D – LOAD CALCULATION EXAMPLES

BANK (120/240-VOLT SINGLE-PHASE), Fig. 11–25.

(1) Air Conditioning [Article 440, 430–24, and Table 430–148]
10-horsepower, 230 volts × 50 amperes × 1.25 =   14,375 VA

(2) Heat [220–15 and 220–21] 10 kW omit 220–21

(3) Lighting [Table 220–3(c), and 220–10(b)]
30,000 square foot × 3.5 VA =   105,000 VA × 1.25 =   131,250 VA
Actual Lighting, Omit
200 units × 1.65 ampere × 120 volts × 1.25 = 49,500 VA

(4) Motor [Table 430–148]
5-horsepower, 230 volts × 28 amperes =   6,440 VA
(Some exams use 240 volts × 28 amperes = 6,720 VA)

(5) Receptacles [220–13]
Actual Load 400 receptacles × 180 VA = 72,000 VA

```
                                    72,000 VA
                                  – 10,000 VA  × 1.00 =   10,000 VA
                                    62,000 VA  × .50  = + 31,000 VA
Net computed load                                         41,000 VA
```

Chapter 3 Advanced NEC Calculations And Code Questions — Unit 11 Commercial Load Calculations

(6) Separate Circuits (non-continuous)
30 circuits × 20 amperes × 120 volts = 72,000 VA

(7) Sign [220–3(c)(6), 220–10(b) and 600–5(b)(3)]
1,200 VA × 1.25 = 1,500 VA

SUMMARY	Overcurrent Protection	Conductor	Neutral
(1) Air conditioner	14,375 VA	17,969 VA *1	0 VA *2
(2) Lighting	131,250 VA	131,250 VA	105,000 VA *2
(4) Motor	16,100 VA *3	6,440 VA	0 VA
(5) Receptacles	41,000 VA	41,000 VA	41,000 VA
(6) Separate Circuits	72,000 VA	72,000 VA	72,000 VA
(7) Sign	+ 1,500 VA	+ 1,500 VA	+ 1,200 VA *2
	276,225 VA	**270,159 VA**	**219,200 VA**

*1 Includes 25% for the largest motors (14,375 × 1.25)
*2 The neutral reflects the feeder conductor load at 100 percent, not 125 percent!
*3 Feeder protection is sized according to the largest motor short-circuit ground-fault protection device [430–63].
  28 amperes × 2.5 = 70 amperes, 70 amperes × 230 volts = 16,100 VA

1. Feeder/service protection device
I = VA/E, I = 276,225 VA /240 volts = 1,151 amperes

2. Feeder and service conductor
I = VA/E, I = 270,159 VA/240 volts = 1,125 amperes

3. Neutral Amperes
I = VA/E, I = 219,200/240 volts = 913 amperes

The neutral conductor is permitted to be reduced according to the requirements of Section 220–22.

Neutral demand load =	913 amperes	
First 200 amperes	– 200 amperes × 1.00 =	200 amperes
Remainder	713 amperes × .70 =	+ 499 amperes
		699 amperes

❏ **Summary Example**
The service overcurrent protection device must be sized no less than 1,151 amperes. The service conductors must have an ampacity no less than the overcurrent protection device rating [240–3(c)]. The grounded (neutral) conductor must not be less than 699 amperes.

➤ What size service overcurrent protection device is required?
   (a) 800 amperes   (b) 1,000 amperes   (c) 1,200 amperes   (d) 1,600 amperes
   • Answer: (c) 1,200 amperes [240–6(a)].

➤ What size service THHN conductors are required in each raceway if the service is parallel in four raceways?
   (a) 250 kcmil   (b) 300 kcmil   (c) 350 kcmil   (d) 400 kcmil
   • Answer: (c) 350 kcmil
     1,200 amperes/4 raceways = 300 amperes per raceway, sized based on 75°C terminal rating [110–14(c)(2)]
     350 kcmil rated 310 amperes × 4 = 1,240 amperes [240–3(c) and Table 310–16].

Note. 300 kcmil THHN has an ampacity of 320 amperes at 90°C, but we must size the conductors at 75°C, not 90°C!

➤ What size service grounded (neutral) conductor is required in each of the four raceways?
   (a) No. 1/0   (b) No. 2/0   (c) 250 kcmil   (d) 350 kcmil
   • Answer: (b) No. 2/0
     The grounded conductor must be sized no less than:
     (1) 12½ percent area of the line conductor, 350,000 × 4 = 1,400,000 × .125 = 175,000 [250–23(b)], 175,000/4 = 43,750 circular mil, No. 3 per raceway [Chapter 9, Table 8]. **(2)** When paralleling conductors, no conductor smaller than No. 1/0 is permitted [310–4]. **(3)** The service neutral conductor must have an ampacity of at least 699 amperes, 699 amperes/4 = 175 amperes, Table 310–16, No. 2/0 is required at 75°C [110–14(c)].

356 Unit 11 Commercial Load Calculations   Chapter 3 Advanced NEC Calculations And Code Questions

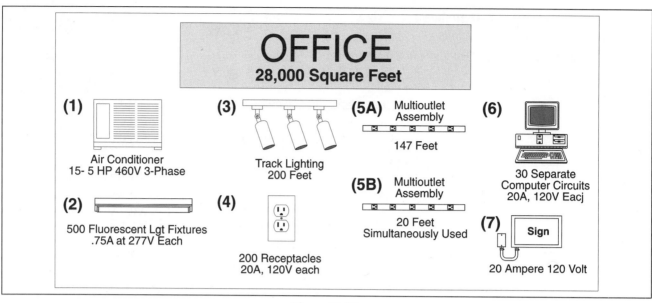

Figure 11-26

Office Example

→ What size grounding electrode conductor is required to a concrete encased electrode?

(a) No. 4 (b) No. 1/0 (c) No. 2/0 (d) No. 3/0

• Answer: (a) No. 4 [250–94 Exception No. 1b]

OFFICE BUILDING (480Y/277-VOLT THREE-PHASE), Fig. 11–26.

(1) Air Conditioning (Table 430–150 and 440–34)
460 volts × 7.6 amperes × 1.732 =   6,055 VA × 15 units =   90,826 VA
25 percent of the largest + VA = 6,055 × .25 =   1,514 VA

(2) Lighting (NEC) [Table 220–3(b) and 220–10(b)]
28,000 square feet × 3.5 VA = 98,000 VA × 1.25 = 122,500 VA (omit)

(2) Lighting Electric Discharge (Actual) [220–10(b)]
500 lights × .75 amperes × 277 volts =   103,875 VA × 1.25 =   129,844 VA

(3) Lighting, Track (Actual) [220–3(c)(5), 220–10(b) and 410–102] 150 VA per 2 feet
200 feet/2 feet = 100 sections x 150 VA =   15,000 VA × 1.25 =   18,750 VA

(4) Receptacles [220-13](Actual)
200 receptacles × 180 VA =   36,000 VA
    − 10,000 VA × 100% =   10,000 VA
    26,000 VA × 50% =   + 13,000 VA
            23,000 VA

(5) Multioutlet Assembly [220–3(c), Exception]
(A) 147 feet/5 feet = thirty 5-foot sections, 30 × 180 VA =   5,400 VA
(B) Simultaneously used 20 feet × 180 VA =   3,600 VA

(6) Separate Circuits (noncontinuous)
120 volts × 20 amperes = 2,400 VA × 30 circuits =   72,000 VA

(7) Sign [220–10(b) and 600–6]
1,200 VA × 1.25 =   1,500 VA

SUMMARY	Overcurrent Protection	Conductor	Neutral	
(1) Air conditioning	92,339 VA	92,339 VA	0 VA	
(2) Lighting (actual)	129,844 VA	129,844 VA	103,875 VA	*1
(3) Lighting (track)	18,750 VA	18,750 VA	15,000 VA	*1
(4) Receptacles (actual)	23,000 VA	23,000 VA	23,000 VA	
(5) Multioutlet Assembly	9,000 VA	9,000 VA	9,000 VA	
(6) Separate Circuits	72,000 VA	72,000 VA	72,000 VA	
(7) Sign	+ 1,500 VA	+ 1,500 VA	+ 1,200 VA	*1
	346,434 VA	346,434 VA	234,875 VA	

*1 The neutral reflects the feeder load at 100 percent, not 125 percent!

1. Feeder/service protection device

I = VA/(E × $\sqrt{3}$), I = 346,434 VA/(480 × 1.732) = 416 amperes

2. Feeder and service conductor amperes

I = VA/(E × $\sqrt{3}$), I = 346,434 VA/(480 × 1.732) = 416 amperes

3. Neutral conductor

I = VA/(E × $\sqrt{3}$), I = 234,875 VA/(480 volts × 1.732) = 283 amperes*.

❑ **Summary Example**

The service overcurrent protection device and conductor must be sized no less than 416 amperes, and the grounded (neutral) conductor must not be less than 283 amperes.

➤ What size service overcurrent protection device is required?

(a) 300 amperes   (b) 350 amperes   (c) 450 amperes   (d) 500 amperes

• Answer: (c) 450 amperes [240–6(a)]. Since the total demand load for the overcurrent device is 416 amperes, the minimum service permitted is 450 amperes.

➤ What size service conductors are required if the total calculated service demand load is 416 amperes?

(a) 300 kcmil   (b) 400 kcmil   (c) 500 kcmil   (d) 600 kcmil

• Answer: (d) 600 kcmil rated 420 amperes [240–3(b) and Table 310–16]
The service conductors must have an ampacity of at least 416 amperes, protected by a 450 ampere protection device [240–3(b)], and the conductors must be selected based on 75ºC insulation rating [110–14(c)(2)].

➤ What size service grounded (neutral) conductor is required for this service?

(a) No. 1/0   (b) 3/0   (c) 250 kcmil   (d) 350 kcmil

• Answer: (d) 350 kcmil

The grounded conductor must be sized no less than:

(1) The required grounding electrode conductor [250–23(b) and Table 250–94], = No. 1/0. **(2)** An ampacity of at least 283 amperes, Table 310–16, 350 kcmil is rated 310 amperes at 75ºC [110–14(c)].

* **Note.** Because of 125 amperes of electric discharge lighting [103,875 VA/(480 × 1.732)], the neutral conductor is not permitted to be reduced for that portion of the load in excess of 200 amperes [220-22].

RESTAURANT (STANDARD) (208Y/120 VOLT, THREE-PHASE SYSTEM), Fig. 11–27.

(1) Air conditioning [Table 430–150, and 440–34]
208 volts × 59.4 amperes × 1.732 × 1.25 =   26,749 VA

(2) Kitchen equipment [220–20]

(A) Ovens	10 kW × 2 units =	20.0 kW	
(B) Mixer	3 kW × 2 units =	6.0 kW	
(C) Water heaters	7.5 kW × 1 unit =	+ 7.5 kW	
Total connected		33.5 kW × .7 =	23.45 kW

(3a) Lighting NEC [Table 220–3(b) and 220–10(b)]
8,400 square feet × 2 VA =16,800 VA × 1.25 = 21,000 VA, omit - less than actual.

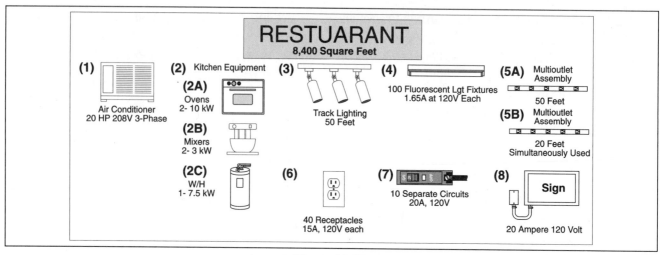

Figure. 11-27

Restaurant Standard Method Example

(3b) Lighting, Track [220–3(c)(5), 220–10(b) and 410–102]
150 VA per two feet, 50 feet/2 = 25 sections
25 sections × 150 VA =   3,750 VA × 1.25 =   4,688 VA

(4) Lighting (actual) [220–10(b)]
100 lights × 1.65 amperes × 120 volts =   19,800 × 1.25 =   24,750 VA

(5) Multioutlet Assembly [220–3(c), Exception]
5,400 VA (1,800 VA + 3,600 VA)
(A) 50/5 = ten 5-foot sections, 10 sections × 180 VA =   1,800 VA
(B) Simultaneous used 20 feet × 180 VA =   3,600 VA

(6) Receptacles (Actual) [220–13] 40 receptacles × 180 VA =   7,200 VA

(7) Separate Circuits [220–10(b)]
120 volts × 20 amperes = 2,400 VA × 10 circuits =   24,000 VA

(8) Sign [220–10(b) and 600–6] 1,200 VA × 1.25 =   1,500 VA

SUMMARY	Overcurrent Protection	Conductor
(1) Air Conditioning	26,749 VA	26,749 VA
(2) Kitchen equipment	23,450 VA	23,450 VA
(3) Lighting (track)	4,688 VA	4,688 VA
(4) Lighting (actual)	24,750 VA	24,750 VA
(5) Multioutlet Assembly	5,400 VA	5,400 VA
(6) Receptacles (actual)	7,200 VA	7,200 VA
(7) Separate Circuits	24,000 VA	24,000 VA
(8) Sign	+ 1,500 VA	+ 1,500 VA
	117,737 VA	117,737 VA

1. Feeder and service protection device

$I = VA/(E \times \sqrt{3})$

$I = 117{,}737 \text{ VA}/(208 \times 1.732) = 327$ amperes

2. Feeder and service conductor

$I = VA/(E \times \sqrt{3})$

$I = 117{,}737 \text{ VA}/(208 \times 1.732) = 327$ amperes

❏ Summary Example

The service overcurrent protection device must be sized no less than 327 amperes. The service conductors must be no less than 327 amperes.

↠ What size service overcurrent protection device is required?

(a) 300 amperes (b) 350 amperes (c) 450 amperes (d) 500 amperes

- Answer: (b) 350 amperes [240–6(a)]

   Since the total demand load for the overcurrent device is 327 amperes, the minimum service permitted is 350 amperes [240–6(a)].

↠ What size service conductors are required?

(a) No. 4/0 (b) 250 kcmil (c) 300 kcmil (d) 400 kcmil

- Answer: (d) 400 kcmil

   The service conductors must have an ampacity of at least 327 amperes, and be protected by a 350 ampere protection device [240–3(b)] next size up is OK, and the conductors must be selected according to Table 310–16, 400 kcmil based on 75ºC insulation rating [110–14(c)(2)] is rated 335 amperes at 75ºC.

Unit 11 – Commercial Load Calculations Summary Questions

Part A – General

11–2 Conductor Ampacity [Article 100]

1. The _____ of a conductor is the rating in amperes a conductor can carry continuously without exceeding its insulation temperature rating. The allowable ampacities as listed in Table 310–16 are affected by ambient temperature, current flow, conductor insulation and conductor bunching or bundling [310–10].
   (a) load rating   (b) ampacity   (c) demand load   (d) continuous factor

11–3 Conductor Overcurrent Protection [240–3]

2. The purpose of conductor _____ is to protect the conductors against excessive or dangerous temperatures. If the ampacity of a conductor does not correspond with the standard ampere rating of a fuse or circuit breaker, the next size up protection device is permitted. This applies only if the conductors do not supply multioutlet receptacles and if the next size overcurrent protection device does not exceed 800-amperes.
   (a) short-circuit protection   (b) ground-fault protection
   (c) overload protection   (d) overcurrent protection

3. The following is a list of some of the standard ampere ratings for overcurrent protection devices (fuses and inverse time circuit breakers): 15, 25, 35, 45, 80, 90, 110, 175, 250 and 350-amperes.
   (a) True   (b) False

11–4 Voltages [220–2]

4. Unless other voltages are specified, branch-circuit, feeder, and service loads shall be computed at a nominal system voltage of _____.
   (a) 600/347   (b) 240/120   (c) 208Y/120   (d) Any of these

11–5 Fraction Of An Ampere

5. There are no NEC rules for rounding when a calculation results in a fraction of an ampere, but it does contain the following note, "except where the computations result in a _____ of an ampere or larger, such fractions may be dropped."
   (a) .05   (b) .5   (c) .49   (d) .51

Part B – Loads

11–6 Air Conditioning

6. What is the feeder or service demand load required for the air conditioning of a six unit office building? Each unit contains one air conditioner rated 3-horsepower, 230 volts.
   (a) 13 kVA   (b) 24 kVA   (c) 45 kVA   (d) 51 kVA

11–7 Dryers

7. What size branch circuit conductor and overcurrent protection is required for a 6-kW dryer rated 240 volts, located in the laundry room of a multifamily dwelling?
   (a) No. 12 with a 20-ampere breaker
   (b) No. 10 with a 20-ampere breaker
   (c) No. 12 with a 30-ampere breaker
   (d) No. 10 with a 30-ampere breaker

8. What is the feeder and service demand for eight 6.75-kW dryers?
   (a) 70 kW   (b) 54 kW   (c) 35 kW   (d) 27 kW

11–8 Electric Heat

9. • What size conductor and protection is required for a single-phase 15-kW, 480-volt heat strip that has a 1.6-ampere blower motor?
   (a) No. 10 THHN with 30-amperes protection
   (b) No. 6 THHN with 45-amperes protection
   (c) No. 10 THHN with 40-amperes protection
   (d) No. 4 THHN with 70-amperes protection

10. What is the feeder and service demand load for a building that has four 20-kW, 230-volt single-phase heat strips that have a 5.4-ampere blower motor for each unit?
    (a) 100 kVA (b) 50 kVA (c) 125 kVA (d) 85 kVA

11–9 Kitchen Equipment

11. What is the branch circuit demand load (in amperes) for one 11.4-kW oven rated 240 volts?
    (a) 60 amperes (b) 27 amperes (c) 48 amperes (d) 33 amperes

12. • What is the feeder and service demand load for the following?
    Water heater .... 9 kW
    Booster heater .... 12 kW
    Mixer .... 3 kW
    Oven .... 2 kW
    Dishwasher .... 1.5 kW
    Disposal .... 1 kW
    (a) 19 kW (b) 21 kW (c) 29 kW (d) 12 kW

13. What is the feeder and service demand load for the following?
    Water heater .... 14 kW
    Booster heater .... 11 kW
    Mixer .... 7 kW
    Oven .... 9 kW
    Dishwasher .... 1.5 kW
    Disposal .... 3 kW
    (a) 25 kW (b) 30 kW (c) 20 kW (d) 45 kW

11–10 Laundry Equipment

14. What is the feeder and service demand load for seven washing machines at 1,500 VA?
    (a) 10,500 VA (b) 15,000 VA (c) 1,125 VA (d) none of these

11–11 Lighting – Demand Factors [220–11]

15. What is the general lighting feeder or service demand load for a 24 room motel, 685 square feet each?
    (a) 10 kVA (b) 15 kVA (c) 20 kVA (d) 25 kVA

16. What is the general lighting and receptacle feeder or service demand load for a 250,000 square foot storage warehouse that has 200 receptacles?
    (a) 25 kVA (b) 55 kVA (c) 95 kVA (d) 105 kVA

11–12 Lighting Without Demand Factors [Table 220–3b) and 220–10(b)]

17. What is the feeder and service (protection device) general lighting demand load for a 3,200 square foot dance club?
    (a) 3,200 VA (b) 6,400 VA (c) 8,000 VA (d) 12,000 VA

18. What is the feeder and service general lighting load for a 90,000 square foot school?
    (a) 238 kVA (b) 338 kVA (c) 90 kVA (d) 270 kVA

11–13 Lighting – Miscellaneous

19. How many 2 x 4 fluorescent fixtures, each rated 277 volts, .8-amperes, can be connected to a 20-ampere circuit? The four lamps are rated 40 watts each and the fixture is to be on for more than 3 hours.
    (a) 5   (b) 7   (c) 9   (d) 20

20. How many 360 watt, 120-volt incandescent fixtures can be connected to a 20-ampere 120-volt circuit continuously?
    (a) 5   (b) 7   (c) 9   (d) 12

21. • What is the VA demand feeder and service load for 130 feet of show window lighting?
    (a) 26 kVA   (b) 33 kVA   (c) 9 kVA   (d) 11 kVA

11–14 Multioutlet Receptacle Assembly [220–3(c) Exception No. 1]

22. What is the feeder demand load for 50 feet of multioutlet assembly and 10 feet of multioutlet assembly simultaneously used?
    (a) 3,600 VA   (b) 12,000 VA   (c) 7,200 VA   (d) 5,500 VA

11–15 Receptacle VA Load [220–3(c)(6)]

23. How many receptacle outlets are permitted on a 20-ampere 120-volt circuit?
    (a) 10   (b) 13   (c) 15   (d) 20

24. What is the service demand load for one-hundred-ten 15- or 20-ampere general use receptacles in a commercial building?
    (a) 5 kVA   (b) 10 kVA   (c) 15 kVA   (d) 20 kVA

25. What is the service demand load for the general lighting and receptacles, for a 30,000 square foot bank?
    (a) 100 kVA   (b) 150 kVA   (c) 200 kVA   (d) 250 kVA

26. • What is the service general lighting demand and receptacle load for a 10,000 square foot office building with 75 receptacles?
    (a) 25 kVA   (b) 35 kVA   (c) 45 kVA   (d) 55 kVA

11–17 Signs [220–(c)(6), 600–5(b)(3), And 220–10(b)]

27. What is the feeder demand load for sizing the overcurrent protection device for one electric sign?
    (a) 1,400 VA   (b) 1,500 VA   (c) 1,920 VA   (d) 2,400 VA

11–18 Neutral Calculations [220–22]

28. What is the neutral current for a balanced 150-ampere 3-wire 208Y/120-volt feeder?
    (a) 0 amperes   (b) 150 amperes   (c) 250 amperes   (d) 300 amperes

Part C – Load Calculations

Marina [555–5]

29. A marina has the following: Twenty-four slips (20-amperes 240-volt receptacles) and thirty slips designed for boats over 20 feet (30-ampere 240-volt receptacles). What size service is required for the marina?
    (a) 400 amperes   (b) 550 amperes   (c) 900 amperes   (d) 1,000 amperes

30. What size conductor is required for the service if the total demand load per phase is 570-amperes and the service is parallel in two raceways?
    (a) 4/0   (b) 250 kcmil   (c) 300 kcmil   (d) 350 kcmil

Mobile Home Park [550–22]

31. What is the feeder and service load for a mobile home park that has the facilities for forty-two sites? The system is 120/240-volt single-phase.
    (a) 650 amperes   (b) 510 amperes   (c) 660 amperes   (d) 730 amperes

Chapter 2 Advanced NEC Calculations And Code Questions   Unit 11 Commercial Load Calculations 363

Recreational Vehicle Park [551–73]

32. A recreational vehicle park has seventeen sites with only 20-ampere 240-volt receptacles, thirty-five sites with 20- and 30-ampere 240-volt receptacles, and ten sites with 50-ampere 240-volt receptacles. The feeder and service demand load is approximately _____ amperes.
    (a) 355   (b) 450   (c) 789   (d) 1,114

Restaurant – Optional Method [220–36]

33. A restaurant has a total connected load of 400 kVA and is all electric. What is the demand load for the service?
    (a) 363 kVA   (b) 325 kVA   (c) 298 kVA   (d) 275 kVA

34. A restaurant has a total connected load of 400 kVA and is *not* all electric. What is the demand load for the service?
    (a) 378 kVA   (b) 325 kVA   (c) 300 kVA   (d) 275 kVA

School – Optional Method [220–34]

35. A 28,000 square foot school has a total connected load of 590,000 VA. What is the total demand load?
    (a) 350 kVA   (b) 400 kVA   (c) 450 kVA   (d) 500 kVA

Challenge Questions

General Commercial Calculations

36. • If the service high leg conductor only supplies three-phase loads, what size service conductor would be required for the high leg? Three-phase loads: 10-horsepower, 230-volt 3-phase A/C and 10-kW of heat, 230-volt 3-phase.
    (a) 10 TW   (b) 12 THW   (c) 10 THHN   (d) No. 8 THW

37. • A commercial office building has forty-six 250 watt, 120-volt lights installed on a 208Y/120-volt three–phase system. If the fixtures are on continuously, how many 20 ampere, 120-volt circuits would be required?
    (a) 9 circuits   (b) 7 circuits   (c) 5 circuits   (d) 4 circuits

Conductor Sizing

38. • Three parallel service raceways are installed. Each raceway contains three 500-kcmil THHN conductors. What size bonding jumper is required for each service raceway?
    (a) 250,000   (b) No. 2/0   (c) No. 1/0   (d) No. 4/0

39. • If each service raceway contains 300-kcmil conductors, what size bonding jumper is required for the raceway?
    (a) No. 1   (b) No. 2   (c) No. 1/0   (d) No. 4

11–9 Kitchen Equipment [220–20]

40. What is the kitchen equipment demand load for: one 14-kW range, one 1.5-kW water heater, one .75-kW mixer, one 2-kW dishwasher, one 3-kW booster, and one 3-kW coffee machine?
    (a) 17 kW   (b) 14 kW   (c) 15 kW   (d) 24 kW

11–11 Lighting – Demand Factors [Table 220–3(b) And 220–11]

41. • Each unit of a one-hundred unit hotel is 12 × 15 feet. In addition, there is an office that is 60 × 20 feet, and there are hallways of 120 square feet. What is the general lighting demand load?
    (a) 29 kVA   (b) 37 kVA   (c) 23 kVA   (d) 25 kVA

11–15 Receptacle VA Load [220–3(c)(7) And 220–13]

42. If, in the hall and other areas of a motel, there are one hundred receptacle outlets (not in the motel rooms), the demand load added to the service for these receptacles would be _____ VA.
    (a) 0   (b) 10,000   (c) 14,000   (d) 20,000

43. What is the receptacle demand load for a 20,000 square foot office building?
    (a) 10,000 VA (b) 40,000 VA (c) 30,000 VA (d) 15,000 VA

11–16 Banks And Offices – General Lighting And Receptacle

44. What is the general lighting and general use receptacle load for a 30,000 square foot bank?
    (a) 220 kVA (b) 200 kVA (c) 175 kVA (d) 150 kVA

11–18 Neutral Calculations [220–22]

A 208Y/120-volt 3-phase service has a total connected load of 1,450-amperes.
600-amperes of these are phase-to-phase loads
300-amperes are balanced 120-volt fluorescent lighting
550-amperes of other 120-volt loads

45. • The neutral demand for this service is _____ amperes.
    (a) 1,150 (b) 650 (c) 745 (d) 420

Part C – Load Calculations

Marina Calculations [555]

46. A marina shore power facility has twenty 20-ampere, 240-volt receptacles, seventeen 30-ampere 240-volt receptacles, and seven 50-ampere 240-volt receptacles. After applying demand factors, the service demand load for the shore power boxes is _____ amperes.
    (a) 1,260 (b) 630 (c) 1,160 (d) 625

Mobile/manufactured Home Parks [550]

47. A seventy-five site mobile home park is designed to contain mobile homes that have a 14,000 VA load per site. The service demand load for the park is _____ kVA.
    (a) 1,200 (b) 264 (c) 222 (d) 201

Motel

48. A forty unit motel (300 square feet in each unit) includes 3-kVA air conditioning, 4-kW heat, and every two units share one 1.5-kW water heater. What is the demand load for the motel?
    (a) 281 kVA (b) 202 kVA (c) 226 kVA (d) 251 kVA

Recreational Vehicle Park [Article 551]

49. A recreational vehicle park has forty-two sites. The minimum feeder demand load for these sites would be _____ kVA.
    (a) 151 (b) 101 (c) 139 (d) 56

Restaurant Calculations

50. • The branch circuit rating required for a 12-kW range in a restaurant would be _____ amperes.
    (a) 50 (b) 45 (c) 35 (d) 30

51. • A new restaurant has a total connected lighting load of 30 kVA. The kitchen equipment includes two gas stoves, one gas grill, three gas ovens, one 75-gallon gas water heater, one 5-kW dishwasher, two 2-kW coffee makers, five 2-kW kitchen appliances on their own circuit, and ten 1.5-kVA small appliance circuits. Using the optional method, the service demand load would be closest to _____ kVA.
    (a) 64 (b) 50 (c) 45 (d) 38

School – Optional Method [220–34]

52. • Using the optional method, what is the demand load (VA per square foot) for a 10,000 square foot school that has a total connected load of 320 kVA?
    (a) 15.75 VA per square foot
    (b) 18.75 VA per square foot
    (c) 29.00 VA per square foot
    (d) 12.75 VA per square foot

53. Using the optional method, what is the demand load (kVA) for a 20,000 square foot school that has a total connected load of 160 kVA?
    (a) 135 kVA   (b) 106 kVA   (c) 120 kVA   (d) 112 kVA

NEC Random Questions, From Article 90 Through Chapter 9

54. Each doorway leading into a vault from the building interior shall be provided with a tight fitting door having a minimum fire rating of _____ hours.
    (a) 2   (b) 4   (c) 5   (d) 3

55. The maximum operating temperature of asbestos-covered, type AF heat resistant wire is _____ ºC.
    (a) 124   (b) 150   (c) 194   (d) 220

56. Conduits of 1½ inches and smaller entering an explosion proof enclosure for circuit breakers, motor controllers, and switches intended to interrupt current in the normal performance of the function may produce arcs or sparks and shall not be required to be sealed if the current-interrupting contacts are within a chamber hermetically sealed against the entrance of gases and vapors.
    (a) True   (b) False

57. The power supply to contact conductors of a crane in a Class III location shall be _____.
    (a) isolated from all other systems
    (b) equipped with an acceptable ground detector
    (c) have an alarm in the case of a ground fault
    (d) all of these

58. • Wiring methods permitted in Class I, Division 2 locations include _____.
    (a) threaded rigid metal conduit
    (b) threaded steel intermediate metal conduit
    (c) general purpose boxes and fittings containing no arcing devices
    (d) all of these

59. Switches, circuit breakers, motor controllers, and fuses, including push buttons, relays, and similar devices installed in Class I Division 1 locations shall be provided with approved explosion-proof enclosures, and they shall be approved as a complete assembly for Class I locations.
    (a) True   (b) False

60. The emergency controls for attended self-service stations must be located no more than _____ feet from the dispensers.
    (a) 20   (b) 50   (c) 75   (d) 100

61. • A Class III Division _____ location is where easily ignitable fibers or combustible flying material is stored or handled but not manufactured.
    (a) 1   (b) 2   (c) 3   (d) all of these

62. A Class I, Division 1 location is a location in which _____.
    (a) ignitable concentrations of flammable gases or vapors can exist under normal operating conditions
    (b) ignitable concentrations of such gases or vapors may exist frequently because of repair or maintenance operations or because of leakage
    (c) breakdown or faulty operation of equipment or processes might release ignitable concentrations of flammable gases or vapors, and might also cause simultaneous failure of electric equipment.
    (d) all of these

63. Equipment installed in hazardous locations must be approved and shall be marked to show the _____.
    (a) Class
    (b) Group
    (c) operating temperature reference to 40°C ambient
    (d) all of these

64. Wireways are sheet metal troughs with _____ that are used to form a raceway system.
    (a) removable covers   (b) hinged covers   (c) a and b   (d) none of these

65. Single-pole breakers utilizing listed handle ties shall not be permitted to be used for the circuit disconnect of dispensing equipment.
    (a) True   (b) False

66. Fixtures supported by suspended ceiling systems shall be securely fastened to the ceiling framing member by mechanical means, such as _____.
    (a) bolts   (b) screws   (c) rivets   (d) any of these

67. A disposal is permitted to be cord and plug connected. The cord must not be less than 18 inches or more than _____ inches and must be protected from physical damage.
    (a) 30   (b) 36   (c) 42   (d) 48

68. Each transformer shall be provided with a nameplate giving the name of the manufacturer, rated kVA, frequency, primary and secondary voltage, and the impedance of transformers _____ kVA and larger.
    (a) 112½   (b) 25   (c) 33   (d) 50

69. Hazardous locations are classified depending on the properties of the _____ which may be present and the likelihood that a flammable or combustible concentration or quantity is present.
    I. flammable vapors   II. flammable gases or liquids   III. combustible dusts or fibers
    (a) I only   (b) II only   (c) I or II   (d) I, II, or III

70. A lighting and appliance branch-circuit panelboard shall be provided with physical means to prevent the installation of more _____ devices than that number for which the panelboard was designed, rated, and approved.
    (a) overcurrent   (b) equipment   (c) breaker   (d) all of these

71. A Class I Division 2 location usually includes locations where volatile flammable liquids or flammable gases or vapors are used, but which, in the judgment of the authority having jurisdiction, would become hazardous only in case of an accident or of some unusual operating condition.
    (a) True   (b) False

72. Fixtures shall not be used as a raceway for circuit conductors, except fixtures designed for end-to-end assembly to form a continuous raceway, or fixtures connected together by recognized wiring methods shall be permitted to carry through conductors of 2-wire or _____ branch circuit supplying the fixtures.
    (a) small appliance   (b) appliance   (c) multiwire   (d) industrial

73. Where more than one motor disconnecting means is provided in the same motor branch circuit, only one of the disconnecting means is required to be readily accessible.
    (a) True   (b) False

74. Metal boxes over _____ cubic inches in size shall be constructed so as to be of ample strength and rigidity.
    (a) 50   (b) 75   (c) 100   (d) 125

75. Conductors and busbars on a switchboard, panelboard, or control board shall be so located as to be free from _____ and shall be held firmly in place.
    (a) obstructions   (b) physical damage   (c) both a and b   (d) none of these

76. The bottom of sign and outline lighting enclosures shall not be less than _____ feet above areas accessible to vehicles.
    (a) 12   (b) 14   (c) 16   (d) 18

77. Recessed incandescent fixtures shall have _____ protection and shall so be identified as thermally protected.
    (a) physical   (b) corrosion   (c) thermal   (d) all of these

78. • Fixture wires shall be permitted (1) for installation in lighting fixtures and in similar equipment where enclosed or protected and not subject to _____ in use, or (2) for connecting lighting fixtures to the branch-circuit conductors supplying the fixtures.
    (a) bending or twisting (b) knotting
    (c) stretching or straining (d) none of these

79. The patient care area is any portion of a health care facility wherein patients are intended to be _____.
    (a) examined (b) treated (c) moved (d) a or b

80. A Group "F" atmosphere contains carbon black, charcoal, coal or coke dusts that has been sensitized by other materials so that they present an explosive hazard.
    (a) True (b) False

81. Switchboards shall be so placed as to reduce to a minimum the probability of communicating _____ to adjacent combustible materials.
    (a) sparks (b) backfeed (c) fire (d) all of these

82. Enclosures that are not over _____ cubic inches in size and that have threaded entries or have hubs identified for the purpose, and that do not contain devices or support fixtures, shall be considered to be adequately supported where two or more conduits are threaded wrenchtight into the enclosure or hubs, and where each conduit is supported within 3 feet of the enclosure on two or more sides so as to provide the rigid and secure installation.
    (a) 50 (b) 75 (c) 100 (d) 125

83. Flexible metal conduit can be installed exposed or concealed where not subject to physical damage.
    (a) True (b) False

84. Hazards may be reduced, or hazardous (classified) locations limited or eliminated, by adequate positive-pressure ventilation from a source of clean air in conjunction with effective safeguards against ventilation failure.
    (a) True (b) False

85. Pull boxes or junction boxes that have any dimension over _____ feet, shall have all conductors cabled or racked up in an approved manner.
    (a) 3 (b) 6 (c) 9 (d) 12

86. Lead wires on weatherproof lampholders shall not be less than No. _____ wire.
    (a) 12 (b) 14 (c) 16 (d) 10

87. Four conditions must be met for fixtures to be permitted to be installed in cooking hoods of nonresidential occupancies. One of those conditions is that the fixture shall be identified for use within _____ cooking hoods and installed so that the temperature limits of the material used are not exceeded.
    (a) nonresidential (b) commercial (c) multifamily (d) all of these

88. Emergency circuit wiring must be designed and located in such a manner so as to minimize the hazards that might cause failure because of _____.
    (a) flooding (b) fire (c) icing (d) all of these

89. Branch-circuit conductors within _____ inch(es) of a ballast within the ballast compartment shall have an insulation temperature rating not lower than 90ºC, such as Types THW and XHHW.
    (a) 1 (b) 2 (c) 3 (d) 4

90. Switchboards that have any exposed live parts shall be located in permanently _____ locations and then only where under competent supervision and accessible only to qualified persons.
    (a) dry (b) mounted (c) supported (d) all of these

91. • Receptacles installed indoors or outdoors protected from the weather or other damp locations shall be in an enclosure that is _____ when the receptacle is covered.
    (a) raintight (b) weatherproof (c) rainproof (d) weathertight

92. Cabinets or cutout boxes installed in wet locations shall be _____.
    (a) waterproof (b) raintight (c) weatherproof (d) watertight

93. Because an auxiliary gutter is used to supplement wiring space, it is not a raceway and shall not extend a distance greater than _____ feet beyond the equipment that it supplements.
    (a) 50   (b) 30   (c) 10   (d) 25

94. Splices and taps shall not be located within fixture _____.
    (a) arms or stems
    (b) bases or screw shells
    (c) both a and b
    (d) either a or b

95. • Receptacles mounted in boxes that are set back of the wall surface shall be installed so that the mounting _____ of the receptacle is held rigidly at the surface of the wall.
    (a) screws or nails   (b) yoke or strap   (c) face plate   (d) none of these

96. Snap switches shall not be grouped or ganged in enclosures unless they can be so arranged that the voltage between adjacent switches does not exceed _____, or unless they are installed in enclosures equipped with permanently installed barriers between adjacent switches.
    (a) 100   (b) 200   (c) 300   (d) 400

97. Which of the following switches are required to indicate whether they are in the on or off position?
    (a) General use switches
    (b) Motor circuit switches
    (c) Circuit breakers
    (d) All of these

98. Incandescent lamp fixtures shall be marked to indicate the maximum allowable _____ of lamps.
    (a) voltage   (b) amperage   (c) rating   (d) wattage

99. Where an electric discharge fixture is mounted directly over an outlet box, the fixture must provide access to the conductor wiring within the outlet box.
    (a) True   (b) False

100. • Transformers with nonflammable dielectric fluid rated over 35,000 volts installed in a vault shall be furnished with a _____ when installed indoors.
    (a) fluid confinement area
    (b) pressure relief vent
    (c) means for absorbing or venting any gasses generated by arcing
    (d) all of these

101. In a location where flammable anesthetics are employed, the area is a Class I, Division 1 location which shall extend _____.
    (a) upward to the structural ceiling
    (b) upward to a level 8 feet above the floor
    (c) upward to a level 5 feet above the floor
    (d) 10 feet in all directions

102. • An underground rigid nonmetallic raceway must be not less than _____ feet from the inside wall of the pool or spa, unless space limitations prevent otherwise.
    (a) 8   (b) 10   (c) 5   (d) 25

103. • Class 1 control circuit conductors shall be protected against overcurrent _____.
    (a) in accordance with the values specified in Table 310–16 through 310–31 for No. 14 and larger
    (b) shall not exceed 7-amperes for No. 18, not exceed 10-amperes for No. 16
    (c) and derating factors do not apply
    (d) all of these

104. Edison base type fuseholders must be installed so as to accept _____ fuses by the use of adapters.
    (a) Edison base   (b) medium base   (c) heavy-duty base   (d) Type S

105. The ampacity of a conductor can be different along the length of the conductor. The higher ampacity is permitted to be used for the lower ampacity, if the lower ampacity is no more than _____ feet or no more than _____ percent of the length of the higher ampacity.
    (a) 10/20   (b) 20/10   (c) 10/10   (d) 15/15

106. The earth can be used as the sole equipment grounding conductor.
    (a) True   (b) False

107. Where flexible metal conduit is installed in a wet location, which of the following conductor types are required to be used?
 (a) THWN (b) XHHW (c) THW (d) any of these

108. Alternating current general use snap switches can control _____.
 (a) resistive loads that do not exceed the ampere and voltage rating of the switch
 (b) inductive and tungsten filament (120 volt) loads that do not exceed the ampere and voltage rating of the switch
 (c) motor loads (two horsepower or less) that do not exceed 80% of the ampere rating of the switch
 (d) all of these

109. Cablebus is a(n) _____ assembly of insulated conductors with fittings and conductor terminations in a completely enclosed ventilated protective metal housing.
 (a) factory (b) preassembled (c) complicated (d) approved

110. Bushings or adapters are required at electrical nonmetallic tubing terminations to protect the conductors from abrasion.
 (a) True (b) False

111. Steel or aluminum cable tray systems shall be permitted to be used as equipment grounding conductors provided four requirements are met. One of those requirements is that all cable tray sections and fittings shall be _____ marked to show the cross-sectional area of metal in channel cable trays, or cable trays of one-piece construction, and the total cross-sectional area of both side rails for ladder or trough cable trays.
 (a) legibly (b) durably (c) a or b (d) a and b

112. Class 2 and 3 control conductors shall be installed in a(n) _____ when used in hoistways.
 I. rigid metal conduit
 II. electrical metallic tubing
 III. intermediate metal conduit
 (a) I (b) II (c) III only (d) I, II, or III

113. All boxes and conduit bodies, covers, extension rings, plaster rings, and the like shall be durably and legibly marked with the manufacturer's name or trademark.
 (a) True (b) False

114. Embedded deicing and snow-melting equipment cables, units, and panels shall not be installed where they bridge _____ unless provision is made for expansion and contraction.
 (a) roads (b) over-water spans (c) runways (d) expansion joints

115. Where there are multiple equipment grounding conductors present in a receptacle outlet box, all equipment grounding conductors must be spliced and a pigtail brought out for the receptacle.
 (a) True (b) False

116. Single conductors are only permitted when installed as part of a wiring method listed in Chapter _____.
 (a) 4 (b) 3 (c) 2 (d) 9

117. Service-lateral conductors are required to be insulated except the grounded conductor when it's _____.
 (a) bare copper installed in a raceway
 (b) a bare copper and part of the cable assembly is identified for underground use
 (c) copper-clad aluminum with individual insulation
 (d) a and b

118. Where nails are used to mount knobs for the support of open wiring on insulators, they shall not be smaller than _____ penny.
 (a) 6 (b) 8 (c) 10 (d) none of these

119. • Where fuses are used for motor overload protection, a fuse shall be inserted in each ungrounded conductor and also in the grounded conductor if the supply system is _____ with one conductor grounded.
 (a) 2-wire, 3-phase direct current
 (b) 2-wire, 3-phase alternating current
 (c) 3-wire, 3-phase direct current
 (d) 3-wire, 3-phase alternating current

120. • Where motors are provided with a terminal housing, the housing shall be of _____ and be of substantial construction.
 (a) steel (b) iron (c) metal (d) copper

121. Flat cable assemblies shall be permitted only as branch circuits to supply suitable tap devices for _____ loads.
 I. lighting II. small power III. small appliance
 (a) I only (b) II only (c) III only (d) I, II, and III

122. There shall be a minimum of one _____ ampere branch circuit for the laundry outlet(s) in a dwelling unit.
 (a) 15 (b) 20 (c) 30 (d) b and c

123. Conductors shall be _____ unless otherwise provided in the Code.
 (a) lead (b) stranded (c) copper (d) aluminum

124. A given box has three equipment grounding conductors. When determining the number of conductors in a box for box fill calculations, the three grounding conductors are counted as _____ conductor(s).
 (a) 3 (b) 6 (c) 1 (d) 0

125. FCC cable can be installed only on _____ floors.
 (a) dry or damp (b) continuous (c) sound and smooth (d) all of these

126. In cellular concrete floor raceways, a "cell" shall be defined as a single, enclosed _____ space in a floor made of precast cellular concrete slabs, the direction of the cell being parallel to the direction of the floor member.
 (a) circular (b) oval (c) tubular (d) hexagonal

127. • Where communications conductors and electric light or power conductors are supported by the same pole, they shall _____.
 (a) not be attached to a crossarm that carries electric light or power conductors
 (b) have a vertical clearance of not less than 8 feet from all points of roofs above which they pass
 (c) both a and b
 (d) none of these

128. Intermediate metal conduit shall be firmly fastened within _____ of each outlet box, junction box, device box, fitting, cabinet, or other conduit termination.
 (a) 12 inches (b) 18 inches (c) 2 feet (d) 3 feet

129. Receptacle-type tap connectors for nonmetallic extensions shall be of the _____.
 (a) protected-type (b) grounding-type (c) locking-type (d) all of these

130. Article _____ contains the requirements for the wiring of occupancy locations used for service and repair operation in connection with self-propelled vehicles (including passenger automobiles, buses, trucks, tractors, etc.) in which volatile flammable liquids are used for fuel or power.
 (a) 500 (b) 501 (c) 511 (d) 514

131. No overcurrent device shall be connected in series with any conductor that is intentionally grounded, except where the overcurrent device opens all conductors of the circuit, including the _____ conductor, and is so designed that no pole can operate independently.
 (a) ungrounded (b) grounding (c) grounded (d) none of these

132. A one-family or two-family dwelling-unit requires _____ GFI receptacle(s) located outdoors.
 (a) zero (b) one (c) two (d) three

133. Wiring from emergency source or emergency source distribution overcurrent protection to emergency loads shall _____ other wiring.
 (a) not enter the same raceway or box with
 (b) not enter the same cable or cabinet with
 (c) be kept entirely independent of all
 (d) all of these

134. When service entrance conductors exceed 1100 kcmil for copper, the required grounded conductor for the service must be sized not less than _____ percent of the area of the total area of the largest phase conductor.
 (a) 15 (b) 19 (c) $12\frac{1}{2}$ (d) 25

135. Raceways on exterior surfaces of buildings shall be _____ and arranged to drain.
 (a) raintight (b) rainproof (c) watertight (d) waterproof

136. Where a feeder supplies _____ in which equipment grounding conductors are required, the feeder shall include or provide a grounding means to which the equipment grounding conductors of the branch circuits shall be connected.
    (a) equipment disconnecting means
    (b) electrical systems
    (c) branch circuits
    (d) electrical discharge equipment

137. A listed fixture or a listed fixture assemble shall be permitted to be cord-connected if located _____ the outlet box and the cord is continuously visible for its entire length outside the fixture and is not subject to strain or physical damage.
    (a) within   (b) directly below
    (c) directly above   (d) adjacent to

138. Overcurrent protection for conductors and equipment is provided to _____ the circuit if the current reaches a value that will cause an excessive or dangerous temperature in conductors or conductor insulation.
    (a) open   (b) close   (c) interrupt   (d) blow

139. Backfill used for underground wiring must not _____.
    (a) damage the wiring method
    (b) prevent compaction of the fill
    (c) contribute to the corrosion of the raceway
    (d) all of these

140. Junction boxes for pool lighting shall not be located less than _____ feet from the inside wall of a pool unless separated by a fence or wall.
    (a) 3   (b) 4   (c) 6   (d) 8

141. Feeder conductors for new restaurants shall not be required to be of _____ ampacity than the service-entrance conductors.
    (a) greater   (b) lesser   (c) equal   (d) none of these

142. An infrared heating lamp used in a medium base lampholder must be rated _____ watts or less.
    (a) 150   (b) 300   (c) 600   (d) 750

143. Electric space heating cables shall be furnished complete with factory-assembled non-heating leads at least _____ in length.
    (a) 6 inches   (b) 18 inches   (c) 3 feet   (d) 7 feet

144. The hand hole of metal fixture poles can be omitted for metal poles _____ feet or less above finish grade. This is only permitted if the pole is provided with a hinged base and the grounding terminal is accessible within the hinged base.
    (a) 10   (b) 18   (c) 20   (d) none of these

145. Electrical equipment installed in hazardous (classified) locations must be constructed for the class, division, and group. A group "D" atmosphere contains _____.
    (a) gasoline   (b) methane   (c) propane   (d) all of these

146. A device intended for the protection of personnel that functions to de-energize a circuit within an established period of time when a current to ground exceeds some predetermined value less than that required to operate the overcurrent device of the supply circuit is a(n) _____.
    (a) duel element fuse   (b) inverse time breaker
    (c) ground fault circuit interrupter   (d) safety switch

147. • No seal is required if a conduit (with no unions, couplings, boxes, or fittings) passes through a Class I, Division 1 location if the termination points of the unbroken conduit are in at least _____ inches in the unclassified location.
    (a) 6   (b) 12   (c) 18   (d) 24

148. Lighting track is a manufactured assembly designed to support and _____ lighting fixtures that are capable of being readily repositioned on the track.
    (a) connect   (b) protect   (c) energize   (d) all of these

149. Where it is unlikely that two dissimilar loads will be in use simultaneously, it shall be permissible to omit the smaller of the two in computing the total load to a feeder. This is known as _____.
    (a) dissimilar load factor   (b) coincident loads

(c) noncoincident loads (d) simultaneous load rule

150. Equipment bonding jumpers are not required for listed devices that have mounting screws that provide the grounding continuity between the metal yoke and the flush box (self-grounding receptacles).
 (a) True (b) False

151. Each circuit leading to or through a dispensing pump shall be provided with a switch or other acceptable means to disconnect _____ from the source of supply all conductors of the circuit, including the grounded neutral, if any.
 (a) automatically (b) simultaneously (c) manually (d) individually

152. If festoon lighting exceeds a span of _____ feet, the conductors shall be supported by messenger wires.
 (a) 20 (b) 30 (c) 40 (d) 50

153. Using standard load calculations, the feeder demand factor for five household clothes dryers is _____ percent.
 (a) 70 (b) 80 (c) 50 (d) 100

154. Class II locations are those that are hazardous because of the presence of _____.
 (a) combustible dust
 (b) easily ignitable fibers or flyings
 (c) flammable gases or vapors
 (d) flammable liquids or gases

155. Health care low voltage equipment frequently in contact with bodies of persons shall not exceed _____ volts.
 (a) 50 (b) 115 (c) 230 (d) 10

156. Liquidtight flexible metal conduit listed for grounding is permitted to serve as the equipment grounding conductor (with specific limitations) if the ground return path of the liquidtight flexible metal conduit does not exceed _____ feet.
 (a) 3 (b) 4 (c) 6 (d) 8

157. In nonstructural mounting of boxes and fittings, it shall be permissible to make a _____ installation in existing covered surfaces where adequate support is provided by clamps, anchors, or fittings.
 (a) temporary (b) workmanlike (c) permanent (d) flush

158. In a multiple-occupancy building where electric service and electrical maintenance are provided by the building management and where these are under continuous building management supervision, the service disconnecting means supplying more than one occupancy shall be permitted to be accessible to authorized _____ only.
 (a) inspectors
 (b) tenants
 (c) management personnel
 (d) none of these

159. In a multiple-occupancy building, each occupant shall have access to the _____.
 (a) overcurrent protection devices
 (b) receptacles
 (c) service-entrance assembly
 (d) panelboard

160. Hazards often occur because of _____.
 (a) overloading of wiring systems by methods or usage not in conformity with the Code
 (b) initial wiring not providing for increases in the use of electricity
 (c) a and b
 (d) none of these

161. A space of _____ inches shall be maintained between the resistor and reactors and any combustible material.
 (a) 24 (b) 12 (c) 14 (d) 20

162. The radius of the inner edge of any bend in Type MI cable shall not be less than five times the external diameter of the metallic sheath for cable not more than _____ inch in external diameter.
 (a) ¼ (b) ½ (c) ¾ (d) 1

163. Ground-fault protection of equipment shall be provided in accordance with the provisions of Section 230–95 for solidly grounded wye electrical systems of more than 150 volts to ground, but not exceeding 600 volts phase-to-phase for each building or structure main disconnecting means rated _____ amperes or more.
 (a) 1,000 (b) 1,500 (c) 2,000 (d) 2,500

Chapter 2 Advanced NEC Calculations And Code Questions   Unit 11 Commercial Load Calculations 373

164. The use of electrical metallic tubing shall be permitted for _____ work.
(a) exposed (b) concealed (c) a and b (d) none of these

165. The number of circuits in a single enclosure should be _____ to minimize the effects from a short-circuit or ground fault in one circuit.
(a) counted (b) limited (c) color coded (d) all of these

166. Each section of a manufactured wiring system shall be marked to identify _____.
(a) its location (b) its type of cable or conduit
(c) the size of the wires installed (d) suitability for wet or damp locations

167. Where individual open conductors are not exposed to the weather, the conductors shall be mounted on _____ knobs.
(a) door (b) insulated metal (c) glass or porcelain (d) none of these

168. Where individual open conductors are exposed to _____, the conductors shall be mounted on insulators or on insulating supports attached to racks, brackets, or other approved means.
(a) a corrosive environment (b) the weather (c) the general public (d) any inspector

169. • A spa or hot tub is a hydromassage pool, or tub, designed for the immersion of people. They are not generally designed or intended to have their contents drained or discharged after each use.
(a) True (b) False

170. So constructed or protected that dust will not interfere with its successful operation is called _____.
(a) dusttight (b) dustproof (c) dust rated (d) all of these

171. Electric pipe organ circuits shall be so arranged that all conductors shall be protected from overcurrent by an overcurrent device rated at not more than _____ amperes.
(a) 20 (b) 15 (c) 6 (d) none of these

172. • Switches or other equipment operating at 600 volts, nominal, or less, and serving only equipment within the high-voltage vault, room or enclosure shall be permitted to be installed in the _____ enclosure, room, or vault if accessible to qualified persons only.
(a) restricted (b) low-voltage (c) dedicated (d) high-voltage

173. Unless specified otherwise, live parts of electrical equipment operating at _____ volts or more shall be guarded.
(a) 12 (b) 15 (c) 50 (d) 24

174. Conductors supplying outlets for arc and Xenon projectors of the professional type shall not be smaller than No. _____.
(a) 12 (b) 10 (c) 8 (d) 6

175. • Exposed live parts of motors and controllers operating at _____ volts or more between terminals shall be guarded against accidental contact by enclosure or by location.
(a) 600 (b) 300 (c) 150 (d) 50

176. It is permissible to run unbroken lengths of surface metal raceways through dry _____.
I. walls II. partitions III. floors
(a) I only (b) II only (c) I and II (d) I, II, or III

177. Unless specifically permitted in Section 400–7, flexible cords and cables shall not be used where _____.
(a) run through holes in walls, ceilings, or floors
(b) run through doorways, windows, or similar openings
(c) attached to building surfaces
(d) all of these

178. Interior locations, protected from weather but subject to moderate degrees of moisture, such as some basements, some barns, some cold-storage warehouses and the like, the partially protected locations under canopies, marquees, roofed open porches, and the like, shall be considered to be _____ locations.
(a) damp (b) wet (c) hazardous (d) dry

179. Table 310–70 provides the ampacities of insulated single aluminum conductor isolated in air. If the conductor size is 8 AWG, and the voltage range is 2,001 to 5,000, then the ampacity is _____ amperes.
(a) 64 (b) 85 (c) 115 (d) 150

180. Conduits or raceways, including their end fittings, shall not rise more than _____ inches above the bottom of a switchboard enclosure.
(a) 3 (b) 4 (c) 5 (d) 6

181. The temperature rating associated with the ampacity of a _____ shall be so selected and coordinated as to not exceed the lowest temperature rating of any connected termination, conductor, or device.
(a) wire (b) conductor (c) fuse holder (d) circuit breaker

182. A single receptacle is a _____ contact device with no other contact device on the same yoke.
(a) dual (b) single (c) multiple (d) live

183. The purpose of receptacle placement in dwelling units is to _____ the use of cords across doorways, fireplaces, and similar openings.
(a) minimize (b) discourage (c) preclude (d) exempt

184. In locating receptacles in a marina, consideration should be given to _____.
(a) the maximum tide level
(b) the minimum tide level
(c) wave action
(d) a and c

185. The alternating current resistance for 1,000 feet of No. 4/0 aluminum conductor in a steel raceway is _____ ohms.
(a) 0.063 (b) 0.10 (c) 10.0 (d) 0.11

186. Support of Rigid metal conduit provided by a bored or punched hole in a _____ is intended to meet the support requirements of the Code.
(a) wall (b) truss (c) rafter (d) framing member

187. Circuit breakers shall be marked with a voltage rating no less than the nominal system voltage that is indicative of their capability to _____ fault currents between phases or phase to ground.
(a) break (b) interrupt (c) clear (d) detect

188. Each motor shall be provided with an individual controller.
(a) True (b) False

189. A receptacle shall be considered to be in a location protected from the weather where located under roofed open porches, canopies, marquees, and the like, and will not be subjected to _____.
(a) spray from a hose
(b) a direct lightning hit
(c) beating rain or water run-off
(d) falling or wind-blown debris

190. Auxiliary gutters shall be supported at intervals not exceeding _____ feet.
(a) 3 (b) 4 (c) 5 (d) 6

191. NM cable shall not be permitted for use _____.
(a) in moist, damp, or corrosive locations
(b) as service entrance cable
(c) in storage battery rooms
(d) all of these

192. In general, the minimum size phase, neutral or grounded conductor permitted for use in parallel is No. _____.
(a) 10 (b) 1 (c) 1/0 (d) 4

193. A _____ switch is a manually operated device used in conjunction with a transfer switch to provide a means of directly connecting load conductors to a power source, and of disconnecting the transfer switch.
(a) transfer (b) motor-circuit (c) general-use snap (d) bypass isolation

194. Which of the following is not a standard classification for a branch circuit supplying several loads?
(a) 20 amperes (b) 25 amperes (c) 30 amperes (d) 50 amperes

195. A secondary tie of a transformer is a circuit operating at _____ volts or less between phases that connects two power sources or power supply points.
(a) 600 (b) 1,000 (c) 12,000 (d) 35,000

Chapter 2 Advanced NEC Calculations And Code Questions   Unit 11 Commercial Load Calculations 375

196. Where connected to a branch circuit supplying _____ or more receptacles or outlets, a receptacle shall not supply a total cord- and plug-connected load in excess of the maximum specified in Table 210–21(b)(2).
 (a) two (b) three (c) four (d) five

197. Receptacles connected to circuits having different voltages, frequencies, or types of current (alternating or direct current) on the _____ shall be of such design that the attachment plugs used on these circuits are not interchangeable.
 (a) building (b) interior (c) same premises (d) exterior

198. The metal underground water pipe electrode must have a supplemental electrode. The supplemental electrode must consist of _____.
 (a) a plate made electrode
 (b) a ground ring of No. 2 copper
 (c) a metal frame of building that is grounded
 (d) any electrode listed in Sections 250–81 or 250–83

199. Electric equipment with a metal enclosure or with a nonmetallic enclosure listed for the use and having _____ shall be permitted to be installed in other space used for environmental air unless prohibited elsewhere in the Code.
 (a) adequate fire-resistant characteristics
 (b) adequate low-smoke-producing characteristics
 (c) wiring material suitable for the ambient temperature
 (d) all of these

200. • Enclosures that are not over 100 cubic inches in size and that have threaded entries or have hubs identified for the purpose, and that support fixtures or contain devices, or both, shall be considered adequately supported where two or more conduits are threaded wrenchtight into the enclosure or hubs and where each conduit is supported within _____ inches of the enclosure so as to provide a rigid and secure installation.
 (a) 12 (b) 18 (c) 24 (d) 36

Unit 12

Delta/Delta And Delta/Wye Transformer Calculations

OBJECTIVES

After reading this unit, the student should be able to briefly explain the following concepts:

- Current flow
- Delta transformer voltage
- Delta high leg
- Delta primary and secondary Line currents
- Delta primary or secondary phase Currents
- Delta phase versus line
- Delta current triangle
- Delta transformer balancing
- Delta transformer sizing
- Delta panel schedule in kVA
- Delta panelboard and conductor sizing
- Delta neutral current
- Delta maximum unbalanced load
- Delta/delta example
- Wye transformer voltage
- Wye voltage triangle
- Wye transformer current
- Line current
- Phase current
- Wye phase versus line
- Wye transformer loading and balancing
- Wye transformer sizing
- Wye panel schedule in kVA
- Wye panelboard and conductor sizing
- Wye neutral current
- Wye maximum unbalanced load
- Delta/wye example
- Delta versus wye

After reading this unit, the student should be able to briefly explain the following terms:

- Delta connected
- kVA rating
- Line
- Line current
- Line voltage
- Phase
- Phase current
- Phase load – delta
- Phase load – wye
- Phase voltage
- Ratio
- Balanced load
- Unbalanced load
- Delta secondary
- Wye secondary
- Winding
- Wye connected
- Primary/secondary voltage – amperes
- Three-phase load
- Single-phase load

INTRODUCTION

This unit deals with three-phase delta/delta and delta/wye transformer sizing and balancing. The system voltages used in this Unit were selected because of their common industry configuration.

DEFINITIONS

Delta Connected

Delta-connected means the windings of three single-phase transformers are connected in series to form a closed circuit. A line can be traced from one point of the delta system, through all transformers, then back to the original starting point (series). A Delta connected transformer is represented by the Greek letter delta Δ. Many call it a delta high leg system because the voltage from one conductor to ground is 208 volts, Fig. 12–1.

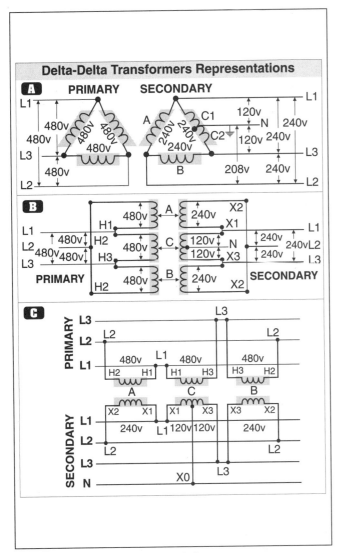

Figure 12-1

Different Ways of Drawing Delta Transformers

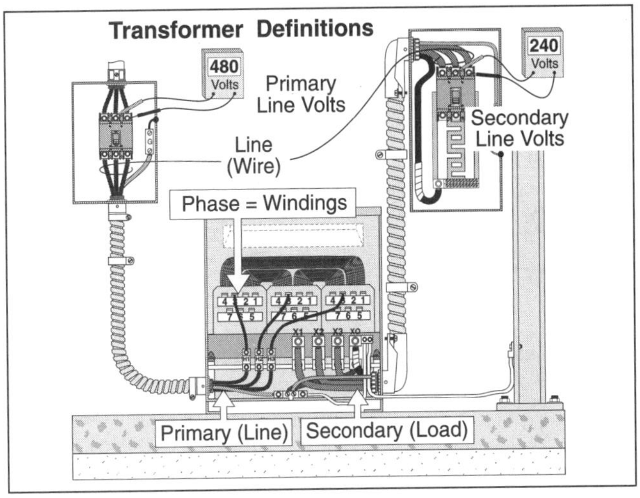

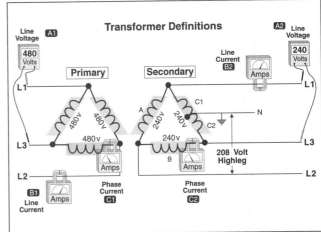

Figure 12-2

Transformer Definitions

Figure 12-3

Transformer Definitions

kVA Rating

Transformers are rated in *kilovolt-amperes* (kVA) and are sized according to the VA rating of the loads.

Note. One kilovolt-ampere is equal to 1,000 volt-amperes.

Line

The *line* is considered the electrical system supply (hot, ungrounded) conductors, Fig. 12–2 and Fig. 12–3.

Line Current

The *line current* for both delta and wye systems is the current on the line (hot, ungrounded) conductors, calculated according to the following formulas, Fig. 12–3, B1 and B2.

Single-Phase Line Current = VA_{Line}/E_{Line}

Three-Phase Line Current = $VA_{Line}/(E_{Line} \times \sqrt{3})$

Line Voltage

The line voltage is the voltage that is measured between any two line (hot ungrounded) conductors. It is also called line-to-line voltage. Line voltage is greater than phase voltage for wye systems and line voltage is the same as *phase voltage* for delta systems, Fig. 12–3, A1 and A2.

Phase

The *phase* winding is the coil shaped conductors that serve as the primary or secondary of the transformer.

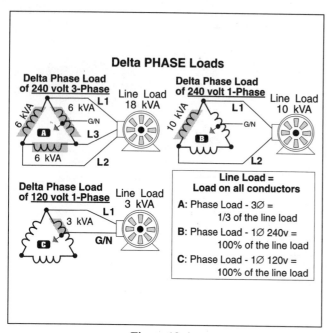

Figure 12-4
Delta Phase Loads

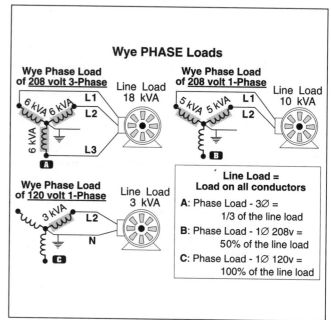

Figure 12-5
Wye Phase Loads

Phase Current

The *phase current* is the current of the transformer winding. For delta systems, the phase current is less than the line current. For wye systems, the phase current is the same as the line current, Fig. 12–3, C1 and C2.

Phase Load Delta

The *phase load* is the load on the transformer winding, Fig. 12–4.

The phase load of a three-phase, 240-volt load = line load/3.

The phase load of a single-phase, 240-volt load = line load.

The phase load of a single-phase, 120-volt load = line load.

Phase Load Wye

The phase load is the load on the transformer winding, Fig. 12–5.

The phase load of a three-phase, 208-volt load = line load/3.

The phase load of a single-phase, 208-volt load = line load/2.

The phase load of a single-phase, 120-volt load = line load.

Phase Voltage

The *phase voltage* is the internal transformer voltage generated across any one winding of a transformer. For a delta secondary, the phase voltage is equal to the line voltage. For wye secondaries, the phase voltage is less than the line voltage, Fig. 12–6.

Ratio (Voltage)

The *ratio* is the relationship between the number of primary winding turns as compared to the number of secondary winding turns. The ratio is a comparison between the primary phase voltage to the secondary phase voltage. For typical delta/delta systems, the ratio is 2:1. For typical wye systems, the ratio is 4:1, Fig. 12–7.

Unbalanced Load (Neutral Current)

The *unbalanced load* is the load on the secondary grounded (neutral) conductors.

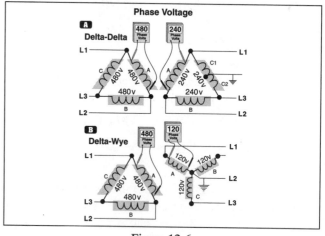

Figure 12-6
Phase Voltage

Delta Secondary – Unbalanced Current

The unbalanced load is calculated using:

$I_{Line\ 1} - I_{Line\ 3}$

Note. L_2 is called the high leg and its voltage to ground is approximately 208 volts, therefore no neutral loads are connected to this line.

Wye Secondary – Unbalanced Current

The unbalanced load is calculated using:

$\sqrt{L_1^2 + L_2^2 + L_3^2 - (L_1 \times L_2 + L_2 \times L_3 + L_1 \times L_3)}$

Winding

The *primary winding* is the winding(s) on the input side of a transformer and the *secondary winding* is the output side of a transformer.

Wye Connected

Wye-connected means a connection of three single-phase transformers of the same rated voltage to a common point (neutral) and the other ends are connected to the line conductors, Fig. 12–8.

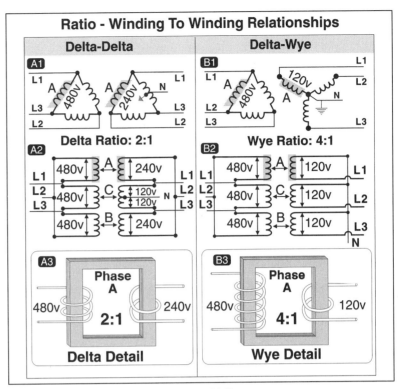

Figure 12-7

Ratio – Winding to Winding Relationships

12–1 CURRENT FLOW

When a load is connected to the secondary of a transformer, current will flow through the secondary conductor windings. The current flow in the secondary creates an electromagnetic field that opposes the primary electromagnetic field. The secondary flux lines effectively reduce the strength of the primary flux lines. As a result, less *counter-electromotive force* is generated in the primary winding conductors. With less *cemf* to oppose the primary applied voltage, the primary current automatically increases in direct proportion to the secondary current, Fig. 12–9.

Note. The primary and secondary line currents are inversely proportional to the voltage ratio of the transformer. This means that the winding with the most number of turns will have a higher voltage and lower current as compared to the winding with the least number of turns, which will have a lower voltage and higher current, Fig. 12–10.

The following Tables show the current relationship between kVA and voltage for common size transformers.

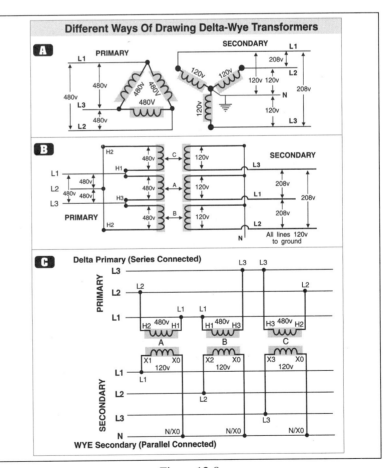

Figure 12-8

Different Ways of Drawing Delta/Wye Transformers

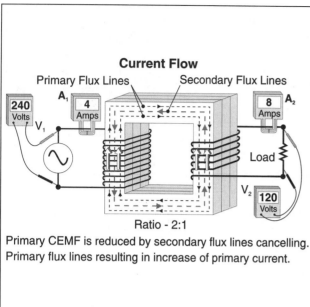

Figure 12-9
Current Flow

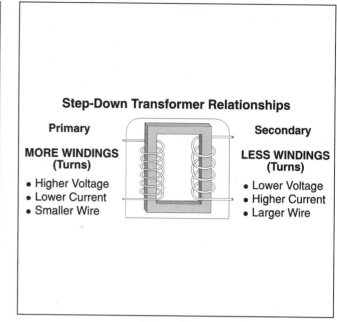

Figure 12-10
Step-Down Transformer Relationships

Single-Phase Transformers I=kVA/E			
kVA Rating	Current at 208 Volts	Current at 240 Volts	Current at 480 Volts
7.5	63 amperes	31 amperes	16 amperes
10	83 amperes	42 amperes	21 amperes
15	125 amperes	63 amperes	31 amperes
25	208 amperes	104 amperes	52 amperes
37.5	313 amperes	156 amperes	78 amperes

Three-Phase Transformers I=kVA/(E x √3)			
kVA Rating	Current at 208 Volts	Current at 240 Volts	Current at 480 Volts
15	42 amperes	36 amperes	18 amperes
22.5	63 amperes	54 amperes	27 amperes
30	83 amperes	72 amperes	36 amperes
37.5	104 amperes	90 amperes	45 amperes
45	125 amperes	108 amperes	54 amperes
50	139 amperes	120 amperes	60 amperes
75	208 amperes	180 amperes	90 amperes
112.5	313 amperes	271 amperes	135 amperes

PART A – *DELTA/DELTA TRANSFORMERS*

12–2 *DELTA TRANSFORMER VOLTAGE*

In a delta configured transformer, the *line voltage* equals the *phase voltage*, Fig. 12–11.

$E_{Line} = E_{Phase}$

Chapter 3 Advanced NEC Calculations And Code Questions   Unit 12 Delta/Delta And Delta/Wye Transformers 381

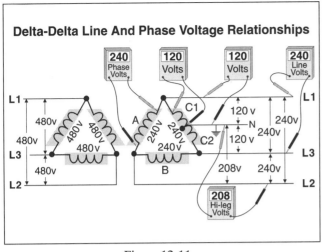

Figure 12-11
Delta/Delta Line/Phase Relationship

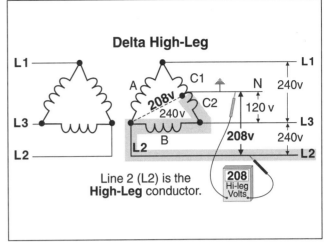

Figure 12-12
Delta High leg

Primary Delta Voltage

LINE Voltage	PHASE Voltage
L_1 to L_2 = 480 volts	Phase Winding A = 480 volts
L_2 to L_3 = 480 volts	Phase Winding B = 480 volts
L_3 to L_1 = 480 volts	Phase Winding C = 480 volts

Secondary Delta Voltage

LINE Voltage	PHASE Voltage	NEUTRAL Voltage
L_1 to L_2 = 240 volts	Phase Winding A = 240 volts	Neutral to L_1 = 120 volts
L_2 to L_3 = 240 volts	Phase Winding B = 240 volts	Neutral to L_2 = 208 volts
L_3 to L_1 = 240 volts	Phase Winding C = 240 volts	Neutral to L_3 = 120 volts

12–3 DELTA HIGH LEG

The term *high leg* (bastard leg, identified with an orange cover [384–3(e)]) is used to identify the conductor that has a higher voltage to ground. The high leg voltage is the vector sum of the voltage of transformer "A" and "C_1", or transformers "B" and "C_2," which equals 120 volts × 1.732 = 208 volts (for a 120/240 volt secondary). If the secondary voltage is 230/115 volts, the high leg voltage to ground (or neutral) would be 115 volts × 1.732 = 199.18 volts.

Note. The actual voltage is often less than the nominal system voltage because of voltage drop, Fig. 12–12.

12–4 DELTA PRIMARY AND SECONDARY LINE CURRENTS

In a delta configured transformer, the line current does not equal the phase current.
The primary or secondary *line current* of a transformer can be calculated by the formula:

$$I_{Line} = \frac{VA_{Line}}{E_{Line} \times \sqrt{3}}$$

❑ **Delta Primary Line Current Example**
What is the primary line current for a 150-kVA, 480- to 240/120-volt three-phase transformer, Fig. 12–13, Part A?

(a) 416 amperes   (b) 360 amperes   (c) 180 amperes   (d) 144 amperes

• Answer: (c) 180 amperes

$I_{Line} = VA_{Line}/(E_{Line} \times \sqrt{3})$

$I_{Line} = 150{,}000 \text{ VA}/(480 \text{ volts} \times 1.732)$

$I_{Line} = 180 \text{ amperes}$

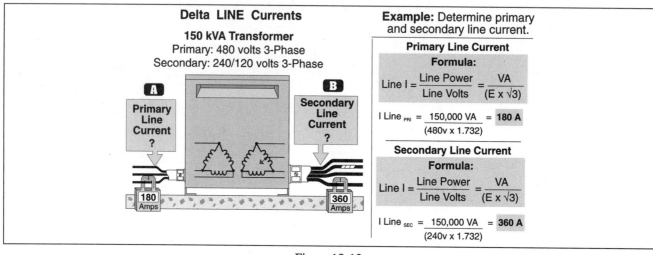

Figure 12-13
Delta Primary/Secondary Line Currents

☐ **Delta Secondary Line Current Example**
What is the secondary line current for a 150-kVA, 480- to 240-volt three-phase transformer, Fig. 12–13, Part B?

(a) 416 amperes (b) 360 amperes (c) 180 amperes (d) 144 amperes

• Answer: (b) 360 amperes

$I_{Line} = VA_{Line}/(E_{Line} \times \sqrt{3})$

$I_{Line} = 150{,}000 \text{ VA}/(240 \text{ volts} \times 1.732)$

$I_{Line} = 360$ amperes

12–5 DELTA PRIMARY OR SECONDARY PHASE CURRENTS

The *phase current* of a transformer winding is calculated by dividing the *phase VA** by the *phase volts*:

$I_{Phase} = VA_{Phase}/E_{Phase}$

The phase load of a three-phase 240-volt load = line load/3.
The phase load of a single-phase 240-volt load = line load.
The phase load of a single-phase 120-volt load = line load.

Note. Remember that it takes 3 phases or windings to create a 3-phase system.

☐ **Delta Primary Phase Current Example**
What is the primary phase current for a 150-kVA, 480- to 240/120-volt three-phase transformer, Fig. 12–14, Part A?

(a) 416 amperes (b) 360 amperes (c) 180 amperes (d) 104 amperes

• Answer: (d) 104 amperes

$I_{Phase} = \dfrac{VA_{Phase}}{E_{Phase}}$ $VA_{Phase} = \dfrac{150{,}000 \text{ VA}}{3 \text{ Phases}} = 50{,}000 \text{ VA}$ $E_{Phase} = 480$ volts

$I_{Phase} = 50{,}000 \text{ VA}/480$ volts

$I_{Phase} = 104$ amperes

☐ **Delta Secondary Phase Current Example**
What is the secondary phase current for a 150-kVA, 480 to 240/120 volt three-phase transformer, Fig. 12–14, Part B?

(a) 416 amperes (b) 360 amperes (c) 208 amperes (d) 104 amperes

• Answer: (c) 208 amperes

$I_{Phase} = \dfrac{VA_{Phase}}{E_{Phase}}$

Phase power = 150,000 VA/3 = 50,000 VA

$I_{Phase} = 50{,}000 \text{ VA}/240$ volts

$I_{Phase} = 208$ amperes

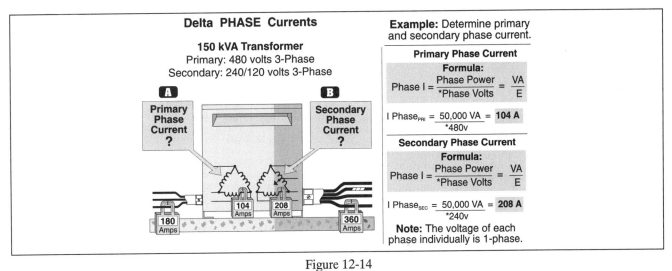

Figure 12-14

Primary/Secondary Phase Relationships

12-6 DELTA PHASE VERSUS LINE

Since each line conductor from a delta transformer is actually connected to two transformer windings (phases), the effects of loading on the line (conductors) can be different than on the phase (winding).

☐ Delta Phase VA, Three-Phase Example

A 36-kVA, 240-volt three-phase load has the following effect on a delta system, Fig. 12–15:

LINE: Total line power = 36 kVA

Line current = I_L = $VA_{Line}/(E_{Line} \times \sqrt{3})$

I_L = 36,000 VA/(240 volts × $\sqrt{3}$)

I_L = 87 amperes, or

$I_L = I_P \times \sqrt{3}$

I_L = 50 amperes × 1.732

I_L = 87 amperes

PHASE: Phase power = 12 kVA (winding)

Phase current = I_P = VA_{Phase}/E_{Phase}

I_P = 12,000 VA/240 volts = 50 amperes, or

$I_P = I_L/\sqrt{3}$

I_P = 87 amperes/1.732 = 50 amperes

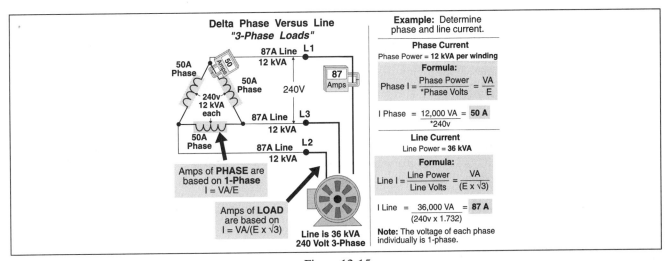

Figure 12-15

Delta Secondary Phase versus Line

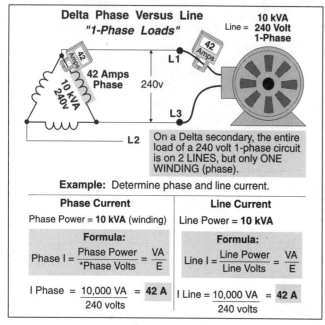

Figure 12-16
Delta Secondary Phase versus Line

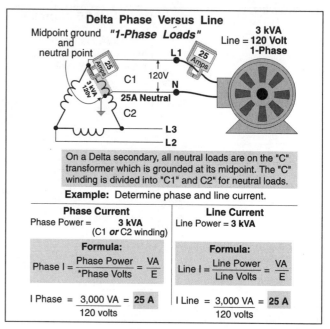

Figure 12-17
Delta Secondary Phase versus Line

❏ **Delta Phase VA, Single-Phase 240 Volt Example**

A 10-kVA, 240-volt single-phase load has the following effect on a delta/delta system, Fig. 12–16:

LINE: Total line power = 10 kVA
Line current = I_L = VA_{Line}/E_{Line} VA
I_L = 10,000 VA/240 volts
I_L = 42 amperes

PHASE: Phase power = 10 kVA (winding)
Phase current = I_P = VA_{Phase}/E_{Phase}
I_P = 10,000 VA/240 volts = 42 amperes

❏ **Delta Phase VA, Single-phase 120 Volt Example**

A 3-kVA, 120 volt single-phase load has the following effect on the system, Fig. 12–17:

LINE: Line power = 3 kVA
Line current = I_L = VA_{Line}/E_{Line} VA
I_L = 3,000 VA/120 volts
I_L = 25 amperes

PHASE: Phase power = 3 kVA (C_1 or C_2 winding)
Phase current = I_P = VA_{Phase}/E_{Phase}
I_P = 3,000 VA/120 volts = 25 amperes

12–7 DELTA CURRENT TRIANGLE

The three-phase line and phase current of a delta system are not equal, the difference is the square root of 3 ($\sqrt{3}$).

$I_L = I_P \times \sqrt{3}$ $I_P = I_L/\sqrt{3}$

The delta triangle (Fig. 12–18) can be used to calculate delta three-phase line and phase currents. Place your finger over the desired item and the remaining items show the formula to use.

12–8 DELTA TRANSFORMER BALANCING

To properly size a delta/delta transformer, the transformer phases (windings) must be *balanced*. The following steps are used to balance the transformer.

Step 1 → Determine the VA rating of all loads.

Step 2 → Balance three-phase loads: ⅓ on Phase A, ⅓ on Phase B, and ⅓ on Phase C.

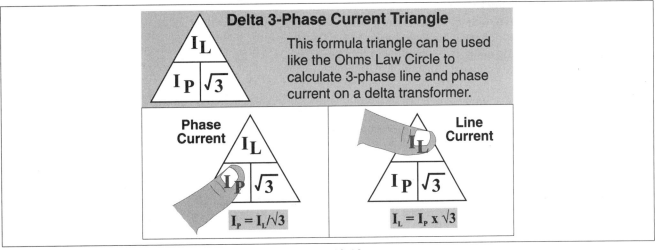

Figure 12-18
Delta Current Triangle

Step 3 → Balance single-phase 240 volt loads (largest to smallest): 100 percent on Phase A or Phase B. It is permissible to place some 240 volt single-phase load on Phase C when necessary for balance.

Step 4 → Balance the 120 volt loads (largest to smallest): 100 percent on C_1 or 100 percent on C_2.

❑ **Delta Transformer Balancing Example**

Balance and size a 480- to 240/120-volt three-phase transformer for the following loads: One 36 kVA three-phase heat strip, two 10 kVA single-phase 240-volt loads, and three 3 kVA 120-volt loads, Fig. 12–19.

	Phase A (L_1 and L_2)	Phase B (L_2 and L_3)	C_1 (L_1)	C_2 (L_3)	Line Total
(1) 36 kVA 240 volt, 3Ø	12 kVA	12 kVA	6 kVA	6 kVA	36,000 VA
(2) 10 kVA 240 volt, 1Ø	10 kVA	10 kVA			20,000 VA
(3) 3 kVA 120 volt, 1Ø			6 kVA *	3 kVA*	9,000 VA
	22 kVA	22 kVA	12 kVA	9 kVA	65,000 VA

*Indicates neutral load.

12-9 DELTA TRANSFORMER SIZING

Once you balance the transformer, you can size the transformer according to the load of each phase. The "C" transformer must be sized using two times the highest of "C_1" or "C_2." The "C" transformer is actually a single unit. If one side has a larger load, that side determines the transformer size.

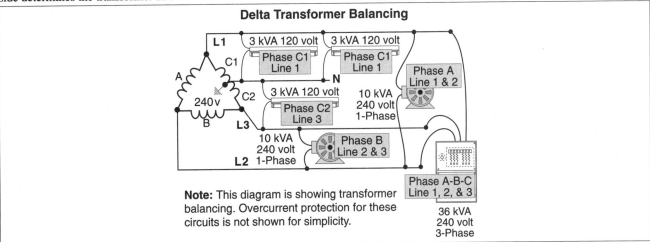

Figure 12-19
Delta Transformer Balancing

❏ **Delta Transformer Sizing Example**

What size 480- to 240/120-volt three-phase transformer is required for the following loads: One 36 kVA three-phase heat strip, two 10 kVA single-phase 240-volt loads, and three 3 kVA 120-volt loads?

(a) three 1Ø 25 kVA transformers (b) one 3Ø 75 kVA transformer
(c) a or b (d) none of these

- Answer: (c) a or b
  Phase winding A = 22 kVA
  Phase winding B = 22 kVA
  Phase winding C = (12 kVA of C_1 × 2) = 24 kVA

12–10 DELTA PANEL SCHEDULE IN kVA

When balancing a panelboard in kVA be sure that three-phase loads are balanced ⅓ on each line and 240-volt single-phase loads are balanced ½ on each line.

❏ **Delta Panel Schedule in kVA Example**

Balance a 240/120-volt three-phase panelboard for the following loads in kVA: one 36 kVA three-phase heat strip, two 10 kVA single-phase 240-volt loads, and three 3 kVA 120-volt loads.

	Line 1	Line 2	Line 3	Line Total
36 kVA 240 volt, 3Ø	12 kVA	12 kVA	12 kVA	36,000 VA
10 kVA 240 volt, 1Ø	5 kVA	5 kVA		10,000 VA
10 kVA 240 volt, 1Ø		5 kVA	5 kVA	10,000 VA
(3) 3 kVA 120 volt, 1Ø	+ 6 kVA *		3 kVA *	9,000 VA
	23 kVA	22 kVA	20 kVA	65,000 VA

*Indicates neutral load.

12–11 DELTA PANELBOARD AND CONDUCTOR SIZING

When selecting the sizing of *panelboards* and *conductors*, we must balance the line loads in amperes.

❏ **Delta Conductors Sizing Example**

Balance a 240/120-volt three-phase panelboard for the following loads in amperes: one 36 kVA three-phase heat strip, two 10 kVA single-phase 240-volt loads, and three 3 kVA 120-volt loads.

	Line 1	Line 2	Line 3	Line Amperes
36 kVA 240 volt, 3Ø	87 Amps	87 Amps	87 Amps	36,000 VA/(240 volts × 1.732)
10 kVA 240 volt, 1Ø	42 Amps	42 Amps		10,000 VA/240 volts
10 kVA 240 volt, 1Ø		42 Amps	42 Amps	10,000 VA/240 volts
3 – 3 kVA 120 volt, 1Ø	+ 50 Amps *		25 Amps *	3,000 VA/120 volts
	179 Amps	171 Amps	154 Amps	

*Indicates neutral load.

Why balance the panel in amperes? Why not take the VA per phase and divide by phase voltage?

- Answer: Line current of a three-phase load is calculated by the formula I_L = VA/(VA Line × $\sqrt{3}$). In our examples 36,000 VA/(240 volts × 1.732) = 87 amperes per line. If we took the per line power of 12,000 VA and divided by one line voltage of 120 volts, we come up with an incorrect line current of 12,000 VA/120 volts = 100 amperes.

12–12 DELTA NEUTRAL CURRENT

The *neutral current* for a delta secondary is calculated as $Line_1$ neutral current less $Line_3$ neutral current.

❏ **Delta Neutral Current Example**

What is the neutral current for the following loads: one 36 kVA three-phase heat strip, two 10 kVA, single-phase 240-volt loads, and three 3 kVA 120-volt loads?
In our continuous example, $Line_1$ neutral current = 50 amperes and $Line_3$ neutral current = 25 amperes.

(a) 0 amperes (b) 25 amperes (c) 50 amperes (d) none of these

- Answer: (b) 25 amperes.
  Neutral current = 50 amperes – 25 amperes = 25 amperes

Chapter 3 Advanced NEC Calculations And Code Questions   Unit 12 Delta/Delta And Delta/Wye Transformers 387

	Line 1	Line 2	Line 3	Ampere Calculation
36 kVA 240 volts, 3Ø	87 amps	87 amps	87 amps	36,000 VA/(240 volts × 1.732)
10 kVA 240 volts, 1Ø	42 amps	42 amps		10,000 VA/240 volts
10 kVA 240 volts, 1Ø		42 amps	42 amps	10,000 VA/240 volts
3 kVA 120 volts, 1Ø	50 amps *		25 amps *	3,000 VA/120 volts

* Indicates neutral (120-volt) loads

12–13 DELTA MAXIMUM UNBALANCED LOAD

The *maximum unbalanced load* (neutral) is the largest phase 120 volt neutral current. The neutral current is the actual current on the neutral conductor.

☐ **Delta Maximum Unbalanced Load Example**

What is the maximum unbalanced load for the following loads: one 36-kVA three-phase heat strip, two 10-kVA single-phase 240-volt loads, and three 3-kVA 120-volt loads?

(a) 0 amperes   (b) 25 amperes   (c) 50 amperes   (d) none of these

• Answer: (c) 50 amperes

Maximum unbalanced current equals the largest line neutral current.

	Line 1	Line 2	Line 3	Ampere Calculation
36 kVA 240 volt, 3Ø	87 amps	87 amps	87 amps	36,000 VA/(240 volts × 1.732)
10 kVA 240 volt, 1Ø	42 amps	42 amps		10,000 VA/240 volts
10 kVA 240 volt, 1Ø		42 amps	42 amps	10,000 VA/240 volts
3 kVA 120 volt, 1Ø	50 amps *		25 amps *	3,000 VA/120 volts

*Indicates neutral (120-volt) loads

12–14 DELTA/DELTA EXAMPLE

One – Dishwasher 4.5 kW, 120v
One – 10 HP A/C motor 240-volt, 3Ø
One – 18 kW 240-volt 3Ø heat strip
Two – 14 kW ranges, 240-volt 1Ø

Two – 3 HP motors, 240-volt, 1Ø
One – 10 kW water heater, 240-volt, 1Ø
Eight – 1.5 kW 120-volt lighting circuits

Motor VA

VA 1Ø = Table Volts × Table Amperes, Table 430–148
3 HP, VA = E × I
VA = 230* volts × 17 amperes = 3,910 VA, Table 430–150
10 HP, VA = 230 volts* × 28 amperes × 1.732 = 11,154
VA 3Ø = Table Volts × Table Amperes × $\sqrt{3}$, Table 430–150

Note. Some exam testing agencies use actual volts (240 volts) instead of table volts.

	Phase A (L$_1$ & L$_2$)	Phase B (L$_2$ & L$_3$)	C$_1$ (L$_1$)	C$_2$ (L$_3$)	Line Total
Heat 18 kW (A/C omitted)	6,000 VA	6,000 VA	3,000 VA	3,000 VA	18,000 VA
14 kW Ranges 240 volt, 1Ø	14,000 VA	14,000 VA			28,000 VA
10 kW Water Heater 240 volt, 1Ø	10,000 VA				10,000 VA
3 HP 240 volt, 1Ø		3,910 VA			3,910 VA
3 HP 240 volt, 1Ø		3,910 VA			3,910 VA
Dishwasher 4.5 kW, 120 volt			4,500 VA *		4,500 VA
Lighting (8 – 1.5 kW), 120 volt (3 on L$_1$, 5 on L$_3$)			4,500 VA *	7,500 VA *	12,000 VA
	30,000 VA	27,820 VA	12,000 VA	10,500 VA	80,320 VA

*Indicates neutral (120 volt) loads.

Note. The phase totals (30,000 VA, 27,820 VA, 22,500 VA) should add up to the Line total (80,320 VA). This is done as a check to make sure all items have been accounted for and added correctly.

10 horsepower, VA = Table Volts × Table Amperes × 1.732
VA = 230 volts × 28 amperes × 1.732 = 11,154 VA (omit)

➤ Transformer Size Question

What size transformers are required?

(a) three 30 kVA single-phase transformers (b) one 90 kVA three-phase transformer

(c) a or b (d) none of these

- Answer: (c) a or b

  Phase A = 30 kVA

  Phase B = 28 kVA

  Phase C = 24 kVA (12 kVA × 2)

➤ High leg Voltage Question

What is the high leg voltage to the ground?

(a) 120 volts (b) 208 volts (c) 240 volts (d) None of these

- Answer: (b) 208 volts

  120 volts × 1.732 = 208 volts, in effect, the vector sum of the voltage of transformer "A" and "C_1" or transformers "B" and "C_2." 240 volts + 120 volts vectorial = 208 volts

➤ Neutral kVA Question

What is the maximum kVA on the neutral?

(a) 3 kVA (b) 6 kVA (c) 7.5 kVA (d) 9 kVA

- Answer: (d) 9 kVA

  If Phase C_2 loads are not on; then phase C_1 neutral loads would impose a 9 kVA load to neutral. (4.5 kW + 4.5 kW). This does not include the 6 kVA on transformer "C" from the heat load since there are no neutrals involved.

➤ Maximum Unbalanced Current Question

What is the maximum unbalanced load on the neutral?

(a) 25 ampere (b) 50 ampere (c) 75 amperes (d) 100 amperes

- Answer: (c) 75 amperes

  I = P/E = 9,000 VA/120 volts = 75 amperes.

  The neutral must be sized to the carry maximum unbalanced neutral current, which in this case is 75 amperes.

➤ Neutral Current Question

What is the current on the neutral?

(a) 0 amperes (b) 13 amperes (c) 25 amperes (d) none of these

- Answer: b) 13 amperes

  C_1 = 9,000 VA/120 volts = 75 amperes

  C_2 = 7,500 VA/120 volts = 62.5 amperes

  Neutral current = 75 amperes less 62.5 amperes = 12.5 amperes

➤ Phase VA Load, Single-phase Question

What is the phase load of each 3-horsepower 240-volt single-phase motor?

(a) 1,855 VA (b) 3,910 VA (c) 978 VA (d) none of these

- Answer: (b) 3,910 VA

  VA = 230 volts × 28 amperes. On a delta system, a 240-volt single-phase load is entirely on one transformer phase.

➤ Phase VA Load, Three-phase Question

What is the phase load for the three-phase, 10-horsepower motor?

(a) 11,154 VA (b) 3,718 VA (c) 5,577 VA (d) none of these

- Answer: (b) 3,718 VA

  VA = Volts × Amperes × 1.732

  VA = 230 volts × 28 amperes × 1.732, VA = 11,154 VA

  11,154 VA/3 phases = 3,718 VA per phase

➤ High leg Conductor Size Question
If only the three-phase 18 kW load is on the high leg, what is the current on the high leg conductor?
(a) 25 amperes   (b) 43 amperes   (c) 73 amperes   (d) 97 amperes

- Answer: (b) 43 amperes
  To calculate current on any conductor:
  $I = P/(E \times \sqrt{3})$ (three-phase)
  $I = 18{,}000 \text{ VA}/(240 \text{ volts} \times 1.732) = 43$ amperes

➤ Voltage Ratio Question
What is the phase voltage ratio of the transformer?
(a) 4:1   (b) 1:4   (c) 1:2   (d) 2:1

- Answer: (d) 2:1
  480 primary phase volts to 240 secondary phase volts

❑ Panel Loading VA Example
Balance the loads on the panelboard in VA.

	Line 1	Line 2	Line 3	Line Total
18 kW Heat 3Ø	6,000 VA	6,000 VA	6,000 VA	18,000 VA
14 kW Ranges 240 volts, 1Ø	7,000 VA	7,000 VA		14,000 VA
14 kW Ranges 240 volts, 1Ø		7,000 VA	7,000 VA	14,000 VA
Water Heater, 240 volts, 1Ø	5,000 VA	5,000 VA		10,000 VA
3 HP 240 volts, 1Ø Motor		1,955 VA	1,955 VA	3,910 VA
3 HP 240 volts, 1Ø Motor		1,955 VA	1,955 VA	3,910 VA
Dishwasher 1Ø 120V	4,500 VA*			4,500 VA
Lighting 1Ø 120V	+ 4,500 VA*		7,500 VA *	4,500 VA
	27,000 VA	28,910 VA	24,410 VA	80,320 VA

* Indicates neutral (120 volt) loads.

❑ Panel Sizing, Amperes Example
Balance the loads from the previous example on the panelboard in amperes.

	Line 1	Line 2	Line 3	Ampere Calculation
18 kW Heat 3Ø	43 amps	43 amps	43 amps	18,000 VA/(240 volts × 1.732)
14 kW Ranges 240 volts, 1Ø	58 amps	58 amps		14,000 VA/240 volts
14 kW Ranges 240 volts, 1Ø		58 amps	58 amps	14,000 VA/240 volts
Water Heater, 240 volts, 1Ø	42 amps	42 amps		10,000 VA/240 volts
3 HP 240 volts, 1Ø Motor		17 amps	17 amps	FLC
3 HP 240 volts, 1Ø Motor		17 amps	17 amps	FLC
Dishwasher 1Ø 120 volts	38 amps *			4,500 VA/120 volts
Lighting 1Ø 120 volts	+ 38 amps *		63 amps *	7,500 VA/120 volts
(3 loads on L₁ and 5 loads on L₃)	219 amps	235 amperes	198 amps	

*Indicates neutral (120 volt) loads.

PART B – DELTA/WYE TRANSFORMERS

12-15 WYE TRANSFORMER VOLTAGE

In a wye configured transformer, the *line voltage* does not equal the *phase voltage*, Fig. 12–20.

> **Reminder:** In a delta configured transformer, line current does not equal phase current.

Primary Voltage (Delta)

LINE Voltage	PHASE Voltage
L₁ to L₂ = 480 volts	Phase Winding A = 480 volts
L₂ to L₃ = 480 volts	Phase Winding B = 480 volts
L₃ to L₁ = 480 volts	Phase Winding C = 480 volts

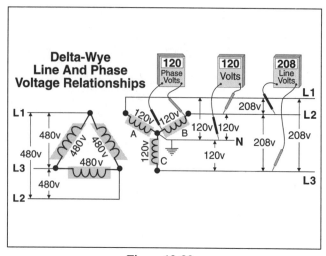

Figure 12-20
Delta/Wye Voltage Relationships

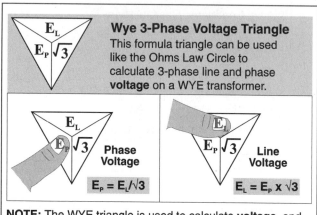

Figure 12-21
Wye Voltage Triangle

Secondary Voltage (Wye)

LINE Voltage	PHASE Voltage	NEUTRAL Voltage
L_1 to L_2 = 208 volts	Phase Winding A = 120 volts	Neutral to L_1 = 120 volts
L_2 to L_3 = 208 volts	Phase Winding B = 120 volts	Neutral to L_2 = 120 volts
L_3 to L_1 = 208 volts	Phase Winding C = 120 volts	Neutral to L_3 = 120 volts

12–16 WYE VOLTAGE TRIANGLE

The *line voltage* and *phase voltage* of a wye system are not the same. The difference is a factor of the $\sqrt{3}$.
$E_{Line} = E_{Phase} \times \sqrt{3}$ or $E_{Phase} = E_{Line}/\sqrt{3}$

The wye voltage triangle (Fig. 12–21) can be used to calculate wye three-phase line and phase voltage. Place your finger over the desired item, and the remaining items show the formula to use. These formulas can be used when either one of the phase or line voltages is known.

12–17 WYE TRANSFORMERS CURRENT

In a wye configured transformer, the three-phase and single-phase 120 volt *line current* equals the *phase current*, Fig. 12–22.
$I_{Phase} = I_{Line}$

12–18 WYE LINE CURRENT

The **line current** of both delta and wye transformers can be calculated by the following formula:
$I_{Line} = VA_{Line}/(E_{Line} \times \sqrt{3})$

❏ **Primary Line Current Question**

What is the primary line current for a 150 kVA 480- to 208Y/120-volt three-phase transformer, Fig. 12–23?

   (a) 416 amperes   (b) 360 amperes
   (c) 180 amperes   (d) 144 amperes

• Answer: (c) 180 amperes.
$I_{Line} = VA_{Line}/(E_{Line} \times \sqrt{3})$
$I_{Line} = 150{,}000 \text{ VA}/(480 \text{ volts} \times 1.732)$
$I_{Line} = 180 \text{ amperes}$

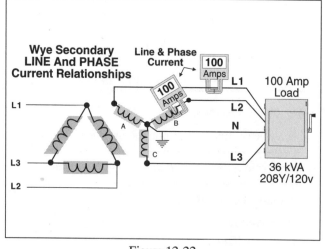

Figure 12-22
Wye Secondary Line and Phase Current

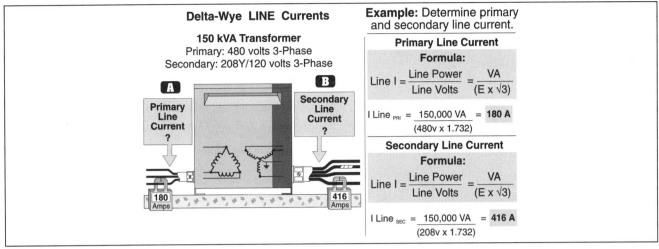

Figure 12-23
Delta/Wye Primary/Secondary Line Current

☐ **Secondary Line Current Question**
What is the secondary line current for a 150-kVA, 480- to 208Y/120-volt, three-phase transformer, Fig. 12–23?

(a) 416 amperes (b) 360 amperes (c) 180 amperes (d) 144 amperes

• Answer: (a) 416 amperes

$I_{Line} = VA_{Line}/(E_{Line} \times \sqrt{3})$

$I_{Line} = 150{,}000 \text{ VA}/(208 \text{ volts} \times 1.732)$

$I_{Line} = 416 \text{ amperes}$

12–19 WYE PHASE CURRENT

The phase current of a wye configured transformer winding is the same as the line current. The phase current of a delta configured transformer winding is less than the line current by the $\sqrt{3}$.

☐ **Primary Phase Current Question**
What is the primary (delta) phase current for a 150-kVA, 480- to 208Y/120-volt, three-phase transformer, Fig. 12–24, Part A?

(a) 416 amperes (b) 360 amperes (c) 180 amperes (d) 104 amperes

• Answer: (d) 104 amperes

$I_{Line} = \dfrac{VA_{Line}}{E_{Line} \times \sqrt{3}}$

$I_{Line} = 150{,}000 \text{ VA}/(480 \text{ volts} \times 1.732)$

$I_{Line} = 180 \text{ amperes}$

$I_{Phase} = \dfrac{VA_{Phase}}{E_{Phase}}$

$I = 50{,}000 \text{ VA}/480 \text{ volts}$

$I_{Phase} = 104 \text{ amperes, or}$

$I_{Phase} = I_{Line}/\sqrt{3} = 108/1.732 = 104 \text{ amperes}$

☐ **Secondary Phase Current Question**
What is the secondary (wye) phase current for a 150-kVA, 480- to 208Y/120-volt, three-phase transformer, Fig. 13-24, Part B?

(a) 416 amperes (b) 360 amperes (c) 208 amperes (d) 104 amperes

• Answer: (a) 416 amperes

$I_{Line} = \dfrac{VA_{Line}}{(E_{Line} \times \sqrt{3})}$

$I_{Line} = 150{,}000 \text{ VA}/(208 \text{ volts} \times 1.732)$

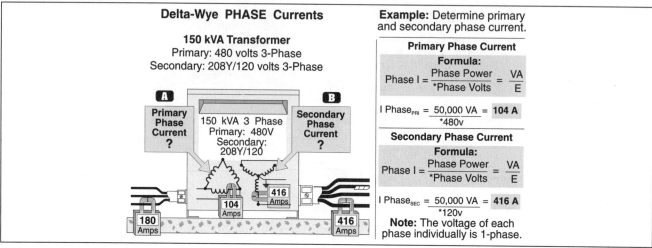

Figure 12-24
Delta/Wye Primary/Secondary Phase Current

I_{Line} = 416 amperes, or

$I_{Phase} = \dfrac{VA_{Phase}}{E_{Phase}}$

I_{Phase} = 50,000 VA/120 volts

I_{Phase} = 416 ampere

12–20 WYE PHASE VERSUS LINE

Since each line conductor from a wye transformer is connected to a different transformer winding (phase), the effects of three-phase loading on the line are the same as the phase, Fig. 12–25.

☐ **Phase VA Load, Three-Phase Example**

A 36-kVA, 208 volt, three-phase load has the following effect on the system:

LINE: Line power = 36 kVA

Line current = I_L = VA/(E × $\sqrt{3}$)

I_L = 36,000 VA/(208 volts × $\sqrt{3}$)

I_L = 100 amperes

PHASE: Phase power = 12 kVA (any winding)
Phase current = I_P = VA_{Phase}/E_{Phase}
I_P = 12,000 VA/120 volts = 100 amperes

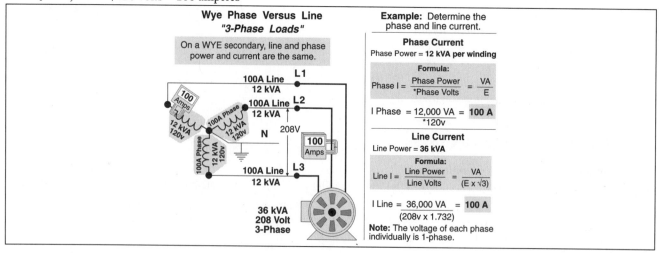

Figure 12-25
Wye Secondary Phase vs. Line

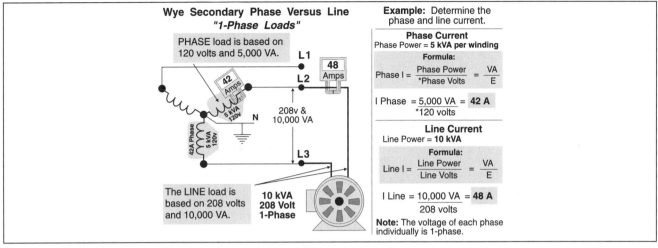

Figure 12-26
Wye Secondary Phase vs. Line

❑ **Phase VA Load, Single-Phase 208 Volt Example**
A 10-kVA, 208-volt, single-phase load has the following effect on the system, Fig. 13-26:
LINE: Line power = 10 kVA
Line current = I_L = VA/E
I_L = 10,000 VA/208 volts
I_L = 48 amperes
PHASE: Phase power = 10 kVA (winding)
Phase current = I_P = VA_{Phase}/E_{Phase}
I_P = 5,000 VA/120 volts
I_P = 42 amperes

❑ **Phase VA Load, Single-phase 120 Volt Example**
A 3-kVA, 120-volt single-phase load has the following effect on the system, Fig. 12–27:
LINE: Line power = 3 kVA
Line current = I_L = VA/E
I_L = 3,000 VA/120 volts
I_L = 25 amperes
PHASE: Phase power = 3 kVA (any winding)
Phase current = I_P = VA_{Phase}/E_{Phase}
I_P = 3,000 VA/120 volts
I_P = 25 amperes

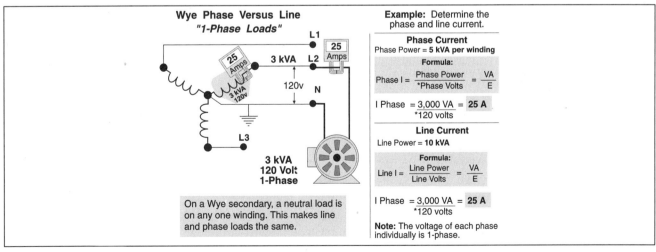

Figure 12-27
Wye Secondary Phase vs. Line

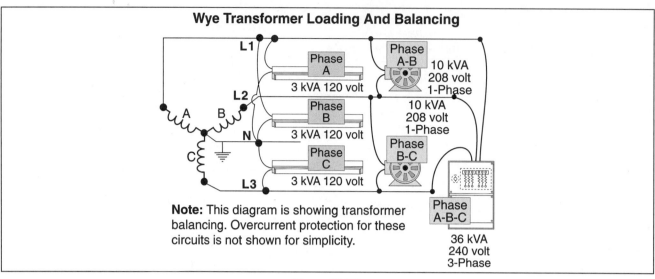

Figure 12-28
Wye Transformer Loading and Balancing

12-21 WYE TRANSFORMER LOADING AND BALANCING

To properly size a delta/wye transformer, the secondary transformer phases (windings) or the line conductors must be balanced. Balancing the panel (line conductors) is identical to balancing the transformer for wye configured transformers! The following steps should be helpful to balance the transformer.

Step 1: → Determine the VA rating of all loads.
Step 2: → Balance 3Ø loads: ⅓ on Phase A, ⅓ on Phase B, and ⅓ on Phase C.
Step 3: → Balance single-phase 208-volt loads (largest to smallest): 50 percent on each phase (A to B, B to C, or A to C).
Step 4: → Balance the 120-volt loads (largest to smallest): 100 percent on any phase.

❏ **Transformer Loading and Balancing Example**

Balance and size a 480- to 208Y/120-volt three-phase phase transformer for the following loads: One 36 kVA, three-phase heat strip, two 10-kVA, single-phase 208-volt loads, and three 3 kVA, 120-volt loads, Fig. 12-28.

	Phase A (L$_1$)	Phase B (L$_2$)	Phase C (L$_3$)	Line Total
36 kVA 208 volt, 3Ø	12 kVA	12 kVA	12 kVA	36 kVA
10 kVA 208 volt, 1Ø	5 kVA	5 kVA		10 kVA
10 kVA 208 volt, 1Ø		5 kVA	5 kVA	10 kVA
(3) 3 kVA 120 volt, 1Ø	+ 6 kVA *		3 kVA *	9 kVA
	23 kVA	22 kVA	20 kVA	65 kVA

* Indicates neutral (120 volt) loads.

12-22 WYE TRANSFORMER SIZING

Once you balance the transformer, you can size the transformer according to the load of each phase.

❏ **Transformer Sizing Example**

What size 480- to 208Y/120-volt, three-phase phase transformer is required for the following loads: one 36-kVA, three-phase heat strip, two 10-kVA single-phase 208-volt loads, and three 3-kVA 120-volt loads?

(a) three 1Ø 25 kVA transformer (b) one 3Ø 75 kVA transformer

(c) a or b (d) none of these

• Answer: (c) a or b
  Phase A = 23 kVA
  Phase B = 22 kVA
  Phase C = 20 kVA

	Phase A (L₁)	Phase B (L₂)	Phase C (L₃)	Line Total
36 kVA 208 volts, 3Ø	12 kVA	12 kVA	12 kVA	36 kVA
10 kVA 208 volts, 1Ø	5 kVA	5 kVA		10 kVA
10 kVA 208 volts, 1Ø		5 kVA	5 kVA	10 kVA
(3) 3 kVA 120 volts, 1Ø	+ 6 kVA *		3 kVA *	9 kVA
	23 kVA	22 kVA	20 kVA	65 kVA

* Indicates neutral (120 volt) loads.

12–23 WYE PANEL SCHEDULE IN kVA

When balancing a panelboard in kVA be sure that three-phase loads are balanced ⅓ on each line and 208 volt 1Ø loads are balanced ½ on each line.

❏ Panel Schedule, kVA Example

Balance and size a 208Y/120-volt, three-phase phase panelboard in kVA for the following loads: one 36-kVA, three-phase heat strip, two 10-kVA, single-phase 208-volt loads, and three 3-kVA, 120-volt loads.

	Line 1	Line 2	Line 3	Line Total
36 kVA 208 volt, 3Ø	12 kVA	12 kVA	12 kVA	36,000 VA
10 kVA 208 volt, 1Ø	5 kVA	5 kVA		10,000 VA
10 kVA 208 volt, 1Ø		5 kVA	5 kVA	10,000 VA
3 kVA 120 volt, 1Ø	6 kVA *			6,000 VA
3 kVA 120 volt, 1Ø			3 kVA *	3,000 VA
	23 kVA	22 kVA	20 kVA	65,000 VA

* Indicates neutral (120 volt) loads.

12–24 WYE PANELBOARD AND CONDUCTOR SIZING

When selecting and sizing the panelboard and conductors, we must balance the line loads in amperes.

❏ Panelboard Conductors Sizing Example

Balance and size a 208Y/120-volt, three-phase panelboard in amperes for the following loads: one 36-kVA, three-phase heat strip, two 10-kVA, single-phase 208-volt loads, and three 3-kVA, 120-volt loads.

	Line 1	Line 2	Line 3	Ampere Calculation
36 kVA 208 volt, 3Ø	100 amps	100 amps	100 amps	36,000/(208 volts × 1.732)
10 kVA 208 volt, 1Ø	48 amps	48 amps		10,000 VA/208 volts
10 kVA 208 volt, 1Ø		48 amps	48 amps	10,000 VA/208 volts
2 – 3 kVA 120 volt, 1Ø	50 amps *			6,000 VA/120 volts
3 kVA 120 volt, 1Ø	+		25 amps *	3,000 VA/120 volts
	198 amps	196 amps	173 amps	

Why balance the panel in amperes? Why not take wattage or VA per phase and divide by one phase voltage?

- Answer: Line current of a single-phase load is calculated by the formula, I_L = VA Line/ELine.

  In our examples, I_L = 10,000 VA/208 volts = 48 amperes per line.

  If we took the per line power of 5,000 and divided by one line voltage of 120 volts, we come up with an incorrect line current of 5,000 VA/120 volts = 42 amperes.

12–25 WYE NEUTRAL CURRENT

To determine the neutral current of a wye system, we use the following formula:

I neutral = $\sqrt{L_1^2 + L_2^2 + L_3^2 - (L_1 \times L_2 + L_2 \times L_3 + L_1 \times L_3)}$

❏ Neutral Current Example

Balance and size the neutral current for the following loads: one 36-kVA three-phase heat strip, two 10-kVA single-phase 208-volt loads, and three 3-kVA 120-volt loads. L_1 = 50 amperes, L_2 = 0 amperes, L_3 = 25 amperes.

(a) 0 amperes   (b) 25 amperes   (c) 35 amperes   (d) 50 amperes

- Answer: (c) 35 amperes
  Based on the previous example

$I_{Neutral} = \sqrt{L_1^2 + L_2^2 + L_3^2 - (L_1 \times L_2 + L_2 \times L_3 + L_1 \times L_3)}$

$I_{Neutral} = \sqrt{(50^2 + 0^2 + 25^2) - [(50 \times 0) + (0 \times 25) + (50 \times 25)]}$

$I_{Neutral} = \sqrt{2,500 + 625 - 1250} = \sqrt{1,875} = 43.3$ amperes neutral current

12–26 WYE MAXIMUM UNBALANCED LOAD

The maximum unbalanced load (neutral) is the largest phase 120 volt neutral current with other lines off.

❏ **Maximum Unbalanced Load Example**
Balance and size the maximum unbalanced load for the following loads: one 36-kVA, three-phase heat strip, two 10-kVA, single-phase 208-volt loads, and three 3-kVA, 120-volt loads. L1 = 50 amperes, L2 = 0 amperes, L3 = 25 amperes.

(a) 0 amperes   (b) 25 amperes   (c) 50 amperes   (d) none of these

- Answer: (c) 50 amperes
  The maximum unbalanced current equals the largest line neutral current.

	Line 1	Line 2	Line 3	Ampere Calculation
36 kVA 208 volts, 3Ø	100 amps	100 amps	100 amps	36,000 VA/(208 volts × 1.732)
10 kVA 208 volts, 1Ø	48 amps	48 amps		10,000 VA/208 volts
10 kVA 208 volts, 1Ø		48 amps	48 amps	10,000 VA/208 volts
2 – 3 kVA 120 volts, 1Ø	50 amps *			3,000 VA/120 volts
3 kVA 120 volts, 1Ø			25 amps *	3,000 VA/120 volts
	198 amps	196 amps	173 amps	

* Indicates neutral (120 volt) loads.

12–27 DELTA/WYE EXAMPLE

One – Dishwasher 4.5 kW, 120-volts   Two – 3 HP motors, 208-volts, 1Ø
One – 10 HP A/C motor 208-volts, 3Ø   One 10 kW water heater 208-volts, 1Ø
One – 18 kW 208-volts, 3Ø heat strip   Eight 1.5 kW, 120-volts, lighting circuits
Two – 14 kW ranges 208-volts, 1Ø

	Phase A (L1)	Phase B (L2)	Phase C (L3)	Line Total
Heat (A/C omitted *)	6,000 VA	6,000 VA	6,000 VA	18.0 kVA
14 kW Ranges 208 volt, 1Ø	7,000 VA	7,000 VA		14.0 kVA
14 kW Ranges 208 volt, 1Ø		7,000 VA	7,000 VA	14.0 kVA
10 kW Water Heater 208 volt, 1Ø	5,000 VA		5,000 VA	10.0 kVA
3 HP 208 volt, 1Ø Motor	1,945 VA	1,945 VA		3.89 kVA
3 HP 208 volt, 1Ø Motor		1,945 VA	1,945 VA	3.89 kVA
Dishwasher 4.5 kW, 120 volt			4,500 VA *	4.5 kVA
Lighting (8 – 1.5 kW circuits)	6,000 VA *	3,000 VA *	3,000 VA *	12.0 kVA
	25,945 VA	26,890 VA	27,445 VA	80.28 kVA

Note. The phase totals (26 kVA, 27 kVA and 27.5 kVA) should add up to the line total (80.5 kVA). This method is used as a check to make sure all items have been accounted for and added correctly.

Motor VA
VA 1Ø = Table Volts × Table Amperes
3 HP, VA = E × I, VA = 208 volts × 18.7 amperes = 3,890 VA/2 = 1,1945 VA per phase
VA 3Ø = Table Volts × Table Amperes × √3

10 HP A/C, VA = Table Volts × Table Amperes × 1.732
10 HP A/C, VA = 208 volts × 30.8 amperes × 1.732 = 11,096 VA (omitted - smaller than 18 kW heat)

➤ Transformer Sizing Question
What size transformers are required?

(a) three 30 kVA single-phase transformers (b) one 90 kVA three-phase transformer
(c) a or b (d) none of these

- Answer: (c) a or b
  Phase A = 26 kVA
  Phase B = 27 kVA
  Phase C = 28 kVA

➤ Maximum kVA on Neutral Question
What is the maximum kVA on the neutral?

(a) 3 kVA (b) 6 kVA (c) 7.5 kVA (d) 9 kVA

- Answer: (c) 7.5 kVA Phase "C"
  Count 120-volt loads only, do not count phase to phase (208-volt) loads

➤ Maximum Neutral Current Question
What is the maximum current on the neutral?

(a) 50 ampere (b) 63 ampere (c) 75 amperes (d) 100 amperes

- Answer: (b) 63 amperes
  I = P/E = 7,500 VA/120 volts = 63 amperes. The neutral must be sized to carry the maximum unbalanced neutral current which in this case is 63 amperes.

➤ Neutral Current Question
What is the current on the neutral?

(a) 25 amperes (b) 33 amperes (c) 50 amperes (d) 63 amperes

- Answer: (b) 33 amperes

Formula

$I_{Neutral} = \sqrt{L_1^2 + L_2^2 + L_3^2 - (L_1 \times L_2 + L_2 \times L_3 + L_1 \times L_3)}$

L_1 = 6,000 VA/120 volts = 50 amperes
L_2 = 3,000 VA/120 volts = 25 amperes
L_3 = 7,500 VA/120 volts = 63 amperes

$\sqrt{(50^2 + 25^2 + 63^2) - [(50 \times 25) + (25 \times 63) + (50 \times 63)]}$

$\sqrt{(2,500 + 625 + 3,969) - (1,250 + 1,575 + 3,150)}$

$\sqrt{7,094 - 5,975}$ $\sqrt{1,119}$ = 33 amperes

➤ Phase VA, Three-Phase Question
What is the per phase load of each 3-horsepower 208-volt, single-phase motor?

(a) 1,945 VA (b) 3,910 VA (c) 978 VA (d) none of these

- Answer: (a) 1,945 VA
  VA 1Ø = Table Volts × Table Amperes
  3 HP, VA = E × I, VA = 208 volts × 18.7 amperes = 3,890 VA
  On a wye system, a 208 volt 1Ø load is on two transformer phases
  3,890 VA/2 phases = 1,945 VA

➤ Phase VA, Single-Phase Question
What is the per phase load for the three-phase, 10-horsepower A/C motor?

(a) 11,154 VA (b) 3,698 VA (c) 5,577 VA (d) None of these

- Answer: (b) 3,698 VA
  10-horsepower, VA = Table volts × Table amperes × 1.732

VA = 208 volts × 30.8 amperes × 1.732 = 11,095 VA
11,095 VA/3 phases = 3,698 VA per phase

→ **Voltage Ratio Question**
What is the phase voltage ratio of the transformer?
(a) 4:1   (b) 1:4   (c) 1:2   (d) 2:1

- Answer: (a) 4:1
  480 primary phase volts to 120 secondary phase volts

☐ **Balance Panel VA Example**
Balance the loads on the panelboard in VA.

	Line 1	Line 2	Line 3	Line Total
Heat 18 kW(A/C 11 kW/ omitted)	6,000 VA	6,000 VA	6,000 VA	18.0 kVA
14 kW Ranges 208 volts, 1Ø	7,000 VA	7,000 VA		14.0 kVA
14 kW Ranges 208 volts, 1Ø		7,000 VA	7,000 VA	14.0 kVA
10 kW Water Heater 208 volts, 1Ø	5,000 VA		5,000 VA	10.0 kVA
3 HP 208 volts, 1Ø Motor	2,000 VA	2,000 VA		4.0 kVA
3 HP 208 volts, 1Ø Motor		2,000 VA	2,000 VA	4.0 kVA
Dishwasher 4.5 kW 120 volt			4,500 VA *	4.5 kVA
Lighting (8 – 1.5 kW circuits) (4 on L₁, 2 on L₂, 2 on L₃)	+ 6,000 VA *	3,000 VA *	3,000 VA *	12.0 kVA
	26,000 VA	27,000 VA	27,500 VA	80.5 kVA

☐ **Panelboard Balancing and Sizing Example**
Balance the previous example loads on the panelboard in amperes.

	Line 1	Line 2	Line 3	Ampere Calculations
18 kW Heat 3Ø	50.0 amps	50.0 amps	50.0 amps	18,000 VA/(208 volts × 1.732)
14 kW Ranges 208 volts, 1Ø	67.0 amps	67.0 amps		14,000 VA/208 volts
14 kW Ranges 208 volts, 1Ø		67.0 amps	67.0 amps	14,000 VA/208 volts
Water Heater, 208 volts, 1Ø	48.0 amps		48.0 amps	10,000 VA/208 volts
3 HP 208 volts, 1Ø Motor	18.7 amps	18.7 amps		Motor FLC
3 HP 208 volts, 1Ø Motor		18.7 amps	18.7 amps	Motor FLC
Dishwasher 1Ø 120 volt			38.0 amps	4,500 VA/120 volts
Lighting 1Ø 120 volt	+ 50.0 amps	25.0 amps	25.0 amps	1,500 VA/120 volts
	233.7 amps	246.4 amps	246.7 amps	

12–28 DELTA VERSUS WYE

Wye Truisms:
Phase CURRENT is the SAME as line current.
$I_{Phase} = I_{Line}$
$I_{Line} = I_{Phase}$

Wye Phase VOLTAGE is DIFFERENT than line voltage.
Line Voltage is greater than Phase Voltage by the square root of 3 ($\sqrt{3}$).
$E_{Line} = E_{Phase} \times \sqrt{3}$
$E_{Phase} = E_{Line} / \sqrt{3}$

Delta Truisms:
Phase VOLTAGE is the SAME as line voltage.
$E_{Phase} = E_{Line}$
$E_{Line} = E_{Phase}$

Delta phase CURRENT is DIFFERENT than line current.
Line current is greater than Phase Current by the square root of 3 ($\sqrt{3}$).
$I_{Line} = I_{Phase} \times \sqrt{3}$
$I_{Phase} = I_{Line} / \sqrt{3}$

Unit 12 – Delta/Delta And Delta/Wye Transformers Summary Questions

Definitions

1. Delta connected means the windings of three single-phase transformers (same rated voltage) are connected in _____. A delta-connected transformer is represented by the greek letter delta Δ.
   (a) series (b) parallel (c) series-parallel (d) a and b

2. Which system is called the high leg system because the voltage from one conductor to ground is between 190 and 208-volts-to-ground?
   (a) delta (b) wye

3. The _____ is the electrical system ungrounded conductors.
   (a) load (b) line (c) phase (d) system

4. The _____ voltage is the voltage measured between any ungrounded conductors.,
   (a) load (b) line (c) phase (d) system

5. • Line voltage of a _____ system is greater than phase voltage and line voltage of a _____ system is the same as phase voltage.
   (a) delta, delta
   (b) wye, wye
   (c) delta, wye
   (d) wye, delta

6. The _____ is the coil shape conductors that serve as the primary or secondary of the transformer.
   (a) line (b) load (c) phase (d) none of these

7. The phase current is the current of the transformer winding. For _____ systems, the phase current is less than the line current. For _____ systems, the phase current is the same as the line current.
   (a) wye, wye
   (b) delta, delta
   (c) wye, delta
   (d) delta, wye

8. The phase load is the load on the transformer winding. For delta systems the phase load is _____.
   (a) 3Ø 240-volt load = line load/3
   (b) 1Ø 240-volt load = line load
   (c) 1Ø 120-volt load = line load
   (d) all of these

9. The phase load is the load on the transformer winding. For wye systems the phase load is _____.
   (a) 3Ø 208-volt load = line load/3
   (b) 1Ø 208-volt load = line load/2
   (c) 1Ø 120-volt load = line load
   (d) all of these

10. The phase voltage is the internal transformer voltage generated across any one "winding" of a transformer. For _____ systems, the phase voltage is equal to the line voltage. For _____ systems the phase voltage is less than the line voltage.
    (a) wye, wye
    (b) delta, delta
    (c) wye, delta
    (d) delta, wye

11. The ratio is the relationship between the number of primary winding turns compared to the number of secondary winding turns. The ratio is a comparison between the primary phase voltage and the secondary phase voltage. For typical delta/delta systems, the ratio is ____, and for typical wye systems the ratio is _____.
    (a) 1:2, 1:4 (b) 2:1, 4:1 (c) 4:1, 2:1 (d) none of these

400 Unit 12 Delta/Delta And Delta/Wye Transformers   Chapter 2 Advanced NEC Calculations And Code Questions

12. What is the turns ratio of a 480- to 208Y/120-volt transformer?
    (a) 4:1   (b) 1:4   (c) 1:2   (d) 2:1

13. • The unbalanced load is the load on the secondary grounded conductors. For _____ systems the formula is:
    $\sqrt{L_1^2 + L_2^2 + L_3^2 - (L_1 \times L_2 + L_2 \times L_3 + L_1 \times L_3)}$
    (a) delta   (b) wye   (c) a or b   (d) none of these

14. The unbalanced (neutral) current for a _____ system can be calculated by: $I_{Line\,1} - I_{Line3}$.
    (a) delta   (b) wye   (c) a or b   (d) none of these

15. _____-connected means a connection of three single-phase transformers to a common point (neutral) and the other end is connected to the line conductors.
    (a) delta   (b) wye   (c) a or b   (d) none of these

12–1 Current Flow

16. When a load is connected to the secondary of a transformer, current will flow through the secondary conductor winding. The current flow in the secondary creates an electromagnetic field that opposes the primary electromagnetic field. The secondary flux lines _____.
    (a) effectively reduce the strength of the primary flux lines
    (b) less cemf is generated in the primary winding conductors
    (c) primary current automatically increases in direct proportion to the secondary current
    (d) all of these

17. The primary and secondary line currents are directly proportional to the ratio of the transformer.
    (a) True   (b) False

Part A – Delta/Delta Transformer

12–2 Delta Transformers Voltages

18. In a delta configured transformer, the Line Voltage equals the Phase Voltage.
    (a) True   (b) False

12–3 Delta High Leg

19. If the secondary voltage of a delta/delta transformer is 220/110-volts, the high leg voltage to ground (or neutral) would be _____ volts.
    (a) 190   (b) 196   (c) 202   (d) 208

12–4 Delta Primary And Secondary Line Currents

20. In a delta configured transformer, the line current equals the phase current.
    (a) True   (b) False

21. What is the primary line current for a 45-kVA, 480- to 240/120-volt 3Ø transformer?
    (a) 124 amperes   (b) 108 amperes   (c) 54 amperes   (d) 43 amperes

22. What is the secondary line current for a 45-kVA, 480- to 240-volt 3Ø transformer?
    (a) 124 amperes   (b) 108 amperes   (c) 54 amperes   (d) 43 amperes

12–5 Delta Primary Or Secondary Phase Currents

23. The phase current of a transformer winding is calculated by dividing the phase load by the phase volts: $I_{Phase} = VA_{Phase}/E_{Phase}$.
    (a) True   (b) False

24. What is the primary phase current for a 15-kVA, 480- to 240/120-volt 1Ø transformer?
    (a) 124 amperes   (b) 62 amperes   (c) 45 amperes   (d) 31 amperes

Chapter 2 Advanced NEC Calculations And Code Questions   Unit 12 Delta/Delta And Delta/Wye Transformers 401

25. What is the secondary phase current for a 15-kVA, 480- to 240/120-volt 1Ø transformer?
    (a) 124 amperes (b) 62 amperes (c) 45 amperes (d) 31 amperes

12–6 Delta Phase Versus Line

26. A 15-kVA 240-volt 3Ø load has the following effect on a delta system:
    (a) I_L = 15,000/(240 x $\sqrt{3}$), I_L = 36 amperes
    (b) I_P = 5,000/240, I_P = 21 amperes
    (c) I_L = I_P x $\sqrt{3}$, I_L = 21 x 1.732, I_L = 36 amperes
    (d) All of the above

27. • A 5-kVA 240-volt 1Ø Load has the following effect on a delta/delta system:
    (a) Line Power = 5 kVA
    (b) Phase Power = 5 kVA
    (c) I_P = 5,000 VA/240-volts = 21 amperes
    (d) All of the above

28. • A 2-kVA 120-volt 1Ø load has the following effect on a delta system:
    (a) Line Power = 2 kVA
    (b) I_L = 2,000 VA/120-volts, I_L = 17 amperes
    (c) Phase Power = 3 kVA (C1 or C2 winding)
    (d) a and b

12–7 Delta Current Triangle

29. The line and phase current of a delta system are not equal. The difference between line and phase current can be described by which of the following equations?
    (a) I_L = I_P x $\sqrt{3}$
    (b) I_P = $I_L/\sqrt{3}$
    (c) a and b
    (d) none of these

12–8 And 12–9 Delta Transformer Balancing And Sizing

30. To properly size a delta/delta transformer, the transformer must be balanced in kVA. Which of the following steps should be used to balance the transformer?
    (a) Balance 3Ø loads, ⅓ on Phase A, 1/3 on Phase B, and 1/3 on Phase C.
    (b) Balance 1Ø 240-volt loads, 100 percent on Phase A or Phase B.
    (c) Balance the 120-volt loads, 100 percent on Phase C1 or C2.
    (d) All of these

31. • Balance and size a 480- to 240/120-volt 3Ø delta transformer for the following loads: One 18-kVA 3Ø heat strip, two 5-kVA 1Ø 240-volt loads, and three 2-kVA 120-volt loads. What size transformer is required?
    (a) Three 1Ø 12.5-kVA transformers
    (b) One 3Ø 37.5-kVA transformer
    (c) a or b
    (d) None of these

12–10 Delta Panel Schedule In kVA

32. When balancing a panelboard in kVA be sure that 3Ø loads are balanced 1/3 on each line and 240-volt 1Ø loads are balanced ½ on each line.
    (a) True (b) False

33. • Balance a 120/240-volt delta 3Ø panelboard in kVA for the following loads: One 18-kVA 3Ø heat strip, two 5-kVA 1Ø 240-volt loads, and three 2-kVA 120-volt loads.
    (a) The largest Line load is 13 kVA.
    (b) The smallest Line load is 10 kVA.
    (c) Total Line load is equal to 34 kVA.
    (d) All of these.

12–11 Delta Panelboard And Conductor Sizing

34. When selecting and sizing panelboards and conductors, we must balance the line loads in amperes. Balance a 120/240-volt 3Ø panelboard in amperes for the following loads: One 18-kVA 3Ø heat strip, two 5-kVA 1Ø 240-volt loads, and three 2-kVA 120-volt loads.
    (a) The largest Line is 98 amperes.
    (b) The smallest Line is 81 amperes.
    (c) Each Line equals to 82 amperes.
    (d) a and b.

12–12 Delta Neutral Current

35. What is the neutral current for: One 18-kVA, 3Ø heat strip, two 5-kVA 1Ø 240-volt loads, and three 2-kVA 120-volt loads?
    (a) 17 amperes (b) 34 amperes (c) 0 amperes (d) A and B

12–13 Delta Maximum Unbalanced Load

36. • What is the maximum unbalanced load in amperes for: One 18-kVA, 3Ø heat strip, two 5-kVA 1Ø 240-volt loads, and three 2-kVA, 120-volt loads?
    (a) 25 amperes (b) 34 amperes (c) 0 amperes (d) None of these

37. What is the phase load for a 18-kVA, 3Ø heat strip?
    (a) 18 kVA (b) 9 kVA (c) 6 kVA (d) 3 kVA

38. • What is the transformer winding phase load for a 5-kVA, 1Ø 240-volt load?
    (a) 5 kVA (b) 2.5 kVA (c) 1.25 kVA (d) None of these

39. If only a 3Ø, 12-kW load is on the high leg, what is the current on the high leg conductor?
    (a) 29 amperes (b) 43 amperes (c) 73 amperes (d) 97 amperes

40. What is the ratio of a 480- to 240/120-volt transformer?
    (a) 4:1 (b) 1:4 (c) 1:2 (d) 2:1

Part B Delta/Wye Transformers

12–15 Wye Transformers Voltages

41. In a wye configured transformer, the line voltage equals the phase voltage.
    (a) True (b) False

12–17 Wye Transformers Current

42. In a wye configured transformer, the line current equals the phase current.
    (a) True (b) False

12–18 Line Current

43. The line current of both delta and wye 3-phase transformers can be calculated by the formula: $I_{Line} = VA_{Line}/(E_{Line} \times \sqrt{3})$.
    (a) True (b) False

44. What is the primary line current for a 22-kVA, 480- to 208Y/120-volt 3Ø transformer?
    (a) 61 amperes (b) 54 amperes (c) 26 amperes (d) 22 amperes

45. What is the secondary line current for a 22-kVA, 480- to 208Y/120-volt 3Ø transformer?
    (a) 61 amperes (b) 54 amperes (c) 27 amperes (d) 22 amperes

12–19 Phase Current

46. The phase current of a wye configured transformer winding is the same as the line current. The phase current of a delta configured transformer winding is less than the line current by the $\sqrt{3}$.
    (a) True (b) False

47. • What is the primary phase current for a 22-kVA, 480- to 208Y/120-volt 3Ø transformer?
    (a) 15 amperes (b) 22 amperes (c) 36 amperes (d) 61 amperes

48. • What is the secondary phase current for a 22-kVA, 480- to 208Y/120-volt 3Ø transformer?
    (a) 15 amperes (b) 22 amperes (c) 36 amperes (d) 61 amperes

12–20 Wye Phase Versus Line

49. Since each line conductor from a wye transformer is connected to a different transformer winding (phase), the effects of loading on the line are the same as the phase. A 12-kVA, 208-volt 3Ø load has the following effect on the system:
    (a) Line Current = I_L = VA/(E x $\sqrt{3}$), I_L = 12,000/(208 x $\sqrt{3}$), I_L = 33 amperes
    (b) Phase Power = 4 kVA
    (c) Phase Current = I_P = VA_{Phase}/E_{Phase}, I_P = 4,000/120 = 30 amperes
    (d) All of these

50. A 5-kVA, 208-volt 1Ø load has the following effect on the system:
    (a) Line Power = 5 kVA
    (b) Line Current = I_L = VA/E, I_L = 5,000/208, I_L = 24 amperes
    (c) Phase Power = 5 kVA
    (d) Phase Current = I_P = VA_{Phase}/E_{Phase}, I_P = 2,500/120 = 21 amperes
    (e) All of these

51. A 2-kVA, 120-volt 1Ø load has the following effect on the system:
    (a) Line Power = 2 kVA
    (b) Line Current = I_L = VA/E, I_L = 2,000/120, I_L = 17 amperes
    (c) Phase Power = 2 kVA
    (d) Phase Current = I_P = VA_{Phase}/E_{Phase}, I_P = 2,000/120 = 17 amperes
    (e) All of these

52. • What is the phase load of a 3-horsepower, 208-volt, 1Ø motor?
    (a) 1,945 VA   (b) 3,890 VA   (c) 978 VA   (d) none of these

53. • What is the per phase load for a 3Ø, 7.5-horsepower, 208-volt motor?
    (a) 2,906 VA   (b) 8,718 VA   (c) 4.359 VA   (d) none of these

12–21 And 12–22 Wye Transformer Balancing And Sizing

54. To properly size a delta/wye transformer, the secondary transformer phases (windings) or the line conductors must be balanced. Balancing the panel (line conductors) is identical to balancing the transformer for wye configured transformers. Which of the following steps should be used to balance the transformer?
    (a) balance 3Ø loads, 1/3 on each phase
    (b) balance 1Ø 208-volt loads 50 percent on each phase
    (c) balance the 120-volt loads 100 percent on any phase
    (d) all of these

55. • Balance and size a 480- to 208Y/120-volt 3Ø transformer for the following loads: One 18-kVA, 3Ø heat strip, two 5-kVA, 1Ø 208-volt loads, and three 2-kVA, 120-volt loads. What size transformer is required?
    (a) three 1Ø 12.5-kVA transformers   (b) one 3Ø 37.5-kVA transformer
    (c) a or b   (d) none of these

12–23 Wye Panel Schedule In kVA

56. When balancing a panelboard in kVA be sure that 3Ø loads are balanced 1/3 on each line and 208-volt 1Ø loads are balanced 1/2 on each line. Balance a 208Y/120-volt 3Ø panelboard in kVA for the following loads: One 18-kVA, 3Ø heat strip, two 5-kVA, 1Ø 208-volt loads, and three 2-kVA, 120-volt loads.
    (a) The largest line is 12.5 kVA.
    (b) The smallest line is 10.5 kVA.
    (c) Each line equals to 11.33 kVA.
    (d) all of these.

12–24 Wye Panelboard And Conductor Sizing

57. • When selecting and sizing panelboards and conductors, we must balance the line loads in amperes. Balance a 208Y/120-volt 3Ø panelboard in amperes for the following loads: One 18-kVA, 208-volt 3Ø heat strip, two 5-kVA, 1Ø 208-volt loads and three 2-kVA, 120-volt loads.
   (a) The largest Line current is 108 amperes
   (b) The smallest Line is 91 amperes
   (c) a and b
   (d) None of these

12–25 Wye Neutral Current

58. To determine the neutral current for a 3-wire wye system we must use the following formula:
   I neutral = $\sqrt{(L_1^2 + L_2^2 + L_3^2) - (L_1 \times L_2 + L_2 \times L_3 + L_1 \times L_3)}$
   (a) True
   (b) False

59. • What is the neutral current for: One 18-kVA, 3Ø heat strip, two 5-kVA, 1Ø 208-volt loads and three 2-kVA, 120-volt loads.
   (a) 47 amperes
   (b) 29 amperes
   (c) 34 amperes
   (d) A and B

12–26 Wye Maximum Unbalanced Load

60. • What is the maximum unbalanced load in amperes for: One 18-kVA, 3Ø heat strip, two 5-kVA, 1Ø 208-volt loads, and three 2-kVA, 120-volt loads.
   (a) 17 amperes
   (b) 34 amperes
   (c) 0 amperes
   (d) a and b

12–27 Delta Versus Wye

61. Which of the following statements are true for wye systems?
   (a) Phase current is the same as line current: $I_P = I_L$
   (b) Phase voltage is the same as the line voltage.
   (c) $E_L = E_P \times \sqrt{3}$.
   (d) a and c

62. Which of the following statements are true for delta systems?
   (a) Phase voltage is the same as line voltage: $E_P = E_L$
   (b) Phase current is the same as the line current.
   (c) Line current is greater than Phase Current: $I_L = I_P \times \sqrt{3}$.
   (d) a and c

Challenge Questions

Delta/Delta Questions

The following statement applies to the next seven questions:

The following loads are to be connected to a three-phase delta/delta 480-volt to 240-volt transformer.
1 – 25-horsepower, 240-volt, three-phase synchronous motor, Air conditioning
2 – 1½-horsepower, 240-volt, single-phase motors
2 – 2-horsepower, 120-volt motors
3 – 1900 watt, 120-volt lighting loads
18 kW, 240-volt, 3-phase heat
Balance the above loads as closely as possible, then answer the next seven questions.

63. • Three single-phase transformers are connected in series; what size transformers are required for the above loads?
   (a) 10/10/10 kVA
   (b) 15/15/15 kVA
   (c) 10/10/20 kVA
   (d) 15/15/10 kVA

64. • The line VA load of the 25-horsepower three-phase synchronous motor would be _____ VA.
   (a) 9,700
   (b) 21,113
   (c) 32,031
   (d) 28,090

65. • After balancing the loads, the maximum unbalanced neutral current is _____ amperes.
    (a) 27   (b) 1.5   (c) 48   (d) 148

66. • The transformer phase load of one 1½-horsepower motor would be _____ VA.
    (a) 1,380   (b) 2,760   (c) 2,300   (d) 1,150

67. • The line load of one of the 1½-horsepower motors would be _____ VA.
    (a) 1,380   (b) 2,760   (c) 2,300   (d) 1,150

68. • The transformer phase VA load of the 25-horsepower three-phase synchronous motor is closest to _____ VA.
    (a) 3,520   (b) 7,038   (c) 21,100   (d) 32,100

69. • If the only load on the high leg is the three-phase 25-horsepower motor, the high leg conductor would have to be sized a minimum of _____ amperes.
    (a) 13   (b) 66   (c) 18   (d) 22

The following statement applies to the next two questions:

A three-phase 37.5-kVA delta/delta transformer has a primary of 480-volts and a secondary of 240/120-volts.

70. • The primary line current is closest to _____ amperes.
    (a) 45   (b) 90   (c) 50   (d) 65

71. • The maximum overcurrent device size for the 37.5-kVA delta/delta transformer is _____ amperes.
    (a) 40   (b) 45   (c) 50   (d) 60

The following statement applies to the next question:

An office building contains the following loads:
1 – 5-horsepower 240-volt three-phase motor
7 – 1,500 watt, 120-volt heaters
2 – 3-horsepower 240-volt motors
5 – 3 kW 240-volt, single-phase heaters
30 kW 120-volt small appliance
10 – 5 kW 240-volt, single-phase machines
40 kW 120-volt lighting

72. • If the only loads on the center tapped transformer (C_1 and C_2) are the 120-volt loads, what is the total kW load on this transformer?
    (a) 30–40 kW   (b) 50–60 kW   (c) 80–90 kW   (d) over 100 kW

73. • The total load on the center tapped 240/120-volt transformer is 100 kW, which consists of 80 kW of 120-volt balanced loads, and 20 kW of 240-volt loads. What is the maximum unbalanced neutral current?
    (a) 167 amperes   (b) 0 amperes   (c) 192 amperes   (d) 333 amperes

74. • On a delta/delta three-phase transformer, a 30,000 VA, 240-volt three-phase load would have a phase current of _____ amperes on the secondary winding.
    (a) 125   (b) 72   (c) 42   (d) none of these

75. • What is the secondary line current of a three-phase, 45-kVA, delta/delta transformer that has a turn ratio of 2:1 and a primary voltage of 460?
    (a) 113 amperes   (b) 55 amperes
    (c) 36 amperes   (d) cannot be determined

76. • A 2,200/220-volt three-phase transformer is connected delta/delta. The secondary phase current is 300 amperes and the secondary line current is 240 volts. What is the primary line current?
    (a) 40 amperes   (b) 20 amperes   (c) 50 amperes   (d) 30 amperes

77. • For a given three-phase transformer, the connection giving the highest secondary voltage is _____.
    (a) wye primary, delta secondary   (b) delta primary, delta secondary
    (c) wye primary, wye secondary   (d) delta primary, wye secondary

Part B – Wye Transformers

78. • A delta/wye, 480/208, 120-volt three-phase transformer has a secondary line current of 200 amperes. What is the primary line current?
    (a) 31 amperes (b) 72 amperes (c) 87 amperes (d) 53 amperes

79. • If the secondary phase voltage of a delta/wye connected transformer is 277 volts, and a 10-kVA three-phase load was connected to that secondary, the line current would be _____ amperes.
    (a) 7 (b) 12 (c) 32 (d) none of these

The following statement applies to the next two questions:

1 – 15-horsepower, 208-volt three-phase induction motor
2 – 1.5-kW, 208-volt dryers
4 – 3-kW, 120-volt lighting banks
1 – 8-kW, 208-volt cooktops
5 – 2.5-kW, 208-volt heat strips

80. • Each 2.5-kW heat strip has a line load of _____ .
    (a) 1,250 watts (b) 12,500 watts (c) 2,500 watts (d) no load

81. • The load per line of the three-phase motor can be determined by which of these formulas?
    (a) 3 × V × I (b) (V × I)/3 (c) V × I (d) no phase load

82. • If a three-phase motor has a total load of 12,000 VA, and is connected to a delta/wye transformer, the load on one phase of the secondary will be _____ VA.
    (a) 12,000 (b) 4,000 (c) 3,000 (d) 1,500

83. • A three-phase 480Y/208-120-volt transformer provides power to nine 2,000 VA, 120-volt lights and nine 2,000 watt, 208-volt heaters. The maximum neutral current is _____ amperes.
    (a) 50 (b) 75 (c) 87 (d) 95

84. • There are five 2,500 watt, 208-volt heat strips on the secondary of a three-phase 208Y/120-volt transformer. What is the load per phase for each heat strip?
    (a) 1,403 watts (b) 1,250 watts (c) 2,500 watts (d) 12,500 watts

The following statement applies to the next question:

The system is 208Y/120 Volt, three-phase. The loads are:
1 – 10-horsepower, three-phase induction motor
4 – 1,500 VA, 120-volt circuits
5 – 1,500 VA, 120-volt lights
1 – 6,000 VA, three-phase water heater
1 – 3,000 watt, 208-volt single-phase industrial oven
1 – ¾-horsepower, 208-volt single-phase motor

85. • The phase load of the 10-horsepower three-phase motor is closest to _____ VA.
    (a) 3,700 (b) 5,500
    (c) 4,300 (d) 11,100

The following statement applies to the next question:

The system is three-phase 208Y/120-volt and the loads are:
9 – 2-kVA 120-volt lighting loads
5 – 3-kVA 120-volt loads
2 – 2.5 kW 208-volt heating units

86. • The neutral conductor must be size to carry _____ amperes.
    (a) 10 (b) 25
    (c) 90 (d) 100

NEC Questions From Article 90 Through Chapter 9

87. The Code covers the requirements for the installation of wiring and equipment of fire alarm signaling systems operating at _____ volts or less.
    (a) 50 (b) 300 (c) 600 (d) 1,000

88. • The _____ of any system is the ratio of the maximum demand for a system, or part of a system, to the total connected load of a system under consideration.
    (a) load (b) demand factor (c) minimum load (d) computed factor

89. • A _____ receptacle without GFCI protection is permitted to be located in a dwelling unit garage for one appliance if located within the dedicated space for the appliance.
    (a) multioutlet (b) duplex (c) single (d) a and b only

90. Each system _____ conductor, wherever accessible, shall be identified by separate color coding, marking tape, tagging, or other equally effective means. This only applies to multiwire branch circuits when there is more than one system voltage in a building.
    (a) grounded (b) ungrounded (c) grounding (d) all of these

91. A storage battery supplying emergency lighting and power shall maintain not less than 87½ percent of full voltage at total load for a period of at least _____ hour(s).
    (a) 2 (b) 1½ (c) 1 (d) ¼

92. Fixtures shall be supported independently of the outlet box where the weight exceeds _____ pounds.
    (a) 60 (b) 50 (c) 40 (d) 30

93. Extreme _____ may cause some nonmetallic conduits to become brittle and therefore more susceptible to damage from physical contact.
    (a) temperatures (b) corrosive conditions
    (c) heat (d) cold

94. Class III locations are those that are hazardous because of the presence of _____.
    (a) combustible dust (b) easily ignitable fibers or flyings
    (c) flammable gases or vapors (d) flammable liquids or gases

95. • For angle or U Pulls, the distance between each cable or conductor entry inside the box (pull and junction boxes for use on systems over 600 volts, nominal) and the opposite wall of the box shall not be less than _____ times the outside diameter, over sheath, of the largest cable or conductor.
    (a) six (b) twelve (c) twenty-four (d) thirty-six

96. The demand percentage used to calculate 45 receptacles on a boat yard feeder is _____ percent.
    (a) 90 (b) 80 (c) 70 (d) 50

97. • It shall be permitted to base the _____ rating of a range receptacle on a single range demand load specified in Table 220-19.
    (a) circuit (b) voltage (c) ampere (d) load

98. Feeders supplying _____ ampere receptacle branch circuits shall be permitted to be protected by a ground-fault-circuit-interrupter in lieu of the GFCI requirements of Section 210-8.
    I. 15   II. 20   III. 30
    (a) I only (b) II only (c) III only (d) I and II

99. Locations of lamps for outdoor lighting shall be below all _____, transformers, or other electric utilization equipment.
    (a) switches (b) feeder circuits (c) branch circuits (d) energized conductors

100. An enclosure that is weatherproof only when the receptacle cover is closed, can be used for receptacles in a wet location, when the receptacle is used for _____ while attended.
    (a) portable equipment (b) portable tools (c) fixed equipment (d) a and b

101. General use branch circuits using flat conductor cable shall not exceed _____ amperes.
    (a) 15 (b) 20 (c) 30 (d) 40

102. Fuseholders for cartridge fuses shall be so designed that it will be difficult to put a fuse of any given class into a fuseholder that is designed for a(n) _____ lower, or _____ higher, than that of the class to which the fuse belongs.
 (a) voltage, amperage
 (b) amperage, voltage
 (c) voltage, current
 (d) current, voltage

103. The individual conductors in a cablebus shall be supported at intervals not greater than _____ feet for horizontal runs.
 (a) 2½ (b) 3 (c) 3½ (d) 4

104. The equipment grounding conductor shall be identified by _____.
 (a) a continuous outer green finish
 (b) being bare
 (c) a continuous outer green finish with one or more yellow stripes
 (d) any of these

105. Equipment or materials included in a list published by an organization acceptable to the authority having jurisdiction, and concerned with product evaluation, are known as _____.
 (a) labeled (b) listed (c) approved (d) identified

106. Conduit installed underground or encased in concrete slabs which are in direct contact with the earth shall be considered a _____ location.
 (a) dry (b) damp (c) wet (d) moist

107. Conductors installed between buildings, structures, or poles as well as wiring and equipment located on or attached to the outside of buildings, structures, or poles must comply with Article 225.
 (a) True (b) False

108. Utilization equipment would be equipment that utilized electricity for _____.
 (a) chemicals (b) heating (c) lighting (d) any of these

109. Grounding-type attachment plugs shall be used only where a(n) _____ ground is to be provided.
 (a) equipment (b) isolated (c) computer (d) branch-circuit

110. The minimum size conductor for operating control and signaling circuits in an elevator is No. _____.
 (a) 20 (b) 16 (c) 14 (d) 12

111. • A _____ shall be located in sight from the motor location and the driven machinery location.
 (a) controller
 (b) protection device
 (c) disconnecting means
 (d) all of these

112. The minimum thickness of the sealing compound in Class I, Division 1 and 2 locations shall not be less than the trade size of the conduit or sealing fitting and in no case less than _____ inches.
 (a) 3/16 (b) 3/8 (c) ½ (d) 5/8

113. One conductor of flexible cords that is intended to be used as a _____ circuit conductor shall have a continuous marker readily distinguishing it from the other conductor or conductors.
 (a) grounded (b) equipment (c) ungrounded (d) all of these

114. In a balanced, 208Y/120 volt, 3-phase, 4-wire system, the grounded (neutral) conductor will carry _____ amperes, if the loads supplied are linear loads and no harmonic currents are present.
 (a) full load (b) zero (c) fault current (d) none of these

115. Where required for the reduction of electric noise for electronic equipment, electrical continuity of the metal raceway is not required and the metal raceway can terminate to a(n) _____ nonmetallic fitting(s) or spacer on the electronic equipment.
 (a) listed (b) labeled (c) identified (d) marked

116. • For installations to supply only limited loads of a single branch circuit, the service disconnecting means shall have a rating of not less than _____ amperes.
 (a) 15 (b) 20 (c) 25 (d) 30

Chapter 2 Advanced NEC Calculations And Code Questions   Unit 12 Delta/Delta And Delta/Wye Transformers 409

117. Where open conductors cross ceiling joists and wall studs and are exposed to physical damage, they shall be protected by a substantial running board; running boards shall extend at least _____ inch outside the conductors, but not more than _____ inches, and the protecting sides shall be at least 2 inches high and at least 1 inch nominal in thickness.
(a) ¾, 1½ (b) ¾, 2 (c) 1, 2½ (d) 1, 2

118. Branch circuit conductors supplying a single motor shall have an ampacity not less than _____ rating.
(a) 125 percent of maximum motor nameplate
(b) 125 percent of the motor full-load current
(c) 125 percent of the motor full locked rotor
(d) 80 percent of the motor full-load current

119. Equipment is required to be approved for the "Class" and for the explosive, combustible, or ignitible properties of the specific _____ that will be present.
(a) gas or vapor (b) dust (c) fiber or flyings (d) all of these

120. An insulated grounded conductor of No. _____ or smaller shall be identified by a continuous white or natural gray outer finish along its entire length.
(a) 3 (b) 4 (c) 6 (d) 8

121. Unless an individual switch is provided for each fixture located over combustible material, lampholders shall be located at least _____ feet above the floor, or shall be so located or guarded that the lamps cannot be readily removed or damaged.
(a) 3 (b) 6 (c) 8 (d) 10

122. Meters, instruments, and relays, including kilowatt-hour meters, instrument transformers, resistors, rectifiers, and thermionic tubes, installed in Class I Division 1 locations must have these devices installed in explosion proof enclosures or purged and pressurized enclosures.
(a) True (b) False

123. • Equipment approved for use in dry locations only, shall be protected against permanent damage from the weather during _____.
(a) floods (b) building construction
(c) building demolition (d) hurricanes

124. Service drops run over roofs shall have a minimum clearance of _____ feet.
(a) 8 (b) 12 (c) 15 (d) 3

125. Sizes No. 18 and No. 16 fixture wires can be used for the control and operating circuits of X-ray equipment when protected by not larger than _____ ampere overcurrent devices.
(a) 15 (b) 20 (c) 25 (d) 30

126. • _____ identified for use on lighting track shall be designed specifically for the track on which they are to be installed.
(a) Fittings (b) Receptacles (c) Lampholders (d) all of these

127. When multiple ground rods are used for a made grounding electrode, they shall be separated not less than _____ feet apart.
(a) 6 (b) 8 (c) 20 (d) 12

128. Conductor overload protection shall not be required where the interruption of the _____ would create a hazard, such as in a material handling magnet circuit or fire pump circuit.
(a) circuit (b) line (c) phase (d) system

129. The circular mil area of a No. 12 conductor is _____.
(a) 10,380 (b) 26,240 (c) 6,530 (d) 6,350

130. Reasonable efficiency of operation can be provided when voltage drop is taken into consideration in sizing the _____ conductors.
(a) service-vertical (b) grounding (c) service-lateral (d) none of these

131. Surface metal raceways and their fittings shall be so designed that the sections can be _____.
(a) electrically coupled together
(b) mechanically coupled together
(c) installed without subjecting the wires to abrasion
(d) all of these

132. When determining the number of conductors that are considered current-carrying for Note 8(a) of Table 310-16, a grounding conductor will be _____.
    (a) counted as one conductor
    (b) considered to be a current-carrying conductor
    (c) considered to be a noncurrent-carrying conductor and is not counted
    (d) counted as one conductor for each ground wire in the raceway

133. Type _____ fuse adapters shall be so designed that when once inserted in a fuseholder they cannot be easily removed.
    (a) A   (b) E   (c) S   (d) P

134. Four conditions must be met for fixtures to be permitted to be installed in cooking hoods of nonresidential occupancies. One of those conditions is that wiring methods and materials supplying the fixture(s) shall not be _____ within the cooking hood.
    (a) protected   (b) exposed   (c) grounded   (d) covered

135. • Open wiring on insulators within _____ feet from the floor shall be considered exposed to physical damage.
    (a) 4   (b) 5   (c) 7   (d) none of these

136. When the derating factors of Note 8(a) of Table 310-16 are used, auxiliary gutters shall not contain more than _____ at any cross section.
    (a) 25 conductors
    (b) 40 current-carrying conductors
    (c) 20 conductors
    (d) no limit on the number of conductors

137. Nonmetallic underground conduit with conductors shall be capable of being supplied on reels without damage or _____ and shall be of sufficient strength to withstand abuse, such as impact or crushing, in handling and during installation without damage to conduit or conductors.
    (a) distortion   (b) breakage   (c) shattering   (d) all of these

138. The _____ clearances of all service-drop conductors shall be based on a conductor temperature of 60ºF, no wind; with final unloaded sag in the wire, conductor, or cable.
    (a) horizontal   (b) lateral   (c) vertical   (d) final

139. A thermal barrier shall be required if the space between the resistors and reactors and any combustible material is less than _____ inches.
    (a) 2   (b) 3   (c) 6   (d) 12

140. When equipment grounding conductor(s) are installed in a metal box, an electrical connection is required between the equipment grounding conductor and the metal box enclosure by means of a _____.
    (a) grounding screw
    (b) listed grounding clip
    (c) listed grounding fittings
    (d) any of these

141. Splices and taps shall be made in _____.
    (a) junction boxes
    (b) outlet and device boxes
    (c) conduit bodies
    (d) any of these

142. Flexible cords used in show windows and _____ shall be Type AFS, S, SE, SEO, SEOO, SJ, SJE, SJEO, SJEOO, SJO, SJOO, SJT, SJTO, SJTOO, SO, SOO, ST, STO, or STOO.
    (a) elevators
    (b) buildings
    (c) run through openings
    (d) show cases

143. There shall be no reduction in the size of the neutral conductor on _____ type lighting loads.
    (a) dwelling unit (b) hospital   (c) nonlinear   (d) motel

144. Sufficient access and _____ shall be provided and maintained about all electric equipment to permit ready and safe operation and maintenance of such equipment.
    (a) ventilation
    (b) cleanliness
    (c) circulation
    (d) working space

145. A wet-niche lighting fixture is intended to be installed in a _____.
    (a) transformer
    (b) forming shell
    (c) hydromassage bathtub
    (d) all of these

146. For each floor area inside a commercial garage, the entire area up to a level of _____ inches above the floor shall be considered to be a Class I, Division 2 location except where the enforcing agency determines that there is mechanical ventilation providing a minimum of four air changes per hour.
    (a) 6
    (b) 12
    (c) 18
    (d) 24

147. A place of assembly is a building, portion of a building, or structure intended for the assembly of _____ or more persons.
    (a) 50
    (b) 100
    (c) 150
    (d) 200

148. Entrances to rooms and other guarded locations containing exposed live parts shall be marked with conspicuous _____ forbidding unqualified persons to enter.
    (a) warning signs
    (b) locks
    (c) both a and b
    (d) neither a nor b

149. • The minimum size conductor permitted in parallel for elevator lighting is No. _____ provided the ampacity is equivalent to a No. 14 wire.
    (a) 14
    (b) 20
    (c) 16
    (d) 1/0

150. Cablebus framework, where _____, shall be permitted as the equipment grounding conductor for branch circuits and feeders.
    (a) adequately bonded
    (b) welded
    (c) protected
    (d) galvanized

151. NM cable shall be permitted for use in _____ which are three floors or less.
    (a) one- and two-family dwellings
    (b) multifamily dwellings
    (c) stores and offices
    (d) all of these

152. Armor cable is limited or not permitted _____.
    (a) in commercial garages
    (b) where subject to physical damage
    (c) in theaters
    (d) all of these

153. Wiring methods and equipment installed behind panels designed to permit access (such as drop ceilings) shall be so arranged and secured to permit the removal of panels to give access to electrical equipment.
    (a) True
    (b) False

154. An impedance heating system that is operating at a(n) _____ greater than 30, but not more than 80, shall be grounded at designated point(s).
    (a) voltage
    (b) amperage
    (c) wattage
    (d) temperature

155. Motors shall be located so that adequate _____ is provided and so that maintenance, such as lubrication of bearings and replacing of brushes, can be readily accomplished.
    (a) space
    (b) ventilation
    (c) protection
    (d) all of these

156. Where livestock is housed, that portion of the equipment grounding conductor run underground to the remote building disconnecting means shall be insulated or covered _____.
    (a) aluminum
    (b) copper
    (c) copper-clad aluminum
    (d) none of these

157. Solid dielectric insulated conductors operated above 2,000 volts in permanent installations shall have _____ insulation and shall be shielded.
    (a) ozone-resistant (b) asbestos (c) hi-temperature (d) perfluoro-alkoxy

158. Where single-conductor MI cables are used, all phase conductors and, where used, the _____ conductor shall be grouped together to minimize induced voltage on the metal sheath.
    (a) grounded (b) neutral c) grounding (d) largest

159. Use of FCC systems shall be permitted on wall surfaces in surface metal _____.
    (a) raceways (b) cable trays (c) busways (d) plenums

160. All conductors of a circuit, including the equipment grounding conductors, must be contained within the same _____.
    (a) raceway (b) cable (c) trench (d) all of these

161. Conductors located above a heated ceiling shall be considered as operating in an ambient temperature of _____ °C.
    (a) 86 (b) 30 (c) 50 (d) 20

162. The overhead service conductors from the last pole or other aerial support to and including the splices if any, are called the _____.
    (a) service-entrance conductors
    (b) service drop
    (c) service conductors
    (d) overhead service

163. The emergency controls for unattended self-service stations must not be less than _____ feet, nor more than _____ feet from the dispensers.
    (a) 10, 25 (b) 20, 50 (c) 20, 100 (d) 50, 100

164. Each continuous-duty motor rated more than _____ horsepower shall be protected against overload by one of four means.
    (a) 1 (b) 2 (c) 3 (d) 4

165. In communication circuits, the bonding together of all separate electrodes shall be permitted with a minimum size jumper of No. _____ copper.
    (a) 10 (b) 8 (c) 6 (d) 4

166. No seal is required if a conduit (with no unions, couplings, boxes, or fittings) passes through a Class I, Division 2 location if the termination points of the unbroken conduit are in at least _____ inches in the unclassified location.
    (a) 6 (b) 12 (c) 18 (d) 24

167. Flexible metal conduit cannot be installed _____.
    (a) underground
    (b) embedded in poured concrete
    (c) where subject to physical damage
    (d) all of these

168. In a(n) _____ that is an apartment or living area in a multifamily building where laundry facilities are provided on the premises that are available to all building occupants, a laundry receptacle shall not be required.
    (a) building
    (b) dwelling unit
    (c) structure
    (d) room

169. • Where installing direct buried cables, a _____ shall be used at the end of a conduit that terminates underground.
    (a) connector
    (b) weatherproof coupling
    (c) bushing
    (d) rubber tape

170. In reference to mobile/manufactured homes, examples of "Appliance, Portable:" could be _____, but only if these appliances are not built-in.
    (a) a refrigerator

(b) gas range equipment
(c) a clothes washer
(d) all of these

171. The minimum and maximum size of EMT is _____ inches, except for special installations.
(a) 5/16 to 3   (b) 3/8 to 4   (c) ½ to 3   (d) ½ to 4

172. Systems and circuit conductors are grounded to _____.
(a) limit voltages due to lightning, line surges, or unintentional contact with higher voltage lines
(b) stabilize the voltage to ground during normal operation
(c) facilitate overcurrent device operation in case of ground faults
(d) a and b

173. The metal frame of the building that is effectively grounded is considered part of the grounding electrode system.
(a) True   (b) False

174. When bonding enclosures, metal raceways, frames, fittings and other metal noncurrent-carrying parts, any nonconductive paint, enamel, or similar coating shall be removed at _____.
(a) contact surfaces   (b) threads   (c) contact points   (d) all of these

175. The lighting outlets provided for illumination about electrical equipment over 600 volts, shall be so arranged that persons changing lamps or making repairs on the _____ will not be endangered by live parts or other equipment.
(a) lighting system
(b) electrical system
(c) electrical equipment
(d) panelboard

176. All panelboard circuits and circuit _____ shall be legibly identified as to purpose or use on a circuit directory located on the face or inside of the panel doors.
(a) manufacturers   (b) conductors
(c) feeders   (d) modifications

177. Where the service disconnecting means is a power-operated circuit breaker, it shall be able to be opened by hand in the event of a _____.
(a) ground-fault   (b) short-circuit
(c) power surge   (d) power supply failure

178. • A reduction of _____ conductors shall be made for outlet box fill for a hickey and two clamps in a box.
(a) one   (b) two   (c) three   (d) zero

179. A wall space shall be permitted to include _____ or more walls of a room (around corners) where unbroken at the floor line.
(a) one   (b) two   (c) three   (d) four

180. The length of lighting track may not be altered by the addition or subtraction of sections of track.
(a) True   (b) False

181. An apparatus enclosed in a case that is capable of withstanding an explosion of a specified gas or vapor that may occur within it; and of preventing the ignition of a specified gas or vapor surrounding the enclosure by sparks, flashes, or explosion of the gas or vapor within; and that operates at such an external temperature that a surrounding flammable atmosphere will not be ignited is defined as a(n) _____.
(a) overcurrent protection device
(b) thermal apparatus
(c) explosionproof apparatus
(d) bomb casing

182. Raceways shall be _____ between pulling points before installing conductors.
(a) completely installed
(b) tested for ground faults
(c) a minimum of 80 percent completed

(d) none of these

183. Flat cable assemblies shall have conductors of No. _____ special stranded copper.
 I. 14 II. 12 III. 10
 (a) I only (b) II only (c) III only (d) I, II, and III

184. Table 310-73 provides ampacities of an insulated triplexed or three single conductor copper cables in isolated conduit in air based on conductor temperature of 90ºC (194ºF) and ambient air temperature of 40ºC (104ºF). If the conductor size is 4 AWG, and the voltage range is 5,001 to 35,000, then the ampacity is _____ amperes.
 (a) 83 (b) 110 (c) 130 (d) 150

185. A Group "E" atmosphere contains combustible metal dusts.
 (a) True (b) False

186. A fused switch shall not have fuses _____.
 (a) in series
 (b) in parallel
 (c) less than 100 amperes
 (d) over 15 amperes

187. The feeder demand load for four 6 kW cooktops is _____ kW.
 (a) 17 (b) 4 (c) 12 (d) 24

188. • A conductor encased within material of composition or thickness that is not recognized by the Code is called a _____ conductor.
 (a) noninsulating (b) bare (c) covered (d) none of these

189. Article 440 applies to electric-driven air-conditioning and refrigeration equipment that have a hermetic refrigerant motor compressor.
 (a) True (b) False

190. Transformer vaults shall be located where they can be ventilated to the outside air without using flues or ducts wherever such an arrangement is _____.
 (a) permitted (b) practicable (c) required (d) all of these

191. A header that attaches to a floor duct shall be installed _____ to the cell.
 (a) parallel (b) straight
 (c) right angle (d) none of these

192. • Where the direct current system consists of a _____, the grounding conductor shall not be smaller than the neutral conductor.
 (a) 2-wire balancer set
 (b) 3-wire balancer set
 (c) balancer winding with overcurrent protection
 (d) b or c

193. The highest current at rated voltage that a device is intended to interrupt under standard test conditions is the _____.
 (a) interrupting rating
 (b) manufacturer's rating
 (c) interrupting capacity
 (d) GFI rating

194. • Circuit breakers shall _____ all ungrounded conductors of the circuit.
 (a) open (b) close (c) protect (d) inhibit

195. • Which of the following are required to be bonded?
 I. All electric equipment within 5 feet of the inside pool
 II. The pool structure steel.
 III. Metal fittings having a dimension of greater than 4 inches within or attached to the pool
 (a) I only (b) II only (c) III only (d) I, II, and III

196. X-ray equipment mounted on a permanent base with wheels for moving while completely assembled is defined as _____.
    (a) portable (b) mobile (c) movable (d) room

197. Service cables, unless continuous from pole to service equipment or meter, shall be equipped with a raintight _____.
    (a) raceway (b) service head (c) cover (d) all of these

198. Two or more electrodes that are effectively bonded together shall be considered as a single electrode in this sense.
    (a) True (b) False

199. Equipment intended to break current at other than fault levels shall have an interrupting rating at system voltage sufficient for the current that must be interrupted.
    (a) True (b) False

200. In other than _____ runs, the cables shall be fastened securely to transverse members of the cable trays.
    (a) horizontal (b) vertical (c) temporary (d) permanent

Index

A

A - Ampere 9
ac
   See Alternating current
AC motors 110, 112
Air conditioner 194
Air conditioning
   branch circuit 339
Alternating current 9, 78
   values 80
Alternator 9
Amapacity 202
   derating factors 198
Ambient temperature 186, 196
   derating factors 199
American wire gauge 250
Ampacity 186, 193, 196, 338
   allowable 196
   conductor 338
Ampacity summary 202
Ampere 9
   fraction of 279
Ampere-turns 84, 116
Apparent power 86
Appliances 285
   cooking 285
   cooking appliance neutral load 292
Armature 15, 78 - 79, 82, 110
   winding 110
Autotransformer 117

B

Back-EMF 82
Battery 8
Box
   calculations 156
   junction 156
   See also outlet box
   pull 156
   size - outlet 154
Box calculation
   angle pull 157
   distance between raceways 157
   distance from removable cover 157
   straight pull 156
   u-pull 157
Branch circuit
   cooking equipment 280
   motor 221
   number required 284
   overcurrent protection 222
   small appliance 280
Bundled 186, 196

C

Calculations
   air conditioning vs. heat 287, 317
   angle pull 157
   appliance demand load 287, 317
   box fill 151
   box size 156
   conductor allowable ampacity 186
   cooking appliance neutral load 292
   cooking equipment 280
   cooking equipment demand load 288, 318
   dryer demand load 288, 318
   dryer neutral load 293
   dryers 341
   dwelling unit examples 286
   dwelling unit-optional method 290
   dwelling unit-standard method 285
   dwelling-unit feeder and service 290
   electric heat 341
   general lighting and receptacles 285
   general lighting VA load 283
   kitchen equipment 342
   lighting - miscellaneous 345
   lighting - other occupancies 344
   lighting demand factors 343 - 344
   marina 349
   mobile home park 350
   motor nameplate amperes 110, 115
   motor review 230

motor-feeder protection 227
motor-short circuit 225
motors- feeder size 226
multifamily - optional method 321
multifamily dwelling-unit 315
multifamily dwelling-units examples 316
multifamily feeders and service 321
multioutlet assembly 345
neutral 292, 349
neutral current 51
neutral reduction not permitted 349
neutral reduction over 200 amps 349
nipple size 148
number of circuits required 284
one counter mt'd cooking unit/ two ovens 282
one wall mt'd oven/one counter cooking unit 281
raceway 148
raceway size 148
recreational vehicles park 350
restaurant-optional method 351
school-optional method 353
series-parallel circuit resistance 49
service conductor size 289, 319
small appliance and laundry circuits 285
straight pull 156
u-pull 157
unit load 285
voltages 339
Capacitance 85
Capacitive reactance 10, 85
Capacitor 85
uses 86
Chapter 9 Tables 143
Charged 85
Circuit
balanced 3-wire 200
existing 260
Circular mils 250
Clamp-on ammeters 16
Closed loop 41, 78
Coil 84
Commutator 110
Condensers 85
Conductance 9
Conductor 249
ampacity 338
branch circuit 221
bundled 198
compact aluminum building 148, 209
cross-sectional area of insulated 146, 209
current carrying 198
current carrying - bundled 200

different size 154
equipment grounding 146, 209
equivalents 152
fill 143, 209
grounding 191
length 258
neutral 292
parallel 190 - 191
properties 148, 184, 209
same size & insulation 144, 146, 209
sizing 257
Conductor resistance 10
Conductors
allowable ampacities 196
allowable ampacity 186
branch circuit 221
bundling 198
current carrying 200
grounded 51
impedance 83
insulation 184
overcurrent protection 193
parallel 190
shape 84
temperature rating 195
voltage drop 191
Conduit body 156 - 157
Contacts 86
Continuous load 338
overcurrent protection 192
Continuous Loads 338
Control circuits 41
Counterelectromotive-force 82 - 83
Cover
removable 157
CPS (cycles per second) 79
Current 9
alternating 8
applied 82
direct 8
fault 192
flow 78
line 377, 381
line - wye 391
neutral 386
neutral - wye 395
opposition to flow 10
phase 378, 381
phase - wye 391
resistor or circuit 43
thru each branch 46
Current carrying conductors

grounded 200
wye 3-wire circuit 201
Cycles per second 79

D

dc
   See Direct current
Degrees 80
Delta/delta 376
Delta/wye 376
Demand factors
   general lighting 344
Demand load
   dryer neutral 293
   unbalanced 292
Device
   next size up - okay 339
Dielectric 85
Direct current 8, 78
   circuits 251
Direct current motor 111
Dryer 285
   branch circuit and OCPD 341
Dwelling-unit calculations
   appliances 285
   clothes dryer 285
   cooking equipment 285
   feeder and service 290
   general lighting/receptacles/small appliance 285
   laundry 285
   standard method 285

E

E - Voltage 9
Eddy currents 83, 118, 251
   skin effect 83
Effective 80
Efficiency 90
Electric field 85
Electric heating 341
Electric motor 110
Electromotive force 9, 78, 251
Electrostatic field 85
EMF 78
Equipment bonding jumpers 152, 154

F

Field winding 110
Flux density 116

Flux lines 84
Formulas
   alternating current resistance 252
   ambient temperature 196
   ambient temperature derating factors 199
   apparent power 86
   circuit resistance 47
   current 46
   current - neutral 52
   current - series 43
   dc current resistance 250
   effective current 80
   effective to peak 81
   efficiency 90
   equal resistor method 47, 50
   extending circuits 260
   horsepower size 114
   inductive reactance 82
   input watts 90
   Kirchoff's law 42
   kVA 118
   motor - branch circuit conductor size 225 - 226
   motor full load current 221
   motor output watts 114
   motor VA 113
   nameplate amps - single phase 115
   nameplate amps - three phase 115
   new ampacity 198
   new conductor ampacity 200
   number of circuits required 284
   output watts 90
   overload protection size 226
   peak current 80
   peak to effective 81
   power consumed 46
   power factor 88
   power to apparent power 88
   reciprocal method 48
   resistance - series 42
   resistance of parallel conductors 253
   resistance parallel conductors 253
   resistance total 49
   transformer current 119
   transposing a formula 7
   true power 89
   VA calculations 232
   voltage 42
   voltage drop 254, 256
   voltage drop - conductor sizing 257
   voltage drop - extending circuits 260
   voltage drop - length of conductors 258
   voltage drop - limit current 260

voltage drop - sizing conductors 257
voltage law 42
Fractions 3
Frequency 79

G

General lighting 283
    load 344
Generator 9
    alternating current 78
Ground-fault 222, 225
Grounded 51
Grounded conductor 51
    overloading 53

H

Harmonic currents 349
Heaters 222
Hertz 79
High leg 381
High voltage 78
Horsepower
    watts 114
    watts per HP 114
Hotel
    general lighting 344
Hystersis losses 118
Hz - Hertz 79

I

I - Ampere 9
Impedance 10, 83
In phase 79
Induced voltage 82
Induction 16, 82
Inductive reactance 10, 82 - 83, 251
Insulation 184
Interrupting rating 192
    short-circuit 192
Iron 118
Iron core 84

K

k - Kilo 4
Kilo 4
Kilo-volt ampere 377
Kirchoff's first law 42
kVA 86
    See kilo-volt ampere
kW 4

L

Lags 79
Leads 79
Lentz's law 82
Lighting - other occupancies
    club example 345
    store example 344
Load
    maximum 260
    maximum unbalanced 387
    maximum unbalanced - wye 396
    phase 378, 382
    unbalanced 378
load calculations
    bank example 354
    restaurant (optional) example 352
    restaurant (standard) example 357
    school (optional) example 353
Loop 116
Low reluctance 118

M

Magnetic cores 84
Magnetic coupling 116
Magnetic field 82, 110 - 111
Magnetic flux lines 78
Magnetomotive force 116
Marina
    calculation 349
    demand factor 349
Megger 17
Megohmer 17
Megohmmeter 17
MMF 116
Mobile home park
    calculation 350
Motel
    general lighting 344
Motors
    a-c 112
    branch circuit conductor 195
    branch circuit protection 222
    calculation review 230
    dual voltage 110, 113
    feeder protection 227
    feeder size 226
    hp/watts 114

induction 112
input 113
nameplate amperes 110, 115
nameplate current rating 223
overload protection 223
reversing a d-c 111
reversing an a-c 112
service factor 223
speed control 110 - 111
synchronous 112
temperature rise 224
universal 112
VA calculations 232
wound rotor 112
Multifamily calculations optional method
    feeders/service calculations 321
    neutral calculations 292
Multifamily calculations standard method
    a/c vs. heat 315
    appliances 316
    feeder and service conductor size 316
    feeders and service calculations 315
    general lighting circuits 315
Multioutlet 345
Multiplier 5
Multiwire circuits
    120/240-volt, 3-wire 51
    dangers 53
    wye 3-wire 52
Mutual induction 116

N

National Electrical Code
    210-19(a), FPN No. 4 191
    210-4 51
    210-52(f) 283
    210-70(a) 283
    220-11 285
    220-16(a)(b) 285
    220-18 341
    220-21 340, 342
    220-3(c) Ex. No. 3 345
    220-3(c)(6) 347
    240-1 FPN 339
    300-5 222
    310-13(b) 51
    370-28 156
    Article 100 51, 338
    Article 430 221
    articles 26
    changes and deletions 26
    Chapter 9, Part B 279
    Chapter 9, Part B Examples 339
    Chapter 9, Table 8 9
    chapters 26
    definitions 26
    Example 8 230 - 231
    exceptions 26
    fine print notes (FPN) 26
    FPN's 210-19(a), 215-2(b), 230-31(c), 310-15 254
    layout 25
    parts 26
    sections and tables 26
    standard size device ratings 192
    superscript letter 26
    Table 310-13 184
    Table 310-16 186
    Table 310-5 187
    Tables 310-16 thru 310-19 186
    terms 25
Neutral 51
Neutral Calculations
    nonlinear loads 349
Neutral load
    cooking appliance 292
Nipple 146
Nonlinear load 53, 201
Nonliner Loads
    demand factor 349

O

Office
    general lighting 347
Ohm 9
Ohm meter 16
Ohm's law 10
Open loop 45
Out of phase 82
Outlet box
    calculations 151
    conductor equivalents 152
    conductors different size 154
    conductors same size 151
    fill 151
    not counted 154
Overcurrent 222
Overcurrent protection 192 - 193, 339
    continuous load 192
    device - next larger 193
    feeder and services 194
    grounded conductor 194
    interrupting rating 192

next size up-okay 339
sizing conductors 193
standard sizes 192, 339
tap conductors 194
Overload 53, 222
heaters 222
number 225
protection 222
Overload protection
motors 223
Overvoltage 53

P

P - Power 10
Panelboard
delta 386
delta - sizing 386
wye - sizing 395
Parallel circuits 45
circuit resistance 47
equal resistors 47, 50
practical uses 45
understanding calculations 46
Parallel conductors 190
Parenthesis 7
Peak 80
Percent increase 5
Percentage 4
Percentage reciprocals 5
Perpendicular 16
Phase 79
degrees 80
PIE circle 12
Pigtailing 51
Pigtails 152, 154
Polarity 15, 79
Polarized 15
Potential difference 51
Power 10, 44
each branch 46
square of the voltage 14
Power factor 86, 88
Power source 8
Power supply 43, 46
Product 48
Pull box 156
PVC
expansion charcteristics 148

R

R - Resistance 9
Raceway
calculations 148
existing 150
Raceway fill
conductor properties 143
Ratio 90, 116, 378
Reactance 10
Receptacle
general-use 283
laundry 283
minimum load 346
Receptacles 338
Reciprocal 5
Recreational vehicles park
calculation 350
Residential calculations
appliance receptacle circuits 280
cooking equipment 280
See also dwelling-unit calculations
feeder & service 280
general requirements 279
laundry circuit 283
lighting & receptacles 283
one counter-mounted cooking unit 281
other Code sections 280
Resistance 9, 83
alternating current 148
Resistor 17
Resonance 86
Restaurant
calculation-optional method 351
optional method 351
RMS 80
Root-mean-square 80
Rotor 78, 110
Rounding off 7, 339
rounding 7

S

Saturated 116
School
calculation-optional method 353
otional method 353
Self-induced voltage 84
Self-induction 82
Series 15, 17
Series circuits 41
current 43

notes 45
power 44
understanding calculations 42
voltage 42
Series-parallel circuits
   review 49
Service Factor 223
   S.F. 223
Short circuit 192, 222, 225
   current 110
Short circuit protection
   branch circuit 225
   ground-fault 225
Shunt 16
Signal circuits 41
Signs 348
Sine wave 79
Skin effect 83, 251
   eddy currents 83
Solenoid 15
Speed control 111
Square root 6
Squaring 6
Stationary 110
Stator 112
Sum 48

T

Tables
   220-10(b) 356
   220-13 347
   220-3(b) 283, 344, 356
   555-5 349
   aluminum - ac vs. dc current resistance 252
   Chapter 9 Table 5 147
   Chapter 9, Table 1 144
   Chapter 9, Table 5 147
   copper, ac vs. dc current resistance 252
   Table 310-13 185
   Table 310-16 186
   Table 310-16 note 8 199
   Table 370-16(a) 151
   Table 373-6(a) 158
   Table 430-152 225
   Table 8 Chapter 9 250
   Table 9 Chapter 9 250
   terminal size 189
Temperature
   ambient 186, 196
   coefficient 250
   correction factors 196

   insulation rating 195
   rating 188
   rise 224
Temperature rise 224
Terminal
   ratings 188
Terminal rating 188
Theory 25
Torque 110
Transformer 116
   auto 117
   conductor resistance loss 117
   current 16, 116, 119
   flux leakage loss 117
   kVA Rating 118
   power loss 117
   primary vs. secondary 116
   secondary voltage 116
   step-down 117
   step-up 117
   turns ratio 118
   voltage variations 78
Transformers
   balancing 384
   delta/delta 376
   delta/wye 376
   size 385
   wye - sizing 394
Transposing formulas 7
True power
   watts 86, 89
Truisms
   wye 398
Turns 84

U

Unbalanced current 51
   miltiwire circuit 51
Ungrounded 51

V

V - Voltage 9
VA 86
Volt
   phase 382
Volt-amperes 86
Voltage 78, 279
   120/240, 208Y/120, etc. 279
   induced 82
   line 380

line - wye 389
nominal system 339
phase 377 - 378, 380
phase - wye 389
wye triangle 390
Voltage drop 42, 46, 191, 254
   a-c resistance 252
   ac vs. dc resistance 251
   conductor length 250
   conductor resistance 249
   conductor resistance a-c 251
   conductor resistance dc 250
   conductor sizing 187
   considerations 254
   convert cu-al 253
   cross-sectional area 250
   determining circuit 254
   extending circuits to limit 260
   feeder & branch circuit 254
   limiting - conductor length 258
   limiting current 260
   material 249
   prevention - conductors 257
   recommendations 254
   temperature 250
Voltage ratio 118
Voltmeters 17

W

W - Watt 10
Water heater 194
Watts 10, 89
   horsepower/HP 114
   true power 89
Waveform 78
   alternating current 78
   direct current 78
Winding 84, 116
   primary 116
   secondary 116, 379
Windings 379
Wye-connected 379

Z

Z - Impedance 10